Maillard Reactions in Chemistry, Food, and Health

Scientific Committee
Dr John Baynes, *University of South Carolina*
Dr Daniel Gallagher, *University of Minnesota*
Dr Theodore Labuza, *University of Minnesota*
Dr Annette Lee, *Picower Institute*
Dr Vincent Monnier, *Case Western Reserve University*
Dr John O'Brien, *University College Cork*
Dr Gary Reineccius, *University of Minnesota*

Local Organizing Committee
Dr Theodore Labuza
Ms Gwen Reed
Dr Gary Reineccius

Sponsors
The Organizing Committee wishes to express their appreciation to the following corporate sponsors for their support of the meeting:

Aleton Inc.
Boehringer–Mannheim Gnbh.
BSN, Centre de Recherche
General Mills
Marrion Merrell Dow
Mead–Johnson Research Center
Nestle, Switzerland
Roche Diagnostic Systems
The Coca Cola Company
The Quaker Oats Company
The Upjohn Company

Maillard Reactions in Chemistry, Food, and Health

Edited by

Theodore P. Labuza
Department of Food Science and Nutrition, University of Minnesota, Minnesota, USA

Gary A. Reineccius
Department of Food Science and Nutrition, University of Minnesota, Minnesota, USA

Vincent Monnier
Institute of Pathology, Case Western Reserve University, Ohio, USA

John O'Brien
Department of Food Chemistry, University College Cork, Cork, Ireland

John Baynes
School of Medicine, University of South Carolina, Charleston, USA

The Proceedings of the Fifth International Symposium on the Maillard Reaction, held at the University of Minnesota, on the 29 August–1 September 1993.

Special Publication No. 151

ISBN 0-85186-802-9

A catalogue record for this book is available from the British Library

Published by The Royal Society of Chemistry,
Thomas Graham House, Science Park, Cambridge
CB4 4WF

Printed and bound by Hartnolls Ltd., Bodmin

Dedication

This volume is dedicated to the memory of Professor Franz Ledl who until his recent untimely death was Professor of Food Chemistry and Analytical Chemistry at Stuttgart University in Germany.

Professor Ledl will long be remembered for his many contributions to our present understanding of Maillard reaction chemistry. Especially noteworthy are his basic studies (with J. Beck and T. Severin) on the origin of the aromagenic 1-deoxydiketoses from Amadori compounds [Carbohyd. Res 177, 240 (1988)] and on the formation of visible chromophors during more advanced Maillard processes, for example (with T. Severin) [Z. Lebensm. Unters. Forsch. 175, 262 (1982)].

On par with his research contributions was his talent for teaching and communication. Many future scholars of Maillard chemistry will profit greatly from Professor Ledl's lucid and comprehensive reviews such as those appearing in Z. Ernährungswiss 30, 4 (1991) and (with E. Schleicher) in Angew. Chem. Int. Ed. Engl. 29, 565 (1990).

Professor Ledl's career in the Maillard reaction field was truly widespread ranging initially from traditional food chemistry to the now popular area of glycosylated proteins and the chemistry of aging. Pertaining to cross-linked proteins, Professor Ledl's recent study on the chemistry of hydroxyalkyl pyrroles (with E. Klein, W. Bergmüller and T. Severin) [Z. Lebensm. Unters. Forsch. 194, 556 (1992)] provides future insight for unraveling the longtime enigma of melanoidin structure.

It is hoped that Professor Ledl's students will continue in the fine tradition established by their mentor. Scientific excellence is his legacy for us all.

Preface

The 5th International Symposium on the Maillard Reaction was held at the University of Minnesota, in Minneapolis, Minnesota (USA) from August 29 through September 1, 1993. Previous symposia were held in Switzerland in 1989, Japan in 1985, the United States in 1982 and Sweden in 1979.

Nearly 300 scientists attended the four day conference to discuss aspects of chemistry, kinetics, technology and toxicology of the reaction in foods and as it relates to health and aging. In total, 61 oral and 52 poster presentations were delivered. The stage for these presentations was set by 3 plenary lectures in the areas of food, health and pharmaceuticals. The banquet speaker provided a most provocative presentation on how the Maillard reaction may well have lead to the superior development of human beings and the demise of the dinosaurs.

The proceedings which follow are complete manuscripts of the oral presentations (including plenary lectures and banquet address) and abstracts of the poster presentations. The final chapter (immediately preceding the poster abstracts) of this book is both a summary of the conference and some thoughts beyond the meeting.

The conference organizers wish to express their appreciation to the University of Minnesota for making the facilities available, the Department of Food Science and Nutrition for its support with the numerous tasks associated with such a meeting and the host of students, both graduate and undergraduate, who helped make everything run smoothly during the meeting.

Contents

The Role of the Maillard Reaction *In Vivo*

Anthony Cerami

THE PICOWER INSTITUTE FOR MEDICAL RESEARCH, 350 COMMUNITY DRIVE, MANHASSET, NY 11030, USA

Summary
The reaction of glucose with proteins and nucleic acids in vivo has been found to play an important role in the development of the complications of diabetes and aging. In this review the evidence accumulated over the past 17 years on this process is described.

The discovery by Maillard in 1912 (Maillard, 1912) that the heating of amino acids and reducing sugars led to the formation of yellow-brown pigments ushered into the field of chemistry the importance of amino groups in accelerating the rearrangements of reducing sugars. For the next sixty years the Maillard reaction was studied by food chemists extensively since it could be shown to be responsible for the generation of aromatic, flavorful compounds which enhance the taste of food products, as well as explain the decreased nutritional availability of food which occurs with prolonged storage (Reynolds, 1965; Ledl and Schleicher, 1990). The importance of the Maillard reaction in living organisms has been appreciated only recently (Bucala and Cerami, 1992). The initial studies of this process centered on the biochemical abnormalities which occur in diabetes mellitus, an illness characterized by high circulating levels of glucose (Koenig and Cerami, 1975). Recent studies have implicated the Maillard reaction in aging and in the normal, physiological turnover of proteins. In order to distinguish Maillard products that form in vivo under normal physiological conditions of glucose concentration, pH and temperature, from Maillard products that form during exposure to elevated temperature, we introduced the term Advanced Glycosylation Endproducts or AGEs. The following review will outline our current understanding of AGEs and assess their diverse biological effects.

Studies of the non-enzymatic glycosylation of proteins in vivo began with the discovery of elevated amounts of the minor hemoglobin, hemoglobin A_{1c}, in the blood of patients and animals with diabetes (Koenig et al., 1976a; Koenig et al., 1976b). In the mouse hemoglobin A_{1c} was shown to occur as the result of a post-translational modification of hemoglobin A whose rate of formation was dependent on the ambient blood glucose concentration (Koenig and Cerami, 1975). Thus, diabetic animals had a three-fold increase in the rate of HbA_{1c} synthesis and a three-fold increase in the levels of this minor hemoglobin. Structural studies of HbA_{1c} revealed that a glucose Amadori product was attached to the amino-terminal valine of the β chain of HbA (Koenig et al., 1977; Bunn et al., 1979). The presence of Amadori products in vivo was not restricted to hemoglobin but was observed to occur on the amino-terminal amino acid and the ϵ-amino groups of lysine of many different proteins (Bucala et al., 1992). As in the case of hemoglobin, the amount of the Amadori product on these proteins was increased in the diabetic state.

The measurement of the amount of Amadori products on hemoglobin has proven to be a useful marker for the glycemic control of diabetic patients over a four week period (Jovanovic and Peterson, 1981). Figure 1 records the temporal relationship between weekly urinary glucose excretion and HbA_{1c}. HbA_{1c} values fall three weeks after the

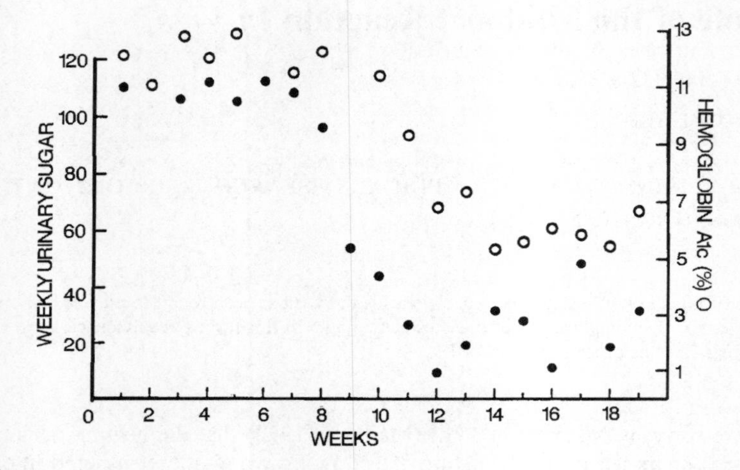

Figure 1. Temporal relationship between weekly urinary glucose excretion and hemoglobin A_{1C}. (Permission granted by the New England Journal of Medicine, 295:417-420, 1976)

initiation of good metabolic control as evidenced by decreased glucose loss in the urine. In effect, the level of HbA_{1c} reflects the average-integrated blood glucose levels for a twenty eight day period. The utility of this assay has recently been highlighted by the report of the Diabetes Control and Complications Trial (Shamoon et al., 1993). In this trial diabetic patients were randomly assigned to two groups which had either regular control or stricter control which was achieved by measuring blood sugar many times a day. In the first group the patients had a HbA_{1c} of approximately 9% while the strict control group averaged approximately 7.5%. This small decrease in HbA_{1c} of the strict control patients was associated with a dramatic decrease (75-90%) in the number of patients who developed complications of diabetes over a nine year period. This study poignantly demonstrates the importance of glucose control in diabetic patients and the long term pathogenicity of glucose.

The identification of Amadori products on hemoglobin immediately opened the door to the possible presence of further Maillard rearrangement products on long-lived proteins. Since the crystallin proteins of the ocular lens have an extremely long half-life (some investigators, in fact, believe that they may never turn over for the life of the individual), this was the first tissue in which AGEs were first described (Monnier and Cerami, 1981). Yellow-brown, fluorescent pigments were noted to accumulate in the lens with age and at an accelerated rate in patients with diabetes. These pigments were also found to be capable of cross-linking proteins, one of the hallmarks of cataract formation. Figure 2 records the characteristic fluorescence excitation spectra of lens crystallins incubated with reducing sugars and the spectra of freshly isolated young and aged human lenses. A similar fluorescent excitation maxima can be observed to have accumulated in the aged lenses and the crystallins incubated with glucose. In addition, the progressive accumulation of AGEs on lens crystallins in vitro was observed to lead to the formation of disulfide and non-disulfide cross-links and the formation of high-molecular weight aggregates which were opalescence and scattered light (Monnier et al., 1979). These aggregates mimic the progressive opacification of cataractous lenses in diabetes.

Connective tissue collagen is another long-lived protein that has been found to accumulate AGEs in vivo (Monnier et al., 1984). Figure 3 demonstrates the results of a study in which the AGE content of human dura collagen was measured in people of increasing age. AGE associated fluorescence was found to increase with the age of the individual. Patients with diabetes had more collagen associated fluorescence than was expected for their chronological age. From model studies it has been observed that the AGEs that are found on collagen are associated with increased cross-linking of collagen molecules (Kohn et al., 1984). This is associated with increased rigidity, decreased solubility and susceptibility to enzymatic digestion. These phenomena have been observed with collagen reacted with reducing sugars in vitro as well as collagen, containing AGEs, isolated from diabetic or aged animals. In addition to AGEs participating in intermolecular cross-linking of collagen molecules together, AGEs that form on long-lived molecules such as collagen can cross-link various serum proteins (Brownlee et al., 1983). Thus, serum albumin, LDL, and immunoglobulins have been found to be covalently trapped on matrix proteins in vitro and in vivo. This trapping of plasma proteins could lead to thickening of basement membrane and the activation of complement that could lead to tissue damage.

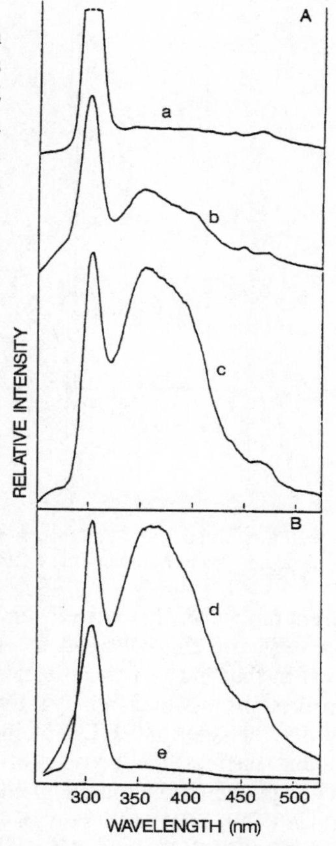

Figure 2. Fluorescence-excitation spectra of human and bovine lens protein digests. (A) Bovine lens crystallins incubated without sugar (a); 5mM glucose (b); and 5 mM glucose-6-phosphate (c). (B) Human lens protein from cataractous (d) and young normal lenses (e). (Monnier and Cerami, 1981).

During the course of our studies of AGE formation in vivo we were struck by the fact that tissues isolated from animals or people have less AGEs than we would have predicted from the incubation in vitro of proteins with glucose. This prompted the hypothesis that a removal system might exist in vivo (Vlassara et al., 1985a). Macrophages, well known scavenger cells of the body, were found to possess specific AGE-receptors that could mediate the uptake and eventual breakdown of AGE-proteins. There are approximately 1.5 x 10^5 receptors per cell with a Ka of 1.7 x 10^7 M^{-1}. Biochemical studies of mouse macrophages have revealed two proteins (60 and 90 kDa) with specific binding activity to AGE-proteins (Vlassara et al., 1985b). AGE-receptors have been noted to occur on a number of othercells as well including endothelial cells, smooth muscle cells, mesangial cells and fibroblasts (Vlassara, 1992). The relationship of the 60 and 90 kDa proteins and the various cell types needs further investigation.

The identification of a macrophage scavenger system for AGE-modified proteins has opened up a new area of investigation into the role of AGE modification in connective tissue homeostasis. Although connective tissue matrix proteins, including collagen, have extended life span in vivo, they are slowly removed and replaced throughout adult life. The monocyte/macrophage plays an important role in this process both by removing the

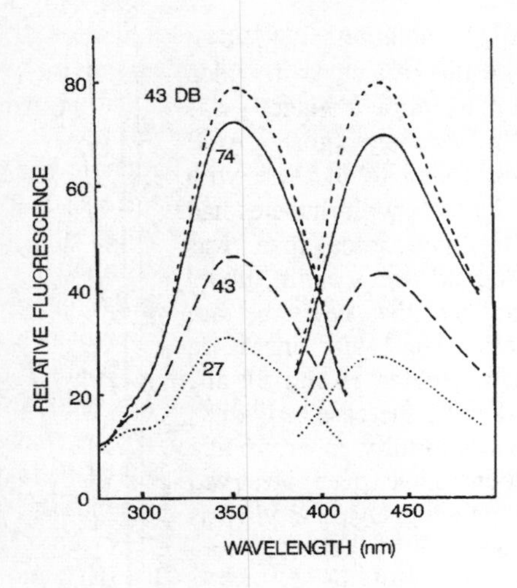

Figure 3. Fluorescence-excitation spectra of the samples of dura collagen. The spectra are differentiated by age as noted. (From Monnier et al., 1984).

senescent molecules that have accumulated AGEs over time and by initiating the steps that lead to new protein synthesis and tissue remodeling. The net accumulation of AGE-proteins in the tissues is a reflection of the balance of the reaction of glucose with the matrix proteins and their removal by the macrophage. The importance of the recognition of the macrophage for AGE-proteins in this remodeling process is best demonstrated by the potent chemotactic activity of AGE-proteins (Kirstein et al., 1990). After the macrophage has found and engulfed the senescent protein, the cell releases a number of cytokines (TNF, IL1) and growth factors [platelet derived growth factor (PDGF), and insulin-like growth factor 1 (IGF-1)] (Doi et al., 1992; Kirstein et al., 1992) . These cytokines and growth factors can recruit additional cells to assist in the removal and resynthesis of the matrix proteins. For example, mesenchymal cells are stimulated to release collagenase and other proteases which degrade matrix proteins while concomitantly, fibroblasts are induced to synthesize new collagen and other matrix proteins. In effect, the AGE-protein induced release of growth factors by macrophages promotes mesenchymal proliferation and the production of connective tissue matrix. This removal and replacement of senescent proteins is a delicate process that can be easily disrupted by the excessive net accumulation of AGEs in the tissues of diabetics and aging individuals. In these states there is not only an accumulation of excess AGE-matrix proteins but an enhanced deposition of matrix proteins, thickening of basement membranes and proliferation of smooth muscle cells as the body tries to repair the damaged tissue (Vlassara, 1992). As will be described below, the injection of AGE-proteins and -peptides into normal animals results in their entrapment on matrix proteins and leads to the abnormal growth response and matrix deposition that is seen to occur in the kidney and vascular walls of animals and people with diabetes and aging.

Over the past 10 years a number of papers have been published which point to the association of accumulated AGEs with various biological and pathological phenomena. In order to define a clearer relationship of cause and effect between the formation of AGEs and the response we initiated a program to find specific inhibitors of AGE formation. The ability of sulfite to inhibit the Maillard reaction in food chemistry was an obvious starting point and a model for what we hoped to achieve. Investigation of the inhibition of AGE formation by sulfite revealed that sulfite trapped a specific product 1-alkyl-2-formyl-3,4-diglycosyl pyrrole (AFGP) that was not a visible chromophore but had the ability to still cross-link amino groups together (Farmar et al., 1988). This, coupled with the fact that sulfite would be toxic to biological systems, ruled out sulfite and, thus, we searched for new compounds. We reasoned that the carbonyl moieties formed during the rearrangement of the sugar-protein adducts would be an important target for intervention. The simplest molecule that we could think of that would have a free and uncharged amino-group at physiological pH was aminoguanidine (Brownlee, 1986). The addition of aminoguanidine to proteins incubated in the presence of reducing sugars prevented the formation of AGEs on the proteins and also importantly prevented the cross-linking of proteins. An inhibition of AGE formation and cross-linking of aminoguanidine was also observed in diabetic rats. It is important to point out that aminoguanidine does not interfere with the formation of Amadori compounds but inhibits subsequent rearrangements which are essential for cross-linking reactions. Studies of a model reaction of aminoguanidine and the Amadori product of glucose and propylamine has revealed two major products shown in Figure 4. In these studies the 1,4,dideoxyglucosone product reacts with two aminoguanidines or forms a stable cyclic derivative (Chen and Cerami, 1993). Further studies are needed to determine if this accounts for the inhibition of AGE formation in vivo.

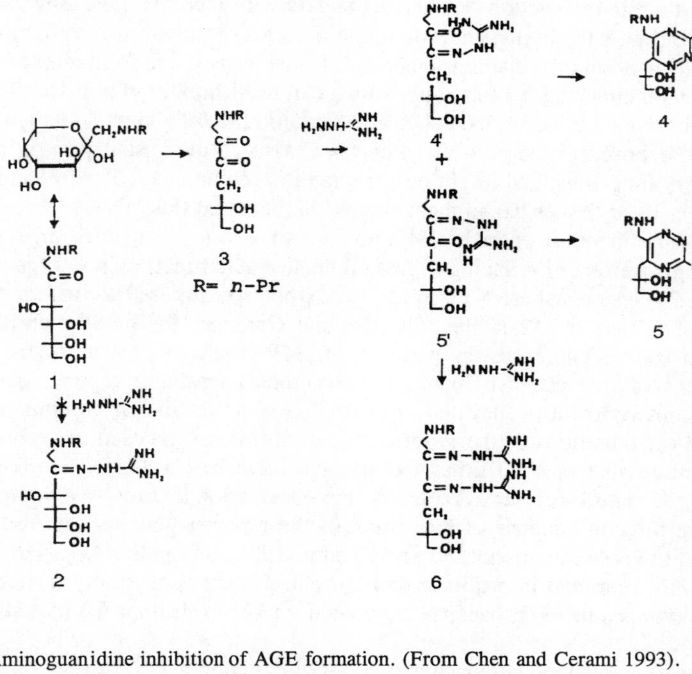

Figure 4. Aminoguanidine inhibition of AGE formation. (From Chen and Cerami 1993).

Since the identification of aminoguanidine as an AGE inhibitor, a number of groups around the world have confirmed the ability of aminoguanidine to inhibit AGE formation and cross-linking in experimental animals as well as prevent a number of the sequelae associated with diabetes, including nephropathy, neuropathy, and retinopathy (Ellis and Good, 1991; Kihara et al., 1991; Hammes et al., 1991). Pilot clinical studies of aminoguanidine for twenty eight days in patients with diabetes have shown that the formation of AGEs on hemoglobin is significantly inhibited (Makita et al., 1992). These encouraging preliminary results have stimulated the initiation of larger clinical studies on the efficacy and safety of aminoguanidine which are currently underway. The clinical endpoint that will be addressed in the first double-blind, multicenter trial is the progression of the loss of kidney function in diabetic patients.

During the course of our studies of the role of the Maillard reaction on proteins in vivo, we became intrigued with the possibility that the amino groups of the bases of DNA could also react with reducing sugars. The incubation of either DNA or single nucleotides which contained amino groups (adenine, guanine, and cytidine) with reducing sugars was observed to produce absorbance and fluorescence changes similar to those seen on AGE-proteins (Bucala et al., 1984). When biologically active DNA from either a bacteriophage (f1) or plasmid (pBR322) were incubated with reducing sugars there was a time- and sugar-dependent loss of biological activity when the AGE-DNA was introduced into the host, E.coli. Of note was the fact that the AGE-DNA had been damaged permanently by the induction of mutations (Bucala et al., 1985). Many of these mutations came about as a result of the attempt by the bacteria to repair the AGE-DNA since bacteria which lacked repair mechanisms did not give rise to mutated DNA.

Although the rate of reaction of reducing sugars with DNA is quite slow, the rate of damage can be accelerated by the addition of amino acids or polyamines to the incubation. In fact the preincubation of reducing sugars with lysine leads to the formation of a reactive intermediate, containing sugar and lysine, which can react rapidly within 15 minutes with either single or double stranded DNA (Lee and Cerami, 1987a). Thus, although the DNA of a cell is not exposed to the high glycemia that exists in the vasculature of a diabetic, the DNA is very long-lived and might be expected to accumulate AGEs. Examination of DNA from long-lived tissues has in fact revealed the fact that this DNA is cross-linked to itself and proteins in older animals (Bojanovic et al., 1970). In order to evaluate the potential damaging effect of reducing sugars on nucleic acid function in vivo several years ago we devised a system where a plasmid DNA would be exposed to different amounts of glucose-6-phosphate (G6P) in the cell (Lee and Cerami, 1987b). Several strains of bacteria which had mutations in the pathway of G6P could be grown under conditions which accumulated this reactive material. Plasmids containing reporter genes were exposed to an environment of elevated or normal G6P levels for an 18 hour period and then evaluated for mutations of the reporter genes. Plasmids exposed to an intracellular G6P concentration that was 30 times the normal value had a 15 fold increase in the mutation rate. Perhaps more remarkable was the observation that the vast majority of the mutations were the consequence of disruption of the reporter gene by the insertion of a host derived transposable element (Lee and Cerami, 1991). Why the AGE-DNA plasmid causes the mobilization and insertion of this transposable element is not understood at all since many previous studies in bacteria have been unable to prompt the moving of these elements.

Recently we have extended our studies of AGE damage of DNA to mammalian cells as well. Utilizing a shuttle vector that can propagate in either mammalian cells or bacteria we have observed that the introduction of AGE-shuttle vectors into mammalian cells is also associated with the induction of mutations in reporter genes (Bucala et al., 1993). Again, one of the common types of mutations that we have observed is the incorporation of a mammalian cell piece of DNA that has some of the properties of a mammalian transposable element. Again, the mechanism for this process is obscure. The obvious question that we wish to know is whether AGE-DNA damage is occurring in mammals. We have begun to address this by utilizing transgenic mice for the lacI gene. This gene is present in all cells of the mouse and can be used as a reporter gene to reflect any damage that might have occurred in the mouse. The two situations we are evaluating are the induction of mutations in various tissues during normal aging and the induction of mutations in embryos that are being carried by diabetic mice. The latter experiments are prompted by the fact that the incidence of birth defects in children born to diabetic mothers is increased many fold. The time of occurrence and the type of abnormalities reflect a damage in DNA metabolism. Preliminary data from both studies have revealed a time dependent increase in mutations of the reporter gene in the spleen and kidney of aging animals and an increased rate of mutations in the embryos that have been exposed in utero to hyperglycemia (Lee et al.[submitted]; Lee et al., 1993). Further studies are underway to determine the nature of the mutations that have occurred in these animals.

Patients with diabetes who loose their kidney function and are forced to go on dialysis face a very bad prognosis. Within 18 to 24 months after going on dialysis more than half of the patients have died primarily as a result of vascular disease including stroke and heart attacks. The pathogenic mechanism for their very rapid demise is unknown. Several years ago we began to evaluate the possible role of AGE damage as an underlying pathogenic process in this group of patients. Analysis of the coronary vessels of these patients revealed an increased amount of AGEs attached to vascular matrix compared to diabetics without kidney dysfunction (Makita et al., 1991). Of interest, patients without diabetes but who had lost kidney function for other reasons had more AGEs than normal age matched controls. Further studies revealed that patients on dialysis had increased amounts of AGE-peptides in the plasma which could not be effectively removed by the dialysis procedure. The AGE-peptides have a molecular weight of 2 to 6 thousand and retain the ability to cross-link with other proteins in vitro and in vivo. Presumably, these peptides represent the uptake and breakdown of senescent proteins by the macrophage system. Evaluation of high flux and peritoneal dialysis procedures also reveal an ineffectiveness at normalizing AGE-peptide levels; the only procedure that normalized the AGE-peptide levels was the transplantation of a normal kidney. Within hours after transplantation the AGE-peptide levels fell to values seen in diabetic patients with normal kidney function.

The finding of the cross-linking properties of AGE-peptides to matrix proteins in vitro suggested that a similar process could be occurring in vivo as well. AGE-peptides were prepared in vitro by the enzymatic digestion of AGE-albumin (Vlassara et al., 1992). These AGE-peptides or normal peptides were injected into normal animals. Examination of the matrix proteins from a number of different sites within the animal receiving the AGE-peptides revealed that there was an accumulation of AGEs which could be prevented if the animals were simultaneously given the AGE inhibitor aminoguanidine. Daily injection of AGE-peptides into rats led to the progressive accumulation of AGEs in the tissue and the development over several months of pathological changes that mimicked those sequelae seen in diabetic animals. It thus appears that not only is glucose an important factor in the complications of diabetes but also AGE-peptides are pathogenic. The ability to induce in normal animals complications which are characteristic of diabetes further strengthens the hypothesis that the Maillard reaction is important in the pathogenesis of the complications of diabetes and aging. In the years ahead we can look forward to further evaluation and strengthening of this hypothesis.

Acknowledgements
The work described in this review represents the considerable efforts of a large number of colleagues over nearly twenty years of work. I dedicate this review to them and thank them for their dedication to furthering our understanding of diabetic complications. The support of the NIH and the Brookdale Foundation are also gratefully acknowledged.

References

Bojanovic, J.J., Jevtovic, A.D., Pantic, V.S., Dugandzic, S.M., and Javonovic, D.S. (1970). Thymus histones in young and old rats. *Gerontologia* 16, 304-312.

Brownlee, M., Pongor, S. and Cerami, A. (1983). Covalent attachment of soluble proteins by nonenzymatically glycosylated collagen: role in the *in situ* formation of immune complexes. *J Exp Med* 158, 1739-1744.

Brownlee, M., Vlassara, H., Kooney, T., Ulrich, P. and Cerami, A. (1986). Aminoguanidine prevents diabetes-induced arterial wall protein crosslinking. *Science* 232, 1629-1632.

Bucala, R., Lee, A.T., Rourke, L., and Cerami, A. (1993) Transposition of an *Alu*-containing element induced by DNA-advanced glycosylation endproducts. *Proc Natl Acad Sci USA* 90, 2666-2670.

Bucala, R., Model, P. and Cerami, A. (1984). Modification of DNA by reducing sugars: A possible mechanism for nucleic acid aging and age-related dysfunction in gene expression. *Proc Natl Acad Sci USA* 81, 105-109.

Bucala, R., Model, P., Russel, M. and Cerami, A. (1985). Modification of DNA by glucose-6-phosphate induces DNA rearrangements in an Escherichia coli plasmid. *Proc Natl Acad Sci USA* 82, 8439-8442.

Bucala, R., Vlassara, H. and Cerami, A. (1992). Advanced Glycosylation Endproducts. In: *Post-Translational Modifications of Proteins*, Harding, J.J., James, M. and Crabbe, C., (eds), CRC Press Inc. Boca Raton, FL.

Bunn, H.F., Shapiro, R., McManus, M., Garrick, L., McDonald, M.J., Gallop, P.M. and Gabbay, K.H. (1979). Heterogeneity of human hemoglobin A_0 due to non-enzymatic glycosylation. *J. Biol. Chem.* 254, 3892-3898.

Chen, Candy Hauh-Jyun and Cerami, A. (1993). Mechanism of Inhibition of Advanced Glycosylation by Aminoguanidine *In Vitro*. *J Carbohydrate Chemistry* 12, 731-742.

Doi, T., Vlassara, H., Kirstein, M., Yamada, Y., Striker, G.E., Striker, L.J. (1992). Receptor-specific increased messangial cell extracellular matrix production is mediated by PDGF. *Proc. Natl. Acad. Sci. USA* 89, 2873-2877.

Ellis, E.N., and Good, B.H. (1991). Prevention of Glomerular Basement Membrane Thickening by Aminoguanidine in Experimental Diabetes Mellitus. *Metab. Clin. Exp.* 40, 1016-1019

Farmar, J.G, Ulrich, P.C. and Cerami, A. (1988). Novel pyrroles from sulfite-inhibited Maillard reactions: insight into the mechanism of inhibition. *J Org Chem* 53, 2346-349.

Hammes, H.P., Martin, S., Federlin, K.L., Geisen, K. and Brownlee, M. (1991). Aminoguanidine Treatment Inhibits the Development of Experimental Diabetic Retinopathy. *Proc. Natl. Acad. Sci. USA* 88, 11555-11558.

Jovanovic, L. and Peterson, C.M. (1981). The clinical utility of glycohemoglobins. *Am. J. Med.* 70, 331.

Kihara, M., Schmelzer, J.D., Low, P.A., Poduslo, J.F., Curran, G.L., and Nickander, K.K. (1991). Aminoguanidine on Nerve Blood Flow, Vascular Permeability, Electrophysiology, and Oxygen Free Radicals. *Proc. Natl. Acad. Sci. USA* 88, 6107-6111.

Kirstein, M., Aston, C., Hintz, R., Vlassara, H. (1992). Receptor-specific induction of insulin-like growth factor I (IGF-I) in human monocytes by advanced glycosylation endproduct-modified proteins. *J. Clin. Invest.* 90, 439-446.

Kirstein, M., Brett, J., Radoff, S., Stern, D., and Vlassara, H. (1990). Advanced Glycosylation Endproducts selectively induce monocyte migration across intact endothelial cell monolayers, and elaboration of growth factors: Role in aging and diabetic vasculopathy. *Proc. Natl. Acad. Sci. USA* 87, 9010-9014.

Koenig, R.J., Araujo, D.C. and Cerami, A. (1976a). Increased hemoglobin A_{1C} in diabetic mice. *Diabetes* 25, 1-5.

Koenig, R.J., Blobstein, S.H. and Cerami, A. (1977). Structure of carbohydrate of hemoglobin A_{1C}. *J Biol Chem* 252, 2992-2997.

Koenig, R.J. and Cerami, A. (1975). Synthesis of hemoglobin A_{1C} in normal and diabetic mice: Potential model of basement membrane thickening. *Proc Natl Acad Sci* USA 72, 3687-3691.

Koenig, R.J., Peterson, C.M., Jones, R.L., Saudek, C., Lehrman, M. and Cerami, A. (1976b). Correlation of glucose regulation and hemoglobin A_{1C} in diabetes mellitus. *N Engl J Med* 295, 417-420.

Kohn, R.R, Cerami, A. and Monnier, V.M. (1984). Collagen aging *in vitro* by nonenzymatic glycosylation and browning. *Diabetes* 33, 57-59.

Lee, A.T. and Cerami, A. (1987a). The formation of reactive intermediate(s) of glucose-6-phosphate and lysine capable of rapidly reacting with DNA. *Mutation Res* 179, 151-158.

Lee, A.T. and Cerami, A. (1987b). Elevated glucose-6-phosphate levels are associated with plasmid mutations *in vivo*. *Proc Natl Acad Sci USA* 84, 8311-8314.

Lee, A.T. and Cerami, A. (1991). Induction of gamma-delta transposition in response to elevated glucose-6-phosphate levels. *Mutation Research* 249, 125-133.

Lee, A., Cerami, A., and Bucala, R. Glucose-mediated DNA Damage and Mutations: *in vitro* and *in vivo*. *Proceedings Book of the 5th Int'l Symposium on the Maillard Reaction* p 249.

Lee, A.T., Plump, A., Cerami, A., and Bucala, R. (1993). Diabetes-Induced Teratogenesis: Role of DNA Damage in a Transgenic Mouse Model. *Diabetes* 42 sppl. 1, 85A.

Ledl, F. and Schleicher, E. (1990). New aspects of the Maillard reaction in foods and in the human body. *Angew. Chem.* 6, 565-706.

Maillard, L.C. (1912). Action des acides amines sur les sucres; formation des melanoidines par voie Méthodique. *C. R. Hebd. Seances Acad. Sci.* 154, 66-68.

Makita, Z., Radoff, S., Rayfield, E.J., Yang, Z., Skolnik, E., Friedman, E.A., Cerami, A., and Vlassara, H. (1991) Advanced Glycosylation Endproducts in patients with diabetic nephropathy. *N Engl J Med* 325, 836-842.

Makita, Z., Vlassara, H., Rayfield, E., Cartwright, K., Friedman, E., Rodby, R., Cerami, A. and Bucala, R. (1992) Hemoglobin-AGE: A circulating marker of advanced glycosylation. *Science* 258, 651-653.

Monnier, V.M. and Cerami, A. (1981). Nonenzymatic browning *in vivo*: Possible process for aging of long-lived proteins. *Science* 211, 491-494.

Monnier, V.M., Kohn, R.R. and Cerami, A. (1984). Accelerated age-related browning of human collagen in diabetes mellitus. *Proc Natl Acad Sci USA* 81, 583-587.

Monnier, V.M., Stevens, V.J. and Cerami, A. (1979). Nonenzymatic glycosylation, sulfhydryl oxidation and aggregation of lens proteins in experimental sugar cataracts. *J Exp Med* 150, 1098-1107.

Reynolds, T. M. (1965). Chemistry of nonenzymatic browning. II. *Adv. Food Res.* 14, 167-283.

Shamoon, H., et.al. (1993). The Effect of Intensive Treatment of Diabetes on the Development and Progression of Long-Term Complications in Insulin-Dependent Diabetes-Mellitus *N Engl J Med* 329, 977-986.

Vlassara, H. (1992a). Receptor-Mediated Interactions of Advanced Glycosylation Endproducts with Cellular Components within Diabetic Tissues. *Diabetes* 41(2), 52-57.

Vlassara, H. (1992b). Monocyte/macrophage receptors for proteins modified by advanced glycation endproducts: role in normal tissue remodeling and in pathology. In: *Mononuclear phagocytes Biology of monocytes and macrophages* van Furth, Ralph (ed), Kluwer Academic Publishers, Dordrecht, The Netherlands. 26, 193-201.

Vlassara, H., Brownlee, M. and Cerami A. (1985a). Recognition and uptake of human diabetic peripheral nerve myelin by macrophages. *Diabetes* 34, 553-557.

Vlassara, H., Brownlee, M. and Cerami, A. (1985b). High-affinity receptor-mediated uptake and degradation of glucose-modified proteins: A potential mechanism for the removal of senescent macromolecules. *Proc Natl Acad Sci USA* 82, 5588-5592.

Vlassara, H., Fuh, H., Makita, Z., Krungkrai, S., Cerami, A., Bucala, R. (1992). Exogenous Advanced Glycosylation Endproducts Induce Complex Vascular Dysfunction In Normal Animals; A Model For Diabetic And Aging Complications. *Proc. Natl. Acad. Sci. USA* 89, 12043-12047.

The Maillard Reaction in Foods

George P. Rizzi

THE PROCTER AND GAMBLE COMPANY, MIAMI VALLEY LABORATORIES, CINCINNATI, OH 45239-8707, USA

Summary

Maillard or non-enzymic browning reactions can play a pivotal role in food acceptance through the ways they influence quality factors such as flavor, color, texture and nutritional value. An overview of Maillard chemistry is presented and an attempt is made to show a chemical basis for the major quality factors.

Introduction

The Maillard reaction in foods is actually a complex network of chemical reactions which usually take place during food processing or storage. Products of Maillard reactions are significant in foods since they often influence food quality and acceptance.

First recognized by Louis Maillard in 1912, the amine-catalyzed degradations of reducing sugars have been extensively studied and by the 1950's they were firmly established as causative factors for flavor, color and textural changes in foods. Since the 1950's, largely as a result of analytical advances, the scope of food-related Maillard chemistry has expanded greatly (Fig. 1). Nowadays it's realized that primary Maillard products also react with endogenous food ingredients like lipids, flavonoids, terpenes and fermentation or metabolic products. Today, the outcome of these reactions is viewed both from the standpoint of aesthetics as well as quality where quality also includes nutrition and safety issues.

The purpose of this paper is to serve as an overview of the Maillard reaction and how it can influence food acceptability. Limited space does not permit a full coverage of the subject, but references to key research papers and reviews are provided.

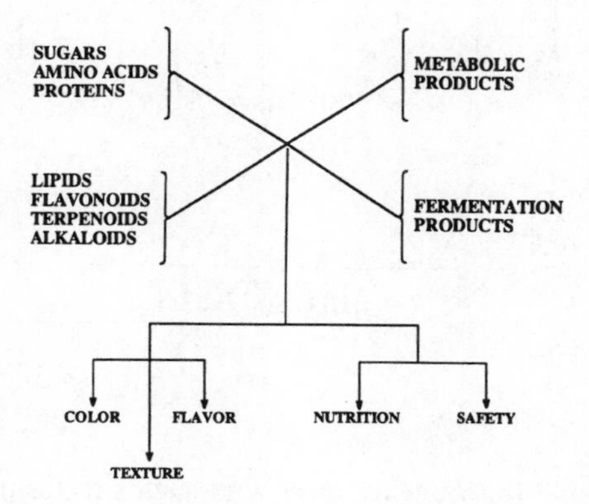

Figure 1. Current view of the Maillard reaction in foods.

Nature of the Reaction

Sugars, amino acids and proteins react to form a wide variety of products ranging from gases at room temperature to water-insoluble melanoidins. Volatile and soluble products contribute to aroma and taste while less soluble materials associate more with color and texture.

Like any chemical process the Maillard reaction is subject to control variables such as structure, concentration, pH, temperature, pressure, time, water activity, metal ions, light and inhibitors (Ellis, 1959; Namiki, 1988). In general Maillard reactions are accelerated at higher temperatures and high pH and pendant nitrogens in protein i.e. the ϵ-amino group in lysine seem especially prone to react. External pressure was shown to slow the rate of browning without affecting the formation of colorless intermediates (Tamaoka et al., 1991).

Major Reaction Pathways

Initial Maillard reactions are reasonably well defined, however later reactions and formation of melanoidins are still obscure (Ledl and Schleicher, 1990). Major pathways leading ultimately to melanoidins are shown in Fig. 2. Reactions of aldehydo-sugars are used for illustration, however similar reactions take place with keto-sugars.

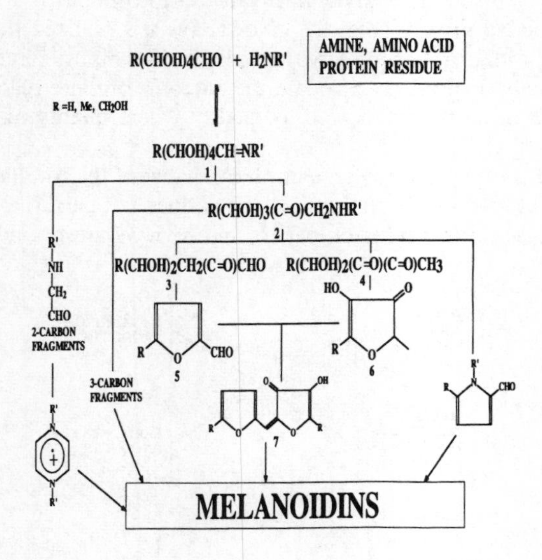

Figure 2. Basic Maillard reaction scheme.

Molecules with amine functionality react with sugars to form an imine (Schiff base) **1** which partially isomerizes to a somewhat more stable amino ketone (Amadori compound) **2**. Further reactions of compounds **1** and **2** are strongly affected by pH. Under alkaline conditions (>pH 7) compounds **1** and **2** undergo mainly chain fragmentation to form 2- and 3-carbon fragments which quickly react further to form melanoidins (Hayashi and Namiki, 1986). At strongly acidic pH (< pH 5) the Amadori compound **2** will eliminate the original amine ($R'NH_2$) to generate a 3-deoxyosone **3**. At intermediate pH's (pH 5-7) a similar fragmentation with loss of amine leads to an isomeric 1-deoxyosone **4**. The deoxyosones are highly reactive Maillard reaction intermediates and their further reactions lead to flavor and color formation in addition to higher molecular weight products.

Flavor Compound Formation

Most studies on Maillard flavors deal with volatile aroma components, while only few reports exist on taste. Maillard-derived aromas are extremely complex and many components are formed in trace amounts by side-reactions and obscure pathways (Baltes, 1980). Certain generalities have been observed, i.e. in the case of sugar derivatives with intact C-skeletons and for reactions of sugar fragmentation products. The deoxyosones (derived from Amadori or Heyns compounds) are considered to be primary sources of aroma volatiles.

Deoxyosones undergo cyclization/dehydration to produce flavor important furan derivatives and different types of furans are formed depending on osone structure. For example monosaccharides xylose, rhamnose and glucose can generate furfural, 5-methylfurfural and 5-hydroxymethylfurfural **5**(R=H, CH_3, and CH_2OH) via cyclization of respective 1-deoxyosones, **3**. In a similar way 3-deoxyosones lead to the 3-furanones **6**. Some 3-furanones e.g. (**6** R=CH_3) have extremely low aroma thresholds (0.04 PPM) and exhibit characteristic sweet, caramel aroma notes. In some foods like meat the

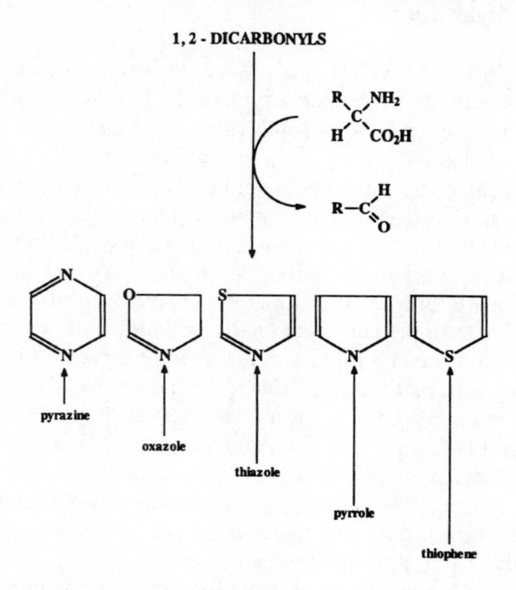

Figure 3. The Strecker degradation.

3-furanones react further to produce a different flavor quality. Thus, reaction with ambient sulfur i.e. H_2S leads to a wide range of sulfur-containing heterocycles which smell like cooked meat (van den Ouweland and Peer, 1975). Some of the sulfur compounds like 2-methyl-3-furanthiol and its related disulfide have odors which are also considered to be highly characteristic of cooked meat (Gasser and Grosch, 1988).

Chain fragmentation of deoxyosones **3** and **4** also leads to important flavor consequences, principally by generating reactants for the Strecker degradation (Figure 3). Retroaldol reactions of deoxyosones produces lower molecular weight 1,2-dicarbonyls like pyruvaldehyde which can react with ambient amino acids to form flavor volatiles. Strecker products include aldehydes (RCHO) derived from parent amino acids and a wide variety of heterocyclic compounds like pyrazines, oxazoles, thiazoles, pyrroles and thiophenes. Many volatile heterocycles occur in foods and their mechanistic origins and flavor properties are both well documented (Vernin, 1982). Some compounds like alkylpyrazines occur almost universally in all processed foods while impact compounds like 3-furanthiols seem to occur in particular foods like roasted meat or coffee. More work is needed to understand the chemistry of specific impact compounds in Maillard-derived flavors.

Some foods produce unique Maillard flavors derived from specific precursor compounds. For example, meats containing lipids generate longchain heterocycles i.e. 2-pentylpyridine in roast lamb (Buttery et al., 1977) and in related model studies (Whitfield, 1992). Also, cereal-based foods contain relatively large amounts of proline which extensive model reactions have shown can react with sugars to produce bitter tasting [dipyrrolidinohexosereductones](Pabst et al., 1985) and corny, baked bread aromas [pyridines, pyrroles](Tressl et al., 1981).

Determination of Flavor Significance

A problem facing modern flavor technologists is how to determine which compounds are flavor significant among thousands of Maillard products. For volatile compounds the aromavalue (AV) concept has proven to be of some value. A dimensionless number (AV) is defined as the ratio of a volatile component's concentration in a food to its threshold concentration. The basic assumption is that the magnitude of (AV) or more typically, log (AV) is a predictor of flavor significance for a single compound in a complex mixture. From the standpoint of psychophysics the assumption is at best a crude approximation, but on the other hand, it works! Workers at WRRC (Albany, CA) used aroma values to determine sensory significance among the volatiles of nineteen different foodstuffs (Teranishi et al., 1991). For example, in tomato paste volatiles fourteen compounds had $\log(AV) > 0$, suggesting they were significant. Subsequently, seven of these compounds were combined to reproduce tomato paste aroma. The aromavalue concept is the underlying hypothesis in other GC-based sensory analyses such as CHARManalysis™ (Acree et al., 1984) and Aroma Extract Dilution Analysis (Ullrich and Grosch, 1987). A noteworthy example of Aroma Extract Dilution Analysis is the characterization of a key aroma impact compound, 2-acetyl-1-pyrroline in wheat bread crust (Schieberle, 1990). Further refinements are needed to more accurately correlate analytical data with sensory significance. Unfortunately, the algorithms needed to relate human perception of aroma quality to GC data and especially to understand how changes in aroma quality varies with concentration still seem as distant as science fiction.

Melanoidin Structure

A complete picture of melanoidin structure still eludes food technologists mainly due to the inhomogeneous nature of the brown polymers. Published structural information is mainly based on melanoidins isolated from model systems. A tetrameric repeating structure was proposed for melanoidin generated from reducing sugars, butylamine and ammonia (Kato and Tsuchida, 1981). The authors concluded that osone-imines are likely precursors of the polymer, but that aromatic structures like pyrroles are probably by-products and not involved to a major extent in polymer formation. Later spectroscopic evidence on glucose-glycine melanoidins seemed to support a non-aromatic polymer structure (Kato et al., 1986), however other workers noted a buildup of pyrrole and/or indole N in melanoidins with reaction time (Benzing-Purdie and Ratcliffe, 1986). More recently, reactions of 6-aminohexanoic acid and sugars moderated with bisulfite produced pyrroles containing nearly intact carbohydrate residues (Farmar et al., 1988). These pyrroles are a clue to some type of melanoidin structure since they are colored, fluorescent and they possess reactive sites for further reactions. Thus, it has been proposed that nucleophilic displacement at pyrrole hydroxymethyl groups can lead to cross-linked structures and possibly to polymers (Klein et al., 1992). Cross-linked biopolymers should be strongly considered as components of food melanoidins, especially in view of their ubiquitous occurrence in living organisms.

Another type of cross-linked products are the pyrazinium radical cationic structures derived from 2-carbon fragments (Figure 2)(Namiki, 1988). Polymers containing stable free radicals may also be an indication of the well known antioxidant activity of food melanoidins. Conceivably, the first-formed condensation products of 2-carbon fragments i.e. dihydropyrazines may act as one electron reducing agents i.e. like ascorbic acid (Liao and Seib, 1988) to inhibit autoxidation. In the proposed reaction, the dihydropyrazines will be oxidized to stable radical cations which have been observed.

In retrospect, it appears that future use of alternate model systems such as peptides plus sugars or amino acids plus polysaccharides may be useful to help unraveling food melanoidin structure.

Colored Maillard Products

Compounds responsible for Maillard reaction colors are still largely unknown, especially the high molecular weight, water insoluble products. Work to date has mainly been confined to relatively low molecular weight unsaturated species containing extended double bond conjugation. For example the 3-furanone **6**(R=H) was observed to react sequentially with furfural by aldol condensation/dehydration to form a yellow product **7** and an orange tricyclic compound (Figure 2)(Ledl and Severin, 1978). Similar compounds have been isolated from sugar-amino acid reactions (Ledl and Severin, 1978; Ames et al., 1993) and predictably, based on available reactivity, many more remain to be elucidated.

A higher molecular weight red pigment is obtained in reactions of proteins containing ϵ-amino groups and dehydroascorbic acid [DHA] (Hayashi et al., 1985). Earlier studies on DHA and amino acids established the probable chromophor in the insoluble pigment as a structure similar to the barbituric acid-derived purple pigment, murexide (Kurata et al., 1973) and the purple/red pigment, Ruhemann's purple formed by ninhydrin-amine reactions. Similar reactions of amino acids or proteins with sugar-derived dehydroreductones may provide yet another source of Maillard pigments.

Nutrition

Nutritional effects due to Maillard reaction products (MRP) have received much attention in recent years (O'Brien and Morrissey, 1989; Hurrell, 1990). The presence of MRP in the diet has been linked to two broad areas, namely: reduced bioavailability of protein and antinutritive effects. MRP may be the indigenous materials associated with normal food preparation or they can be materials produced separately and added i.e. as flavor supplements etc. The presence of MRP in foods is not considered to be a significant problem except in restricted diets like infant formulas or in intravenous feeding.

Maillard reactions affect protein bioavailability by derivatizing protein-bound, dietary limiting amino acids such as lysine, arginine and histidine and possibly tryptophane and cysteine. For example, the pendant e-amino groups of lysine react with reducing sugars to form pendant Amadori compounds. For fructose-lysine, both the protein-bound and the free Amadori compound are not bioavailable as lysine in the rat and in human infants 71% of the bound form is excreted (Hurrell, 1990).

Protein bioavailability is also affected by cross-linking phenomena similar to that observed *in vivo* (Pongor et al., 1984). Cross-linking of 2-carbon fragments (Figure 2) (Namiki, 1988) or further reactions of pendant Amadori compounds or pyrraline (Klein et al., 1992) can lead to indigestible forms of protein. Also, other *in vivo* results suggest that multiple amino acid residues i.e. lysine plus arginine can be involved in cross-linked proteins via pentosidine formation (Sell and Monnier, 1989).

MRP exhibit antinutritive effects by mechanisms involving complexation with micronutrients, destruction of vitamins and by acting as inhibitors of digestive enzymes. Lysinoalanine, a product of protein bound lysine and cysteine is known to form stable metal ion complexes (Hayashi, 1982) and similar complexes with MRP may be the cause of observed urinary excretion of zinc, copper and iron (Hurrell, 1990). Vitamins with reactive functionality i.e. NH_2 in thiamine or CHO in pyridoxal can be destroyed in thermally induced amine/carbonyl reactions. The effect of MRP on digestive enzymes has recently been reviewed (O'Brien and Morrissey, 1989) and inhibition was reported for trypsin, carboxypeptidases A and B, aminopeptidase N and intestinal disaccharidases.

Safety and Toxicity

Meaningful studies on the toxicity of MRP have been limited by the purity of Maillard reaction mixtures which tend to be complex and undefined (O'Brien and Morrissey, 1989). The development of *in vitro* screening assays, in particular the Ames test have focused attention on mutagenicity as a major criterion for selecting mixtures or compounds for further study. Concern over mutagenicity in cooked foods originated when mutagenic substances were observed in pyrolyzed proteins and amino acids (Namiki, 1988).

Cooked muscle foods seem to contain more mutagens than other foods. A particularly potent class of mutagens, the imidazoquinolins (IQs) were first observed in broiled fish and also isolated from heated beef extract and hamburger (Kasai et al., 1980). IQ compounds are moderately carcinogenic *in vivo*, but relevance to human cancers is unknown. A mechanism for IQ formation was proposed involving creatinine, and Maillard reaction products (Jägerstad et al., 1983) which is supported by recent results (Milic et al., 1993). Interestingly, the formation of IQ compounds in model systems can also be suppressed by the addition of pre-formed MRP (Yen and Chau, 1993).

Overview and Outlook for Future Work

Research on the Maillard reaction has been strongly driven by developments in analytical chemistry since the 1950's. As a result thousands of compounds have been identified in processed foods and in model systems. However, more work is needed to understand how these materials can influence the quality and acceptance of foods. Progress has been made in the sensory area using the aromavalue concept, but better tools are needed to relate human subjective response to objective data.

Structures of melanoidins and color bodies remain elusive due to the inhomogeneous, complex nature of high molecular weight Maillard products. Perhaps model systems based on different starting materials like peptides/protein plus sugar or amino acid plus polysaccharide will lead to melanoidins more amenable to structural analysis. For example, recent model studies using polylysine plus xylose produced cross-linked products with interpretable structure (Pongor et al., 1984).

Nutrition and human safety are indispensable consequences of Maillard reactions in foods. Recent reports have focused on the occurrence and possible significance of mutagenic/carcinogenic MRP. Based on published risk/exposure data for foodstuffs and environmental hazards, the human carcinogen hazard of MRP appears to be relatively small (Ames et al., 1987). In addition, recent evidence showed that lipid oxidation *in vivo* is a major cause of DNA damage probably far exceeding damage from external sources (Ames and Gold, 1991). It would appear that MRP with their well documented antioxidative properties may be doing more good than harm.

References

Acree, T.E., Barnard, J. and Cunningham, D.G. (1984) A procedure for the sensory analysis of gas chromatographic effluents. Food Chem. 14, 273-286.

Ames, B.N. and Gold, L.S. (1991) Endogenous mutagens and the causes of aging and cancer. Mutation Res. 250, 3-16.

Ames, B.N., Magaw, R. and Gold, L.S. (1987) Ranking possible carcinogenic hazards. Science 236, 271-280.

Ames, J. M., Apriyantono, A. and Arnoldi, A. (1993) Low molecular weight coloured compounds formed in xylose-lysine model systems. Food Chem. 46, 121-127.

Baltes, W. (1980) Die bedeutung der Maillardreaktion für die aromabildung in lebensmitteln. Lebensmittelchem. Gerichtl. Chem. 34, 39-47.

Benzing-Purdie, L. M. and Ratcliffe, C. I. (1986) A study of the Maillard reaction by ^{13}C and ^{15}N CP-MAS NMR: influence of time, temperature and reactants on major products. In: Amino-Carbonyl Reactions in Food and Biological Systems: Dev. in Food Sci., Vol. 13 (Fujimaki, M., Namiki, M. and Kato, H., Eds.), Elsevier, New York, pp. 193-205.

Buttery, R. G., Ling, L. C., Teranishi, R. and Mon, T. R. (1977) Roasted lamb flavor: basic volatile components. J. Agr. Food Chem. 25, 1227-1229.

Ellis, G. P. (1959) The Maillard reaction. Adv. Carbohydr. Chem. Vol. 14, pp. 63-134.

Farmar, J. G., Ulrich, P. C. and Cerami, A. (1988) Novel pyrroles from sulfite-inhibited Maillard reactions: insight into the mechanism of inhibition. J. Org. Chem. 53, 2346-2349.

Gasser, U. and Grosch, W. (1988) Identification of volatile flavor compounds with high aroma values from cooked beef. Z. Lebensm.- Unters. Forsch. 186, 489-494.

Hayashi, T., Terao, A., Ueda, S. and Namiki, M. (1985) Red pigment formation by the reaction of oxidized ascorbic acid and protein in a food model system of low moisture content. Agric. Biol. Chem.(Tokyo) 49, 3139-3144.

Hayashi, T. and Namiki, M. (1986) Role of sugar fragmentation in an early stage browning of amino-carbonyl reaction of sugar with amino acid. Agric. Biol. Chem.(Tokyo) 50, 1965-1970.

Hayashi, R. (1982) Lysinoalanine as a metal chelator, an implication for toxicity. J. Biol. Chem.257, 13896-13898.

Hurrell, R. F. (1990) Influence of the Maillard reaction on the nutritional value of foods. In: The Maillard Reaction in Food Processing, Human Nutrition and Physiology (Finot, P. A., Aeschbacker, H. U., Hurrell, R. F. and Liardon, R., Eds.) Birkhäuser Verlag, Boston, pp. 245-258.

Jägerstad, M., Reutersward, A. L., Öste, R., Dahlqvist, A., Grivas, S., Olsson, K. and Nyhammar, T. (1983) Creatinine and Maillard reaction products as precursors of mutagenic compounds formed in fried beef. In The Maillard Reaction in Foods and Nutrition (Waller, G. R. and Feather, M. S., Eds.) ACS Symp. Series 215, American Chemical Society, Washington, D.C., pp 507-519.

Kasai, H., Yamaizumi, Z., Wakabayashi, K., Nagao, M., Sugimura, T., Yokoyama, S., Miyazawa, T.and Nishimura, S. (1980) Structure and chemical synthesis of ME-IQ, a potent mutagen isolated from broiled fish. Chem. Lett. 1391-1394.

Kato, H. and Tsuchida, H. (1981) Estimation of melanoidin structure by pyrolysis and oxidation. In: Prog. Fd. Nutr. Sci., Vol. 5 (Eriksson, C., Ed.) Pergamon Press, New York, pp. 147-156.

Kato, H., Kim, S. B. and Hayase, F. (1986) Estimation of the partial chemical structures of melanoidins by oxidative degradation and ^{13}C CP-MAS NMR. In: Amino-Carbonyl Reactions in Food and Biological Systems: Dev. in Fd. Sci., Vol. 13 (Fujimaki, M., Namiki, M. and Kato, H., Eds.), Elsevier, New York, pp. 215-223.

Klein, E., Ledl, F., Bergmüller, W. and Severin, T. (1992) Reactivity of Maillard products with a pyrrole structure. Z. Lebensm. Unters. Forsch. 194, 556-560.

Kurata, T., Fujimaki, M. and Sakurai, Y. (1973) Red pigment produced by the reaction of dehydro-L-ascorbic acid with α-amino acid. Agric. Biol. Chem.(Tokyo) 37, 1471-1477.

Ledl, F. and Severin, T. (1978) Braunungsreaktionen von pentosen mit aminen. Z. Lebensm. Unters. Forsch. 167, 410-413.

Ledl, F. and Schleicher, E. (1990) New aspects of the Maillard reaction in foods and in the human body. Angew. Chem. Int. Ed. Engl. 29, 565-594.

Liao, M. -L. and Seib, P. A. (1988) Chemistry of L-ascorbic acid related to foods. Food Chem. 30, 289-312.

Milic, B. Lj., Djilas, S. M. and Canadanovic-Brunet, J. M. (1993) Synthesis of some heterocyclic aminoimidazoazarenes. Food Chem. 46, 273-276.

Namiki, M. (1988) Chemistry of Maillard reactions: recent studies on the browning reaction mechanism and the development of antioxidants and mutagens. Adv. Food Res. 32, 115-184.

O'Brien, J. and Morrissey, P. A. (1989) Nutritional and toxicological aspects of the Maillard browning reaction in foods. Crit. Rev Food. Sci. Nutr. 28, 211-248.

van den Ouweland, G. A. M. and Peer, H. G. (1975) Components contributing to beef flavor. Volatile compounds produced by the reaction of 4-hydroxy-5-methyl-3(2H)-furanone and its thio analog with hydrogen sulfide. J. Agr. Food Chem. 23, 501-505.

Pabst, H. M. E., Ledl, F. and Belitz, H.-D. (1985) Bitter compounds obtained by heating sucrose, maltose and proline. Second report. Z. Lebensm. Unters. Forsch. 181, 386-390.

Pongor, S., Ulrich, P. C., Bencsath, F. A. and Cerami, A. (1984) Aging of proteins: isolation and identification of a fluorescent chromophor from the reaction of polypeptides with glucose. Proc. Natl. Acad. Sci. USA 81, 2684-2688.

Sell, D. R. and Monnier, V. M. (1989) Structure elucidation of a senescence cross-link from human extracellular matrix. J. Biol. Chem. 264, 21597-21602.

Schieberle, P. (1990) Studies on bakers yeast as a source of Maillard-type bread flavour compounds. In: The Maillard Reaction in Food Processing, Human Nutrition and Physiology (Finot, P. A., Aeschbacker, H.U., Hurrell, R. F. and Liardon, R., Eds.), Birkhäuser Verlag, Boston, pp. 187-196.

Tamaoka, T., Itoh, N. and Hayashi, R. (1991) High pressure effect on Maillard reaction. Agric. Biol. Chem.(Tokyo) 55, 2071-2074.

Teranishi, R., Buttery, R. G., Stern, D. J. and Takeoka, G. (1991) Use of odor thresholds in aroma research. Lebensm.-Wiss. Technol. 24, 1-5.

Tressl, R., Grünewald, K. G. and Helak, B. (1981) Formation of flavour components from proline and hydroxyproline with glucose and maltose and their importance to food flavour. In: Flavour '81, Walter de Gruyter & Co., New York, pp. 397-416.

Ullrich, F. and Grosch, W. (1987) Identification of the most intense volatile flavour compounds formed during autoxidation of linoleic acid. Z. Lebensm. Unters. Forsch. 184, 277-282.

Vernin, G. (1982) Chemistry of Heterocyclic Compounds in Flavours and Aromas (Vernin, G., Ed.), John Wiley & Sons, New York.

Whitfield, F. B. (1992) Volatiles from interactions of Maillard reactions and lipids. Crit. Rev. Fd. Sci. Nutr. 31, 1-58.

Yen, G.-C. and Chau, C.-F. (1993) Inhibition by xylose-lysine Maillard reaction products of the formation of MeIQx in a heated creatinine, glycine and glucose model system. Biosci. Biotech. Biochem. 57, 664-665.

Maillard Reaction and Drug Stability

Vijay Kumar and Gilbert S. Banker

PHARMACEUTICS DIVISION, COLLEGE OF PHARMACY, THE UNIVERSITY OF IOWA, IOWA CITY, IOWA 52242, USA

Summary

The great majority of drugs have amine functionality. These drugs when mixed with reducing sugars or other carbonyl containing pharmaceutical adjuvants often produce extensive mottling or discoloration, ("Maillard browning"), of the final dosage form over time. In this article, the results of various studies that have been reported involving the Maillard reaction and pharmaceutical products are reviewed.

Introduction

Pharmaceutical preparations containing reducing sugars or other carbonyl containing compounds such as pharmaceutical adjuvants, and amine drugs, frequently show a non-enzymatic browning ("Maillard") reaction during storage. The occurrence of the Maillard reaction in food products may result in the formation of desirable or undesirable flavor and color, but its occurrence in pharmaceutical formulations is indicative of incompatibilities between the drug(s) and supposedly inert carbonyl containing pharmaceutical adjuvants. In addition to producing mottling, or discoloration of the dosage form, the Maillard reaction can cause a reduction in the effective amounts of the therapeutic agents present, may alter drug bioavailability, and/or may produce compounds that can cause adverse or toxic effects.

The reactions of carbohydrates with amines, causing Maillard browning, are well known and were reviewed by Mauron (1981). Reactions occur with aldehyde groups of the open-chain forms of carbohydrates, to form schiff bases, which in turn undergo cyclization to give the corresponding N-substituted glycosamine. The latter, under mild acidic conditions, undergoes Amadori rearrangement to produce N-substituted 1-amino-1-deoxy-2-ketose forms. This Amadori product, depending on the reaction conditions, undergoes degradation reactions to produce a variety of flavor components and pigments. The various types of sugars that have been implicated in the browning of pharmaceutical products include glucose, lactose, dextrates, dextrose, and sucrose. Glucose is commonly used as a nutritional supplement or sweetening agent in liquid dosage forms, whereas other sugars serve as diluents (or fillers) in tablet formulations. Other commonly used fillers include: mannitol, sorbitol, calcium sulfate, dibasic calcium phosphate NF, tribasic calcium sulfate NF, starch, calcium carbonate, and microcrystalline cellulose. Fillers are important ingredients. They are used in large amounts, and can greatly affect the biopharmaceutic and physicochemical properties of the final dosage form. Carbohydrate fillers are particularly important because of their non-toxic nature, acceptable taste, low cost, reasonable solubility profiles, and their general ability to enhance the mechanical strength of tablet dosage forms. Other ingredients commonly used in tablet formulations, in addition to drugs and fillers, include: binders (dry or wet), lubricants, dissolution enhancers, dissolution retardants, antiadherants, glidants, wetting agents, antioxidants, preservatives, coloring agents, and flavoring agents.

Maillard Browning of Lactose-Amine Drug Systems

Lactose is the most commonly and widely used diluent/filler in tablets. Its solubility and sweetness are much lower than other sugars. It is obtained from whey, a milk by-product during cheese manufacture, and is available commercially in both crystalline and spray-dried forms. Spray-dried lactose was developed largely for its special utility in tablet making (excellent powder flow, may be directly compressed following simple admixture with many drugs, and good compression characteristics). However, spray-dried materials have been found to exhibit the greatest tendency to discolor on storage. Brownley and Lachman (1964) and Koshy et al. (1965) have shown that there is a direct relationship between the degree of browning and the amounts of 5-hydroxymethylfurfural (5-HMF) present in the lactose. Since samples of crystalline and spray-dried lactose from different suppliers show different properties, they cannot be used interchangeably in pharmaceutical formulations.

The browning of pharmaceutical dosage forms containing lactose and amine drugs, in addition to other ingredients, has been studied by several investigators (Blaug and Huang, 1972; Duvall et al., 1965a and 1965b; Hammouda and Salakawy, 1971a, 1971b and 1972; Henderson and Bruno, 1970). Duvall et al. (1965b) investigated the browning of dextroamphetamine sulfate solutions, (pH 8), containing lactose, and found that the solution becomes progressively darker on storage at 50°C with the formation of a brownish black precipitate. A new spectral peak with maximum at 320 nm in the ultraviolet-visible spectrum of the solution was noted. On the basis of infrared data and thin-layer chromatographic results, they concluded that the product is of a Schiff base type, formed from a reaction between the primary amine of the dextroamphetamine and the carbonyl group of the lactose.

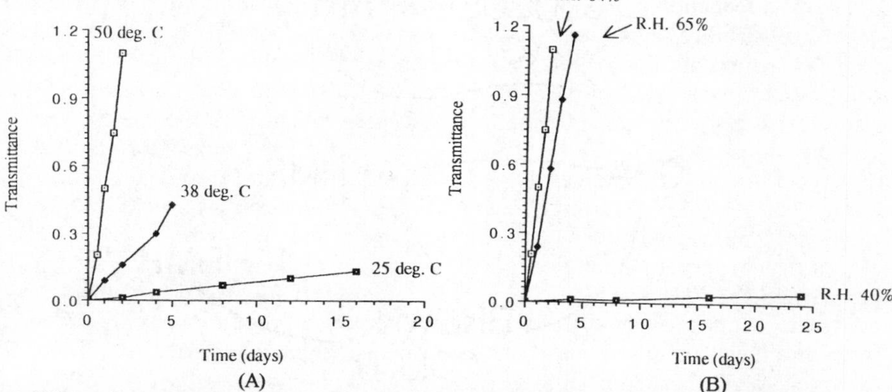

Figure 1 Effect of temperature (A) and relative humidity (B) on the browning of Antacid tablets (initial pH of tablets 8.25)

Salakawy and Hammouda (1972) studied the effect of temperature, pH and relative humidity on the browning of antacid tablets that contained calcium carbonate, lactose, glycine, gelatin, and talc as ingredients. Three other formulations, one without lactose, (initial pH 8.25), and the other two formulations without calcium carbonate, (one was adjusted to an initial pH of 7 and the other to pH 6.0), were used for comparison purposes. Their results indicated that the browning of antacid tablets is lactose induced, and it increases with an increase in the temperature (Figure 1A) and relative humidity (Figure 1B).

Calcium carbonate enhances Maillard browning through its effect of increasing the initial pH of the tablets. The pH of the tablets decreased as the browning reaction progressed. It was suggested that the initial pH of the tablets was the basic factor for the browning of the antacid tablets. Neomycin tablets containing lactose and other ingredients have been reported to exhibit similar effects of temperature, relative humidity and pH (Hammouda and Salakawy, 1971b).

Wu et al. (1970) investigated the browning of isoniazid-lactose system in solid state using diffuse reflectance spectroscopy (DRS). The rates of browning at 95, 100, 105, and 110°C were determined by measuring the reflectance at 450 nm. From the Arrhenius plot (log k versus 1/T), it was suggested that the lactose would require an infinitive time to turn brown at 25°C, whereas the isoniazid-lactose system would discolor in approximately four years. Brownley and Lachman (1964) reported that lactose remained white after storage for 36 months under ambient conditions. The plot between remission function, $f(r_\infty)$, which was calculated according to the equation described by Wendlandt and Hecht (1966), and the heating time indicated that the browning of lactose-isoniazid occurs in two stages. Initially, the browning reaction, represented by a sigmoidal curve, involves dehydration of lactose as well as the chemisorption of isoniazid on the surface of lactose. The water liberated from the lactose produces a true solution phase. The dehydrated lactose and the isoniazid then proceed to react to form a condensation product and various carbonyl compounds, (e.g., 5-hydroxymethylfurfural), through a stepwise degradation of lactose. During the second stage, the browning reaction follows an apparent zero order rate law. A mechanism for the browning of the lactose-isoniazid system, as proposed by Wu et al. (1970), is shown in Figure 2.

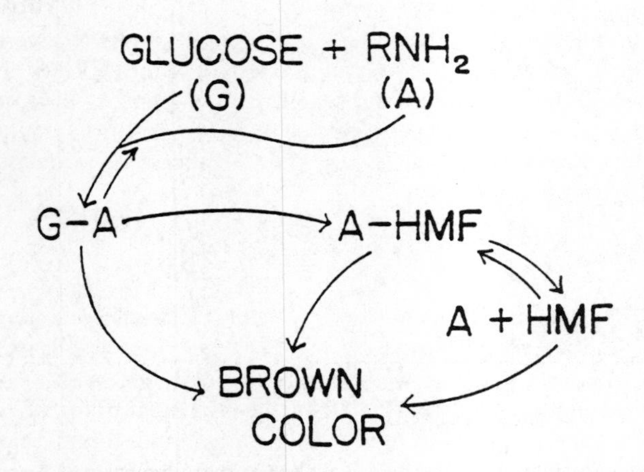

Figure 2 Mechanism for the browning of lactose-isoniazid system

Blaug and Huang (1972) studied the browning of spray-dried lactose and dextroamphetamine sulfate solid-solid mixtures using DRS, and found that the browning of dextroamphetamine sulfate-lactose follows the same mechanisms as suggested by Wu et al. (1970), for the isoniazid and lactose system. The browning rate constants, determined at 25°C by extrapolation of an Arrhenius-type plot of log k versus 1/T, were reported as approximately 50.7 and 79.3 days for samples containing 10 and 15 mg of dextroamphetamine sulfate per gram of spray-dried lactose, respectively.

Maillard Browning of Dextrates-Amine Drug Systems

Dextrates suitable for use in pharmaceutical formulations are obtained by controlled hydrolysis of starch. One dextrate product consists of 92% dextrose and 8% higher saccharides and is useful as a direct compression excipient. Henderson and Bruno (1970) reported that, like lactose, dextrates also show a tendency to discolor on storage at different temperatures and humidities. This browning of dextrates is due to dextrose. Duvall et al. (1965b) found that it is linearly related to the formation of the 5-HMF, a degradation product of dextrose. Compared to lactose, dextrose is relatively less susceptible to browning (Duvall et al., 1965a). However, in the presence of amines, dextrose shows more browning than lactose (Duvall et al., 1965b). It has been suggested that the browning is predominantly due to a primary amine-carbonyl (Maillard) reaction.

Blaug and Huang (1973 and 1974) investigated the browning of dextrates and dextroamphetamine sulfate in solid state and in solution phase, (in citrate-phosphate buffer, pH 8), by DRS and absorption spectroscopy, respectively. The rate of discoloration of both solid mixtures and solutions increased on storage as temperature was increased. The rate of browning of solutions increased with an increase in temperatures and pH and decreased with increasing dextroamphetamine sulfate concentration. The rate of browning of solid dextrates-dextroamphetamine sulfate mixtures was found to be considerably slower than that reported for the spray-dried lactose-dextroamphetamine solid mixtures (Blaug and Huang, 1972). The estimated times for browning of dextrate-dextroamphetamine solid mixtures containing 10 mg and 15 mg of dextroamphetamine per gram of dextrate (determined by the Arrhenius-type plots), at 25°C, were approximately 845 and 793 days, whereas for the solutions containing 0.75%, 1%, and 1.25% dextroamphetamine sulfate and 10% dextrates were 33, 41, and 43 days, respectively. On the basis of elemental analysis, infrared data and thin-layer chromatography, the brown solid isolated from solutions was characterized as a condensation product of amphetamine and 5-HMF.

Miscellaneous Examples

Simmon et al. (1970) analyzed the benzocaine content in fifteen commercial throat lozenge products and found that only four products met the label claim for the actual benzocaine content. The other eleven products were found to contain benzocaine N-glucoside as an impurity. Kabasakalian et al. (1969) studied the incompatibility of benzocaine with various additives commonly used in throat lozenges and found that benzocaine is incompatible with corn syrup, citric acid, and natural cherry flavor. Corn syrup contains glucose, whereas both reducing sugars and aldehydes are present in natural cherry flavor. It was suggested the benzocaine reacts with glucose to form N-arylglycosylamine, which then undergoes hydrolysis as well as Amadori rearrangement to form amino ketones called isoglucosamines.

Kigazawa et al. (1973) reported discoloration of the sugar-coated tablets of norphenylephrine hydrochloride when stored in an oven for 24 hours at 25°C. They attributed the browning of the tablets to the reactions between norphenylephrine hydrochloride and decomposition products of sugar, 5-HMF and levulinic acid.

Nystatin is a polyene-amino sugar (mycosamine: 3-amino-3,6-dideoxy-D-mannose) containing antifungal antibiotic. It is available in varying degree of crystallinity, has poor aqueous solubility and is extremely sensitive to heat, atmospheric oxidation, and pH (outside the pH range of 6 to 9). These properties make it difficult to formulate this drug into a pharmaceutical dosage form with optimum efficacy and physicochemical stability. Ward (1975) reported that nystatin tablets containing anhydrous citric acid and other ingredients, when stored at room temperature, 40°C and 50°C, showed extensive mottling after 12 months, 5 months, and 6 weeks, respectively. Solid-state interactions of mannosamine hydrochloride with anhydrous acids (citric acid, oxalic acid, ascorbic acid, and tartaric acid) over the temperature range 20-150°C resulted in the formation of color, whereas similar solid-state reactions of acids with fillipin (a sugar-free neutral polyene antibiotic) did not produce any color. This led to the conclusion that the nystatin in the presence of acid undergoes hydrolysis of the glycosidic bond to produce the free amino-sugar, which then undergoes a series of complex reactions (carmelisation and/or Maillard reaction), to produce discoloration of the tablets.

El-Kheir et al. (1991 and 1975), Metwally et al. (1991) and Despande and Shirolkar (1987) reported the browning of various commercial liquid oral formulations containing sugars and amine drugs in addition to other adjuvants. Oxidized celluloses which contain aldehyde, carboxylic, and/or ketone anhydroglucose units, in addition to glucopyranose rings, have also been found to produce extensive mottling or discoloration when used with amine drugs in solid and liquid formulations (Kumar and Banker, unpublished work).

Inhibitors of the Maillard Reaction

Inorganic bisulfites have been commonly used for the stabilization of pharmaceuticals (Schroeter, 1961). Their use in preventing Maillard browning of pharmaceutical dosage forms was first suggested by Griffin and Banker (1967). The effect of concentration of sodium bisulfite on the browning rate of neomycin-lactose tablets, as reported by Hammouda and Salakawy (1971b), is shown in Figure 3.

Showa (1982) reported that browning of white amino acid additives in the presence of sugars in pharmaceuticals and foods can be prevented by coating the surface of the amino acid particles with soluble high molecular weight substances (e.g., dextrin, gelatin, araroba gum, CMC, etc.). A mixture of dextrin coated glycine particles and D-glucose remained white after storage at 50°C for 30 days, whereas a control mixture containing uncoated glycine particles and glucose turned brown.

The use of flavonoid glycoside to prevent discoloration of ascorbate in pharmaceuticals, cosmetics and food has been reported by Inoue and Akiyama (1992). Drakoff (1983) revealed that the browning of aqueous cosmetic formulations containing amino or nitro groups, polyhydroxy compounds or sugars and having a pH of about 7.0 or higher can be prevented by using sodium, potassium, ammonium, and other cationic salts of hydroxymethane sulfonate.

Other potential inhibitors that could be used in pharmaceutical preparations are flavonoid containing extract of Scutellaria baicalensis (Morisaki and Inoue, 1991), ascorbic acid tocopheryl diesters (Inoue, 1989), quinoline and oxophenoxazine derivatives (Inoue, 1991a), anthranilate derivatives (Inoue, 1991b), aminoguanidines (Onada et al., 1989; Oonada et al. 1989; Ohuchida et al., 1989a, 1989b, and 1989c), carbazoyls (Ohuchida et al. 1989d), hydrazide and semicarbazide derivatives (Ohuchida et al., 1989e), malonamide, cyanoacetic acid hydrazide, 2,4-pentanedione, and methylsulfonyl-acetonitrile (Ulrich and

Cerami, 1991), and benzopyrans (Ohuchida et al. 1990). These agents have been formulated into pharmaceutically acceptable solid, solution, and suspension dosage forms, and used for inhibiting the Maillard reaction in vivo. The Maillard reaction in the human body has been reported to be responsible for the development of diabetic complications and various age related disorders (e.g., cataract, neuropathy, nephropathy, arthrosclerosis and atherosclerosis).

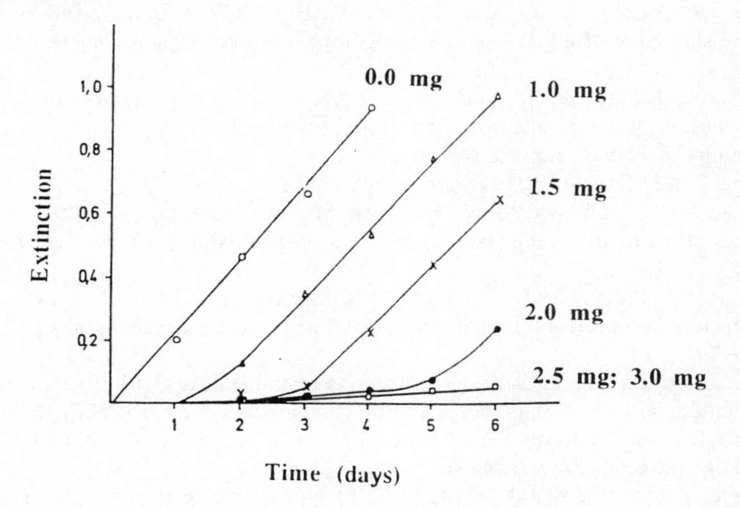

Figure 3 Effect of concentration of sodium bisulfite on the rate of discoloration of neomycin-lactose tablets at 84% relative humidity and 50°C (Tablet composition (mg/tablet): neomycin sulfate 5.0; Lactose 100.0; magnesium stearate 2.0; and sodium bisulfite 0.0-3.0)

Concluding Remarks

Compared to food and biological systems, fewer studies have been reported on the Maillard reaction in pharmaceutical products. Pharmaceutical scientists have largely identified the excipients that are likely to produce this reaction, and formulators typically attempt to avoid ingredients that interact with basic amine drugs to produce the non-enzymatic browning reaction. The pharmaceutical literature is however quite helpful in identifying the range of excipient materials, that are also commonly used in foods and other consumer products, including reducing sugars, polyhydroxy derivatives, or other carbonyl containing compounds which have the potential to exhibit non-enzymatic ("Maillard") browning reactions. The pharmaceutical literature, as surveyed herein, is also quite helpful in reporting on mechanisms and specific agents, (including some materials acceptable for use in foods), capable of inhibiting the Maillard reaction, and thereby stabilizing products.

References

Blaug, S. M., and Huang, W.-T. (1972) Interaction of dextroamphetamine sulfate with spray-dried lactose. *J. Pharm. Sci.*, 61: 1770-1775.

Blaug, S. M., and Huang, W.-T. (1973) Interaction of dextroamphetamine sulfate with dextrates in solution. *J. Pharm. Sci.*, 62: 653-655.

Blaug, S. M., and Huang, W.-T. (1974) Browning of dextrates in solid-solid mixtures containing dextroamphetamine sulfate. *J. Pharm. Sci.*, 63: 1415-1418.

Brownley, C. A., and Lachman, L. (1964) Browning of spray-dried lactose. *J. Pharm. Sci.*, 53: 452-454.

Deshpande, S. G., and Shirolkar, S. (1987) Stability of bromhexine hydrochloride in liquid oral formulations. *The Eastern Pharmacist (December)* 135-138.

Duvall, R. N., Koshy, K. T., and Dashiell, R. E. (1965a) Comparative evaluation of dextrose and spray-dried lactose in direct compression systems. *J. Pharm. Sc.*, 54: 1196-1200.

Duvall, R. N., Koshy, K. T., and Pyles, J. W. (1965b) Comparison of reactivity of amphetamine, metamphetamine, and dimethylamphetamine with lactose and related compounds. *J. Pharm. Sci.*, 54: 607-611.

Drakoff, R. (1983) Cosmetic preservatives. *US 4375465.*

El-Kheir, A. A., El-Bolkiney, M. N., El-Sayed, W., and Metwally, M. M. F. (1991) Study of factors affecting browning reaction in stored cough mixtures. *Egypt. J. Pharm. Sci.*, 32: 175-180.

El-Kheir, A. A., Salem, F. B., Ahmad, A. K. S., and Amer, M. M. (1975) Browning reactions in pharmaceutical preparations containing liquid glucose. *Egypt. J. Pharm. Sci.*, 16: 219-231.

Griffin, J. C., and Banker, G. S. (1967) Reactions of amine drugs with sugars I. Reaction of amphetamine with aldoses, glycosylamines, and bisulfite. *J. Pharm. Sci.*, 56: 1098-1101.

Hammouda, Y., and Salakawy, S. A. (1971a) Lactose-induced discoloration of amino drugs in solid dosage form. *Pharmazie*, 26: 181.

Hammouda, Y., and Salakawy, S. A. (1971b) Non-enzymic browning in solid-dosage forms: lactose-induced discoloration of neomycin tablets. *Pharmazie*, 26: 636-640.

Henderson, N. L., and Bruno, A. J. (1970) Lactose USP (beadlets) and Dextrose (PAF 2011): Two new agents for direct compression. *J. Pharm. Sci.*, 59: 1336-1340.

Inoue, J. (1989) Ascorbic acid tocopheryl diesters for inhibition of Maillard's reaction. *EP 0,430,045 A2.*

Inoue, J. (1991a) Pharmaceutical compositions and methods using quinoline and oxophenoxazine derivatives for inhibition of Maillard's reaction. *CA 2029421AA.*

Inoue, J. (1991b) Maillard reaction inhibitor for treatment of diseases associated with diabetes. *PCT Intl. 11997.*

Inoue, T., and Akiyama, K. (1992) Prevention of ascorbate browning in pharmaceuticals, cosmetics and food. *JP 04-99771.*

Kabasakalian, P., Cannon, G., and Pinchuk, G. (1969) Fractional factorial experimental design study of the incompatibility of benzocaine in throat lozenges. *J. Pharm. Sci.*, 58: 45-47.

Kigazawa, K., Ikari, N., Ohkubo, K., Iimura, H., and Haga, S. (1973) Decomposition and stabilization of Drugs. VIII. Reaction products of norphenylephrine with degradation products of sugars. *Yakugaku Zasshi*, 93: 925-927.

Koshy, K. T., Duvall, R. N., Troup, A. E., and Pyles, J. W. (1965) Factors involved in the browning of spray-dried lactose. *J. Pharm. Sci.*, 54: 549-554.

Mauron, J. (1981) The Maillard reaction in food; a critical review from the nutritional standpoint. *Prog. Fd. Nutr. Sci.*, 5: 5-35.

Metwally, M. M. F., El-Bolkiney, M. N., El-Syed, W., and El-Kheir, A. A. (1991) Identifications of the carbonyl compounds formed in stored cough mixtures. *Egypt. J. Pharm. Sci.*, 32: 787-796.

Morisaki, M. and Inoue, A. (1991) Pharmaceuticals containing inhibitors of Maillard reaction. *JP 03-240725*.

Ohuchida, S., Toda, M., and Miyamoto, T. (1989a) Aminoguanidine derivatives and inhibitory agents on Maillard reaction containing them as active ingredients. *EP 0,325,936 A2*.

Ohuchida, S., Toda, M., and Miyamoto, T. (1989b) Aminoguanidine derivatives. *EP 0,339,496 A2*.

Ohuchida, S., Toda, M., and Miyamoto, T. (1989c) Preparation of N-(arylamino)guanidine derivatives with inhibitory activity of Maillard reaction. *EP 325936*.

Ohuchida, S., Toda, M., and Miyamoto, T. (1989d) Carbazoyl derivatives. *EP 0,323,590 A2*.

Ohuchida, S., Toda, M., and Miyamoto, T. (1989e) Preparation of hydrazide and semicarbazides as Maillard reaction. *EP 323590*.

Ohuchida, S., Toda, M., and Miyamoto, T. (1990) Benzopyran derivatives. *EP 0,387,771 A2*.

Onada, S., Toda, M., and Miyamoto, T. (1989) Inhibitors (e.g., aminoguanidines, etc.) of the Maillard reaction and formulations containing them. *JP 64-56614*.

Oonada, S., Toda, M., and Miyamoto, T. (1989) Sulfonylaminoguanidines as Maillard reaction inhibitors. *JP 64-83059*.

Salakawy, S. A., and Hammouda, Y. (1972) Non-enzymic browning in solid-dosage forms: Factors involved in the browning of antacid tablets containing glycine. *Pharmazie*, 27: 595-599.

Schroeter, L. C. (1961) Sulfurous acid salts as pharmaceutical antioxidants. *J. Pharm. Sci.*, 50: 891-901.

Showa, D. K. K. (1992) Coating of amino acids with polymers for prevention of discoloration in pharmaceuticals and foods. *JP 57-128662*.

Simmons, D. L., Ranz, R. J., and Quan, C. C. D. (1970) Available benzocaine content in throat lozenge products on the Canadian market. *Can. J. Pharm. Sci.*, 5: 85-86.

Ulrich, P. C., and Cerami, A. (1991) Inhibitors of the advanced glycosylation of proteins (Maillard reaction inhibitors), and methods of use therefor. *EP 450598*.

Ward, M. D. (1975) A nystatin-acid solid state interaction study. *J. Pharm. Pharmacol.*, 59: 59P.

Wendlandt, W. W., and Hecht, H. G. (1966). *Reflectance spectroscopy*, New York, NY.: Interscience.

Wu, W.-H., Chin, T.-F., and Lach, J. L. (1970) Interaction of isoniazid with magnesium oxide and lactose. *J. Pharm. Sci.*, 59: 1234-1242.

Anthropological Implications of the Maillard Reaction: An Insight

Luc De Bry

GENERAL BISCUITS BELGIUM DE BEUKELAER-PAREINLAAN 1, B-2200 HERENTALS, BELGIUM

Summary
Investigation into the origins of the Maillard reaction, subsequent to the mastering of fire by humans, some 700,000 years ago, raised the question : "Why do we heat some of our foods?" Attempts to answer this led to the formulation of the following hypothesis: if we roast cocoa beans, bake bread, cook soybeans, etc., and not strawberries or lettuce, is it to obtain the desired Maillard aroma, or is it because we want to obtain safe access to rich sources of kilocalories ? Applying the new heating technology to storage organs of higher plants, and using their organoleptic senses, our ancient ancestors realized a useful action. In essence we find a negative correlation between the activity of anti-nutritional factors and the generation of Maillard compounds from heating ($r = -0.93$ for cocoa and -0.987 for soybean). Analyzing the plant-food-animal co-evolution and the appearance of Angiosperms suggests that, acting as a food safety indicator, non-enzymatic browning caused by heating has given a tremendous nutritional and competitive advantage to *Homo sapiens* over other species, whereas, more than asteroids, volcanoes, oxygen deficit or other yet undetermined factors, it may well be the failure to adapt to the need for developing the Maillard reaction which may have brought the end of the last dinosaur, and stuck apes at the level of the banana. Thus, in the search for safe kilocalories, a major paradigm shift of the human race was to cross over the fire frontier and benefit from its chemical efffects.

Introduction

First the Maillard Reaction or First the Seed ?
Back from Switzerland and the previous Maillard Symposium, a food chemist and a microbiologist concluded their report saying : "God gave us two great gifts : fermentation and the Maillard Reaction!". A plant scientist unexpectedly appeared. "Hey, wait a minute", he said. "Aren't you forgetting something, i.e. the Seed!" "The Seed ??", replied the two other scientists. "Well, yes, the Seed", repeated the plant scientist, "because without seeds, you would have neither fermentation nor the Maillard reaction, would you?"."Well, we never thought about it that way", murmured the two bewildered food scientists. "So, actually, God gave us one excellent thing, the Seed, plus two accessories, i.e. fermentation and the Maillard Reaction. Thus, the plant scientist concluded "First the Seed!" the motto of the American Seed Association.

Is this so? The biological sciences of botany, zoology, anthropology, and food science are different disciplines although complementary. The former are largely observational, centered on interactions between paleo- and living-organisms in their natural habitats, and are carried out in the fields of Nature. By contrast, the latter is experimental, concentrating on interactions at the molecular level, and carried out at the laboratory bench or in the food processing plant with a safe and nutritious consumable food as the ultimate goal.

Table I shows the caloric range for some foods, the first four items which can be eaten raw are low in caloric content. Seeds and tubers, which contain more calories, cannot be eaten raw and require a heating treatment to process them into the edible products listed lower in the table. Table II shows some of the typical problems involved in eating these plant parts raw, and establishes the reasons why food processing, including cooking (heating), may have come about. The toxicological problem arises from the fact that the seed and tuber are the carriers of the genetic material and plants developed chemical defense mechanisms to protect these sex gene carriers. To eat them without any kind of processing could cause severe metabolic disorders leading to death.

Table I. Range of kilocalories contained in plant organs and foods.

PLANT ORGANS and FOODS		RANGE of kcal/100g
GATHERING	Leaves	7 - 15
	Flowers	10 - 20
	Roots	20 - 40
	Fruits	10 - 91
	Tubers	100 - 150
	Seeds	260 - 700
	Legumes	260 - 340
	Cereals	285 - 385
	Oil-seeds	525 - 700
PROCESSING (Heating)	Pastries	300 - 450
	Biscuits	350 - 550
	Chocolates	600 - 650
Average daily need(Human)= 2500 kcal		

Indeed, higher plants do not generate energy-rich seeds or tubers for humans to eat. They are there for promoting new lives upon germination, and, as seeds and tubers cannot run away from predators, plant descendants and kilocalories are protected by anti-nutritional factors, e.g. cyanogens, anti-trypsin, alkaloids, lectins, polyphenols, etc. (Table II). Even wheat, rice and soybean, the three most important crops of the world, are highly toxic (D'Mello et al., 1991; Thompson, 1993). The higher the kilocalorie content, the higher seems to be the level of anti-nutritional activity, eg. cowpeas (Marconi et al, 1993). Maillard reaction studies in foods have generally been concerned with color or flavor or the loss of protein quality upon the heating, or high temperature storage of sugar-amine systems (Parliment et al., 1989; Finot et al., 1990). Non-enzymatic browning thus has become a widely used quality parameter in applications such as drying, frying, extrusion and baking (Nursten, 1986). On a different spectrum, plant toxicologists realize that heating is helpful for inactivating protein inhibitors such as phytohaemagglutinins (D'Mello et al., 1991). The apparent discrepancy between the nutritional loss of essential amino acids during the Maillard reaction and the advantage of eliminating anti-nutritional activity has already been raised (Labuza, 1972). It should be noted that, although studied for almost one hundred years, the kinetics of the Maillard reaction are still not well understood (Labuza and Baisier, 1992) and the implications of the reaction for the aging process of humans are just beginning to be recognized in the medical field (Monnier and Baynes, 1989).

Today, humans seem to be the only animals willing to risk food poisoning from seeds on a scale enlarged by modern agriculture, and regularly applying the Maillard reaction before eating any seed or tuber. For about 700,000 years, humans are also the sole animals capable of mastering fire (Leakey and Lewin, 1992). Hence the question of why historically did humans learn to roast cocoa beans, bake bread, cook soybeans etc., was it really for obtaining Maillard aroma and color, or was it because we wanted to obtain safe access to their rich source of kilocalories?

Table II. Metabolic effects of some anti-nutritional factors from higher plants.

ANTI-NUTRITIONAL FACTORS	HIGHER PLANTS	METABOLIC EFFECTS UPON INGESTION
Cyanogens	Cassava, peanut	Death within one hour
Protease Inhibitors	Cereals, soya	Death within a month or two
Alkaloids	Potato, pepper	Block action of neuro-transmitters, teratogenic
Amino Acids Analogs	Legumes	Osteo- and neuro-lathyrism, abnormalities, paralysis
Lectins (phytohaemagglutinins)	Soya	Aggregation of red blood cells, gastroenteritis, nausea
Saponins	Legumes	Haemolysis of red blood cells
Glucosinolates	Oilseed rape	Goitrogenicity non-alleviable by iodine
Allergens	Wheat gliadins	Altered immunological response
Fibrous Polysaccharides	Plant cell walls	Reduction of mineral bioavailability

Life and death with the Maillard Reaction

This enhancement of nutrition to attain food satiety, through the development of food safety, looks unique to *Homo sapiens*. Indeed, natural plant poisons, ubiquitous among the Angiosperms, are so powerful that alkaloids have even been proposed as potentially responsible for the demise of dinosaurs (Swain, 1977). The proper use of the Maillard reaction would thus promote life, whereas its absence would accelerate death and animal species extinction.

Development of Food Safety and Satiety

By heat treatment, a process referred to as toasting in the soy processing industry, the inferior nutritive value of raw soy protein products can be improved to one nearly comparable to meat and milk. Loss in anti-trypsin activity has been shown to be a function of moisture content, temperature, particle size and time (Rackis, 1965). One can observe that there is a negative correlation (r = - 0.987) between anti-trypsin activity and protein efficiency upon autoclaving (Figure 1) of raw soy beans. Thus, up to a well defined level, heat processing contributes to developing food safety and satiety.

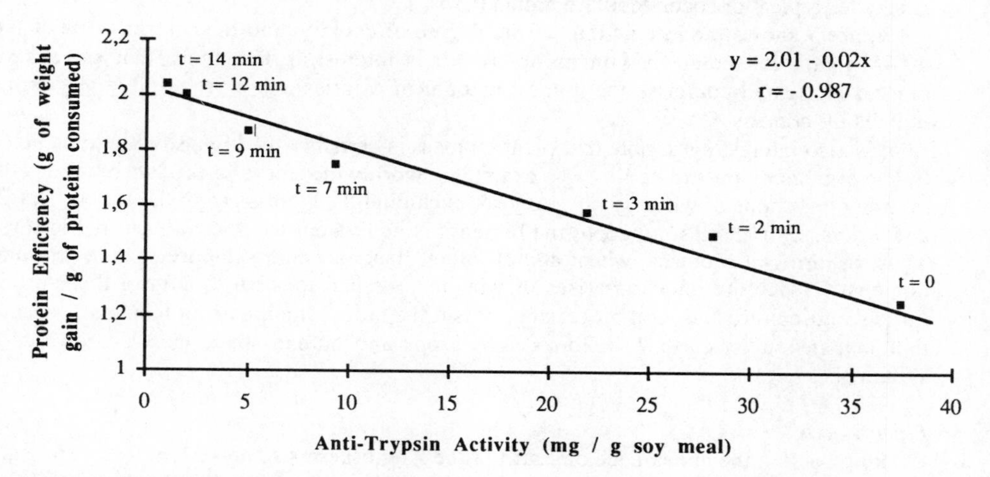

Figure 1. Inactivation of soy anti-trypsin by autoclaving, calculated from data of Rackis, 1965.

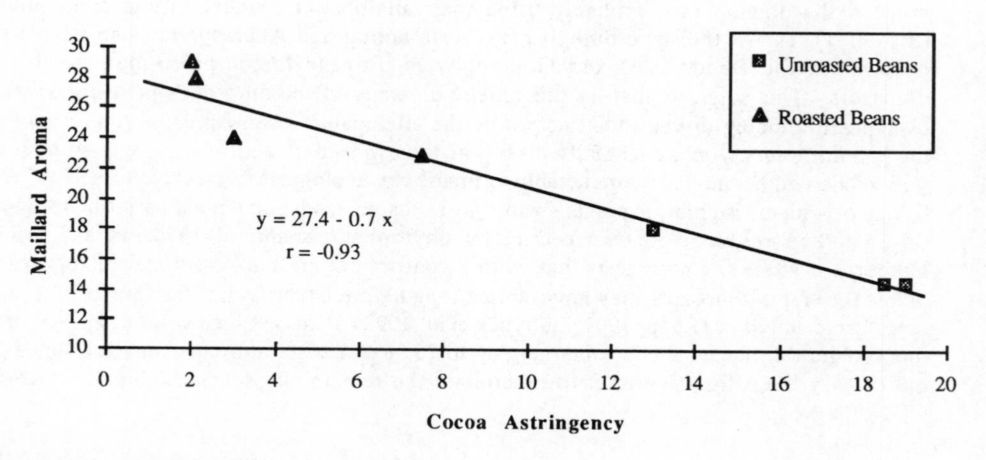

Figure 2. Cocoa Maillard aroma and astringency upon roasting.

When roasting Forastero cocoa beans, one can observe a negative correlation (r=- 0.93) between the loss of astringency and the formation of desirable Maillard aromas (Figure 2). Unfortunately cocoa beans have a fairly high polyphenol content (Pierpoint, 1985) which can elicit an astringent response (Ozawa et al., 1987) and can induce esophageal cancer (Morton, 1978).Astringency which has a negative effect is defined as the tactile effect impacting on the whole mouth epithelium, i.e. tongue, palate, jaws, internal cheeks and lips (Breslin et al., 1993).The loss of astringency would be due to the complexation of polyphenols with cocoa proteins, rendered irreversible upon roasting. Furthermore, as cocoa proteins have been shown to have anti-trypsin activity (Dodo et al., 1992), their irreversible complexation with polyphenols would also inactivate their inhibitory potential. In model systems, the addition of epicatechin has been shown to decrease the formation of aldehydes typical of cocoa Maillard aroma (Cros, 1992).

Figure 3 shows the interaction of the degree of cooking and the caloric value of the major plants consumed by humans today. It is interesting that those that are baked, roasted or toasted comprise the highest amount of calories and exist as the major crops utilized by humans.

It is also interesting to note that plant biomass is certainly the highest volume niche of living organisms on the earth. For example, worldwide, there is produced about 600 million metric tons of wheat grains per year, excluding the biomass from the stalks, straws and roots. For 5 billion people the biomass is equivalent to 350 million metric tons. Thus, in terms of biomass, wheat and all other plants are more important on earth than humans. Hence the question arises of whether we had to learn to control these crops through agriculture and food processing, or is it the crops who are using humans to ensure their expansion on earth ? It looks as if crops and humans have evolved towards a symbiosis.

Paleo-Foods

Looking back to the appearance on Earth of the Angiosperms some 100 million years ago, i.e. the ancestors of all our modern crops, one realizes that they first appeared in little pockets along the coasts. They progressively expanded on the continents, overwhelming most of the other plants, probably using their allelopathic and thus toxic compounds (Swain, 1977). At the same time, it may worth noting that Angiosperms also started to overwhelm the Benettitales and Ginkyophytes, i.e. the major paleo-plant-foods of dinosaurs. This suggests that, as the genetic diversity of the dinosaur's plant-foods was disappearing, being driven to extinction by the allelopathic compounds of Angiosperms, the last dinosaurs would eventually disappear too. Indeed, genetic diversity is correlated with niche width, and partly predictable, primarily by ecological factors (Nevo, 1993). No living organism, including plants, wants to be eaten, and plant poisons seem to have originated in response to overfeeding by phytophagous animals (Coe et al., 1987; Harborne, 1988). This suggests that, with a contribution of anti-nutritional factors from plants the end of dinosaurs may have started long before the arrival of the famous asteroid or volcanic activities (Raup, 1991; Bultynck et al.,1992) Birds evolved with beaks enabling them to get rid of seed shells containing up to 75 % of the antinutrients, and developed a pH of 4 in their digestive tract to precipitate the remaining protease inhibitors (Kolb, 1975).

Conclusion

Evolutionary Concepts in Maillard Reaction Thoughts

Why have plants evolved such powerful chemical defenses? How was this heating process and the correlation between food safety and aroma plus color discovered, and why is it that only humans are benefiting from this useful action? One of the main objects of this presentation was to portray a picture of an evolutionary possible function of the Maillard reaction developed by humans. Maillard reaction volatiles have a positive impact in the identification of safe foods. The molecular structure of one key reactant , i.e. the amino acids, is directly correlated not only to the genetic code, but also to the basic tastes (Siemion et al., 1987). It suggests that olfactory receptors have evolved as a way to recognize food safety, whereas taste buds have evolved to locate kilocalories and essential nutrients. Overheating to completely remove anti-nutritional activities usually results in protein degradation and other deteriorative reactions, giving rise to "burnt reaction products", that cause other adverse effects in animals due to the formation of mutagens and carcinogens (O'Brien and Morrissey, 1989; Weisburger, 1992). Therefore, precise control of the heating process and Maillard reaction kinetics is critical to the preparation of modern safe foods with enhanced nutritional value. The negative correlations (Figures 1 and 2) suggest that it would be the function of the Maillard aroma and color to act as a food safety indicator with the additional role of the expansion of the food kilocalorie supply available to those who learned how to cook (Figure 3).

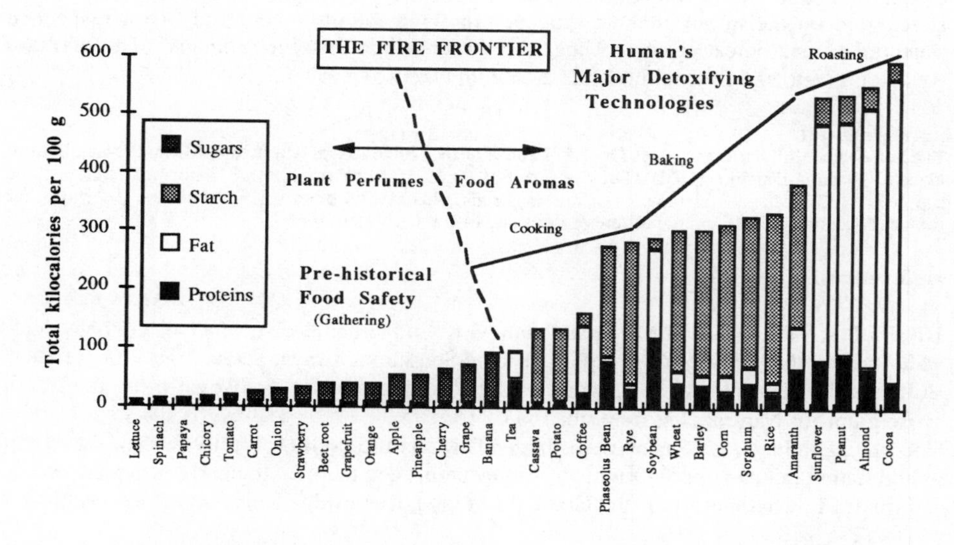

Figure 3. Kilocalorie content of the 33 most important world crops; (compiled from Kloppenburg, 1988; Scherz and Senser, 1989).

The elucidation of functional characteristics is not always obvious and is, therefore, one of the challenging problems presented by and to living organisms. About one hundred million years ago, Angiosperms appeared. *Homo sapiens* appeared one million years ago. In between these two events were the dinosaurs who disappeared and the apes and other mammals, who, not being able to cook, constantly ate fruits and leaves to take in enough calories for life. Evolution, in the context of anti-nutritional factors and the Maillard reaction, implies first that, as a rule, non-functional characteristics do not survive, and secondly that those properties which have evolved are likely to have a decisive advantage for survival. Indeed, both the biosynthesis of plant chemical defenses and the development of the Maillard reaction opened the way to the development of agriculture and have given a tremendous competitive advantage to Angiosperms and humans over other species. Thus, the crossing of the fire frontier was a major advantage to the human race.

Perspectives : "First the Maillard Reaction!"
Looking at the origin of agriculture some 10,000 years ago, one may say "First the Seed!". Continuing 700,000 years further back, scientists are realizing that in order to overcome the efficiency of seed poisons, it was "First the Maillard Reaction!" From this perspective, it is worth noting that the great planner of life has left us with a double dilemma;on the one hand, powerful anti-nutritional factors from plants, e.g. protease inhibitors or alkaloids, can bring not only problems, death and extinction (Swain, 1977), but also benefits (Thompson, 1993), e.g. in our fight against pathogens, such as the HIV virus (Tyms et al., 1987; Paris et al., 1993). On the other hand, years after having consistently contributed to develop a safe food supply and to enhance our nutrition, long-term physiological consequences of the Maillard reaction occurring with glucose and the various tissue proteins in our bodies may be as important or more important than many other age-related mechanisms of bringing us closer to death. It is interesting that, in his early work, Dr. L.C. Maillard himself speculated on the importance of glucose-protein reactions in the body with respect to potential physiological effects. Thus, one can conclude that with humans, "The Maillard Reaction giveth and the Maillard Reaction doth taketh away!"

Acknowledgments
The author gratefully acknowledges Dr. T.P. Labuza of the University of Minnesota, M. Ir. T. Neutelings of General Biscuits Belgium (GBB), Dr. F. Martin of Royal Institute for Natural History of Brussels, and suppliers of GBB for useful samples of raw materials, plus several other scientists around the world, for limitless encouragements, supply of rare reprints and lively discussions.

References

Breslin P.A.S., Gilmore M.M., Beauchamp G.K. and Green B.G. (1993). Psychophysical Evidences that Oral Astringency is a Tactile Sensation. *Chem. Senses* 18 : 405-417.

Bultynck P., Dhondt A.V. and Martin F. (1992). Dinosaurs & Co. Royal Institute of Belgium for Natural History (Publ.); pp. 115.

Coe M.J., Dilcher D.L., Farlow J.O., Jarzen D.M. and Russell D.A. (1992). Dinosaurs and Land Plants.*In* : The Origins of Angiosperms and their Biological Consequences. Friis E.M., Chaloner W.G. and Crane P.R. (eds.); Cambridge University Press (Publ.) UK; 225-258.

Cros E. (1992). Effects of Roasting Parameters on Cocoa Aroma Formation. *In* : The Reactions of Maillard and of Caramelisation. Colloque de l'Alliance 7 - Cedus (ed. and Publ.), Paris; 65-69 (*in French*).

D'Mello J.P.F., Duffus C.M. and Duffus J.H. (eds.) (1991). Toxic Substances in Crop Plants. The Royal Society of Chemistry (Publ.); pp. 339.

Dodo H.W., Fritz P.J. and Furtek D.B. (1992). A Cocoa 21 Kilodalton Seed Protein has Trypsin Inhibitory Activity. *Café Cacao Thé* 36 : 279-284.

Finot P.A., Aeschbacher H.U., Hurrell R.F. and Liardon R. (1990). The Maillard Reaction in Food Processing, Human Nutrition and Physiology. Birkhäuser Verlag (Publ.); pp. 516.

Harborne J.B. (1988). Introduction to Ecological Biochemistry. Academic Press (Publ.); pp. 356.

Kolb E. (1975). Physiology of Domestic Animals. Vigot Frères (eds, Publ.) Paris; pp. 974.
Kloppenburg J.R. Jr. (1988). First the Seed : The Political Economy of Plant Biotechnology. Cambridge University Press (Publ.); pp. 349.
Labuza T.P. (1972). Nutrient Losses during Drying and Storage of Dehydrated Foods. *CRC Critical Rev. Food Technol.* 217-240.
Labuza T.P. and Baisier W.M. (1992). The Kinetics of Nonenzymatic Browning. *In* : Physical Chemistry of Foods. Schwartzberg H.G. and Hartel R.W. (eds.), M. Dekker (Publ.), New York; pp. 595-649.
Leakey R. and Lewin R. (1992). Origins Reconsidered : In Search of What makes us Human. Double Day (Publ.); pp. 375.
Marconi E., Ng N.Q. and Carnovale E. (1993). Protease Inhibitors and Lectins in Cowpea. *Food Chemistry* 47 : 37-40.
Martin R.D. (1993). Primate Origins : Plugging the Gaps. *Nature* 363 : 223-234.
Monnier V. and Baynes J. (1989). Toward a Maillard Reaction Theory of Aging. *In* : Maillard Reaction in Aging. Liss Press (Publ.); New York; 15-22.
Morton J.F. (1978). Economic Botany in Epidemiology. *Econ. Bot.* 32 : 111-116.
Nevo E. (1993). Evolutionary Processes and Theory : The Ecological-Genetics Interface. *Wat. Sci. Tech.* 27 : 489-496.
Nursten H.R. (1986). Maillard Reaction in Dried Foods. *In* : Concentration and Drying of Foods; McCarthy D. (ed.); Elsevier Applied Science (Publ.); 53-68.
O'Brien J. and Morrissey P.A. (1989). Nutritional and Toxicological Aspects of the Maillard Reaction in Foods. *CRC Critical Rev. in Food Sci. and Nutrition* 3 : 211-248.
Ozawa T., Lilley T.H. and Haslam E. (1987). Polyphenol Interactions : Astringency and the Loss of Astringency in Ripening Fruits. *Phytochemistry* 26 : 2937-2942.
Paris A., Strukelj B., Renko M., Turk V., Pukl M., Umek A. and Korant B.D . (1993). Inhibitory Effect of Carnosolic Acid on HIV-1 Protease in Cell-Free Assays. *J. Natural Products* 56 : 1426-1430.
Parliment T.H., McGorrin R.J. and Ho C.T. (eds.), (1989). Thermal Generation of Aromas. ACS (Publ.); pp. 555.
Pierpoint W.S. (1985). Phenolics in Food and Feedstuffs : the Pleasures and Perils of Vegetarianism. *In* : Annual Proceedings of the Phytochemical Society of Europe, Van Sumere C.E. and Lea P.J. (eds.), Clarendon Press (Publ.); 427-451.
Rackis J.J. (1965). Physiological Properties of Soybean Trypsin Inhibitors and their Relationship to Pancreatic Hypertrophy and Growth Inhibition of Rats. *Fed. Proc.* 24 : 1488.
Raup D.M. (1991). Extinction : Bad Genes or Bad Luck ? W.W. Norton & Co. (Publ.); pp. 210.
Scherz H. and Senser F. (1989). Food Composition and Nutrition Tables. Deutsche Forchungsantalt für Lebensmittelchemie, Garching b. München (ed.). WVG mbh (Publ.) Stuttgart; pp. 1032.
Siemion I.Z., Paradwski A., Gòrnicki P. and Wiewiòrowski M. (1987). Branched-Unbranched Amino Acids and the Genetic Code. *Bulletin of the Polish Academy of Sciences Chemistry* 35 : 521-528.
Swain T. (1977). Secondary Compounds as Protective Agents. *Ann. Rev. Plant Physiol.* 28 : 479-501.
Thompson L.U. (1993). Potential Health Benefits and Problems associated with Antinutrients in Foods. *Food Research International* 26 : 131-149.

Tyms A.S., Berrie E.M., Ryder T.A., Nash R.J., Hagarty M.P., Taylor D.L., Mobberley M.A., Davis J.M., Bell E.A., Jeffries D.J., Taylor-Robinson D. and Fellows L.E. (1987). Castanospermine and other Plant Alkaloids Inhibitors of Glucosidase Activity block the Growth of HIV. *Lancet* 2 : 1025-1026.

Weisburger J.H. (1992). Mechanisms of Macronutrient Carcinogenesis. *In* : Macronutrients : Investigating their Role in Cancer. Micozzi M.S. and Moon T.E. (eds.). Marcel Dekker (Publ.) New York; 3-31.

The Maillard Reaction of Disaccharides

Monika Pischetsrieder and Theodor Severin

INSTITUT FÜR PHARMAZIE UND LEBENSMITTELCHEMIE DER UNIVERSITÄT MÜNCHEN, SOPHIENSTRASSE 10, 80333 MÜNCHEN, GERMANY

Summary

The Maillard reaction of disaccharides which are connected by a 1,4-glycosidic link leads to the formation of compounds that differ in structure from those obtained from glucose or fructose. 2,3-Dihydro-3,5-dihydroxy-6-methyl-4*H*-pyran-4-one (1) and 4-hydroxy-2-(hydroxymethyl)-5-methyl-3(2*H*)-furanone (2) can be considered as typical products when glucose reacts with primary or secondary amines. On the other hand, when maltose or lactose is heated under comparable conditions, 4-(glycosyloxy)-2-hydroxy-2-methyl-2*H*-pyran-3(6*H*)-ones (3), 4,5-dihydroxy-2-(glycosyloxy)-5-methyl-2-cyclopenten-1-ones (4) and 4-(glycosyloxy)-5-(hydroxymethyl)-2-methyl-3(2*H*)-furanones (5) are obtained as main products. 3 and 4 are transformed during prolonged heating into stable isomaltol glycosides (6) or after elimination of the sugar residue into maltol (7). All these maltose and lactose derived intermediates react further with primary amines, such as propylamine or α-N-acetyl lysine to give a wide range of heterocyclic compounds. These amines represent models for lysine side chains of proteins. Nitrogen containing products, which have been isolated so far are 1-[3-(glycosyloxy)-1-propyl-2-pyrrolyl]-1-ethanones (8), 2-[1-(propylamino)ethylidenyl]-3-(2*H*)furanones (9) and 4-(glycosyloxy)-2-methyl-propylpyridinium-3-olates (10). Degradation of the pyridinium betaines (10) leads to the formation of 3-hydroxy-2-methyl-1-propyl-4(1*H*)-pyridone (11), 3-glycosyloxy-2-methyl-1-propyl-4(1*H*)-pyridones (12) or 1,4-dihydro-2-methyl-1-propyl-4-(propylimino)-3-pyridinol (13). Polysaccharides that are connected by a 1,4-glycosidic link, such as starch or dextrins react in a similar way. The disaccharide degradation products 11 and 13 can be isolated from a reaction mixture of starch and amine.

Introduction

Carbohydrates that are connected by a 1,4-glycosidic link occur in great abundance in foods. Starch for example is the main ingredient of many vegetable foodstuffs. The two most important disaccharides are maltose, which can be found as a degradation product of starch, and lactose, a component of milk and dairy products. Several investigations have shown that the Maillard reaction of disaccharides leads to the formation of substances that are not known to arise from monosaccharides. Many low molecular weight volatile reaction products of maltose have been identified that are of significance for the flavour of heated foodstuffs. However, these compounds play only a minor role in the overall decomposition and must be considered as by-products. In this review, we describe the pathway of the Maillard reaction of 1,4-glycosidic linked carbohydrates and give a survey of the main products of their degradation.

Materials and Methods

The preparation of 3, 4, 6, 8, 9, 10, 12 and 13 is described by Kramhöller *et al.* (1993a); of 5 by Pischetsrieder and Severin (in press). 7 was purchased from Fluka (Switzerland).

Degradation of starch and preparation of 11 were carried out as documented by Kramhöller et al. (1993b).

Results and Discussion

When glucose, maltose or lactose are heated in the presence of primary or secondary amines, the first products that can be obtained are N-substituted amino-2-deoxyketoses. However, these Amadori compounds are not stable when heated or stored, but they give rise to a range of substances with dicarbonyl structures (Figure 1). The proportion of the deoxyosones depends particularly on the pH-value. Degradation in nearly neutral aqueous solution leads to the formation of the 1-deoxyhexosuloses as main products, the further reactions of which are most interesting for food chemistry and physiology.

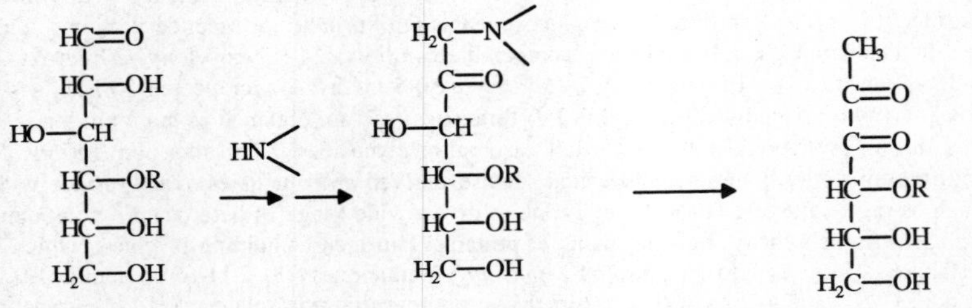

HC=O, HC—OH, HO—CH, HC—OR, HC—OH, H_2C—OH

R = H,glu,gal

HN<

H_2C—N<, C=O, HO—CH, HC—OR, HC—OH, H_2C—OH

N-substituted amino ketoses =Amadori compounds

CH_3, C=O, C=O, HC—OR, HC—OH, H_2C—OH

1-deoxyosones

Figure 1. Initial steps of the Maillard reaction of mono- and disaccharides (I).

The next steps that take place are enolization and cyclization of the 1-deoxyosones, resulting in a five or a six membered ring (Figure 2).

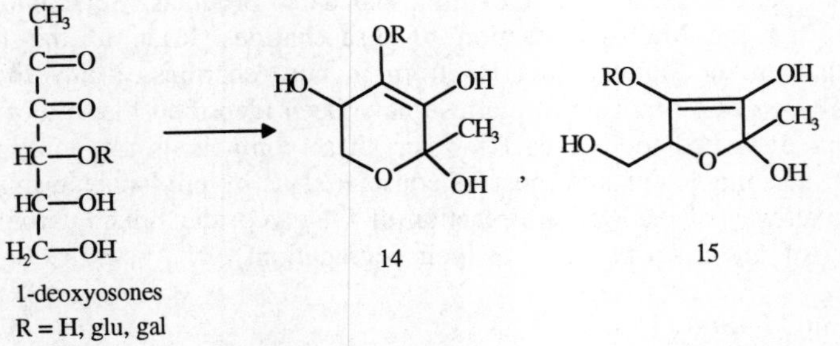

CH_3, C=O, C=O, HC—OR, HC—OH, H_2C—OH

1-deoxyosones
R = H, glu, gal

14 , 15

Figure 2. Initial steps of the Maillard reaction of mono- and disaccharides (II).

From this point on the degradation of monosaccharides and disaccharides follows different pathways. A pyranoid intermediate of general structure 14 is suggested from both mono- and disaccharides. If the R group is a hydrogen atom, the elimination of the α-hydroxyl group is favoured and 2,3-dihydro-3,5-dihydroxy-6-methyl-4*H*-pyran-4-one (1) arises, which is a typical product of the Maillard reaction of glucose (Figure 3).

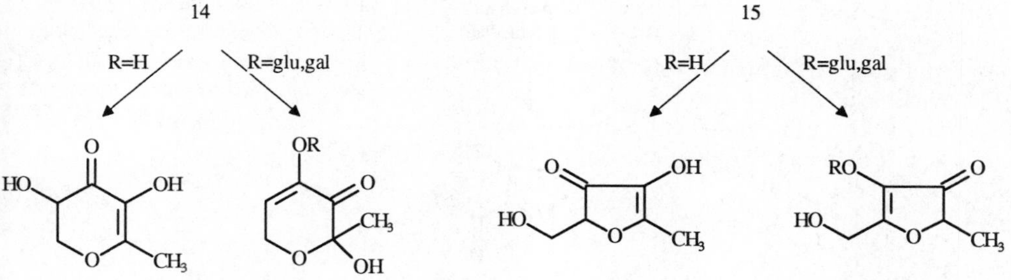

Figure 3. Different pathways of the further decomposition of mono- or disaccharides.

On the other hand, in the disaccharide-derived intermediates the sugar residue is a poor leaving group and the OH group at C-5 is eliminated in accordance with expectation. As a result, 4-(glycosyloxy)-2-hydroxy-2-methyl-2*H*-pyran-3(6*H*)-ones (3) are formed, the galactosyl derivative of which has already been detected in heated milk (Figure 3) (Ledl et al., 1986). In aqueous solution, 3 isomerises to give 4,5-dihydroxy-2-(glycosyloxy)-5-methyl-2-cyclopenten-1-ones (4) until equilibrium is reached (Figure 4). Although the open-chain intermediate appears only in small amounts, it could be isolated after peracetylation. Prolonged heating results in the transformation of 3 and 4 into more stable compounds; dehydration leads to the isomaltol glycosides (6), whereas maltol (7) arises, when the sugar residue is eliminated from 3. Both, 6 and 7 are long known degradation products of disaccharides and, in addition, maltol has particular significance as a naturally occurring flavour enhancer (Figure 4).

The five-membered cyclisation products of the 1-deoxyosones give rise to other compounds. The first steps, that take place are enolization and elimination of water. In the glucose-derived intermediate, the proton is separated from the hydroxyl group in position 4 and, consequently, 4-hydroxy-2-(hydroxymethyl)-5-methyl-3(2*H*)-furanone (2) arises. By contrast, a sugar residue at C-4 prevents this type of reaction and, as a result, 4-(glycosyloxy)-5-(hydroxymethyl)-2-methyl-3(2*H*)-furanones (5) are formed from disaccharides (Figure 3). As an ether of a reductone, 5 can easily be oxidised. So it can be assumed that it contributes to the well-known anti-oxidative effect of Maillard products that protects processed or heated food from oxidative spoilage.

All these early products of the Maillard reaction of disaccharides are quite stable substances that can be isolated from reaction mixtures of amines and maltose or lactose. On the other hand, they also represent intermediates that can undergo further reactions with other components, particularly with proteins or amino acids. A very reactive part of protein that is most likely to attack by sugar degradation products is the ε-amino group of lysine. It could be shown by detailed investigations that other, simpler amines

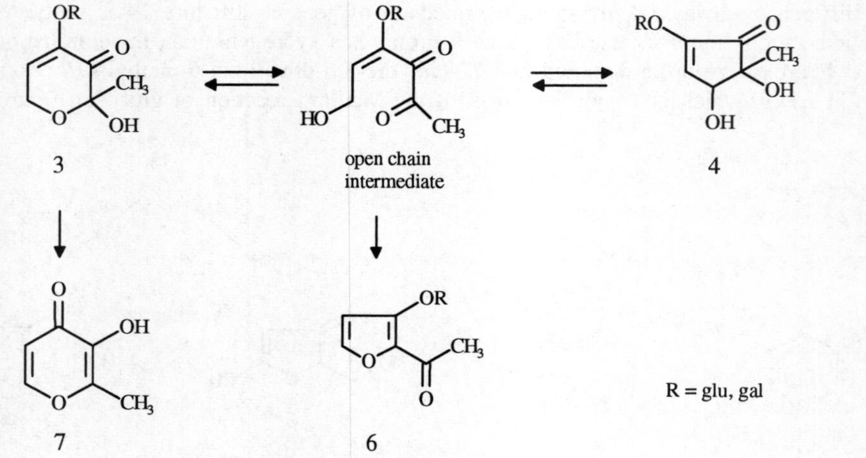

Figure 4. Continuing degradation of maltose and lactose.

such as α-N-acetyl lysine or propylamine behave in a similar way to protein-bound lysine and can, therefore, be used as suitable models.

Condensation reactions of the disaccharide decomposition products with primary amines were thoroughly investigated and a wide range of nitrogen containing heterocycles could be isolated. Heating of 3 with propylamine in neutral aqueous solution leads to the formation of 1-[3-(glycosyloxy)-1-propyl-2-pyrrolyl]-1-ethanone (8) as the nitrogen-containing main product (Figure 5).

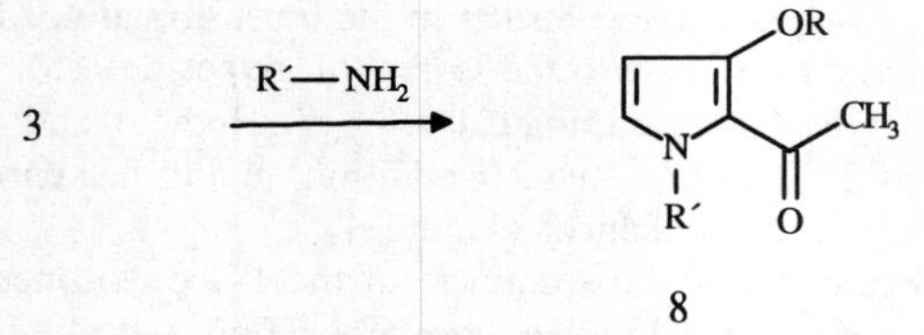

Figure 5. Reaction of β-pyranone derivative with primary amine.

On the other hand, when the isomaltol glycosides (6) are reacted under similar conditions, 2-[1-(propylamino)ethylidenyl]-3-(2*H*)furanone (9) and 4-(glycosyloxy)-2-methyl-propylpyridinium-3-olates (10) are formed (Figure 6). 10 is not a stable product that represents the end of the decomposition of disaccharides, but forms the basis of some new reactions. After prolonged heating, 10 readily eliminates the sugar residue to give 3-hydroxy-2-methyl-1-propyl-4(1*H*)-pyridone (11), a stable compound that can be considered as one of the final products of the Maillard reaction of disaccharides (Figure 7). Particularly under slightly alkaline conditions an intramolecular rearrangement of the sugar residue from the C-4 to the C-3 position can proceed and 3-glycosyloxy-2-methyl-1-propyl-4(1*H*)-pyridone (12) is formed. Finally, substitution of a propylamine for the

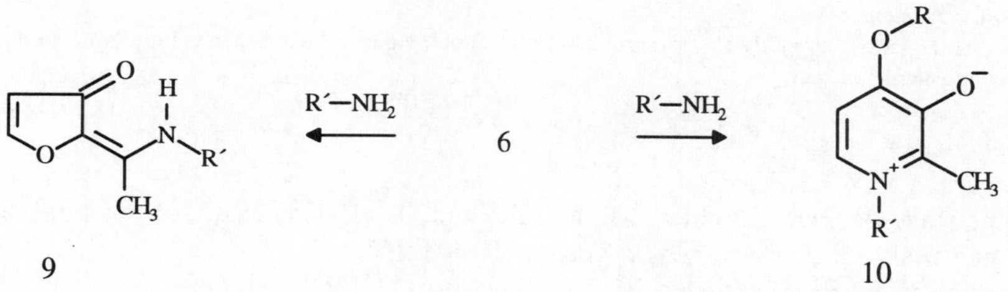

Figure 6. Reactions of the isomaltol glycosides with primary amines.

sugar residue leads to 1,4-dihydro-2-methyl-1-propyl-4-(propylimino)-3-pyridinol (13), a basic compound, which forms stable complexes with several metal ions (Figure 7) (Kramhöller et al., 1993a). Assuming a pyridon imine derivative (13), that contains two molecules of protein-bound lysine, this type of product can indicate a new pathway for the crosslinking of proteins.

Experience shows that polysaccharides are more resistant to thermal decomposition than monosaccharides or reducing disaccharides. Nevertheless, they undergo Maillard reactions when heated with amines.

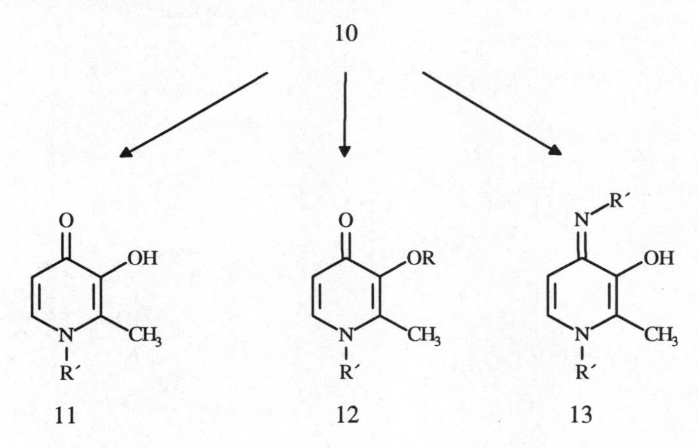

Figure 7. Degradation of pyridinium betaines.

It seems very likely, that the decomposition of polysaccharides that are connected by a 1,4-glycosidic link, such as starch or dextrins follows the pathway of the Maillard reaction of disaccharides and, as a result, similar products should arise, that only differ in the length of the sugar residue. Indeed, from a reaction mixture of starch or dextrins 11 and 13 can be isolated and separated by HPLC (Kramhöller et al., 1993b). 11 and 13 are typical disaccharide degradation products that aren't found in reaction mixtures of monosaccharides.

In consequence, it can be concluded, that the results obtained for disaccharides can be applied to the large number of 1,4-glycosidic linked polysaccharides.

Acknowledgements
This work was supported by grants from Fond der Chemischen Industrie and Deutsche Forschungsgemeinschaft.

References

Kramhöller, B., Pischetsrieder, M., and Severin, T. (1993a) Maillard reactions of lactose and maltose. *J. Agric. Food Chem.* 41: 347-351.

Kramhöller, B., Pischetsrieder, M., and Severin, T. (1993b) Maillard reactions of dextrins and starch. *Z. Lebensm. Unters. Forsch.* 197: 227-229.

Ledl, F., Ellrich, G., Klostermayer, H. (1986) Proof and identification of a new Maillard compound in heated milk. *Z. Lebensm. Unters. Forsch.* 182: 19-24.

Pischetsrieder, M., and Severin,T. (1994) Maillard reaction of maltose - isolation of 4-(glucopyranosyloxy)-5-hydroxymethyl-2-methyl-3-furanone. *J. Agric. Food Chem.* (in press).

Evaluation of a Lysine–Glucose Maillard Model System Using Three Rapid Analytical Methods

M. B. Assoumani, D. Maxime,[1] and N. P. Nguyen

SANDERS ALIMENTS, LABORATOIRE DE BIOTECHNOLOGIE, 17 QUAI DE L'INDUSTRIE F-91200 ATHIS-MONS; [1]SIM MONTPELLIER, FRANCE

Summary
The use of spectrophotometric methods to monitor Maillard browning reactions has been described by many authors. The limitation of such methods is lack of specificity. The present work describes a new approach using an enzyme electrode (immobilized α-L-Lysine oxidase from *Trichoderma viride*) to evaluate lysine loss. The experiments were conducted at 120°C (30 minutes, pH 5) with the lysine-glucose molar ratio varying from 0.2 to 1.0. Lysine loss was highly correlated with spectrophotometric data (A_{287} nm) using least squares polynomial fitting (R^2=0.991). The data obtained the microwave spectral analysis (scanning between 2400 and 2450 MHz) were also highly correlated with lysine loss (R^2=1). This technique is based on dielectric properties of the media to be analyzed. It was able to show the importance of the polar groups that were formed (amplitude of the spectrum) and the structure change (frequency of the spectrum) with the Maillard model system. The experiment was extended to different carbohydrates (pentoses and hexoses) at a fixed molar ratio. The results agreed with those in the literature: lysine loss was higher with pentoses.

Introduction

There is a need for simple and rapid methods to monitor Maillard reactions in the food and pharmaceutical industries. Generally, lysine concentration is measured due to its sensitivity to Maillard reactions through its ε-amino group.

Lysine analysis with a lysine oxidase electrode offers several advantages over conventional chemical methods due to its specificity and sensitivity (Assoumani *et al.*, 1990; Assoumani and Nguyen, 1991). Maillard model systems containing lysine and sugars in solution can be monitored by this technique or by spectrophotometric techniques .

Another possible approach is the use of a microwave spectral analysis technique. This emerging technique, which is currently binding applications in physics and organic chemistry, is based on measurement of the molecular relaxation that follows energy absorption. The behavior of biological materials under an electromagnetic field is characterized by their dielectric properties :

$\varepsilon^* = \varepsilon' - j\varepsilon''$ (Von Hippel, 1954)

where:

ε^*= the relative permittivity of the material (frequency-dependent);

ε'= dielectric constant (a measure of a material's ability to store electrical energy);

ε''= dielectric loss factor (a measure of a material's ability to dissipate electrical energy);

the ratio $\tan \delta = \varepsilon'' / \varepsilon'$ (the loss tangent of the material due to Joule effects resulting from molecular friction).

Under an electromagnetic field, a product behaves according to its dielectric properties which depend on temperature, water activity and polarity (Mudget, 1985): it is said to be a conducting dielectric. The lower the molecular weight of a compound, the greater will be the strength of its interaction with microwave radiation. This phenomenon is known as dielectric polarization. There are 4 types of dielectric polarization: space-charge (audio-frequencies), dipolar orientation (microwaves), ionic (infra-red) and electronic (UV) (Thuery, 1983).

Dielectric properties are based upon transmission properties which, in turn, depend on the extent of energy reflection and transmission at product surfaces and absorption of energy within the product (Mudgett, 1985). Free water molecules can be viewed as dipoles or polar molecules. An electric field of rapidly changing polarity causes a rotation of the dipoles until they are aligned with the direction of the field (Püschner, 1966). Kuntz (1974) reported 3 regions of interest for measuring dielectric relaxation: 10^5-10^7 Hz, 10^8-10^9Hz and ~ 2 x 10^{10} Hz. The mid region (10^8-10^9 Hz) was shown to be the most useful for complex permittivity experiments as in this region, the molecular relaxation depends on the rotation of water molecules. Several authors have investigated microwave dielectric measurements (Pointon and Woodman, 1971; Parkash *et al.*, 1979; Ollivon *et al.*, 1988; Xu and Harmony, 1992; Marstokk and Møllendal, 1992; Schönfeld *et al.*, 1993). The studies were either based on destructive testing (Ollivon *et al.*, 1988) or undertaken at very high frequency in the 20 - 40 GHz region (Xu and Harmony, 1992; Marstokk and Møllendal, 1992). Therefore, the objective of the present work was to develop a real-time non-destructive microwave spectral analysis method (2.4 GHz - 2.45 GHz); to compare it with the L-Lysine oxidase electrode and UV absorbance methods in the study of a Maillard model system at various lysine-glucose ratios, and to extend the model to different monosaccharides at a fixed lysine-sugar ratio.

Material and Methods

Microwave spectral analysis method

A PROTEON^{TM} (Société de Contrôle Moléculaire, F-91200 Athis-Mons) computer-controlled microwave generator was used. A resonator is excited by a continuous frequency wave between 2400 MHz and 2450 MHz with a fixed microwave power (3 x 10^{-4} watts). The resonator is an open short-circuited coaxial cavity (one side of the cavity is open) of cylindrical form (Pointon and Woodman, 1971; Parkash *et al.*, 1979 and Schönfeld *et al.*, 1993). One mL of the sample to be tested is presented in a Teflon capsule (the microwave absorption cell) in order to produce a small perturbation to the free cavity resonance. The absorbed energy provokes the rotation of polar molecules, which is counteracted by chemical bonds. The response is a spectrum characterized by the maximum amplitude (mV) at the resonance frequency (MHz). Ten consecutive measurements were conducted at ambient temperature and averaged to give the final amplitude and frequency. The accuracy of the spectral measurements was better than 50 KHz.

Absorbance measurement

The color intensity of the samples was measured at 287 nm using a Shimadzu UV 160. The samples were individually tested to find the λ_{max} between 200 nm and 800 nm.

L-Lysine determination using an immobilized α-L-Lysine oxidase electrode
The method was described in detail in a previous paper (Assoumani *et al.*, 1990). The membrane enzyme activity was 6.5 IU. The basic principle was the measurement of oxygen consumption following the introduction of sample and its oxidation:

L-Lysine oxidase
L-Lysine $+O_2 + H_2O$ -------------------------> α-keto-ε aminocaproate $+ NH_3 + H_2O_2$
-------------------------> δ-piperidine-2-carboxylate
(Kusakabe *et al.*, 1980).

The oxygen consumption was followed dynamically by measuring the reaction slope (dpO_2/dt) at the inflection point of the acquisition curve. Total analysis time was 60 seconds. The calibration curve was obtained using lysine concentrations varying from 0 to 0.06 g/L.

Reagents
L-Lysine HCl was purchased from Merck (Darmstadt, Germany). D-Glucose, D- and L-arabinose, D- and L-Xylose, D- and L-ribose, D-Mannose, D-Galactose and D-Fructose were purchased from Sigma Chemicals Company (St Louis, MO).

Reaction mixture
Maillard modeling by varying lysine-glucose molar ratios
The mixtures were prepared by dissolving 0.243 g of lysine (1.334 mM) and varying level of glucose in 20 mL of a pH 5 phosphate buffer (M/15 disodium hydrogenphosphate and M/15 potassium dihydrogenphosphate). The glucose concentrations tested (1.33 mM to 5.55 mM) gave the following lysine-glucose molar ratios: 1:5, 1:4, 1:3, 1:2 and 1:1.

Maillard reaction of lysine and different sugars
Solutions with a lysine:sugar molar ratio of 1:5 were prepared by adjusting reactant concentration: 5:55 mM hexose (D-fructose, D-glucose, D-galactose, D-mannose) and 1.11 mM lysine; 6.67 mM pentose (D-arabinose, L-arabinose, D-ribose, D-xylose, L-xylose) and 1.334 mM lysine.

All the samples were placed in 100 mL vials, closed with a rubber stopper, secured with an aluminium clamp and placed in an autoclave at a pressure of 1 bar. The reaction time was 30 minutes at 120°C.

Results

Lysine-glucose Maillard reaction model system
The visual browning intensity of the different solutions varied depending on molar ratios. Browning decreased as the molar ratio of lysine:glucose increased. Figure 1 shows the variation of absorbance and lysine loss with molar ratio. Interestingly, lysine loss measured using the lysine oxidase electrode followed the same pattern: the highest values of glucose concentration resulted in the highest lysine loss. A good correlation was found between lysine loss and absorbance (R^2=0.991). Figure 2 shows the microwave amplitude and frequency for the molar ratios tested: the amplitude increased with increasing molar ratio and reached a plateau at a ratio of 1:2 . By contrast, the frequency decreased with increasing molar ratio. The curve obtained could be compared to those obtained with lysine

loss or absorbance. It is surprising that the correlation between frequency and lysine loss (Figure 3) was ideal (R^2=1) contrary to that between frequency and absorbance (R^2=0.978).

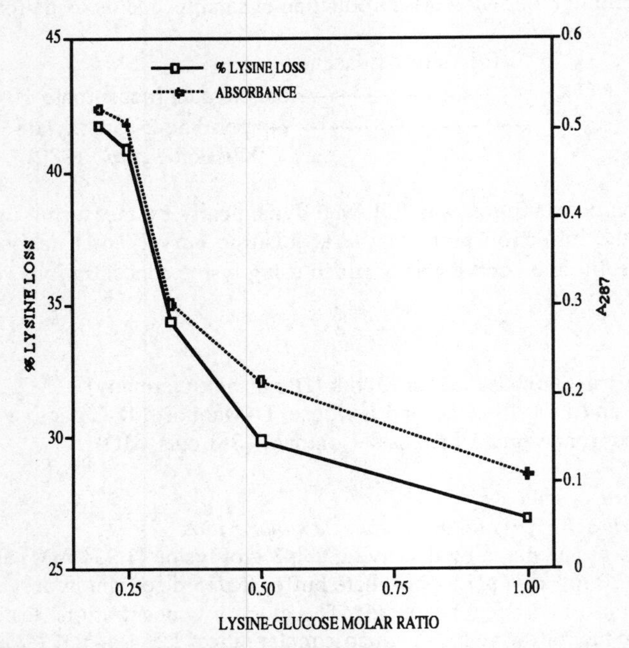

Figure 1. Relationship between the lysine:glucose molar ratio, absorbance at 287 nm and lysine measured using a lysine oxidase electrode.

Maillard reaction of lysine with different sugars

The pentoses gave a more intense browning (black coffee) with a pronounced caramel odor and the reaction resulted in the formation of fine particles in suspension. The solutions containing hexoses were clearer with a caramel type color and a less intense odor. The final pH was more acidic particularly for the pentoses: pH 3.3 compared with pH 3.8 for the hexoses. Figure 4 shows the lysine loss for the tested sugars as measured using the lysine oxidase electrode. Maillard browning intensity was higher for pentoses with an average 34.15 % lysine loss compared to 21.95 % loss for hexoses. The most reactive sugars were: xylose and ribose for the pentoses; galactose and mannose for the hexoses.

Microwave spectral analysis revealed a different behavior for the tested sugars (Figure 5). The hexoses had a higher amplitude and a lower frequency compared with the pentoses.

Discussion

The Maillard reaction of a lysine-glucose system at various molar ratios (1:5 to 1) was followed by 3 rapid analytical methods. The λ_{max} of 287 nm for the treated solutions was in good agreement with previous studies. This was also the case for the increase of lysine loss with excess of glucose (Adrian, 1982; Baisier and Labuza, 1992). The good correlation

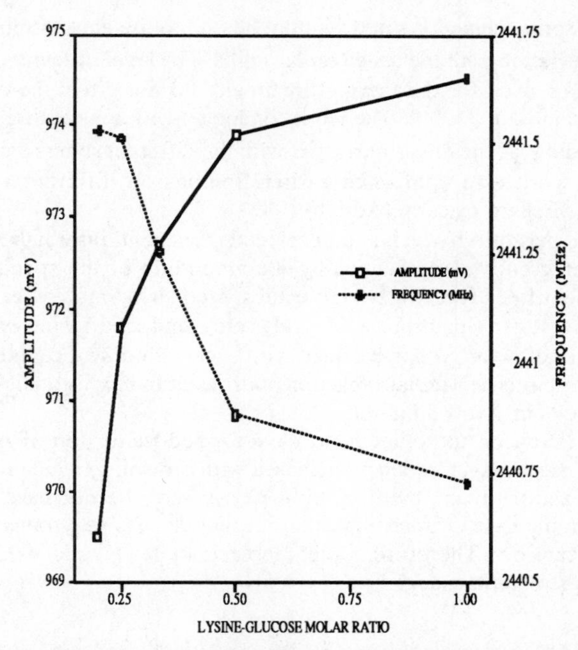

Figure 2. Relationship between the lysine:glucose molar ratio, microwave amplitude and frequency after heat treatment. The microwave values illustrate molecular relaxation resulting from energy absorption.

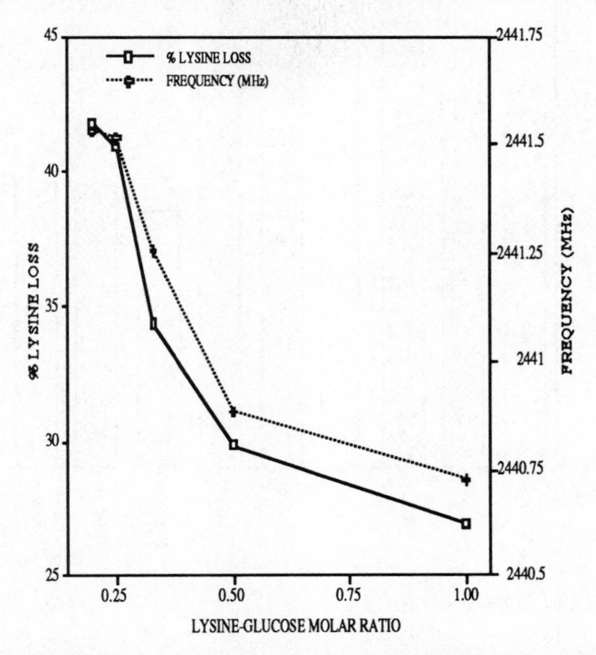

Figure 3. Relationship between lysine-glucose molar ratio, lysine loss (measured using a lysine oxidase electrode) and microwave frequency.

found between absorbance and lysine loss may be due to the chosen model, which didn't show any λ_{max} deviation with the tested molar ratios. The level of levulinic acid produced under the conditions used for the thermal treatment did not affect the enzyme electrode response (Assoumani *et al.*, 1990). The work conducted with the pentoses and the hexoses confirmed this finding as the λ_{max} measured with the different sugars varied from 270 nm to 300 nm. This work also confirmed earlier findings on different sugars and their reactivities in the Maillard reaction (Adrian, 1982).

The microwave permittivity technique provided interesting information concerning the formation of polar groups as evidenced by the amplitude of the spectra. Several polar species can be identified: free water molecules, free sugar molecules, acid molecules produced from the sugar (levulinic acid, NH_2, etc), and free lysine with ionized NH_2 groups. The number of polar groups declined with excess glucose. From this finding it may be concluded that an intense Maillard reaction will result in low formation of polar groups which, in turn, results in a low amplitude.

Frequency variation on the other hand gave a good indication of molecular weight variation. In the present case, frequency increased with browning, which reveals a tendency towards polymerization. It is then possible to say that the decrease in polar groups accompanied by an increase in frequency signifies that those polar groups are involved in a polymerization reaction. Therefore, the increase in frequency can be linked to a compositional or a structural change.

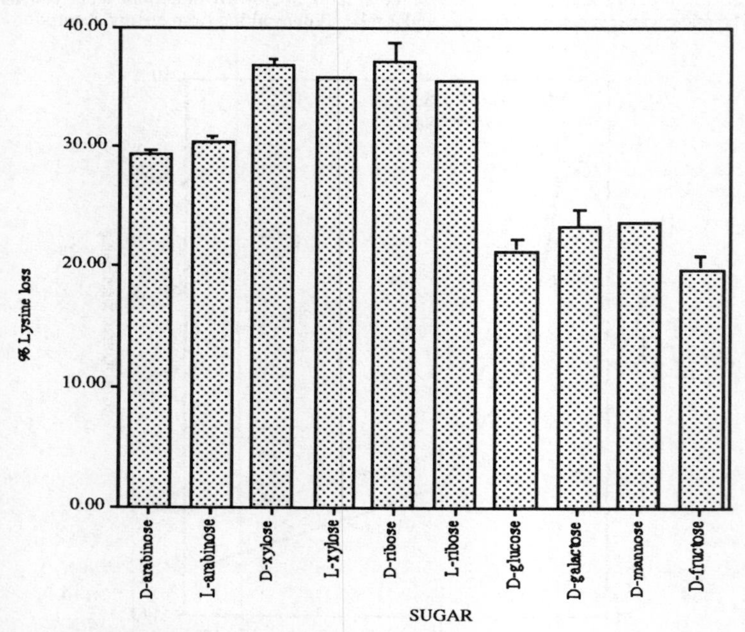

Figure 4. Lysine loss with different sugars.

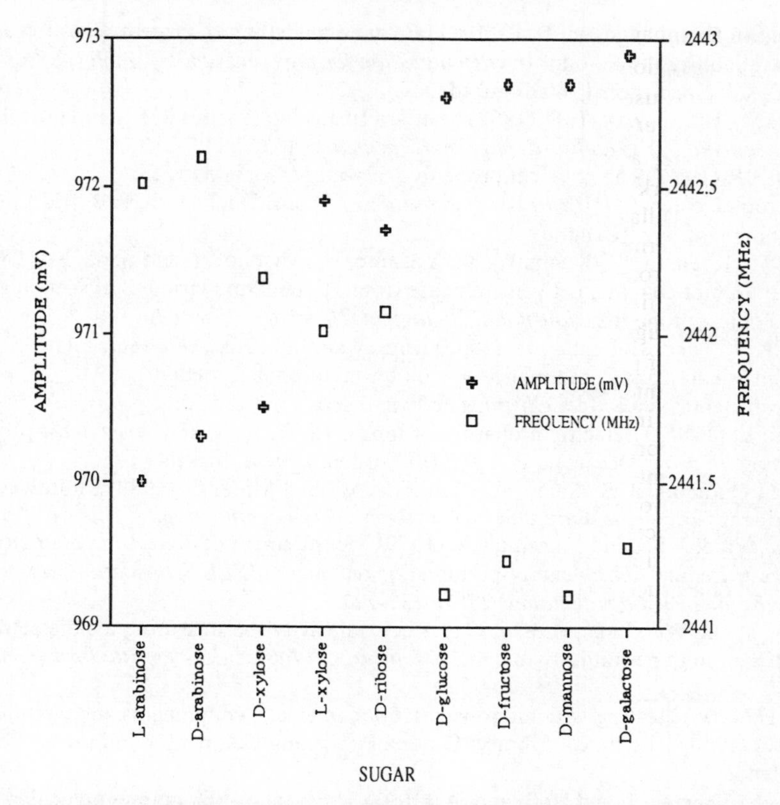

Figure 5. Microwave amplitude and frequency measured after Maillard browning with different sugars solutions at a lysine:sugar molar ratio of 1:5.The values illustrate molecular relaxation resulting from microwave energy absorption.

Conclusion

The combination of a lysine oxidase electrode method and a non destructive microwave spectral analysis method (2.4 GHz - 2.45 GHz) offers potential for kinetic studies of Maillard model system based on lysine.

References

Adrian, J. (1982) The Maillard reaction: 529-608. In: *Handbook of Nutritive value of processed food, Volume 1. Food for human use*. Miloslav Rechcigl Jr. (Ed), CRC Press, Boca Raton (FL).

Assoumani, M. B. Nguyen, N. P., Lardinois, P. F., Van Bree, J., Baudichau, A. and Bruyer, D. C. (1990) Use of a lysine oxidase electrode for lysine determination in Maillard model reactions and in soybean meal hydrolysates. *Lebensmittel Wissenchaft und Technologie*. 23: 322-327.

Assoumani, M. B. and Nguyen, N. P. (1991) Enzyme modelling of protein digestion and L-lysine availability: 86-104. In: *In vitro digestion for pigs and poultry*. Fuller, M.F. (Ed), C.A.B International, Wallingford,U.K.

Baisier, W. M., and Labuza, T. P. (1992) Maillard Browning kinetics in a liquid model system. *Journal of Agricultural and Food Chemistry*. 40 (5): 707-713.

Kuntz, I.D., JR (1974) The physical properties of water associated with biomacromolecules: 93-109. *In Water relations of Foods*. Duckworth, R.B. (Ed.), Academic Press, Inc. (London).

Kusakabe, H., Kodama, K., Kuninaka, A., Yoshino, H., Misono, H. and Soda, K.(1980). A new antitumor enzyme, L-Lysine oxidase from Trichoderma viride. Purification and enzymological properties. *Journal of Biological Chemistry*, 255: 976.

Marstokk, K. M., and Møllendal, H. (1992) Microwave spectrum, intramolecular hydrogen bonding and conformational equilibrium of trans-2-methyl cyclopropanemethanol. *Acta Chemica Scandinavica*. 46: 861-866.

Mudgett, R. E. (1985) Dielectric properties of foods: 15-37. In: *Microwaves in the food processing industry*. Decareau, R. V. (Ed), Academic Press, Inc. (NY).

Ollivon, M., Quinquenet, S., Seras, M., Delmotte, M., and More, C. (1988) Microwave dielectric measurements during thermal analysis. *Thermochimica Acta*. 125: 141-153.

Parkash, A., Vaid, J. K., and Mansingh A. (1979) Measurement of dielectric parameters at microwave frequencies by cavity-perturbation technique. *IEEE Transactions on microwave theory and techniques*. 27(9): 137-222.

Pointon, A. J., and Woodman, K. F. (1971) A coaxial cavity for measuring the dielectric properties of high permittivity materials. *Journal of Physics E: Scientific Instruments*. 4: 208-212.

Püschner, H.(1966) Heating with microwaves. Fundamentals, components and circuit technique. Philips Technical Library. Centrex Publishing Company. Eindhoven. The Netherlands.

Schönfeld, A., Kremer, F. and Hofmann, A. (1993) The b-relaxation in low-molecular-weight and in polymeric side group ferroelectric liquid crystals. *Makromol. Chem*. 194:1149- 1155.

Thuery, J. (1983) (Ed.)Les micro-ondes et leurs effets sur la matière. Applications industrielles, agro-alimentaires et médicales., Lavoisier Technique et Documentation (Paris) et C.D.I.U.P.A (Massy).

Xu, S., and Harmony, M. D. (1992) Microwave study of 3-chloropropionitrile and 3-bromopropionitrile. *Journal of Molecular Structure*. 274: 115-130.

Von Hippel, A. R. (1954) Dielectrics and Waves. MIT Press, Cambridge, MA.

Mechanistic Studies on the Formation of Pyrroles and Pyridines from [1-^{13}C]-D-Glucose and [1-^{13}C]-D-Arabinose

R. Tressl, E. Kersten, C. Nittka, and [1]D. Rewicki

INSTITUT FÜR BIOTECHNOLOGIE, TU BERLIN, SEESTRASSE 13, 13353 BERLIN, GERMANY; [1]INSTITUT FÜR ORGANISCHE CHEMIE, FU BERLIN, TAKUSTRASSE, 14195 BERLIN, GERMANY

Summary

The Maillard reaction of [1-^{13}C]-D-glucose and [1-^{13}C]-D-arabinose with 4-aminobutyric acid (representing peptide bound lysine as well as a Strecker inactive amino acid) and L-isoleucine (representing a Strecker active amino acid) was investigated to get more insight into the reaction pathways involved. The extent and position of the labeling were determined by MS data. The results support 3-deoxyaldoketose as intermediate of N-alkyl-2-formyl-5-hydroxymethyl- and N-alkyl-2-formyl-5-methylpyrroles (**1, 3, 7, 9**) and disqualify 4-deoxy- and 1-deoxydiketose routes to N-alkyl-2-hydroxyacetyl- and N-alkyl-2-acetyl-pyrroles (**2,4,8,10**), respectively. The 1,3-dideoxy-1-amino-2,4-diketose **C** is postulated as a new key intermediate in the formation route to **2, 4, 8,** and **10** from D-glucose. In addition, this β-dicarbonyl route is correlated to the 3-deoxy-aldoketose route by keto-enol tautomerism as demonstrated by Maillard experiments in deuterium-oxide. The D-exchange ratios of compounds **1, 3, 7, 9, 13,** and **14** (which were examined by MS- and ^{1}H NMR spectroscopy) indicated incorporation of D-atoms at C4 of glucose. The β-dicarbonyl pathway of [1-^{13}C]-D-glucose/L-isoleucine (glycine) Maillard experiments generates 2-acetyl-[5-^{13}C]-pyrrole and 2-methyl-3-[6-^{13}C]pyridinol via a Strecker active intermediate. Based on the results of these labeling experiments and on the results of Maillard experiments in deuteriumoxide a revised scheme of the Maillard reaction of D-glucose with amines and α-amino acids is presented.

Introduction

The pathway of the Maillard reaction of D-glucose with amines, primary and secondary α-amino acids has been elucidated by the isolation and characterization of lower molecular weight products. Some of the postulated α-dicarbonyl intermediates (e.g. 3-deoxyaldoketoses, 1-deoxydiketoses, 1,4-dideoxy-2,3-diketoses) were identified as TMS-derivatives by capillary GC/MS (Beck et al., 1988; Huber and Ledl, 1990). The formation pathways to flavor compounds are still obscure and a compound can be generated by different routes from different precursors. Therefore, we investigated the routes to selected flavor compounds with [^{13}C]-labeled sugars in Maillard model systems. The analysis of the labeled compounds by mass spectrometry indicates the distribution, the extent and in (most cases) the position of the [^{13}C]-labeling. By this method, conclusions on the precursors and the reactive intermediates can be drawn, if the sugar skeleton is transformed without splitting off the [^{13}C]-label. In addition, the cleavage of the sugar into labeled and unlabeled fragments can be analyzed. The products formed from these fragments can be detected as mixtures of unlabeled, singly and doubly labeled isotopomers.

For our Maillard experiments we selected 4-aminobutyric acid (GABA) (Tressl et al., 1993a, 1993b) as the most reactive, non Strecker active amino acid (resembling peptide bound L-lysine), L-isoleucine (Tressl et al., 1994) representing a Strecker active primary α-amino acid, and L-proline (L-hydroxyproline) (Tressl et al., 1993c) as a Strecker active secondary α-amino acid. The results of the labeling experiments demonstrate five groups

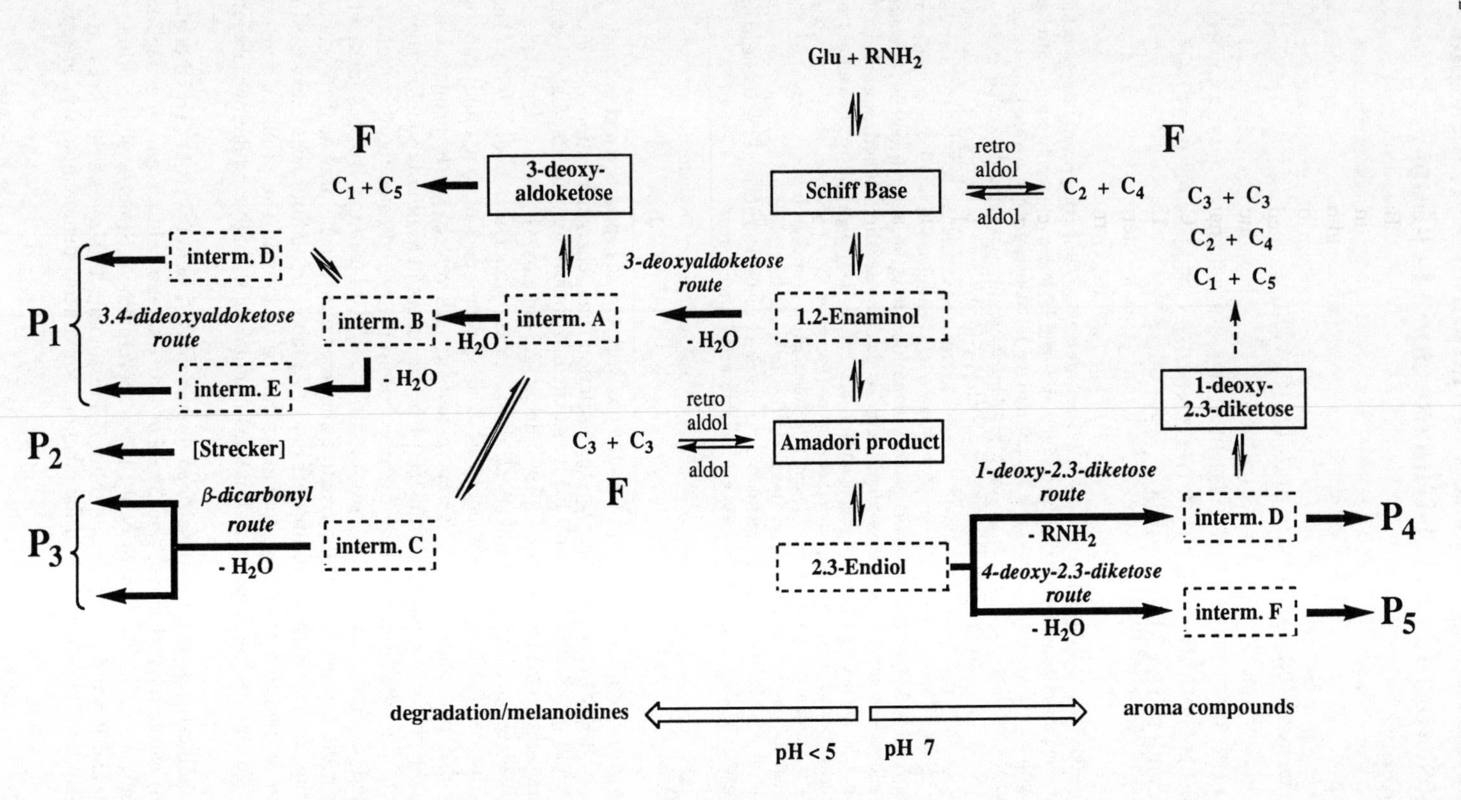

Figure 1. Pathways of the Maillard reaction of D-Glucose (schematic)

(**P_1-P_5**) of C_6-products with an intact sugar skeleton from [1-^{13}C]- and [6-^{13}C]-D-glucose and five characteristic sugar fragmentations **F** (C_1-C_5 ≡ **F_1- F_5**). Based on the results of these [^{13}C]-labeling experiments and on Maillard experiments in deuteriumoxide, we present a revised scheme of the Maillard reaction with amines and α-amino acids (Figure 1). Pyrroles and pyridinoles are generated via allylic dehydration of the 1,2-enaminol to enol **A** from which three distinct groups of pyrroles are generated (P_1-P_3). Nyhammar et al. (1983) demonstrated a Strecker degradation of [1-^{13}C]-D-glucose with glycine into 100% singly labeled pyrroles and pyridinoles, which are summarized as **P_2** products (corresponding to **13-16** in Table I). The investigation of the [^{13}C]-labeled compounds by MS- and NMR spectroscopy showed that the methyl groups in **13-16** are derived from C-6 of glucose. These results support 3-deoxyaldoketose as an intermediate of **13** and **14**, but disqualify 1-deoxydiketose as a precursor of **15** and **16**. Pyrroles and pyridinoles, which were identified in [^{13}C]-D-glucose/4-aminobutyric acid (L-isoleucine) model experiments, are summarized in Figure 2. The observed labeling characteristics support 3-deoxyaldoketose as an intermediate for pyrroles of type **1, 3, 7,** and **9** (**P_1** products) and disqualify 4- and 1-deoxydiketose as intermediates for those of type **2, 4, 8,** and **10** (**P_3** products). The labeling experiments support a β-dicarbonyl intermediate in the Maillard reaction of D-glucose with amines and α-amino acids. In addition, this β-dicarbonyl route is operative in the formation of compounds **15** and **16**. Up to now the formation pathway to **15** and **16** was explained by decarboxylation/dehydration of the Schiff bases (generated from D-glucose and α-amino acids) to 2-deoxyhexose (Nyhammar et al., 1983).

Table I. Formation of pyrroles in reducing sugars/4-aminobutyric acid (isoleucine) Maillard systems (figures = concentration in ppm; + = < 1 ppm; - = not detected).

No.*		component	D-arabinose	L-rhamnose	D-glucose	
R[a)]	R_1[b)]		GABA	GABA	GABA	Ile
1	**7**	figure 2	-	-	3000	4500
2	**8**	figure 2	-	-	380	+
3	**9**	figure 2	+	8250	280	150
4	**10**	figure 2	35	2300	1250	8
5	**11**	figure 2	1870	-	215	55
6	**12**	figure 2	225	-	225	17
13		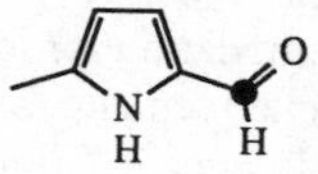	-	-	+	800
14		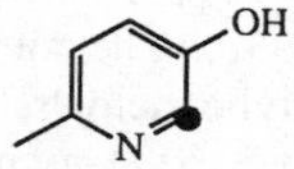	-	-	-	370
15		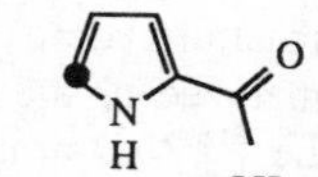	-	-	-	390
16		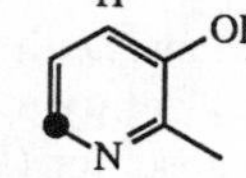	-	-	-	140

* R, R_1: see Figure 2. a) Amino acid: GABA. b) Amino acid: Ile.

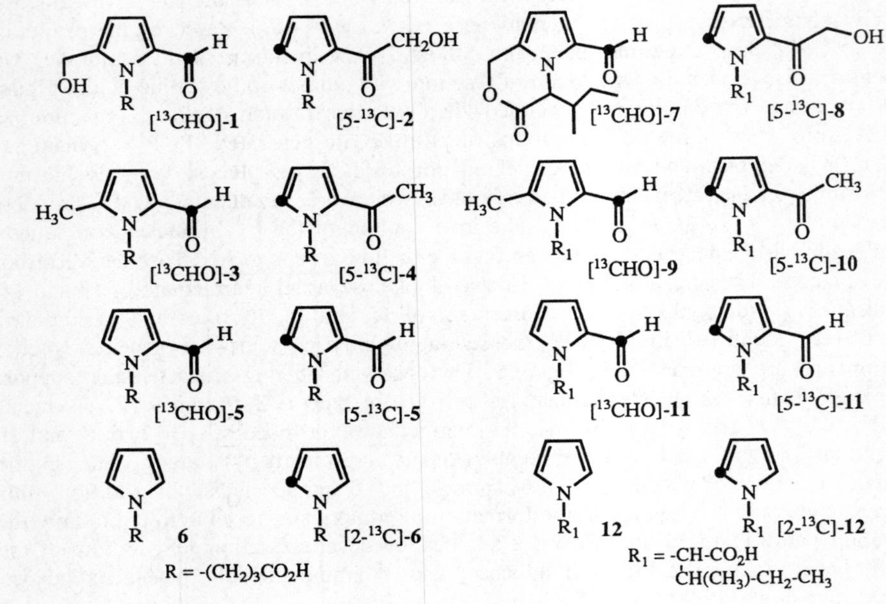

Figure 2. [^{13}C]-labeled pyrroles identified in [1-^{13}C]-D-glucose/4-aminobutyric acid (**1-6**) and [1-^{13}C]-D-glucose/L-isoleucine (**7-12**) Maillard experiments.

Formation Pathways to Pyrroles in [1-^{13}C]-D-Glucose/4-Aminobutyric Acid L-isoleucine Maillard Systems

Recently we identified pyrroles, 2-pyrrolidones, and 4-pyridones as 4-aminobutyric acid specific Maillard products. The described model reactions carried out with [^{13}C]-labeled sugars lead to the corresponding [^{13}C]-labeled products, which were characterized by mass spectrometry (Tressl et al., 1993a, 1993b). In the corresponding L-isoleucine /D-glucose Maillard experiments, compounds **7-12** were identified by MS- and NMR spectroscopy (Tressl et al., 1994). To some extent, L-isoleucine systems formed the analogous pyrroles in comparable amounts to the blocked Strecker GABA model experiments (Table I). In addition, the Strecker degradation products **13-16** (**P_2**) were identified as 100% singly labeled compounds. In agreement with the results of the labeling of compounds **1-4** we postulated a revised scheme (Figure 3) of the Maillard reaction of D-glucose with amines (Tressl et al., 1993b). The routes to **1/3** and **2/4** are correlated by enol **A**, which is transformed via allylic dehydration into **B** and by a vinylogous Amadori rearrangement into **1**. 1,3-Dideoxy-1-amino-2,4-diketose C is postulated as a new key intermediate in the formation route to pyrroles **2/4** from D-glucose. In the isoleucine Maillard system, **7** is analyzed as the corresponding pyrrole lactone and **8** is formed as a trace compound. In the [^{13}C]-D-arabinose/4-aminobutyric acid Maillard system [^{13}CHO]-**5** and [5-^{13}C]-**5** were analyzed as a mixture of 100% singly labeled isotopomers. The mass spectra of the 2-formylpyrroles clearly indicate their formation via intermediate **B** (3,4-dideoxyaldopentose for [^{13}CHO]-5 (60%)), and

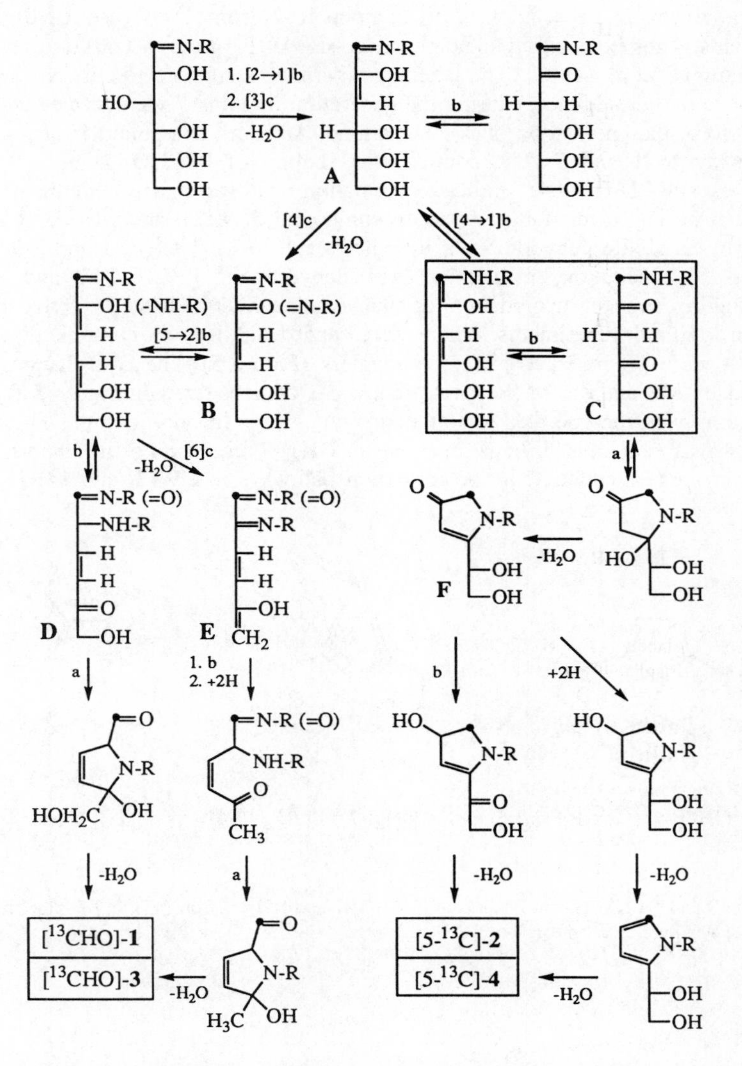

Figure 3. Formation of pyrroles from [1-^{13}C]-D-glucose/GABA (Ile); a=cyclization, b=ketoenol tautomerism, c=allylic dehydration.

intermediate **C** (1,3-dideoxy-1-amino-2,4-diketose for [5-^{13}C]-**5** (40%)). These results demonstrate that the formation of pyrroles via intermediate **C** is a major pathway in glucose and pentose Maillard systems. The formation of 100% singly labeled [^{13}CHO]-**7** in [1-^{13}C]-D-glucose/L-isoleucine model experiments indicates, that the main amount of isoleucine is not degraded via Strecker. The route to compound **7** was further investigated by heating of D-glucose and L-isoleucine in D_2O(comparable to degradation experiments of Amadori compounds (Feather, 1981). The isolated pyrrole lactone was investigated by

MS- and ^{1}H NMR spectroscopy. Some of the results are summarized in Figure 4. The incorporation of **D** in position 4 in compound **7** from D-glucose (and D-fructose) experiments is not expected by the pathway **A→B→D** (Figure 3) according to Anet (1964) and Nyhammar et al. (1983). The labeling experiments support a keto-enol tautomerism of enol **A** and intermediate **C** during the generation of **1** and **7** via intermediate **B** (Figure 6). Therefore, the enolization step 1,2-enaminol/Amadori compound is not important in the pathway to **1** and **7**. The comparable labeling of [^{13}CHO]-**7** and [^{2}H$_4$]-**7** from D-glucose and D-fructose indicate formation pathways via (identical) bisimine intermediates. The formation pathway to compound **3, 9, 13**, and **14** (via **A→B→E→**) is initiated by an allylic dehydration of intermediate **B** to **E**. [^{13}CH0]-**3** and [^{13}CHO]-**9** are generated via reduction, cyclization and dehydration. [^{13}CHO]-**13** and [2-^{13}C]-**14** correspond to Strecker degradation of the bisimine intermediate **E** (Nyhammar et al., 1983). In Maillard experiments, which were carried out in D_2O, compounds of type **3/9** and **13/14** were generated as [^{2}H$_4$]-isotopomers (Figure 5). The D-exchange rates were examined by MS and ^{1}H NMR spectroscopy. As demonstrated in Figure 5 for [^{1}H$_4$]-**13** one D-atom was incorporated into the methyl group. In addition, the D-exchange in position 4 was examined to be comparable to [^{2}H$_4$]-**7**, confirming the importance of the β-dicarbonyl intermediate **C** in the formation pathways to **3,9,13,** and **14** (Figure 7).

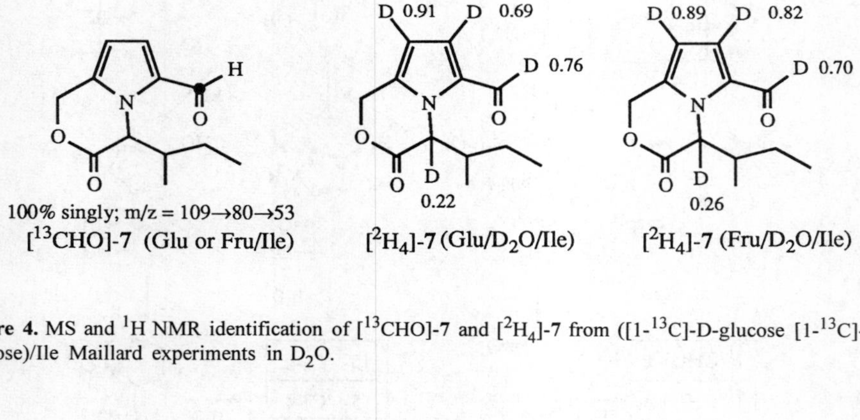

100% singly; m/z = 109→80→53
[^{13}CHO]-**7** (Glu or Fru/Ile) — [^{2}H$_4$]-**7** (Glu/D_2O/Ile) — [^{2}H$_4$]-**7** (Fru/D_2O/Ile)

Figure 4. MS and ^{1}H NMR identification of [^{13}CHO]-7 and [^{2}H$_4$]-7 from ([1-^{13}C]-D-glucose [1-^{13}C]-D-fructose)/Ile Maillard experiments in D_2O.

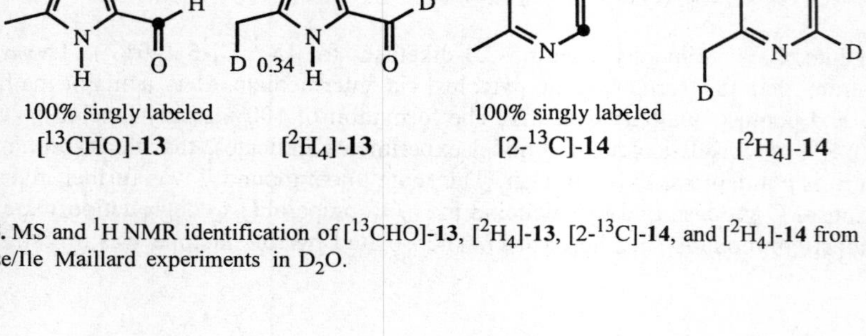

100% singly labeled — 100% singly labeled
[^{13}CHO]-**13** — [^{2}H$_4$]-**13** — [2-^{13}C]-**14** — [^{2}H$_4$]-**14**

Figure 5. MS and ^{1}H NMR identification of [^{13}CHO]-**13**, [^{2}H$_4$]-**13**, [2-^{13}C]-**14**, and [^{2}H$_4$]-**14** from [1-^{13}C]-D-glucose/Ile Maillard experiments in D_2O.

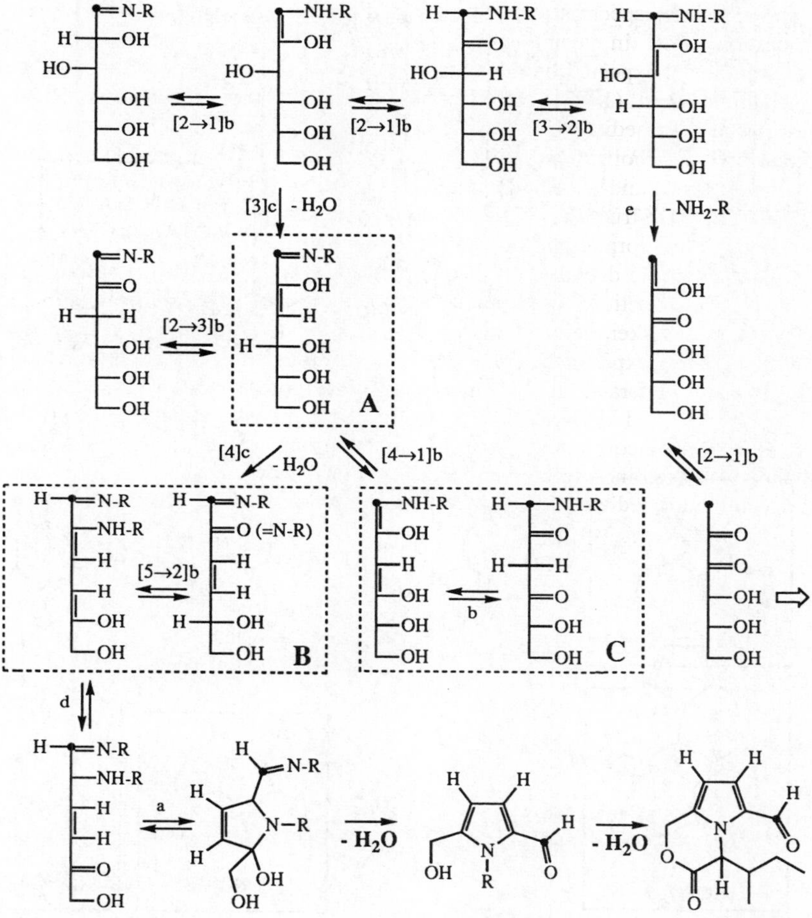

Figure 6. Formation pathway to pyrroles of the type **1/7**; a=cyclization, b=keto-enol tautomerism, c=allylic dehydration, d=vinylogous Amadori rearrangement, e=allylic deamination.

Formation Pathways to Pyrroles 2, 4, 8, and 10 and to the Strecker Degradation Compounds 15 and 16

As key intermediate in the formation of **2** we postulated 1-amino-1,3-dideoxy-2,4-diketose **C**, which is derived from enol **A** via vinylogous Amadori rearrangement (Tressl et al., 1993b). Cyclization, dehydration, and keto-enol tautomerization generate the end products **2** and, by simultaneous reduction, **4** (Figure 3). In the corresponding [1-^{13}C]-D-arabinose Maillard system both, the 3-deoxyaldoketose and the β-dicarbonyl routes generate **5**. The two pathways were examined by [^{13}C]-labeling experiments at a ratio of 60:40. In the D-glucose/L-isoleucine Maillard system the β-dicarbonyl route to [5-^{13}C]-**8** and [5-^{13}C]-**10** was examined as minor pathway whereas [5-^{13}C]-**15** and [6-^{13}C]-**16** were generated as main compounds. These results disqualify 1-deoxydiketose

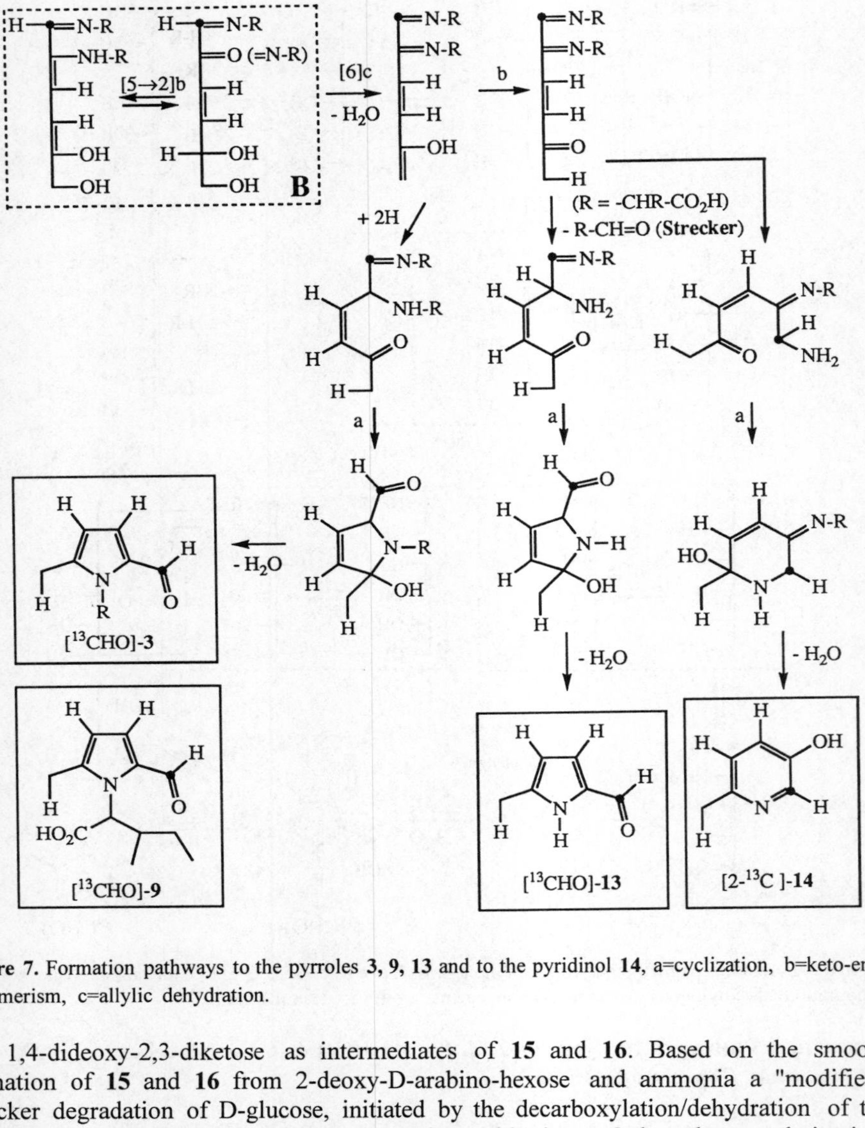

Figure 7. Formation pathways to the pyrroles **3, 9, 13** and to the pyridinol **14**, a=cyclization, b=keto-enol tautomerism, c=allylic dehydration.

and 1,4-dideoxy-2,3-diketose as intermediates of **15** and **16**. Based on the smooth formation of **15** and **16** from 2-deoxy-D-arabino-hexose and ammonia a "modified" Strecker degradation of D-glucose, initiated by the decarboxylation/dehydration of the Schiff base (generated from D-glucose/α-amino acids) into a 2-deoxyhexose derivative, was proposed (Nyhammar et al., 1983). This reaction sequence was not observed with pentoses, tetroses or α-hydroxycarbonyls. Therefore, we postulate a formation pathway to **15** and **16** via intermediate **C** (Figure 8), which is in agreement with [^{13}C]-labeling and D_2O experiments (Figure 9). Keto-enol tautomerization of the intermediate **C**, allylic dehydration and subsequent enolization generate a Strecker active intermediate, which is degraded into 1-amino-1,3,6-trideoxy-2,4-diketose, corresponding to the β-dicarbonyl intermediate **C**. Enolization, cyclization and dehydration form **15/16**.

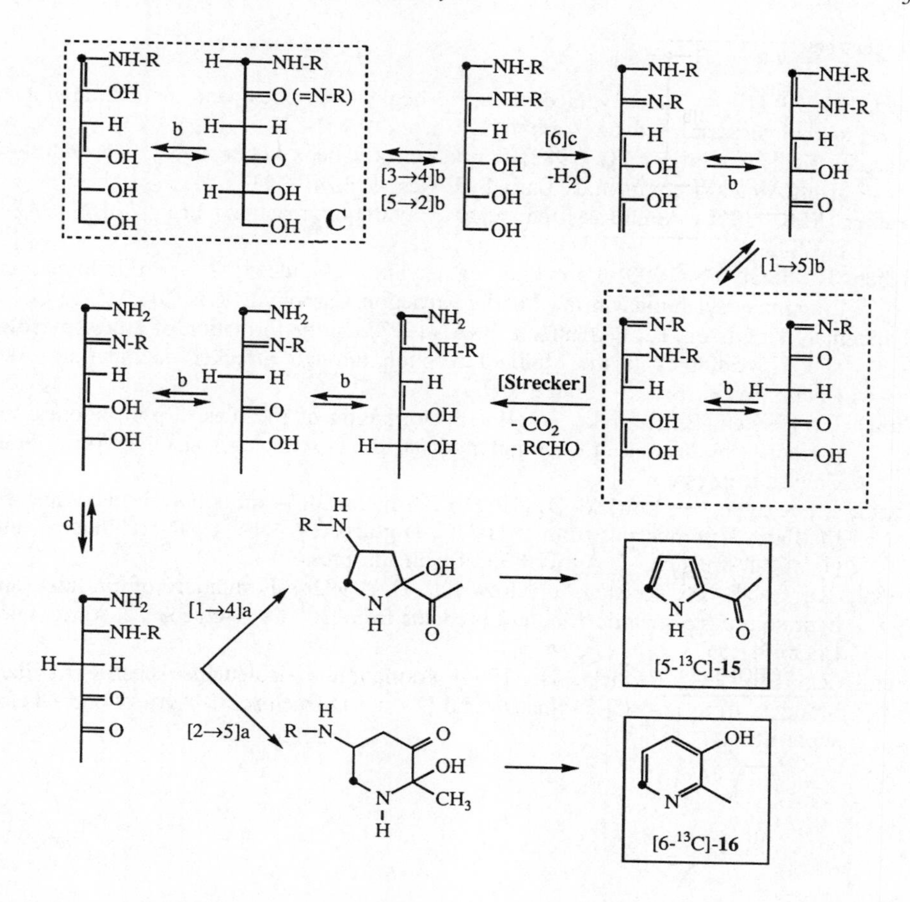

Figure 8. Formation pathway of [5-^{13}C]-**15**/[6-^{13}C]-**16** from [1-^{13}C]-D-glucose/Gly; a=cyclization, b=keto-enol tautomerism, c=allylic dehydration, d=vinylogous Amadori rearrangement.

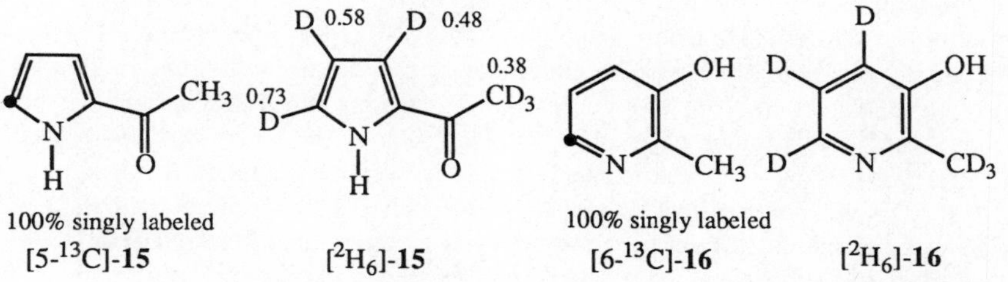

100% singly labeled
[5-^{13}C]-**15**

D 0.58 D 0.48 0.38 CD_3 0.73 D
[$^2H_6$]-**15**

100% singly labeled
[6-^{13}C]-**16**

[$^2H_6$]-**16**

Figure 9. MS and ^{1}NMR identification of [5-^{13}C]-**15**, [$^2H_6$]-**15**, [6-^{13}C]-**16** and [$^2H_6$]-**16** from [1-^{13}C]-D-glucose/Gly Maillard experiments in D_2O.

References

Anet, E.F.L.J. (1964) 3-Deoxyglycosuloses (3-deoxyglycosones) and the degradation of carbohydrates J. Advan. Carbohyd. Chem. 19:181-218.

Beck, J.; Ledl, F.; Severin, T. (1988) Formation of 1-deoxy-D-erythro 2,3-hexodiulose from Amadori compounds Carbohyd. Res. 177:240-243.

Feather, M.S. (1990) Amine-assisted sugar dehydration reactions Prog. Fd. Nutr. Sci. 5:37-45.

Huber, B.; Ledl, F. (1990) Formation of 1-amino-1,4-dideoxy-2,3-hexodiuloses and 2-aminoacetyl-furans in the Maillard reaction Carbohyd. Res. 204;215-220.

Nyhammar, T.; Olsson, K.; Pernemalm, P.-A. (1993) On the formation of 2-acyl-pyrroles and 3-pyridinoles in the Maillard reaction through Strecker degradation. Acta Chem. Scand. Ser. B 37: 879-889.

Tressl, R.; Kersten, E.; Rewicki, D. (1993a) Formation of pyrroles, 2-pyrrolidones and pyridones by heating of 4-aminobutyric acid and reducing sugars. J. Agric. Food Chem., in press.

Tressl, R.; Kersten, E.; Rewicki D. (1993b) Formation of 4-aminobutyric acid specific Maillard products from [1-^{13}C]-D-glucose, [1-^{13}C]-D-arabinose and [1-^{13}C]-D-fructose. J. Agric. Food Chem., in press.

Tressl, R.; Helak, B.; Kersten, E.; Rewicki, D. (1993c) Formation of proline- and hydroxyproline-specific Maillard products from [1-^{13}C]-D-glucose. J. Agric. Food Chem. 41:547

Tressl, R.; Nittka, C.; Rewicki, D. (1994) Formation of isoleucine specific Maillard products from [1-^{13}C]-D-glucose and [1-^{13}C]-D-fructose. J. Agric. Food Chem., submitted.

Mechanistic Studies on the Formation of Maillard Products from [1-^{13}C]-D-Fructose

D. Rewicki,[1] E. Kersten, B. Helak, C. Nittka, and R. Tressl

[1]INSTITUT FÜR ORGANISCHE CHEMIE, FU BERLIN, TAKUSTRASSE 3, 14195 BERLIN, GERMANY; INSTITUT FÜR BIOTECHNOLOGIE, TU BERLIN, SEESTRASSE 13, 13353 BERLIN, GERMANY

Summary

Ketoses are known to react with amino compounds via ketimines to form the so-called Heyns compounds (2-aminoaldoses), which are assumed to undergo subsequent transformations parallel to those observed with the corresponding Amadori compounds. Until now, this assumption was not established by separate mechanistic studies. Therefore, we prepared [1-^{13}C]-D-fructose from [1-^{13}C]-D-glucose by enzymatic methods. In a series of model experiments [1-^{13}C]-D-fructose was heated with 4-aminobutyric acid (Strecker inactive), L-isoleucine (Strecker active), and L-proline (secondary amine type), respectively. The labeled products were analyzed by capillary GC/MS and NMR spectroscopy and the labeling characteristics were examined from MS data. Compared to corresponding experiments with [1-^{13}C]-D-glucose, the significant results are: (1) With 4-aminobutyric acid only trace amounts of 3-deoxyaldoketose products are formed in the D-fructose system, whereas 1-deoxydiketose products were generated in comparable amounts from D-glucose and D-fructose; and (2) With L-isoleucine both D-glucose and D-fructose form 3-deoxyaldoketose- and 1-deoxydiketose products in comparable amounts; but with D-fructose the most effective reaction is the formation of pyrazines initiated by a retro aldol cleavage into C_3+C_3 fragments. This cleavage is also responsible for the formation of mixtures of isotopomeric products in D-fructose systems.

Introduction

Based on the results of [^{13}C]- and [^{2}H]-labeling experiments in Maillard model systems we established a revised reaction scheme of the Maillard reaction of [1-^{13}C]-D-glucose with amines and α-amino acids (Tressl et al., 1993a). Five groups (**P_1- P_5**) of products with the intact glucose carbon skeleton as well as unlabeled or labeled C_1-C_5 fragments (**F_1-F_5**), which are possible precursors of Maillard products (unlabeled, singly and doubly labeled isotopomers), were demonstrated. In order to find out whether this reaction scheme can also be applied to D-fructose Maillard systems, corresponding [1-^{13}C]-D-fructose experiments were started (Tressl et al., 1993b). Up to now there are only a few detailed investigations in this field, although D-fructose is one of the most abundant and most reactive (Bunn and Higgins, 1981) hexoses. Heyns et al. (1961), who described the so-called Heyns rearrangement as a characteristic ketose Maillard reaction for the first time, already found unexpected deviations from the simple rearrangement route depending on the amino compound used. Thus, in D-fructose/L-proline experiments in methanol they characterized not only 30-40% glucose-proline and 10% mannose-proline (HRP's, epimeric Heyns rearrangement products) but also 40-50% fructose-proline (ARP, Amadori rearrangement product). In the corresponding 4-aminobutyric acid/D-fructose systems the aldose amino acids were easily formed, whereas from α-amino acids with space filling side chains (Val, Ile) only minor amounts of HRP's were obtained. With secondary amines the same reductones were formed from D-fructose and D-glucose. Overall, these results pointed to some kind of glucose-fructose equilibration, and this may be one of the reasons for the assumption, that "Amadori and Heyns compounds decompose in a parallel manner" and predominantly via 3-deoxyaldoketose (Ledl and Schleicher, 1990).

Let us, first, briefly consider the initial phase of the Maillard reaction of D-glucose and fructose and the correlation of the ARP and the HRP (Figure 1). Starting with D-glucose, 1,2-enolization of the Schiff base leads to an enaminol, the dehydration of which is the first step on the 3-deoxyaldoketose route. The ARP is an intermediate on the course to the 2.3-endiol, which is easily deaminated to start the 1-deoxydiketose route. In contrast, the 1.2- or 2.3-enolization of the Schiff base of D-fructose immediately leads to corresponding reactive intermediates as starting points for the 3- and 1-deoxyosone routes. The HRP is not necessarily involved, but its formation and transformation into the bisimine is required for the interconversion into the ARP. Thus, one can assume, that whether the Maillard reaction of D-glucose and D-fructose leads to identical products in comparable distribution or not, depends on the relative rates of the

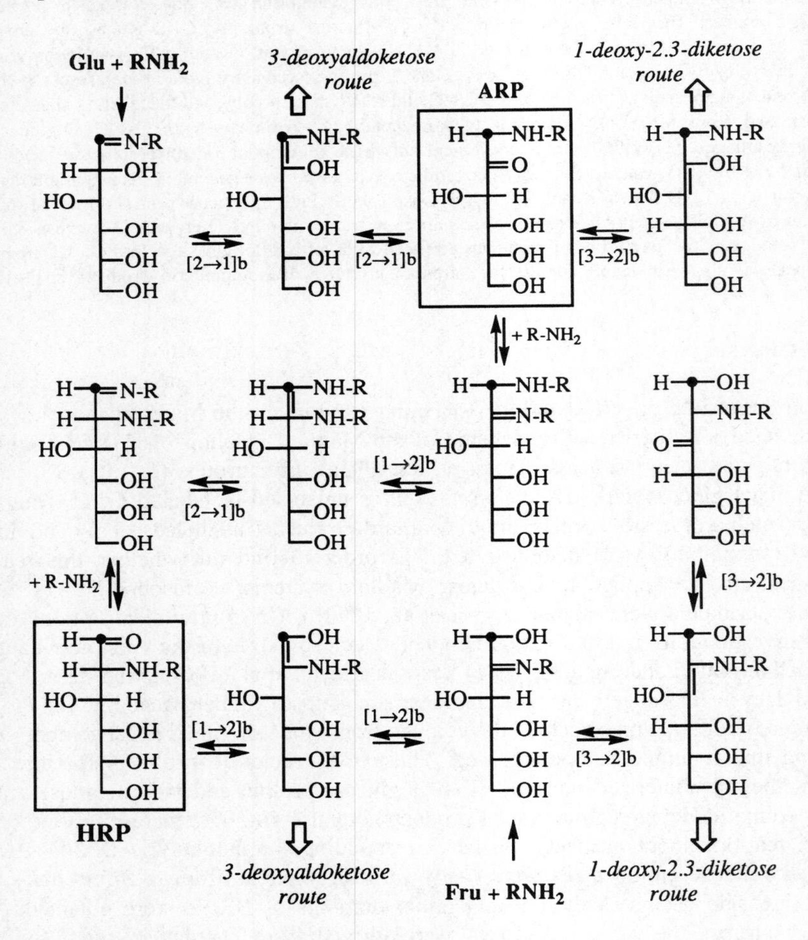

Figure 1. Initial phase of the Maillard reaction of D-glucose and D-fructose (correlation of Amadori and Heyns products); b = keto-enol tautomerization.

Table I. Types and amounts of C_6 compounds (**P_1-P_4**) and of C_3-C_5 fragmentation products (**F_3-F_5**) generated in D-Glucose (D-Fructose)/GABA (Ile, Hyp) Maillard systems.

type of products*	relative amounts** (ppm) formed from					
	D-glucose and			D-fructose and		
	GABA	Ile	Hyp	GABA	Ile	Hyp
P_1	3300	5500	510	25	1400	500
P_2	-	1500	-	-	750	-
P_3	1650	10	-	+	10	-
P_4	600	350	750	250	250	650
F_5	600	90	50	50	80	50
F_4	220	40	50	50	350	50
F_3	70	350	450	20	5700	1000

* According to figure 1 in Lit. (*1*). ** + = < 1 ppm; - = not detectable.

interconversion steps compared to the individual degradation processes. In addition, the products formed may be characteristically influenced by cleavage reactions of early stage intermediates, e.g. C_2+C_4 retro aldol cleavage (Hayashi and Namiki, 1986) in the D-glucose sequence, and C_3+C_3 retro aldol cleavage in the D-fructose sequence, both on the stage of the Schiff base.

To investigate the Maillard reaction of ketoses in more detail, [1-^{13}C]-D-fructose was synthesized from [1-^{13}C]-D-glucose by enzymatic methods as a 91:9 mixture of anomers (Tressl et al., 1993b). [1-^{13}C]-D-Fructose was heated with 4-aminobutyric acid, L-proline, L-hydroxyproline, and L-isoleucine, respectively, for 1.5 h at 160°C in equimolar amounts. The formed compounds were analyzed by capillary GC/MS, identified by MS and NMR, and quantified using internal standards. The extent and position of labeling were derived from the MS data. Selected results are summarized in Table I and are given in more detail in Tables II and III.

[1-^{13}C]-D-Fructose Maillard Systems: 3-Deoxyaldoketose Pathway

Tables I and II indicate differences in the amounts of the amino acid specific **P_1-P_4** products formed from D-fructose and D-glucose: 4-Aminobutyric acid Maillard systems generate only trace amounts of **P_1/P_3** compounds from D-glucose, whereas L-isoleucine, L-hydroxyproline (and L-proline) form comparable amounts of **P_1-P_4** and **F_5** products in D-glucose and D-fructose model experiments. The **F_3** and **F_4** compounds increase strongly in the D-fructose/Ile (Hyp) Maillard systems, corresponding to the retroaldol cleavage of the Schiff bases. The Strecker degradation products **P_2** from [1-^{13}C]-D-glucose/Ile and [1-^{13}C]-D-fructose/Ile were examined as 100% singly labeled isotopomers, comparable to [1-^{13}C]-**7** (Table II). 2,5-Dimethyl-pyrazine (**17**) was examined as mixture of unlabeled, singly and doubly labeled isotopomers (34/50/16) (formed from labeled C_3* and unlabeled C_3-fragments) as well as pyrazine **18** (28/45/27) (obviously formed from C_3*/C_3-, C_2*/C_2- fragments and unlabeled **21**). The unlabeled compounds **19-23** correspond to degradations of L-isoleucine via transamination/reduction (**19,20**) and oxidative decarboxylation/oxidation (**21,22**). The formation of unlabeled **23** is still obscure. The results demonstrate the formation of [^{13}CHO]-**7** as major compound in [1-^{13}C]-D-glucose and [1-^{13}C]-D-fructose Maillard systems. Compounds **1, 3, 7,9, 26,**

Table II. Selected degradation products of [1-^{13}C]-D-glucose and [1-^{13}C]-D-fructose in the Maillard reaction with GABA (Ile, Pro).

No. [a]	Compound	Concentration [ppm] D-glucose			D-fructose		
		GABA	Ile	Pro	GABA	Ile	Pro
1 (P_1) **7** (P_1)	structures see Tressl et al.(1993a)	3000	4500		20	1080	
2 (P_3) **8** (P_3)		380	+		+	-	
3 (P_1) **9** (P_1)		280	150		+	40	
4 (P_3) **10** (P_3)		1250	8		+	10	
5 (F_5) **11** (F_5)		215	55		50	50	
6 (F_4) **12** (F_4)		225	17		20	22	
13 (P_2)		+	800		-	240	
14 (P_2)		-	370		-	80	
15 (P_2)		-	390		-	400	
16 (P_2)		-	140		-	-	
17 (F_3)	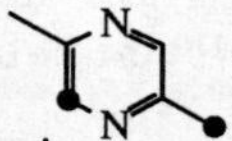		300			5780	
18 (F_3,F_2)	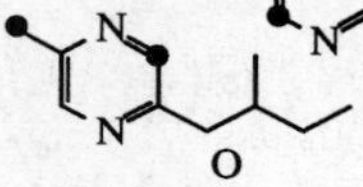		20			330	
19	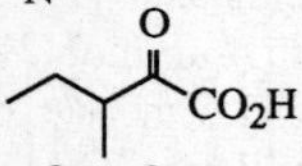		16			18	
20	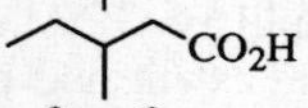		13			2	
21	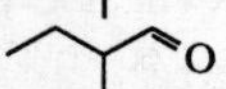		9940			2560	
22	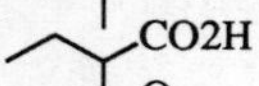		2820			680	
23	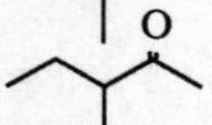		40			13	
24	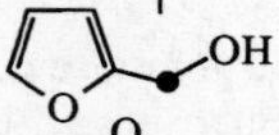	135			37		
25	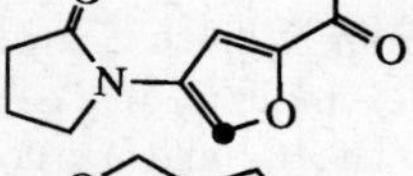	-			50		
26 (P_1)	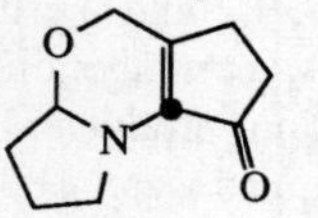			580			500
27 (P_1)	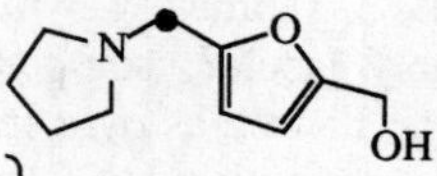			500			500
28 (P_4)	structures see figure 3	215	230	250	200	200	300
32 (P_4)		310	65	+	30	20	+
33 (P_4)		55	55	+	+	25	+

[a] **1-6**: amino acid = GABA; **7-12**: amino acid = Ile.

and **27** were identified from [1-^{13}C]-D-glucose with the intact skeleton (Tressl et al., 1993b,c; 1994). The formation pathways to [^{13}CHO]-**1**/[^{13}CHO]-**7** (Tressl et al., 1993b, 1994), [^{13}CHO]-**3**/[^{13}CHO]-**9** (Tressl et al., 1993b, 1994), and [8a-^{13}C]-**26**/[N^{13}CH$_2$]-**27** (Tressl et al., 1993c) were elucidated. From [1-^{13}C]-D-fructose, compounds **1/7** are also generated as 100% singly labeled compounds, whereas **26** was examined as mixture of unlabeled, singly and doubly labeled isotopomers. In addition, [^{13}CHO]-**1** is formed as a minor compound from D-fructose compared to D-glucose Maillard systems. The formation pathway to **3** and **9** are correlated to retro aldol reaction in D-fructose Maillard systems. Interestingly, **27** was examined with intact skeleton from D-fructose. This might be explained by kinetic effects during Strecker degradation of [1-^{13}C]-D fructose with proline into maltoxazine (**26**) and 5-hydroxymethylfurfuryl-2-pyrrolidine (**27**) (Figure 2). The 3-deoxyosone route was observed as minor pathway in [1-^{13}C]-D-fructose/4-aminobutyric acid Maillard systems (Tressl et al., 1993b).

Table III. Labeling characteristics of selected degradation products of [1-^{13}C]-D-glucose and [1-^{13}C]-D-fructose in the Maillard reaction with GABA (Ile, Pro).

compound	distribution (unlabeled / singly / doubly)					
	D-glucose			D-fructose		
	GABA	Ile	Pro	GABA	Ile	Pro
1 (P_1) **7** (P_1)	0/100/0	0/100/0		0/100/0	0/100/0	
3 (P_1) **9** (P_1)	10/85/5	0/100/0		31/51/18	86/12/2	
13 (P_2)		0/100/0			0/100/0	
14 (P_2)		0/100/0			0/100/0	
15 (P_2)		0/100/0			0/100/0	
16 (P_2)		0/100/0				
24				80/20/0		
25				0/100/0		
26 (P_1)			0/100/0			22/70/8
27 (P_1)			0/100/0			0/100/0
28 (P_4)	0/100/0	0/100/0	0/100/0	0/100/0	0/100/0	0/100/0
32 (P_4)	0/100/0	0/100/0	0/100/0	9/90/1	40/55/5	16/68/16
33 (P_4)	0/100/0	0/100/0	0/100/0	31/60/9	48/50/2	28/55/17

As outlined in Figure 3 the Schiff base of D-fructose can undergo two alternative enolizations into 2,3-enaminol **H** or into 2,1-enaminol **G** leading to the HRP. Allylic dehydration of **G** to enaminol **L** initiates the 3-deoxyosone route. Further dehydration of the bisimine (to **B**) or enolization (to **C**) would open the 3,4-dideoxyaldoketose route as well as the β-dicarbonyl route described for D-glucose (Tressl et al., 1993a), but up to now we cannot explain why these routes do not work in the D-fructose/GABA Maillard system. In this system the pyrrolidone [2'-^{13}C]-**25** was identified as 100% singly labeled product. In the corresponding L-isoleucine system, compounds derived from the intermediate **B** were examined as mixtures of unlabeled, singly, and doubly labeled isotopomers (e.g. [^{13}CHO)-**9**], Table III). These results demonstrate that the reaction pathways of D-fructose Maillard systems depend strongly on the amino acids, catalyzing enolizations and retro aldol reactions.

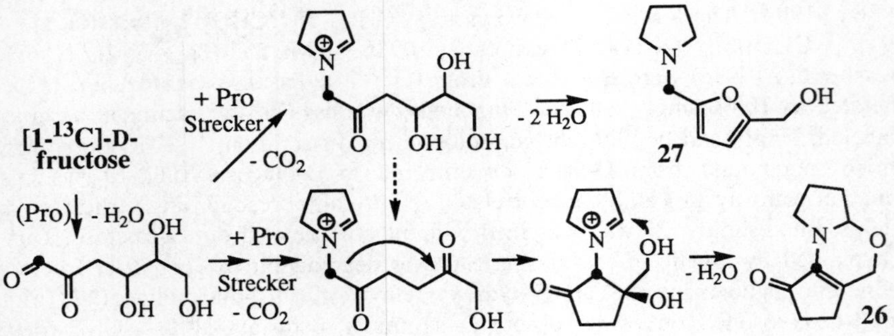

[1-^{13}C]-D-fructose
+ Pro
Strecker
- CO_2
- 2 H_2O
27
(Pro) - H_2O
+ Pro
Strecker
- CO_2
- H_2O
26

Figure 2. Formation pathways of [8a-^{13}C]-**26** and [N-$^{13}CH_2$]-**27** in [1-^{13}C]-D-fructose/Pro Maillard systems.

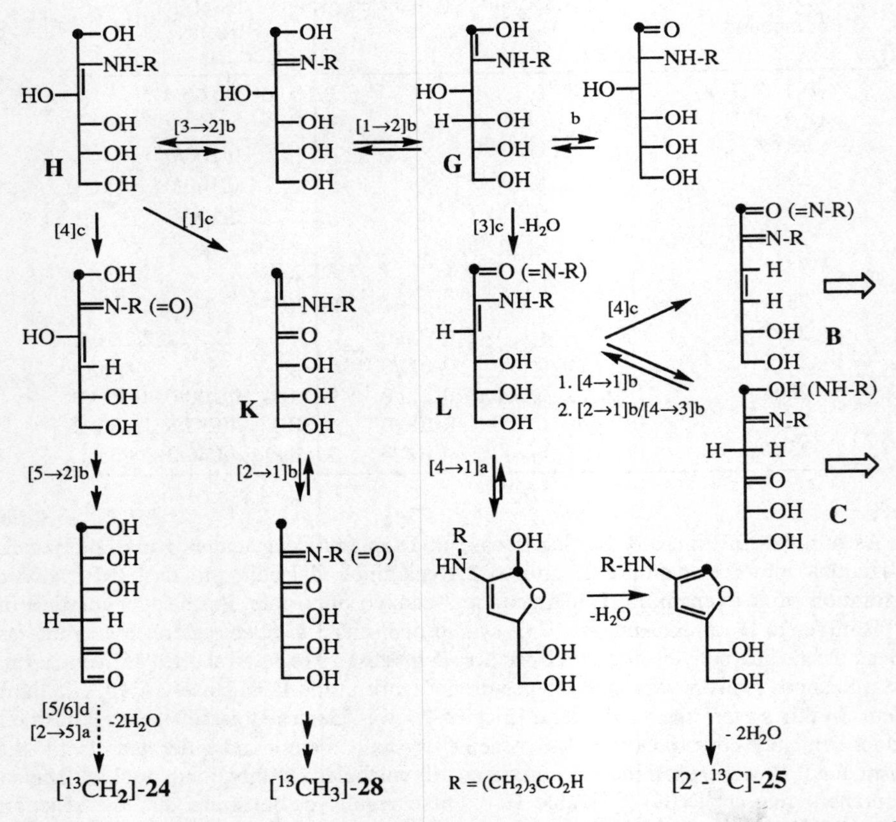

Figure 3. Pathways of the Maillard reaction of D-fructose with GABA, Ile, or Pro; for completion of the scheme (along **B/C**) see Tressl et al. (1993a, Figures 1 and 6); a=cyclization, b=keto-enol tautomerization, c=allylic dehydration.

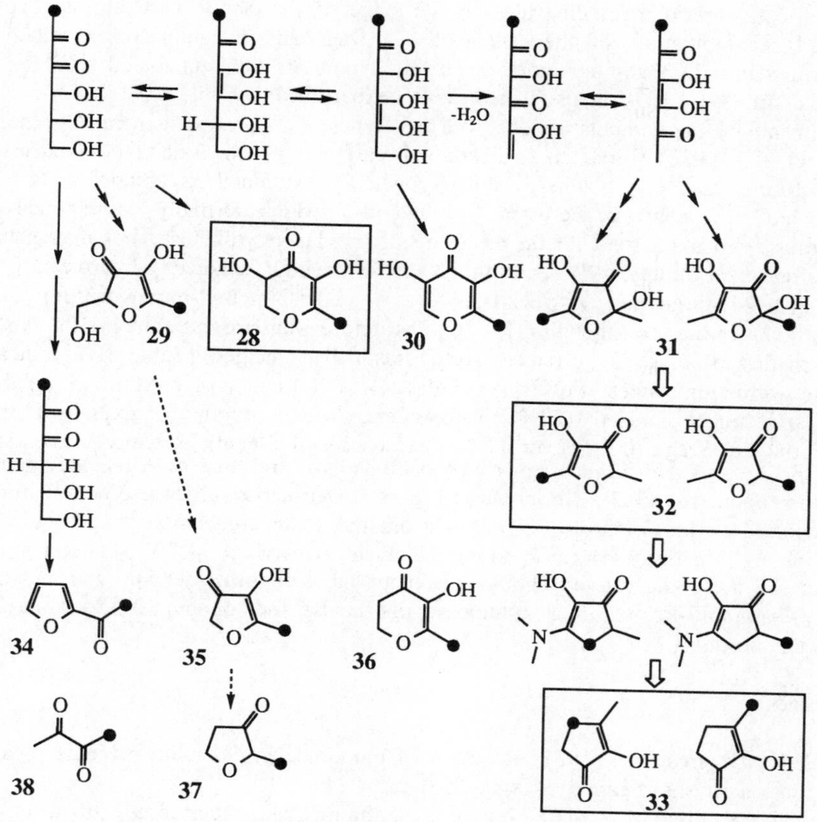

Figure 4. 1-Deoxy-2.3-diketose products generated in [1-^{13}C]-D-fructose/GABA (Ile, Pro) Maillard experiments (only the most abundant isotopomer is shown).

[1-^{13}C]-D-Fructose Maillard Systems: 1-Deoxydiketose Pathway

The degradation of [1-^{13}C]-D-glucose into 1-deoxydiketose products ($\mathbf{P_4}$) has been demonstrated with Pro, Hyp, Cys, GABA and Ile Maillard systems (Figure 4). Overall, these typical $\mathbf{P_4}$ products were also formed in [1-^{13}C]-D-fructose Maillard experiments in comparable amounts. [$^{13}CH_3$]-**28** is generated as 100% singly labeled compound in [1-^{13}C]-D-fructose as well as in [1-^{13}C]-D-glucose Maillard systems. These results confirm the pathway (Figure 3) via intermediate **H**, subsequent dehydration into **K**, forming **28** by [6→2] cyclization. 2,5-Dimethyl-4-hydroxy-3(2*H*)furanone (**32**) and cyclotene (**33**) were examined as mixtures of two 100% singly labeled isotopomers in [1-^{13}C]-D-glucose Maillard systems (Tressl et al., 1993b,c) . In the corresponding [1-^{13}C]-d-fructose/4-aminobutyric acid experiments, a cleavage reaction must be included in the reaction sequence to **32/33** via diacetylformoin as indicated by the formation of mixtures of unlabeled, singly and doubly labeled isotopomers. The formation of isotopomer mixtures is increased in the corresponding isoleucine system, resulting in surprisingly high amounts of unlabeled **32** (40%) and **33** (48%). This might be explained by an effective retro aldol cleavage on the stage of the Schiff base and subsequent preferred recombination of the unlabeled C_3 fragments to aldohexoses which are transformed into **32/33**. In agreement with this hypothesis 86% unlabeled [^{13}CHO]-**9** is generated in [1-^{13}C]-D-fructose/L-isoleucine experiments (Table III). In [1-^{13}C]-D-glucose Maillard experiments we observed two types of $\mathbf{F_5}$ product generating reactions (Tressl et al., 1993b). First, an α-dicarbonyl cleavage of the 3-deoxyaldoketose into labeled formic acid and unlabeled 2-deoxypentose (examined as unlabeled **24**), and, second, a C_1+C_5 cleavage between C-5 and C-6, which formally corresponds to a vinylogous retro aldol reaction on the stage of the 1-deoxydiketose, but may actually occur after cyclization to **29** resulting in 100% to singly labeled 4-hydroxy-5-[^{13}C]-methyl-3(2*H*24)-furanone ([$^{13}CH_3$]-**35**) and 2-[^{13}C]methyl-tetrahydrofuran-3-one ([$^{13}CH_3$]-**37**). In the corresponding [1-^{13}C]-D-fructose Maillard experiments the labeling characteristics of these $\mathbf{F_5}$ products were drastically changed (Table III), indicating different formation routes. Furfurylalcohol (**24**) is formed as a mixture of unlabeled (80%) and singly labeled (20%) isotopomers, which might be explained via a α-dicarbonyl cleavage of the reversed (with respect to the label) 3-deoxy-aldoketose as shown in Figure 3. Norfuraneol (**35**) was examined as a mixture of unlabeled to singly labeled isotopomers (80:20). As shown in Figure 1, D-fructose might also be transformed into 2-deoxy-2-amino-3-ketose derivatives in analogy to the conversion of D-fructose into D-psicose by piperidine or morpholine catalysis (Heyns et al., 1961). Retro aldol cleavage of these derivatives should result in labeled formaldehyde and unlabeled 1-deoxy-1-amino-2-ketoses (e.g. xylulose amino acids), thus opening a new route to $\mathbf{F_5}$ cyclization products.

References

Bunn, H.F.; Higgins, P.J. (1981) Reaction of monosaccharides with proteins: possible evolutionary significance. Science 213: 222-224.

Hayashi, T.; Namiki, M. (1986) Formation of three-carbon sugar fragment at an early stage of the browning reaction of sugar with amines or amino acids. Agric. Biol. Chem. 50: 1959-1964.

Heyns, K.; Paulsen, H.; Schroeder, H. (1961) Die Umsetzung von Ketohexosen mit sekundären Aminosäuren und sekundären Aminen. Tetrahedron 13: 247-257.

Ledl, F.; Schleicher, E. (1990) New aspects of the Maillard reaction in foods and in the human body. Angew. Chem. Int. Ed. Engl. 29: 565-594.

Tressl, R.; Kersten, E.; Nittka, C.; Rewicki D. (1993a) Mechanistic studies on the formation of pyrroles and pyridines from [1-^{13}C]-D-glucose and [1-^{13}C]-D-arabinose, submitted (*this book*)

Tressl, R.; Kersten, E.; Rewicki D. (1993b) Formation of 4-aminobutyric acid specific Maillard products from [1-^{13}C]-D-glucose, [1-^{13}C]-D-arabinose and [1-^{13}C]-D-fructose. J. Agric. Food Chem. (in press).

Tressl, R.; Helak, B.; Kersten, E.; Rewicki, D. (1993c) Formation of proline- and hydroxy-proline-specific Maillard products from [1-^{13}C]-D-glucose. J. Agric. Food Chem. 41:547-553.

Tressl, R.; Nittka, C.; Rewicki, D. (1994) Formation of isoleucine specific Maillard products from [1-^{13}C]-D-glucose and [1-^{13}C]-D-fructose. J. Agric. Food Chem., submitted.

Investigation of the Acyclic Forms of Reducing Sugars and Amadori Products by FTIR Spectroscopy

Varoujan A. Yaylayan, Ashraf A. Ismail, and Alexis Huyghues-Despointes

DEPARTMENT OF FOOD SCIENCE AND AGRICULTURAL CHEMISTRY, MCGILL UNIVERSITY 21, 111 LAKESHORE, STE. ANNE DE BELLEVUE, QUEBEC, CANADA H9X 3V9

Summary

The acyclic forms of reducing sugars and Amadori products were studied by FTIR spectroscopy. The spectra of different aldoses and ketoses were recorded as a function of temperature and pH. Fourier self-deconvolution of the carbonyl region (1700 - 1750 cm^{-1}) has revealed that pentuloses, pentoses, and hexuloses exhibit absorption bands at 1733, 1726 and 1717 cm^{-1} and these bands were tentatively assigned to 3-*keto*, 2-*keto* and *aldehydo* forms of the sugars, respectively, formed through Lobry de Bruyn-Alberda van Ekenstein transformation reactions, whereas, tetruloses showed only two bands at 1733 and 1717 cm^{-1}. In addition, the corresponding 2-*deoxy*-aldoses showed only one peak. The migration of the carbonyl group of the acyclic forms of reducing sugars can proceed through enediol formation; these species absorb in the region of 1620-1700 cm^{-1}. The sugars that exhibited migration of the carbonyl group also showed an absorption band centered at 1650 cm^{-1}. This band was also sensitive to temperature and pH similar to the carbonyl absorption band and, after treatment of the sugar with ozone or passing a stream of air in the presence of $FeCl_3$, the intensity of the band was diminished, indicating its double bond character.

Introduction

Knowledge of the composition of reducing sugars in solution can have considerable practical and theoretical importance. The physical and chemical properties of these sugars in solution depend on the relative concentrations of different tautomeric forms. Their biological properties may also depend on tautomeric equilibria (Angyal, 1984). In non-enzymatic glycation of proteins and amino acids, the concentrations of open-chain forms might be a crucial factor in determining the rate of the reaction if the mutarotation rate is slower than the reaction rate. Although most of the tautomeric forms of the reducing sugars in solution have never been isolated, they can be detected by different techniques, and their concentrations in the equilibrium mixture can be measured. Sugars must undergo a ring opening in order to mutarotate to the high-energy open-chain form, which is usually stabilized by complexation with the solvent molecules. The proportions of the acyclic forms increases with temperature due to the entropy factor (acyclic forms have a greater degree of freedom) and the high enthalpy content (cyclizations are exothermic reactions and hence favored by lower temperatures). Evidence for the presence of the *keto* form of D-fructose in solution has been provided by methods such as polarography and ^{13}C NMR spectroscopy (Angyal, 1992). In a previous study (Yaylayan and Ismail, 1992), we demonstrated that FTIR spectroscopy can be employed in the detection of the carbonyl absorption band of the acyclic form of D-fructose centered at 1728 cm^{-1}. Changes in the intensity of the band at 1728 cm^{-1} allowed the monitoring of the concentration of the acyclic form of D-fructose at different temperatures and pH values. The concentration of the acyclic form was observed to increase with increasing temperature (Figure 1) and was an order of magnitude higher at 80 °C compared to 30 °C. The new equilibrium can be reversed with decreasing temperature. Data obtained (Yaylayan *et al.*, 1993) by the digital integration of the intensities of open-chain carbonyl absorption bands were used to calculate the percentage of the open-chain forms of D-fructose between 25 to 80°C and at acidic and basic pH values, using the molar absorptivity value of the carbonyl peak of 1,3-dihydroxyacetone (1286 L/mol/cm). Generally the values obtained by FTIR were higher than those of ^{13}C NMR studies.

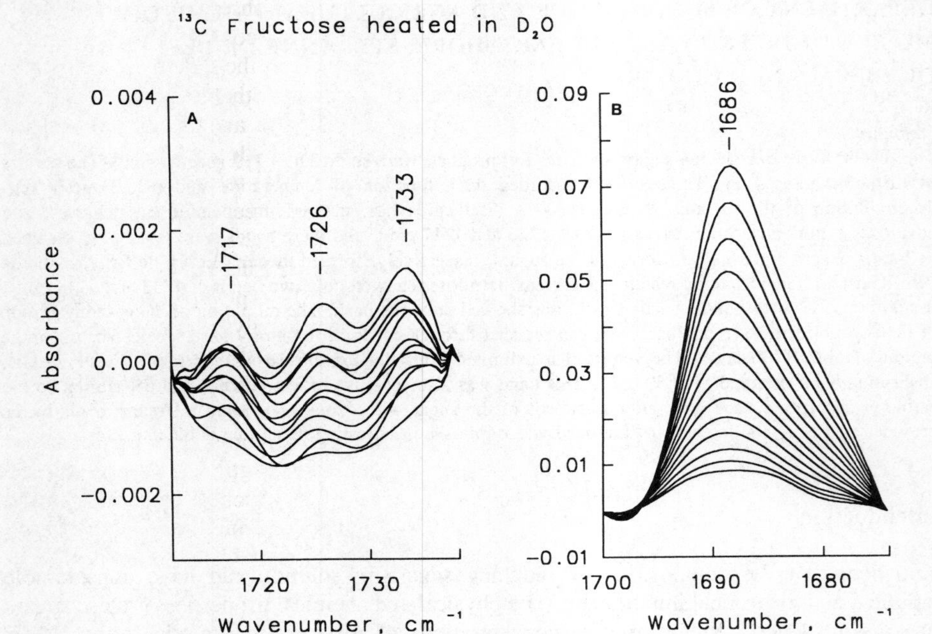

Figure 1. (A) Fourier self-deconvolution result of the 1700-1750 cm^{-1} region of D-[2-^{13}C]fructose. (B) The increase in the carbonyl absorption band of D-[2-^{13}C]fructose (in D_2O) as a function of temperature between 30-85 °C

Material and Methods

All the sugars were obtained from Sigma Chemical Co. D_2O and D-[2-^{13}C]fructose were obtained from MSD Isotopes (Montreal, Canada). Fructosyl proline was synthesized according to literature procedures. Sugar solutions were made up in concentrations ranging from 1.1 M to 2.8 M in D_2O. The solutions were left to stand for a minimum of 24 h at room temperature prior to FTIR measurements. The pHs of the solutions were adjusted by addition of DCl or NaOD.

Temperature Studies.
Sugar solutions in D_2O were placed in a CaF_2 IR cell with a 50-μm Teflon spacer. The temperature of the sample was regulated by placing the IR cell in a temperature-controlled cell holder. Infrared spectra were recorded on a Nicolet 8210 Fourier-transform spectrometer, purged with dry air and equipped with a deuterated triglycine sulfate (DTGS) detector. The initial temperature of the cell was raised by 1 °C per minute, and

every five minutes the temperature was kept constant for 15 minutes to record the spectra. A total of 128 scans at 4-cm^{-1} resolution were coadded.

Results and Discussion

Confirmation of the carbonyl absorption peaks by D-[2-^{13}C]fructose and D-[1-^{13}C]ribose studies.

Generally, ketone and aldehyde carbonyl groups give rise to intense bands in the 1740 - 1710 cm^{-1} region of the spectrum. It is possible to extract a considerable amount of information from the position of the absorption frequency about the electronic and steric effects that arise from the nature of the substituents attached to the carbonyl group. Of special significance to reducing sugars is the substituents at α and α' carbons, since introduction of a halogen at these positions in a ketone or aldehyde, is known to lead to a shift to higher frequencies provided that the halogen can rotate to eclipse the carbonyl group. Most probably this shift arises from a field effect. This effect was demonstrated in carbohydrates by a shift to higher frequency (from 1716 to 1722 cm^{-1}) in the carbonyl absorption band of ribose when compared to the absorption of 2-deoxyribose, due to the presence of the α-hydroxy group. In order to assign carbonyl absorption bands in reducing sugars, a series of reducing *keto* and *aldehydo* sugars were analysed along with ^{13}C - labeled fructose and ribose. Infrared absorption wavelengths are shifted when one of the atoms is replaced with a different isotope. In order to confirm the identity of the peak centered at 1728 cm^{-1} in the spectrum of fructose, the spectrum of D-[2-^{13}C]fructose was compared to that of D-fructose. In unlabeled D-fructose, the carbonyl peak was centered at 1728 cm^{-1} whereas in D-[2-^{13}C]fructose the absorption wavelength was shifted to 1686 cm^{-1} (Figure 1). The expected ratio of $\upsilon_{C\text{-}12}/\upsilon_{C\text{-}13}$ was 1.025 which is very close to the experimentally observed ratio of 1.023 (1728/1686), which confirmed the identity of the peak. Similarly, the peak centered at 1725 cm^{-1} in D-ribose, was shifted to 1684 cm^{-1} in D-[1-^{13}C]ribose.

Enolizations and migration of carbonyl groups in reducing sugars.

Under acid or base catalysis, reducing sugars are known to undergo Lobry de Bruyn-Alberda van Ekenstein transformation reactions (Figure 2); the enediol mechanism of this transformation has been verified by different groups (Cervantes-Laurean *et al.*, 1993; King-Morris and Serianni, 1986). The contribution of 3-*keto*, *aldehydo* and different enediol forms to the total concentration of acylic forms of reducing sugars is not known due to lack of analytical techniques for their detection. In order to verify the presence of different *keto* and *aldehydo* forms of acyclic sugars, Fourier self-deconvolution studies were performed on the carbonyl absorption bands of D-*glycero*-tetrulose (D-erythrulose), D-lyxose, D-xylose, D-ribose, D-ribose-5'-phosphate, 2-*deoxy*-D-ribose, D-ribulose, D-[1-^{13}C]ribose, D-[2-^{13}C]ribose, 2-*deoxy*-D-glucose, D-[2-^{13}C]fructose, D-fructose and fructosyl proline. D-*glycero*-tetrulose, which exist only in acyclic form, showed two distinct bands at 1733 and 1717 cm^{-1} without the need to mathematically enhance the carbonyl band.

D-Lyxose, D-xylose, D-ribose, D-ribulose and D-ribose-5'-phosphate showed three bands in the carbonyl region centered at 1718, 1726 and 1733 cm^{-1}. In D-ribulose, the band at 1726 cm^{-1} was the most prominent whereas in the others the peak centered at 1718 cm^{-1} was the most prominent. 2-*Deoxy*-D-ribose and 2-*deoxy*-D-glucose showed only one band each centered at 1716 and 1718 cm^{-1}, respectively. Fructosyl proline also showed

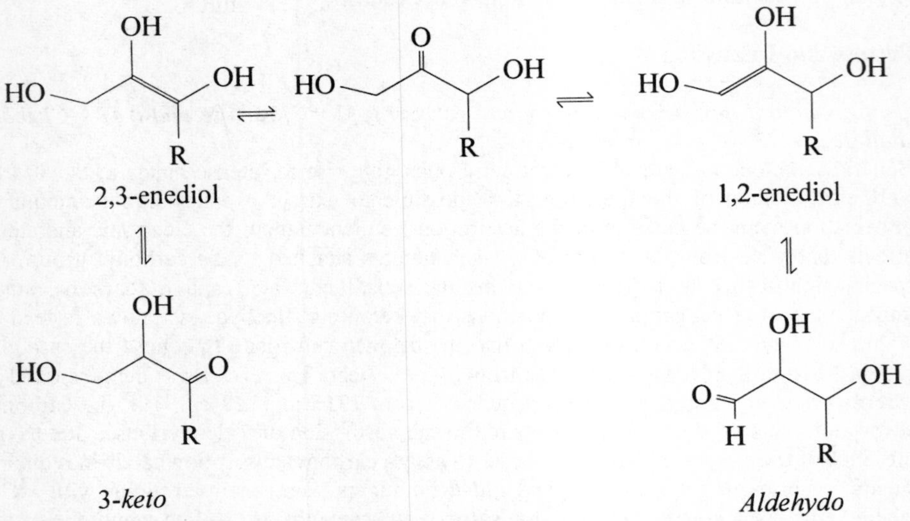

Figure 2. Lobry de Bruyn-Alberda van Ekenstein transformation reactions of reducing sugars.

three distinct bands at 1717, 1726 and 1733 cm^{-1}. The availability of D-[2-^{13}C]fructose and shifting of the 2-*keto* band to 1686 cm^{-1}, allowed the study of the contribution of 3-*keto*, *aldehydo* and 5-*keto* (equivalent of 2-*keto* in the unlabeled sugar) forms of fructose to the total concentration of acyclic forms. The deconvolution of the carbonyl region of D-[2-^{13}C]fructose (Figure 1) revealed the presence of three bands at 1717, 1726 and 1733 cm^{-1}. By comparison with the unlabeled fructose and with other sugars studied, the 1726 band may be assigned to the 5-*keto* form, 1717 to the a*ldehydo* form (glucose) and 1733 to the 3-*keto* form. These three bands in all the sugars studied were sensitive to pH and temperature, particularly in the case of the Amadori compound. At neutral pH, the Amadori compound showed only one peak centered at 1734 cm^{-1}; changing to either acidic or basic pH, catalysed the enolization reactions with the emergence of the other carbonyl bands. The rate of decomposition of different *keto* forms was also influenced by pH when the temperature was increased from 35 to 85 °C (Figure 3). The discrepancy in the calculated values of the open-chain forms of fructose between FTIR and ^{13}C NMR studies, can be partially explained by the fact that ^{13}C NMR measures only the concentration of the 2-*keto* open-chain form, whereas FTIR measures the total carbonyls comprising the open-chain forms of fructose.

Enediol formation.
Alkenes usually absorb in the region between 1620 -1700 cm^{-1}. Interestingly, all the sugars studied that showed migration of the carbonyl group also showed an absorption band centered at 1650 cm^{-1}, where enols commonly absorb. The band was sensitive to temperature (absorption increased with increasing temperature and *vice versa*) and to pH. 2-Deoxy-sugars, which cannot propagate enolizations lacked this band in their spectra. In order to verify the nature of the chemical moiety producing this band, selected sugars such as ribose and fructose were subjected to ozonolysis for 30 min at room temperature and

also to air oxidation in the presence of ferric chloride; in all cases, the intensity of the 1650 cm^{-1} band was diminished, indicating the presence of unsaturation.

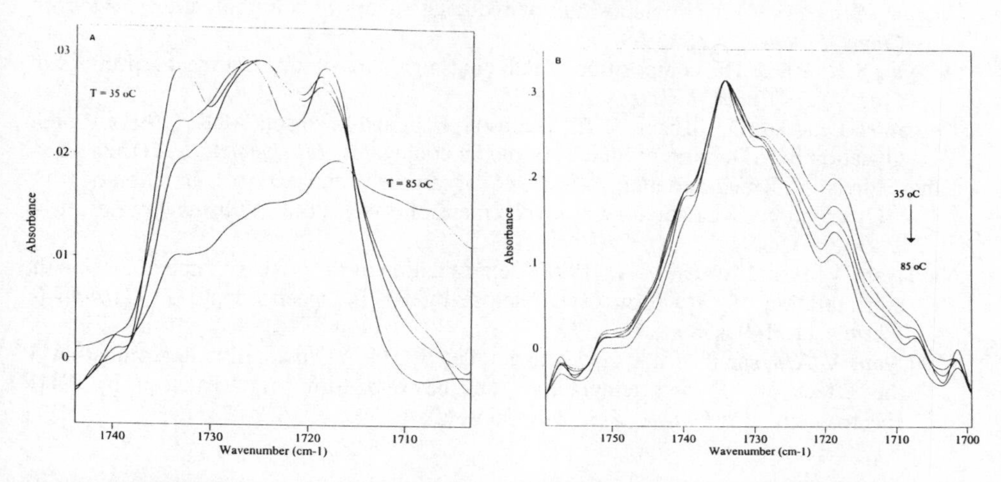

Figure 3. The effect of increasing temperature on the carbonyl absorption band of fructosyl proline A) at pH 9, B) at pH 4.

Conclusions

The time scale of IR spectroscopy allows rapidly exchanging tautomers such as enediols and α-hydroxy ketones to be detected and monitored. This study indicates that the reported values of acyclic forms of reducing sugars based on ^{13}C NMR studies might be underestimations.

Acknowledgments
The authors acknowledge funding for this research by the Natural Sciences and Engineering Research Council of Canada (NSERC). The authors also wish to thank Nicolet Instrument Inc. for their support.

References

Angyal, S.J. (1984) The composition of reducing sugars in solution. *Adv. Carbohydr. Chem. Biochem.* 42: 15-68.

Angyal, S.J. (1992) The composition of reducing sugars in solution: Current aspects. *Adv. Carbohydr. Chem. Biochem.* 49:19-35.

Cervantes-Laurean, D., Minter, D.E., Jacobson, E.L. and Jacobson, M.K. (1993). Protein glycation by ADP-ribose: Studies of model conjugates. *Biochemistry*. 32:1528-1534.

King-Morris, M.J. and Serianni, A.S. (1986) Hydroxide-catalyzed isomerization of D-[1-^{13}C]mannose: Evidence for the involvement of 3,4-enediols. *Carbohydr. Res.* 154: 29-36.

Yaylayan, V.A. and Ismail, A.A. (1992) Determination of the effect of temperature on the concentration of *keto* form of D-fructose by FT IR spectroscopy. *J. Carbohydr. Chem.*, 11: 149-158.

Yaylayan, V.A., Ismail, A.A. and Mandeville, S.(1993) Quantitative determination of the effect of pH and temperature on the *keto* form of D-fructose by FTIR spectroscopy. *Carbohydr. Res.* 248: 355-360.

Naturally Occurring Phenolic Compounds as Inhibitors of Free Radical Formation in the Maillard Reaction

Sonja M. Djilas and Božidar Lj. Milić

ORGANIC CHEMISTRY DEPARTMENT, FACULTY OF TECHNOLOGY, UNIVERSITY OF NOVI SAD, 21000 NOVI SAD, YUGOSLAVIA

Summary

The influence of naturally occurring phenolic compounds, ellagic acid, 2,3,7,8-tetrahydroxy[1]-benzopyrano-[5,4,3-cde][1]-benzopyran-5,10-dione, gallic acid, 3,4,5-trihydroxybenzoic acid, ferulic acid 3-(4-hydroxy-3-methoxyphenyl)-2-propenoic acid and syringic acid, 4-hydroxy-3,5-dimethoxybenzoic acid, as inhibitors of the formation of 1,4-pyrazine cation free radical and pyrazine derivatives in the Maillard reactions between D(+)-glucose and 2-, 3- or 4-aminobutanoic acid, respectively, have been studied by electron spin resonance (ESR) spectroscopy. The effect of phenolic compounds on the formation of amino-imidazoazarenes, a new class of mutagenic heterocyclic amines, which are products of the Maillard reaction in the presence of creatinine, has been studied by HPLC.

Introduction

Several articles on the formation of free radicals in the Maillard reaction between aldoses and amino acids in water solutions, under conditions of base catalysis and high temperature, have been published (Namiki and Hayashi, 1975; Milić et al.,1978). Recently published results (Jagerstad *et al.*, 1986, Milić *et al.*, 1993) showed that pyrazine and pyridine derivatives were an integral part of the mechanism of formation of some heterocyclic amino-imidazoazarenes. In an effort to clarify the mechanism of formation of amino-imidazoazarenes, Milić and coworkers (1993) examined the formation of pyrazine and pyridine free radicals, as intermediates, in the initial step, and pyrazine and pyridine derivatives in the steady-state reaction step between D(+)-glucose and 2-, 3- and 4-aminobutanoic acids, respectively.

The phenolic compounds, as antioxidants, may react with pyrazine free radicals from the Maillard reaction and retard their reactivity (Nawar, 1985). Although methods for the prevention of the Maillard reaction have been extensively studied (Labuza and Schmidl, 1986), information concerning the utilization of naturally occurring phenolic compounds is scarce. Many authors (Fujita *et al.*, 1988; Kikuzaki and Nakatani, 1989; Miura and Nakatani, 1989; Cilliers and Singleton, 1991) have reported that some naturally occurring antioxidant compounds, such as flavones and polyphenoles or phenolic compounds, isolated from seeds (coffee, sesame and soya), leaves (oregano, thyme and rosemary), and those in foods, had a significant influence on the development of cancer when such foods were consumed. Due to such binding, the nature of different free radical species, their reaction pathways and reactivity should be investigated using model and living systems to study the mechanisms of diseases and aging.

An excellent and critical review by Namiki (1990), considered the chemistry of antioxidants as antimutagens, and concluded that the induction of mutations occurred mainly through damage to DNA as a result of active free radical species formation from various types of mutagens and carcinogens. Such radicals, in the presence of oxygen can result in the formation of oxy-radicals which in turn may add to double bond of pyrimidine bases and abstract hydrogen from the deoxyribose moiety resulting in scission of DNA and mutagenesis and carcinogenesis. As free radicals, especially oxy- and peroxy-species are the most important factors of the oxidation of food, it seems reasonable to screen the natural substances that inhibit their activity.

In an effort to clarify the influence of some naturally occurring phenolic compounds on the formation of pyrazine cation free radicals, which may promote the formation of other free radical species and mutagens, the present paper examines the inhibitory effect of ellagic, gallic, ferulic and syringic acids, respectively, on free radical formation.

To establish the effects of such phenolic compounds on the formation of amino-imidazoazarenes, the formation of 2-amino-3,4,8-trimethylimidazo(4,5-f)quinoxaline from the Maillard reactants and creatinine, and from 2,5-dimethylpyrazine, acetaldehyde and creatinine, with and without the examined phenolic compounds was studied. The aim of such experiments was to compare the effects of the four naturally occurring phenolic compounds on the formation of amino-imidazoazarenes from starting products formed by the Maillard reaction.

Materials and Methods

All chemicals used in the present experiments were of analytical purity. Model systems for the ESR determination of pyrazine cation free radicals were obtained by mixing equimolar amounts of D(+)-glucose and the 2-, 3- or 4-aminobutanoic acids separately in an aqueous solution of 0.01 M NaOH at pH 9.0. The final concentrations of both reactants were 1.0M.

Model systems were also prepared with ellagic, gallic, ferulic or syringic acids in the following concentrations 0.005, 0.010, 0.020, 0.040, 0.080 and 0.160 M. Samples for free radical determination were heated to 95 C in the quartz cell of a Bruker 4121 VT-RS high temperature control system for the 20 minutes of the reaction period.

The ESR spectra were recorded on a Bruker 300E Electron spin resonance spectrometer using the following instrument settings: field sweep width,50 G; receiver gain, 1.021 G; microwave power, 6.32×10^{-1} mW and center-field, 3440 G.

Two series of model systems were used to establish the effects of the naturally occurring phenolic compounds on the formation of amino-imidazoazarenes from Maillard reaction products or pyrazine derivatives and acetaldehyde, in the presence of creatinine.

The first series of model systems contained creatinine (1M), D(+)-glucose (1M) and aminobutanoic acid (1M) (2-, 3- or 4-aminobutanoic acids) with and without ellagic, gallic, ferulic and syringic acids (0.020 M); the constituents were dissolved in diethylene glycol containing 11% water (pH 9.0), and refluxed for 3 h at 130°C. The second series of model systems contained creatinine (1M), 2,5-dimethylpyrazine (1M), and acetaldehyde (1M), with and without ellagic, gallic, ferulic and syringic acids (0.020 M), under the same reaction conditions described above. The reaction mixtures were subsequently extracted with 1-butanol, and the extracts were purified on a column of Dowex 50 (H^+ -form) using 0.1 M ethanolic ammonia eluent. The eluates were evaporated to dryness, dissolved in a small amount of methanol and rechromatographed on YMC A-300 and Sep-Pak columns. 2-Amino-3,4,8-trimethylimidazo(4,5-f)quinoxaline was then determined by high-performance liquid chromatography (HPLC), using Waters Associates instrument under the following conditions: first column, YMC A-300 (240 x

5.2 mm i.d.); second column, Sep-Pak (120 x 4.5 mm i.d.) 5μm particle size; mobile phase, methanol/H_3 PO_4 0.01 M/ NaOH 0.01 M (1:0.6:0.6) pH 7.4; flow rate, 0.8 ml/min at 25°C. The total amounts of the total amino-imidazoazarenes were calculated according to the peak area of 2-amino-3,4,8-trimethylimidazo-(4,5-f)quinoxaline.

Results and Discussion

From the 75 ESR spectra obtained (3 spectra of the three model systems of the heated alkaline water solutions of D(+)-glucose or 2-, 3- or 4-aminobutanoic acids and 72 spectra of reaction mixtures of the model systems described above and the added amounts, 0.005, 0.010, 0.020, 0.040, 0.080 and 0.160 mol/dm, of one of the four naturally occurring phenolic compounds: ellagic, gallic, ferulic and syringic acids it was apparent that the phenolic compounds inhibited the formation and reaction of 1,4-pyrazine cation free radicals.

The shape and hyperfine splitting constants of the ESR spectra of free radicals, obtained from the reaction between D(+)-glucose and the three aminobutanoic acid isomers have been reported recently (Milić *et al.*, 1993). The representative ESR spectra of the reaction mixtures of D(+)-glucose and 2-aminobutanoic acid with and without gallic acid at a concentration of 0.020 M, are shown in Figure 1. The complete results of the investigation are presented in Figure 2.

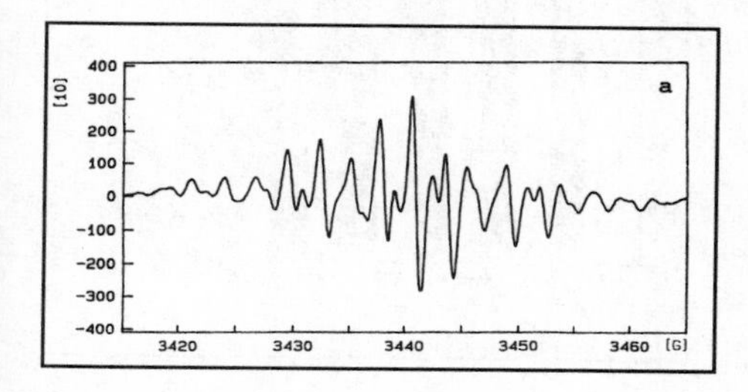

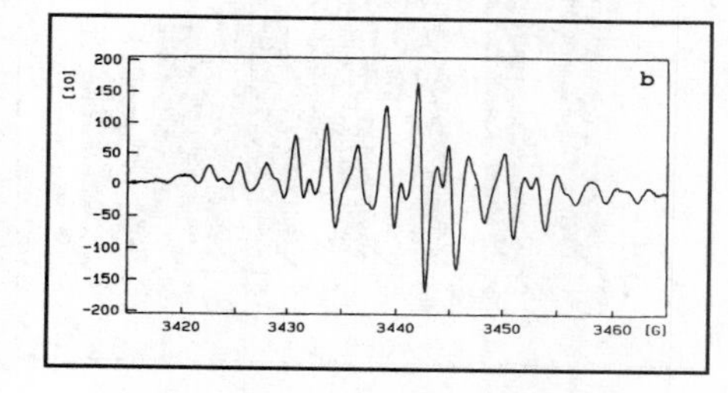

Figure 1. ESR spectra of free radicals obtained from the reaction mixtures: (a), D(+)-glucose and 2-aminobutanoic acid and (b), D(+)-glucose, 2-aminobutanoic acid and gallic acid.

No change in the shape of any of the ESR spectra was detected, but the relative intensity (RI) of ESR signals, corresponding to the concentration of free radicals, was decreased with increasing concentrations of the added phenolic compounds. From the results presented in Figure 2 it is possible to conclude that the inhibitory effect on free radical formation were in the order: ellagic acid (a four hydroxy compound), gallic acid (a three hydroxy compound) and syringic acid (a monohydroxy compound). Minor effects were observed in the case of ferulic acid (a monohydroxy compound).

Although the effect of phenolic compounds on Maillard pathways responsible for the formation of free radicals and volatile compounds has received little attention, studies on related topics suggest several routes by which phenolic compounds may interact in the Maillard reaction. Investigations of the phenolic compounds used in the present study, heated with D(+)-glucose and one of the aminobutanoic acid isomers showed that color intensity decreased with increasing concentration and degree of oxidation of the phenolic compounds.

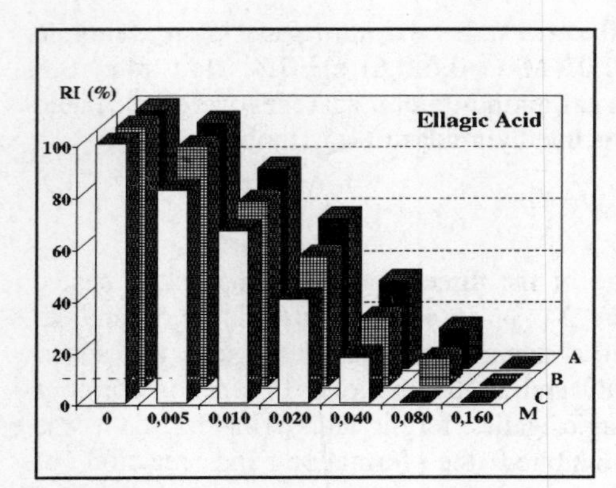

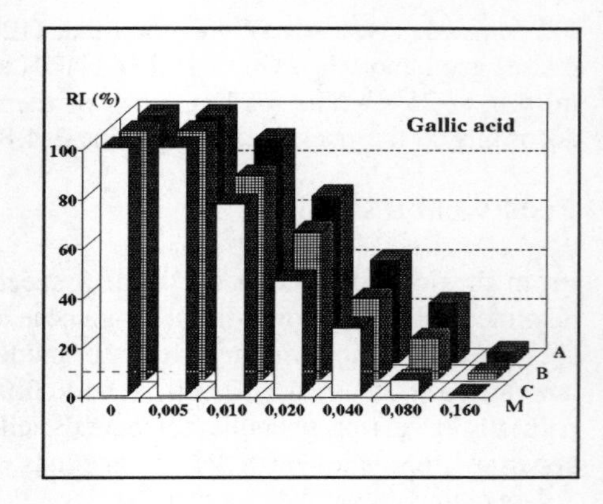

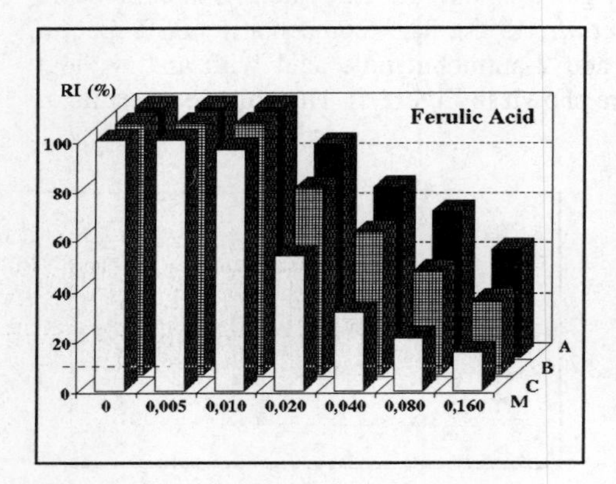

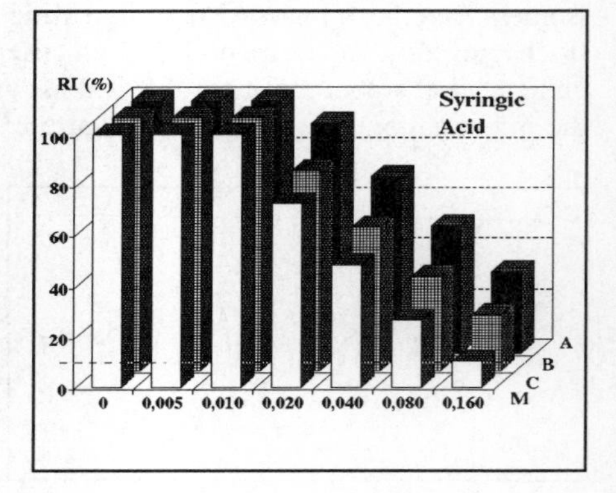

Figure 2. The influence of different amounts of ellagic, gallic, ferulic and syringic acids, on the relative intensity (RI) of signals in the ESR spectra of free radicals in the model system of D(+)-glucose and 2-, 3- or 4-aminobutanoic acids (A,B and C respectively).

Unsaturated carbonyl compounds and hydroxyketones, which are products of the Maillard reaction, also cause browning in the presence of amino compounds, via the formation of Schiff's bases; free radicals derived from such reactions are known to react with the hydroxy groups of phenolic compounds forming semiquinones. If more than one species of free radicals exists in reaction mixtures or if a molecule contains more than one unpaired electron, additional magnetic phenomena arise due to the interaction between them, scalar exchange interactions or dipole-electron interactions. The strength of such interactions allow a reasonable classification into "double-double radicals", "diradicals" and "real multiplet systems". As the spectrum of a diradical in isotropic solution differs from that of the corresponding monoradical, it can be concluded from the unchangeable spectra shapes of Figure 1 that only the 1,4-pyrazine cation radicals in the systems studied exist. Such radicals can react with the hydroxy groups of phenolic compounds presumably with hydroxy groups in the C_4 and C_3 positions. The possibility that two species of free radicals might exist in the reaction mixtures pyrazine radicals and semiquinone oxy-radicals (the latter could be formed under base catalytic conditions from the phenolic compounds and act as good radical scavengers, Fujita *et al.*, 1988) seems to be unlikely. The ESR spectra of such mixtures, especially when the g-values are similar, often consists of a superposition of the contributing spectra that is difficult to analyze. The individual ESR spectra of both free radical species, if they exist in the reaction mixtures, can be recorded separately using the Electron Nuclear Double Resonance(ENDOR) - Induced ESR spectroscopy technique, provided separate signals for the different radicals appear in the spectrum. ENDOR experiments of such system are currently under way.

The Maillard reaction between aldoses and amino compounds leads to the formation of end products of different pyrazine and pyridine derivatives. Milić *et al.*, (1993) showed that these compounds were precursors, and in the reaction with creatinine they form different amino-imidazoazarenes. The results of studies on the effect of phenolic compounds on the formation of amino-imidazoazarenes are presented in Figure 3.

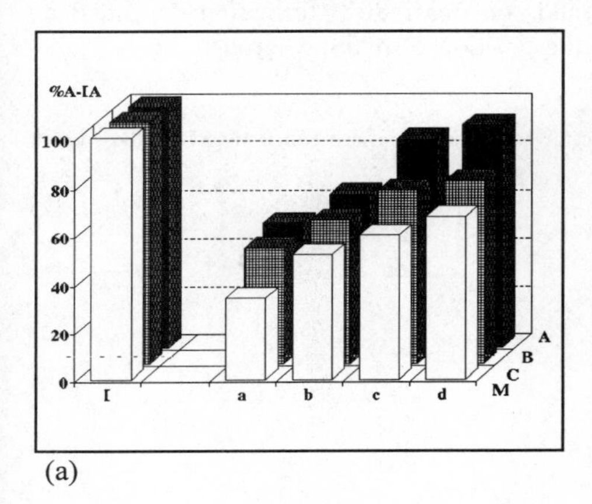

(a)

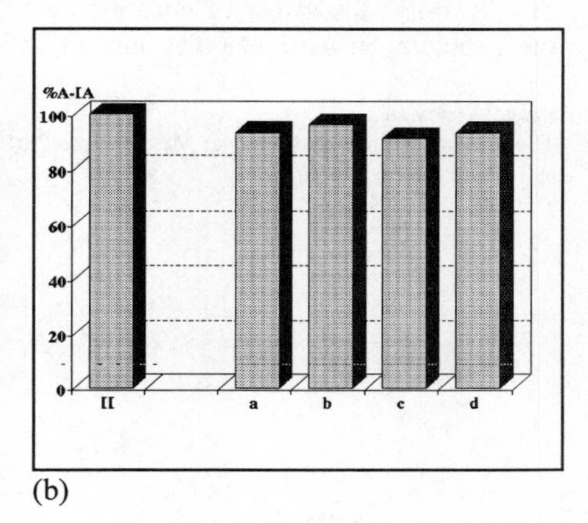

(b)

Figure 3. The influence of ellagic (a), gallic (b), ferulic (c) and syringic acid (d) on the formation of amino-imidazoazarenes (A-IA) in the different model systems:
(a): Model system I: creatinine, D(+)-glucose and 2-, 3- or 4-aminobutanoic acids (A, B and C, respectively). (b): Model system II: creatinine, 2, 3-dimethylpyrazine and acetaldehyde.

In both series of model systems determined 2-amino-3,4,8-imidazo(4,5-f)quinoxaline was assumed to equal the total obtained amino-imidazoazarenes (A-IA).

In the first series of these investigations (Figure 3a), the model system I, containing creatinine, D(+)-glucose and 2-, 3- or 4-aminobutanoic acids, respectively, was assigned a yield of amino-imidazoazarenes of 100%. The other reaction mixtures, Figure 3a, which contained ellagic, gallic, ferulic or syringic acids gave lower yields of amino-imidazoazarenes. The highest inhibitory effect was shown by ellagic acid. In the presence of ellagic acid, total amino-imidazoazarenes ranged from 34%, in the presence of 4-aminobutanoic acid in the model system, to 52% when 2-aminobutanoic acid was used. Inhibitory effects on the formation of amino-imidazoazarenes was also shown by gallic (from 52 to 64%), ferulic (from 60 to 86%) and syringic acids (from 68 to 92%). The second series of model systems (Figure 3b), which contained creatinine, 2,5-dimethylpyrazine, acetaldehyde and ellagic, gallic, ferulic or syringic acid showed no changes in the yields of amino-imidazoazarenes (A-IA).

The phenolic compounds caused a variety of effects on the formation of Maillard reaction products responsible for the formation of amino-imidazoazarenes (i.e. pyrazine and pyridine formation). However, the phenolic compounds had no effect on amino-imidazoazarene formation in the reaction between creatinine, acetaldehyde and pyrazine. The mechanism of action of the phenolic compounds will be the subject of future investigations.

Conclusions

The results described in this paper confirmed that naturally occurring phenolic compounds interact in the Maillard reaction to modify the formation of free pyrazine cation radical formation and degree of the browning produced; marked dissimilarities were shown between the behavior of the four examined phenolic compounds. Significantly, the effect of phenolic compounds on free radical formation depended on the structure, in particular the number and the position of hydroxy groups.

Acknowledgments
The authors express gratitude to Mrs. Nevena Milić for her excellent advice concerning both the content and language of this paper.

References

Cilliers, J.J.L. and Singleton, V.L. (1991) Characterization of the Products of Nonenzymic Autoxidative Phenolic Reactions in a Caffeic Acid Model System. *J. Agric. Food Chem.* 39: 1298.

Fujita, T., Komagoe, K., Niwa, Y., Uehara, I., Hara, R., Mori, H., Okuda, T. and Yoshida, T. (1988) Studies of autoxidation by tannins and flavonoids. III. Inhibition mechanism of tannins isolated from medicinal plants and related compounds on autoxidation of methyl linoleate. *Yakugaku Zasshi* 108: 528.

Jagerstad, M., Grivas, S., Olsson, K., Laser Reutersward, A., Negishi, C. and Sato, S. (1986) *Genetic Toxicology of the Diet* 17: 155-67.

Kikuzaki, H. and Nakatani, N. (1989) Structure of a new antioxidative phenolic acid from Oregano (*Oreganum vulgare L.*). *Agric. Biol. Chem.*, 53: 519.

Labuza, T.P. and Schmidl, M.K. (1986) *Advances in the control of browning reactions in foods.* In "Role of Chemistry in the Quality of Processed Food". (Ed.) O.R. Fennema, W.H. Chang, and C.Y. Lii. p. 65. Food & Nutrition Press, Inc., Westport, CT.

Milić, B.Lj., Djilas, S.M. and Čanadanović-Brunet, J.M. (1993) Synthesis of some heterocyclic amino-imidazoazarenes. *Food Chemistry* 46: 273-276.

Milić, B.Lj., Piletić, M.V., Grujić-Injac, B. and Premović, P.I. (1978) EPR spectral study on the kinetic behaviour of the non-enzymatic browning reaction. *Proc. Fifth International Congress of Food Science and Technology*, Kyoto, Japan, Sept. 1978, p. 208.

Miura, K.and Nakatani, N. (1989) Antioxidative activity of flavonoids from thyme (*Thymus vulgaris L.*). *Agric. Biol. Chem.* 53: 3043.

Namiki, M. (1990) Antioxidants/Antimutagens in Food. *Crit. Rev. Food Sci. Nutr.* 29:273-300.

Namiki, M. and Hayashi, T.(1975) Development of novel free radicals during the amino-carbonyl reaction of sugars with amino acids. *J. Agric. Food Chem.* 23: 487.

Nawar, W.W. (1985) *Lipids.* In "Food Chemistry", (Ed.) O.R. Fennema, 2nd ed.p.139. Marcel Dekker, Inc., New York.

New Insight into the Mechanism of the Maillard Reaction from Studies of the Kinetics of its Inhibition by Sulfite

B. L. Wedzicha, I. R. Bellion, and G. German

PROCTER DEPARTMENT OF FOOD SCIENCE, UNIVERSITY OF LEEDS, LEEDS LS2 9JT, UK

Summary

Systematic kinetic measurements of the rate of reaction of sulfite during its inhibition of the Maillard reaction can provide new fundamental knowledge regarding the factors which control the rate of browning. A detailed kinetic model of the reaction has been derived and critically tested. The turnover of sulfite in glucose-glycine-sulfite mixtures equals the rate of formation of a key intermediate in browning and is shown to predict remarkably well the yield of melanoidin in unsulfited systems. Similarly, the rate of the initial step is inversely correlated to the "induction" time for browning to begin. In practice, the kinetic behavior is complicated by the fact that sulfite species catalyze the first step of the Maillard reaction. However, the contribution of the reaction which does not depend kinetically on sulfite species can easily be resolved. Thus, the role of food components, e.g. acids, bases, amino compounds on the individual steps in the Maillard reaction (in the absence of sulfite) can be investigated without the ambiguity associated with kinetic studies of "browning", however specified. Similar behavior has been identified for the fructose- and maltose-amino acid reactions.

Introduction

Maillard browning involves a series of consecutive reactions whose overall rate is controlled at one or more steps in the sequence. The rates of individual steps are expected to depend on the reaction conditions (e.g. pH, temperature, non-electrolytes, ionic strength, water activity) and the presence of weak acids, bases and specific amino compounds. For the food scientist, the challenge is to be able to identify the factors which determine the rate and extent of browning in a particular food commodity. This has implications in the selection and perhaps manipulation of food for a particular composition or susceptibility towards browning.

The specific role of different amino compounds in browning has been the subject of many investigations, but there is no consensus as to which amino compounds are generally the most important. Several distinct functions of amino compounds are the initiation steps of the early stages of the Maillard reaction, which can be viewed as catalysis, in Strecker degradation and incorporation into the final structure of the melanoidin chromophore. There is no reason to expect that a given amino compound will be equally effective in all functions. Thus, one would suggest that the cocktail of amino compounds present in food might be more important in determining the susceptibility towards browning, rather than suggesting it is due to a single component. Also, the individual stages of the browning reaction might be affected differently by changes in the reaction conditions. Thus, the behavior of any cocktail of amino compounds may be unique to a set of reaction conditions. Given such a complex situation, it is necessary to devise some means of identifying significant rate-determining steps in the Maillard reaction in order to study their kinetics in detail. This possibility is afforded by the use of sulfites as inhibitors of browning.

Theory

The main product of the sulfite ion inhibition of the browning of hexoses is 3,4-dideoxy-4-sulphohexosulose, DSH, formed as a result of nucleophilic attack on the α,ß-unsaturated carbonyl moiety of 3,4-dideoxyhexosulos-3-ene, DDH (McWeeny *et al.*, 1974). The fact that the formation of color is effectively inhibited by this conversion of intermediates in browning to DSH suggests that the chromogens are formed by reactions of the intermediates concerned, after the step at which sulfite interrupts the Maillard reaction sequence.

The concentration of sulfite species, S(IV), in glucose-amino acid-S(IV) systems generally falls with time; the concentration-time profile shows an induction phase (initial rate of loss equals zero) followed by an acceleration of the reaction leading to a constant rate of loss of S(IV) over the main part of the reaction (McWeeny *et al.*, 1969). The least complicated mechanism which accounts for these kinetics is a two-step consecutive reaction (Wedzicha, 1984) which, for the glucose-glycine-S(IV) reaction, takes place as follows:

$$\text{glucose} + \text{glycine} \xrightarrow[k_1]{\text{slow}} I_1 \xrightarrow[k_2]{\text{slow}} I_2 \xrightarrow[+\,S(IV)]{\text{fast}} \text{DSH}$$

When ([glucose] and [glycine])»[S(IV)], k_1 is a pseudo-zero order rate constant for the formation of intermediate I_1 and k_2 is a pseudo-first order rate constant for the conversion of I_1 to intermediate I_2 which is reactive towards S(IV). The identity of I_1 as 3-deoxyhexosulose, DH, has been shown by Wedzicha and Kaban (1986) and Wedzicha and Garner (1991). According to this mechanism, the concentration of I_1 should eventually reach a *steady state*, i.e. rate of formation of I_1 equals the rate of its loss, and the steady state concentration of DH is, then, k_1/k_2. It appears that this situation applies over the major part of the reaction in which S(IV) is lost. Thus, the rate of loss of S(IV) equals k_1, i.e. the rate of the first rate-determining step in the Maillard reaction. Indeed, Wedzicha and Kaputo (1992) showed that the yield of melanoidin in the glucose-glycine reaction can be predicted reasonably well from the value of k_1.

In practice, the measurement of k_1 is complicated by the fact that this reaction step is dependent also on the concentration of S(IV) (Wedzicha and Vakalis, 1988), and the steady state rate of loss of S(IV) is given by,

$$-d[S(IV)]/dt = k_1 + k_1'[S(IV)]$$

where k_1' is the pseudo-first order rate constant for the S(IV)-dependent reaction. The value of k_1 is obtained by extrapolating rate-[S(IV)] data to zero concentration. The value of k_2 may be obtained by extrapolating the linear concentration-time plot, representing the steady-state phase of the reaction, to zero time and subtracting the value of the intercept from the initial S(IV) concentration (Wedzicha and Garner, 1991) to give directly the steady-state concentration of DH, $[DH]_{ss}$. Then, $k_2 = k_1/[DH]_{ss}$.

Identical kinetic behavior has been identified for the maltose-amino acid-S(IV) reaction (Wedzicha and Müller, unpublished) and it is possible that the approach is also useful for studying the Maillard reaction of other disaccharides and oligosaccharides. The fructose-amino acid-S(IV) reaction shows similar kinetic behavior. The main difference from that reported above is that S(IV) concentration-time profiles do not show an induction phase but the same interpretation of the rate of loss of S(IV) is possible (Swales and Wedzicha, 1992). Surprisingly, the ascorbic acid-S(IV) reaction is not amenable to this treatment because the rate of loss of S(IV) is dominated by an S(IV)-dependent reaction (Davies and Wedzicha, 1992).

Results and Discussion

A considerable amount of data on the behaviors of k_1, k_1' and k_2 has now been accumulated and those given here are intended to serve as an illustration of the potential value of the approach being suggested. For glucose-amino acid reactions, the value of k_1' is usually greatest when the amino acid has an overall positive charge (arg and lys), suggesting that there may be an electrostatic effect of the charged group on an approaching sulfite or bisulfite ion. Of greater interest to investigations of the Maillard reaction is the behavior of k_1. It is of first order with respect to both reactants in the glucose-glycine reaction confirming that the reaction step being investigated is the amine-assisted conversion of glucose to DH and that one molecule of each is involved in the rate determining step. This applies also when the medium contains glycerol used to adjust a_w in the range 0.43-1.0.

Table I shows the relationship between k_1, k_2 and the induction time t_i for browning to commence for three glucose-amino acid reaction mixtures measured under the same conditions. Figure 1 illustrates these data plotted as k_1 *vs* the reciprocal of t_i (≡rate) with the line forced through the origin, on the assumption that when the value of k_1 is zero, there is no browning taking place ($t_i = \infty$). Despite the limited amount of data available, the correlation between these quantities is encouraging and suggests that the process described by k_1 controls the production of chromogenic intermediates. On the other hand, there is no correlation between t_i and k_2.

Table I. Values of rate constants k_1 and k_2, and the induction time, t_i, for browning, measured for glucose (1M) + amino acid (0.5 M) in acetate buffer (0.25 M CH_3COONa), pH 5.5, 55 °C (Wedzicha and German, unpublished).

Amino Acid	10^4k_1/mol l^{-1} h^{-1}	$10^2k_2/h^{-1}$	t_i/min
Lysine	2.4	6.6	1250
Arginine	0.8	5.5	5000
Glycine	1.7	1.2	2000

There is much interest in the effects of pH control on the progress of the Maillard reaction. Figure 2 shows concentration-time profiles for S(IV) in reactions with and without pH control. Whilst the rate is clearly different, it is interesting to see that the intercepts, when the linear parts of the concentration-time plots are extrapolated to zero time, are unaffected by pH. This means that the steady state concentration of DH remains unaffected. Thus, the change in pH of unbuffered Maillard reactions affects k_1 and k_2 in the same way, but since k_1 controls the rate of melanoidin formation, pH control leads to an increase in the rate of browning.

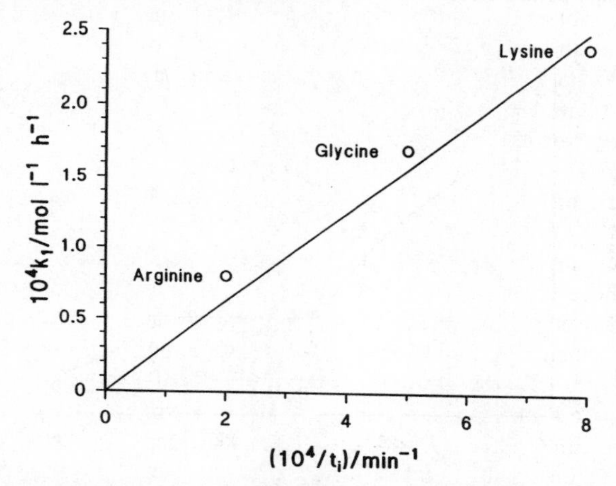

Figure 1. Relationship between k_1 and $1/t_i$ for glucose-amino acid mixtures. Reaction conditions: [glucose] = 1 M, [amino acid] = 0.5 M, acetate buffer (0.25 M), pH 5.5, 55 °C. Data from Table I.

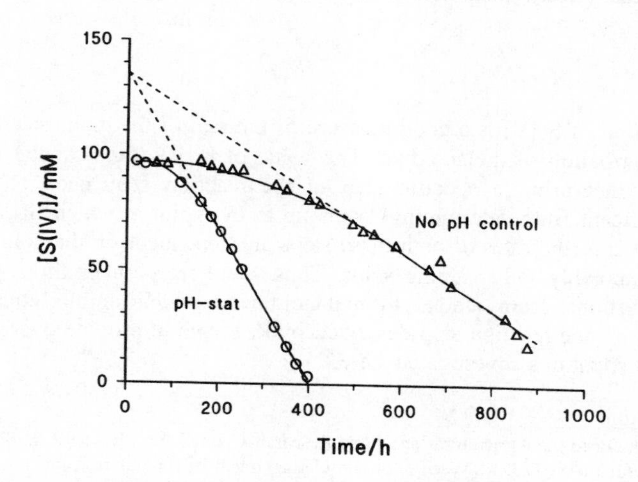

Figure 2. Effect of pH control (pH-stat method) on [S(IV)]-time profiles during the glucose-glycine-S(IV) reaction. Reaction conditions: [glucose] = 1 M, [glycine] = 0.5 M, [S(IV)] ≈ 0.1 M, initial pH 5.5, 55 °C (Bellion & Wedzicha, unpublished).

Figure 3 illustrates the effect of varying a_w on the value of k_1. This measurement represents the specific effect of a_w on the initial reaction step unlike much published work where the effect of a_w on the overall production of color has been reported. It is interesting to compare the behavior of k_1 in the range $0.7>a_w>0.45$ with the extent of browning plotted as a function of a_w by Eichner and Karel (1972). The absorbance of glucose-glycine mixtures at 420 nm increases markedly (3.2-fold) in this range; the increase in k_1 occurs over the same range of a_w values and is of similar magnitude. At this stage the similarity can only be regarded as fortuitous because Eichner & Karel used substantially more concentrated solutions (*c.* 10 M glucose) and no pH control.

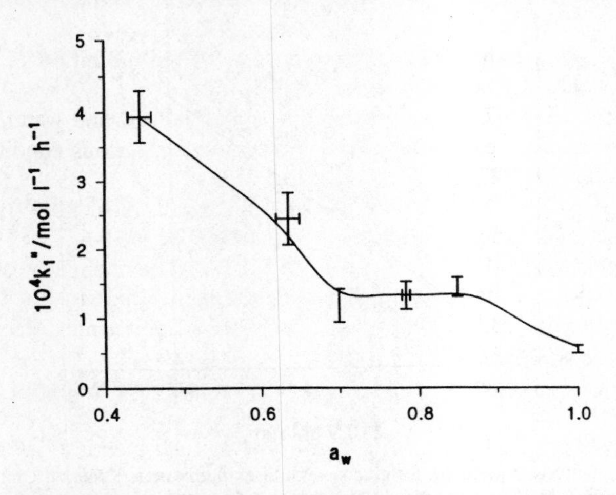

Figure 3. Effect of water activity on the second order rate constant k_1'' defined as k_1/([glucose][glycine]) for the glucose-glycine reaction in the presence of varying amounts of glycerol (Bellion and Wedzicha, 1993).

Conclusions

The rate of reaction of S(IV) is a good measure of the rate of the main reaction pathway leading to the formation of melanoidins. The value of k_1 has fundamental significance with respect to measuring a specific step in the reaction sequence. There are two kinetically significant (rate determining) steps up to the point at which S(IV) intercepts the Maillard reaction; the rates of both these steps are dependent on the concentration of glycine and, presumably, other amino acids. Thus, there are perhaps three points in the sequence of reactions from aldose to melanoidin at which amino compounds are involved. However, the reaction step described by k_1 seems to provide overall control of browning in the situations investigated here.

Acknowledgements
The authors are grateful to the Agricultural and Food Research Council for a Research Fellowship to IRB. They are also grateful to Dr. D.J. McWeeny for stimulating much of this interest and for his continuing encouragement.

References

Bellion, I.R. and Wedzicha, B.L. (1993) Effect of glycerol on the kinetics of the glucose-glycine-sulphite reaction. *Food Chem.*, **47**, 285-288.

Davies, C.G.A. and Wedzicha, B.L. (1992) Kinetics of the inhibition of ascorbic acid browning by sulphite. *Food Add. Cont.*, **9**, 471-477.

Eichner, K. & Karel, M. (1972). The influence of water content and water activity on the sugar-amino browning reaction in model systems under various conditions. *J. Agric. Food Chem.*, **20**, 218-223.

McWeeny, D.J., Biltcliffe, D.O., Powell, R.C.T. and Spark, A.A. (1969) The Maillard reaction and its inhibition by sulphite. *J. Food Sci.*, **34**, 641-643.

McWeeny, D.J., Knowles, M.E. and Hearne, J.F. (1974) The chemistry of non-enzymic browning in foods and its control by sulphites. *J. Sci. Food Agric.*, **25**, 735-746.

Swales, S. and Wedzicha, B.L. (1992) Kinetics of the sulphite-inhibited browning of fructose. *Food Add. Cont.*, **9**, 479-483.

Wedzicha, B.L. (1984) A kinetic model for the sulphite-inhibited Maillard reaction. *Food Chem.* **14**, 173-187.

Wedzicha, B.L. and Garner, D.N. (1991) The formation and reactivity of osuloses in the sulphite-inhibited Maillard reaction of glucose and glycine. *Food Chem.*, **39**, 73-86.

Wedzicha, B.L. and Kaban, J. (1986) Kinetics of the reaction between 3-deoxyhexosulose and sulphur(IV) oxospecies in the presence of glycine. *Food Chem.*, **22**, 209-223.

Wedzicha, B.L. and Kaputo, M.T. (1992) Melanoidins from glucose and glycine: composition, characteristics and reactivity towards sulphite ion. *Food Chem.*, **43**, 359-367.

Wedzicha, B.L. and Vakalis, N. (1988) Kinetics of the sulphite-inhibited Maillard reaction: The effect of sulphite ion. *Food Chem.*, **27**, 259-271.

Chemiluminescence Developed at An Early Stage of the Maillard Reaction

Mitsuo Namiki,[1] Michiko Oka,[2] Miki Otsuka,[2] Teruo Miyazawa,[2] Kenshiro Fujimoto,[2] Kazuko Namiki,[3] Norio Kanamori,[4] and Nobutaka Suzuki[5]

[1]TOKYO UNIVERSITY OF AGRICULTURE, SETAGAYA, TOKYO 156, JAPAN; [2]TOHOKU UNIVERSITY, DEPARTMENT OF FOOD CHEMISTRY, FACULTY OF AGRICULTURE, AOBA, SENDAI 981, JAPAN; [3]SUGIYAMA JOGAKUEN UNIVERSITY, CHIKUSA, NAGOYA 465, JAPAN; [4]TOKUSHIMA UNIVERSITY SCHOOL OF DENTISTRY, TOKUSHIMA 770, JAPAN; [5]SHIMONOSEKI UNIVERSITY OF FISHERIES, SHIMONOSEKI 759-65, JAPAN

Summary

Due to the fact that a fairly stable free radical was observed to form at an early stage of the Maillard reaction, we examined development of chemiluminescence (CL) in the Maillard reaction using a highly sensitive photon counting system. It was demonstrated that marked CL developed generally at an early stage in the Maillard reaction prior to browning. The CL intensity followed the order methyl glyoxal >> glyoxal > xylose > glucose for the carbonyls, and alkylamine > lysine > β-alanine > «-alanine for the amino compounds. CL was observed even at pH 6.0 and increased markedly with increases in pH. These features are almost the same as those observed for free radical formation and browning. CL was negligible in Ar or N_2 and greatly increased in air; it was suppressed by addition of ascorbic acid or cysteine. The mechanism by which CL develops is not yet clear but is assumed to be attributable to the production of electronically excited species through the reaction of oxygen with the free radical product and/or Schiff base product formed at an early stage in the reaction.

Introduction

Concerning the mechanism of the Maillard reaction, the scheme proposed by Hodge (Hodge, 1953), involving Amadori rearrangement as a key step, has been accepted widely as being the most reasonable. However, development of novel free radical products at an early stage of the reaction has been found (Namiki et al., 1973), and a new mechanism has been demonstrated that involves fragmentation of the Schiff base product followed by the formation of a pyrazinium free radical product as well as browning products (Namiki and Hayashi, 1983, Namiki, 1988). In addition, the generation of weak chemiluminescence (CL) during the amino-carbonyl reaction has been reported (Bordalen, 1984), and the effects of various reaction conditions on CL formation have been examined (Kurosaki et al., 1989). Based on our finding on the development of novel free radical products in the Maillard reaction, our research group has also investigated the development and some properties of CL in the Maillard reaction in correlation with free radical formation and browning.

Materials and Methods

Chemiluminescence measurement

CL was measured using the Tokoku Electronic Industry Co (Sendai, Japan) CLD 100 chemiluminescence detector system (sensitivity, 10^{-14} W; wave length region, 300-650 nm). The count per second shown on the vertical axis in the figures is a CL intensity unit measured by the CLD 100 chemiluminescence detector. It is not a direct count of the number of photons but a unit by which one count corresponds to approximately 10 photons.

Experiment A
A solution of an amino compound in phosphate buffer or Merczen buffer and a solution of a carbonyl compound were mixed in a sample cell of the CLD 100 and heated to a given temperature in a sample chamber. CL was recorded as a function of time.

Experiment B
A mixture of amino and carbonyl compounds was heated in a flask and aliquots taken at time intervals during heating were used for the determination of browning by absorption at 420 nm and CL was determined using the CLD 100 system.

Experiment C
CL measurement of the reaction under oxygen-free conditions: Each solution of amino and carbonyl compounds was deaerated and exchanged with pure argon gas by repeated procedure of degassification *in vacuo* of the frozen sample solution followed by saturation with pure argon gas after thawing. These reactant solutions were mixed in a glass cuvette under an argon gas atmosphere and CL was measured instantaneously with heating using the CLD 100 apparatus. All materials used were of reagent grade.

Results and Discussion

CL of Various Maillard Reaction Systems (Experiment A)
Figure 1 shows the results obtained from the various Maillard reaction mixtures at pH 8.0 and 90°C. CL was clearly demonstrated to develop in every amino-carbonyl reaction system very early in the reaction and increased with heating time. As in the glyoxal - methylamine system, CL was observed as soon as the reaction was started and increased very rapidly with heating, giving a first small peak within 100-150 s, then increased again until a maximum of ca. 300 s. This two-step type increase in the development of CL was also observed in other cases including the glucose-β-alanine system, whose initial process is shown in the additional small figure in Figure 1. A comparison was made of the CL intensity at the maximum point for the various amino-carbonyl reaction system.

Concerning the amino compounds, CL developed only with primary amino compounds but was undetectable with secondary and tertiary amino compounds. Among the alkylamines, CL intensity was prim- > sec- > tert-derivative. Among the amino acids, the development followed the order Lys > β-Ala >> Gly and Arg > a-Ala, but it was undetectable in cysteine. Among the carbonyl compounds examined, methylglyoxal was especially reactive and the development of CL was observed in the reaction with most amino acids and proteins at room temperature and 37°C. Other carbonyl compounds were effective in the formation of CL at heating temperatures above 60°C, and the activity followed in the order glycol aldehyde = glyoxal > glyceraldehyde > xylose > fructose = glucose. These effects are almost the same as those observed in the cases of free radical formation and browning (Namiki and Hayashi, 1983). In a previous paper, methylglyoxal was classified as a compound inactive in free radical formation (Hayashi et al., 1977), but this time we were able to detect a weak ESR signal only very early in the reaction with methylamine. This short-life free radical formation seems to be correlated with a high activity in the development of CL.

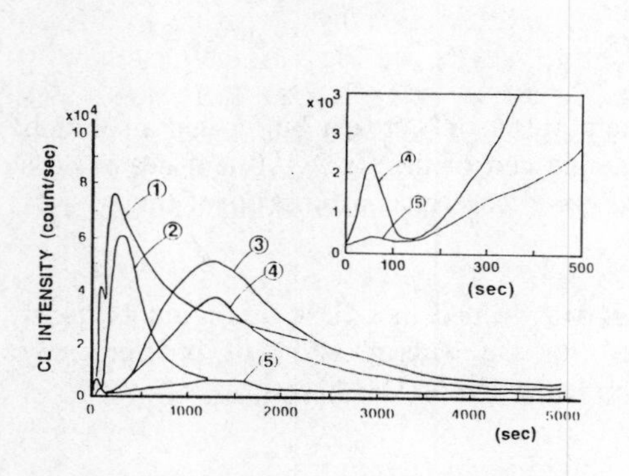

Figure 1. Chemiluminescence in various Maillard Reaction Systems (pH 8.0, 90°C air)
(1) Glyoxal-$MeNH_2$ (25 mM) (2) Xylose-β-Ala (1.0 M) (3) Glycose-Lys (1.0 M) (4) Glucose-β-Ala (1.0 M) (5) Glucose-α-Ala (1.0)

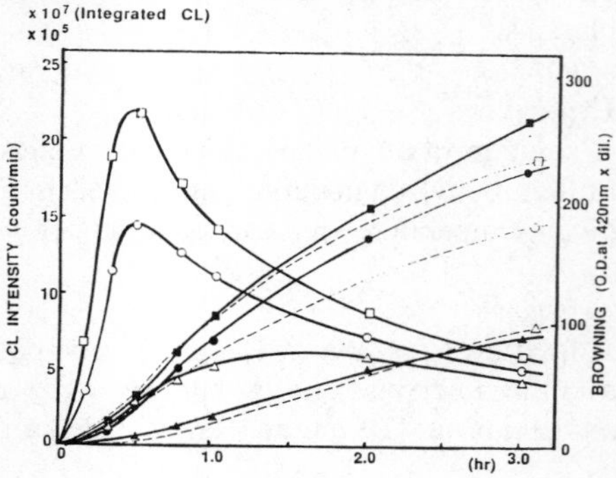

Figure 2. Chemiluminescence and Browning in the Maillard Reaction
Glucose-β-Ala (1.0 M) was heated in air in a boiling water bath. CL was measured at 30°C, and browning was determined at 420 nm at pH 6, 8, and 11, as follows:

pH	6	8	11
CL intensity	△	○	□
Integrated CL	△	○	□
Browning	▲	●	■

CL and Browning in Maillard Reaction (Experiment B)

The relationship between CL and browning was examined with the glucose-β-alanine system at various pHs by method B. As shown in Fig. 2, development of CL was observed at an early stage of the reaction prior to the browning reaction, and the intensity increased rapidly with heating time, especially in the alkaline mixtures, and showed a maximum peak at about 30 min, along with a simultaneous increase of browning. These features in the development of CL are similar to those observed in the free radical formation in the Maillard reaction (Namiki and Hayashi, 1983). Every curve in the integrated CL corresponded well to each curve of the browning at the same pH, indicating that the CL-forming reaction product(s) may directly participate to further the reaction giving rise to melanoidin.

Effect of Oxygen on the Development of CL (Experiment C)

It is commonly recognized that CL is generated from excited molecules such as singlet oxygen and excited carbonyl compounds, and that oxygen plays an important role in the formation of such excited molecules (Campbell, 1988). In fact, Kurosaki et al. (1989) reported that almost all of the CL of the Maillard reaction disappeared under nitrogen gas bubbling. However, the ESR signal of the Maillard reaction can be detected even in an open test tube and disappears rapidly by bubbling of air, indicating that the free radical product is fairly stable in the reaction mixture (Namiki and Hayashi, 1975). On the other hand, browning in the Maillard reaction is known to proceed even in the absence of oxygen, although the effect of oxygen on the reaction remains to be elucidated (Namiki, 1988). Moreover, the development of CL in the Maillard reaction independent of oxygen, e.g., from excited carbonyl or Schiff base groups, is indisputable. Thus, we examined the effect of oxygen on the CL formed in the Maillard reaction by preparing a strictly anaerobic reaction system using Method C. As shown in Fig. 3, the reaction of methylglyoxal-methylamine in this system showed only a little CL at an early stage, and it increased dramatically with the introduction of air. This indicated that almost all of the CL observed in the Maillard reaction is caused by the reaction of oxygen with some reactive reaction product(s) such as a free radical compound and/or a highly oxidizable compound.

Spectral Study of CL Developed in the Maillard Reaction

A visible spectrum of CL developed in the Maillard reaction was measured by two different methods; CL of methylglyoxal-methylamine in aqueous and ethanol solutions at 50°C by the diode array spectrometer (IMUC-7000) and those of glyoxal-methylamine and other systems by use of a filter spectrometer integrating for 10 min from the starting point of the reaction. As shown in Figure 4, in all cases, the spectrum gave a maximum at around 580 nm. A similar type of visible spectrum was also observed by filter spectroscopy in the reaction of glucose-methylamine, glucose-β-alanine, and xylose-β-alanine systems though in the latter two the maximum shifted to around 680 nm. These are the first spectra reporting that there exists a singlet oxygen in the reaction. Thus we tried to detect the near-infrared absorption around 1270 nm in the CL of the methylglyoxal-methylamine system as direct evidence for the presence of singlet oxygen, but were not successful.

Fluorescence spectrum of the reaction mixture of methylglyoxal-methylamine excited at 360 $\pm$ 30 nm showed a maximum around 420 nm at an initial stage then shifted to 580 nm and 630 nm (the later could be an overtone of the excitation light). The reaction mixture did not emit light at 50 ns after excitation, indicating no occurrence of phosphorescence and no contribution of a triplet excited state in the CL development.

Effect of Various Additives on the Development of CL

To investigate the mechanism of the development of CL the effect of various radical scavenging and/or reducing agents on CL in the Maillard reaction was examined. A solution of additive was added to the reaction mixture of glyoxal-methylamine or glucose-β-alanine proceeding at 90°C by method A at around the maximum point of the development of CL. The CL instantaneously and markedly decrease by the addition of ascorbic acid as well as cysteine, then gradually decreased with further heating. The suppressive effect was somewhat stronger with ascorbic acid than with cysteine. The mechanism is not yet clear for either case, but it is assumed to be due to their reducing activity to the oxygen-related excited molecule and/or to the scavenging effect on the free radical product. It should be noted that ascorbic acid is a representative enediol reductone compound and has been known to produce some active oxygen species such as superoxide, hydrogen peroxide, and OH radical by the reduction of the molecular oxygen, especially in the presence of Fe or Cu ion (Fridovich, 1979, Seib, 1982).

However, no clear evidence of formation of singlet oxygen, the most probable origin of CL, in such reaction systems has yet been demonstrated. The fact that the addition of ascorbic acid caused a decrease in CL suggests no generation of singlet oxygen contributing to CL from ascorbic acid. And it was also shown that no detectable

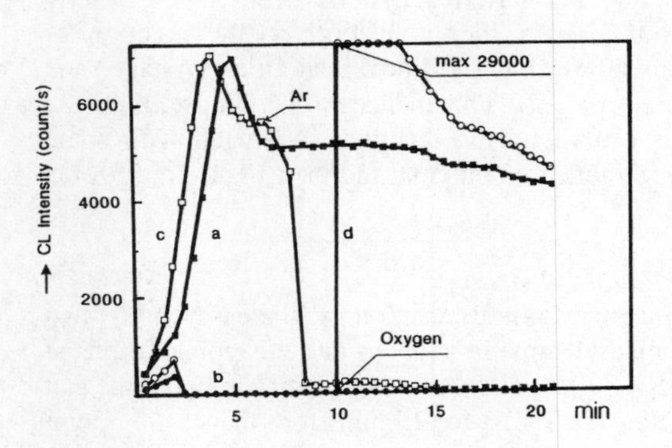

Figure 3. Time curves of chemiluminescence from aqueous solutions of 50 mM methylamine and 50 mM methylglyoxal in aqueous solutions at 35°C.
(a) O_2 gas bubbling,(b) Ar gas bubbling
(c) O_2 gas bubbling and then Ar gas bubbling,
(d) Ar gas bubbling and then
O_2 gas bubbling.

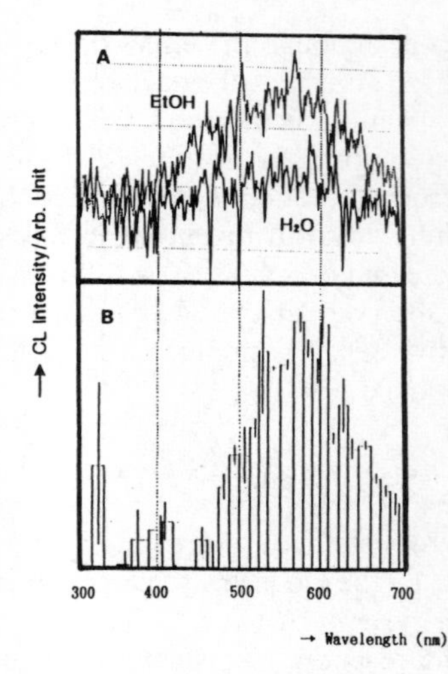

Figure 4. Chemiluminescence Spectra:
A. Methylglyoxal-$MeNH_2$ (50 mM each) were mixed at 50°C in aqueous or ethanolic solutions (Ohtsuka Electronics (IMUC-7000)
B. Glyoxal-$MeNH_2$ (100 mM each) were mixed at 90°C in a buffer solution. (A handmade filter spectrometer)

CL was measured in solution of ascorbic acid (50 mM in phosphate buffer, pH at 8.0, heated in air) with or without the presence of Fe and Cu ions. These facts suggest that no appreciable amount of singlet oxygen giving clear CL is formed by the reaction of a reductone compound with oxygen even in the presence of a metal ion. We also conducted a preliminary experiment to measure CL development during heating of solutions of Amadori product alone and with amino acid using N-(1-deoxy-D-fructosyl)-L-leucine. In this experiment, CL formation was small with the Amadori compound alone as compared with the reaction with amino acid. Thus, we concluded that there is no notable contribution of the Amadori product to the CL in the Maillard reaction to support the mechanism proposed by Kurosaki et al. (1989).

The effect of various additives on CL in the glucose-β-alanine system is summarized in Table I. Addition of L-ascorbic acid, β-carotene, cysteine, Fe-chlorophyllin and ferricyanide showed a significant suppressive effect while addition of NaN_3, SOD, KI, mannitol, and BHT gave a weak effect or no effect.

Further experiments using specific radical scavengers were made on CL in the methylglyoxal-methylamine system in 20% ethanol or water and the results are shown in Table II. No significant effect on CL was observed with NaN_3 and DABCO, suggesting the absence of CL emitted from singlet oxygen in the CL development.

Sensitizers of S_1 (DPA, perylene, and rubrene) and sensitizers of S_1 and T_1 (DBA and dibromoanthracenesulfonic acid) enhanced CL in diluted solutions while they quenched CL in concentrated solutions, indicating the presence of some singlet excited state molecule(s) besides that of oxygen. 9-Acridone-2-sulfonic acid (a sensitizer of S_1 and O_2^-) (Suzuki et al., 1991) enhanced CL, suggesting the presence of S_1 and O_2^-, but no effect of CL was observed by the addition of SOD.

The fact that CL development was inhibited by 2,4,6-tri-tert-butylphenol, and higher concentrations of AIBN and benzoylperoxide, while it was enhanced by diluted solutions of the latter two also suggests the contribution of some radical intermediate(s) in CL development.

Table I. Suppressive effect of additives on CL from glucose-β-alanine (1 M each) in a buffer solution at pH 8.0 and 85-90 °C for 30 min in air.

Additive	Concn. Range (M)	Effect	Remarks
L-Ascorbic Acid	0.1	++	Reducing agent
β-Carotene	0.1	++	$^1O_2$ Scavenger
L-Cysteine	0.1	++	Reducing agent
Fe-chlorophyllin Na	0.1	++	$^-O_2$ Scavenger
Potassium Ferricyanide	0.1	++	Catalyzer
Potassium Ferrocyanide	0.1	+	Catalyzer
KI	0.1	+	·OH Scavenger
NaN_3	0.1	+	Quencher of $^1O_2$
SOD	0.1	±	O_2^- Scavenger
Mannitol	0.2	-	·OH Scavenger
BHT	0.1	-	Radical Scavenger

Table II. Effects of additives on CL from methylamine and methylglyoxal.

Additive	Concn. Range (mM)	Solv.	Effect	Remarks
9,10-Diphenylanthracene (DPA)	10^{-3} - 4 x 10^{-1}	20%EtOH	++	Sensitizer of S_1
Perylene	10^{-3} - 4 x 10^{-1}	20%EtOH	++	Sensitizer of S_1
Rubrene	10^{-3} - 4 x 10^{-1}	20%EtOH	+++	Sensitizer of S_1
9,10-Dibromoanthracene (DBA)	10^{-3} - 4 x 10^{-1}	20%EtOH	±	Sensitizer of S_1 & T_1
9,10-Dibromoanthracene-2-sulfonic Acid	10^{-3} - 4 x 10^{-1}	H_2O	+	Sensitizer of S_1 & T_1
Triethylene Diamine (DABCO)	10^{-1} - 10	H_2O	+	Quencher of $^1O_2$
NaN_3	10^{-1} - 10	H_2O	±	Quencher of $^1O_2$
α,α'-Azobisisobutyronitrile (AIBN)	10^{-1} - 10	20%EtOH	++	Radical Initiator
Benzoyl Peroxide	10^{-1} - 10	20%EtOH	+	Radical Initiator
2,4,6-Tri-tert-butylphenol (TBP)	10^{-1} - 10	20%EtOH	---	Radical Quencher
9-Acridone-2-sulfonic Acid	10^{-1} - 10	H_2O	++	Sensitizer of S_1&O_2

Into a solution of methylamine (1 mL) and 50 mM methylglyoxal (1 mL), an additive (500 μL) was added under O_2 bubbling at 35°C; +, CL enhances; ±, no effect; and -, retarded more than the control.

References

Bordalen, B.E. (1984) Chemiluminescence Method for Estimation of Autoxidation in Foods: Interfering Reactions. In *Analytical Application of Bioluminescence and Chemiluminescence*, pp.577-579. Academic Press, New York.

Campbell, A.K. (1988) Chemiluminescence. Principle and Applications in Biology and Medicine. Ellis Horwood, pp. 15-125, Chichester.

Fridovich, I. (1975) Oxygen Radicals, Hydrogen Peroxide, and Oxygen Toxicity. In *Free Radical in Biology,* (W.A. Pryor, ed.) pp. 239-277, Academic, New York.

Hayashi, T., Ohta, Y., and Namiki, M. (1977) Electron Resonance Spectral Study on the Structure of the Novel Free Radical Products Formed by the Reactions of Sugars with Amino Acids and Amines. *Agric. Food Chem.* 25:1282-1287.

Hodge, J.E. (1953) Chemistry of Browning Reaction in Model Systems. *J. Agric. Food Chem.* 1:928-943.

Kurosaki, Y., Sato, H., and Mizugaki, M. (1989) Extra-weak Chemiluminescence of Drugs. VI. Extra-Weak Arising from the Amino-Carbonyl Reaction. *J. Biolum. Chemilum.* 3:13-19.

Namiki, M. (1988) Chemistry of Maillard Reactions: Recent Studies on the Browning Reaction Mechanism and the Development of Antioxidants and Mutagens. *Adv. Food Res* 32:115-184, Academic, New York.

Namiki, M. and Hayashi, T (183) A New Mechanism of the Maillard Reaction Involving Sugar Fragmentation and Free Radical Formation. *ACS Symp. Ser.* 255:21-46.

Namiki, M., Hayashi, T and Kawakishi, S. (1973) Free Radicals Developed in the Amino-Carbonyl Reaction of Sugars with Amino Acids. *Agric. Biol. Chem.* 37:2935-2937.

Seib, P.A., and Tolbert, B.M. (1982) Ascorbic Acid: Chemistry, Metabolism and Uses. In *Advances in Chemistry*, (P.A. Seib and B.M. Tolbert, eds.) 200:163, American Chemical Society, Washington, DC.

Suzuki, N., Itagaki, T., Goto, A., Yoda, B., Nomoto, T., Mizumoto, I., Inaba, H., and Goto, T. (1991) Studies on the Chemiluminescent Detection of Active Oxygen Species: 9-acridone-2-Sulfonic Acid, A Specific Probe for Superoxide. *Agric. Bio. Chem.* 55:1561-1564.

Effects of Aspirin on Glycation, Glycoxidation, and Crosslinking of Collagen

M-X. Fu, S. R. Thorpe, and J. W. Baynes

DEPARTMENT OF CHEMISTRY AND SCHOOL OF MEDICINE, UNIVERSITY OF SOUTH CAROLINA, COLUMBIA, SC 29208, USA

Summary

Glycation, oxidation and glycoxidation reactions contribute to the chemical modification and crosslinking of proteins by glucose. In this study we show that aspirin, and the related compounds, benzoate and salicylate, inhibit the formation of glycoxidation products and the crosslinking of rat tail collagen *in vitro*, but have negligible effect on glycation of the protein. We conclude that the mechanism of action of aspirin as an inhibitor of advanced Maillard reactions depends on its antioxidant activity in inhibiting oxidation and glycoxidation reactions, rather than inhibition of glycation by acetylation of amino groups in collagen. Our results suggest that aspirin and other non-steroidal anti-inflammatory agents with antioxidant activity may be effective inhibitors of the chemical modification and crosslinking of collagen by glucose *in vivo*.

Introduction

In previous studies on the chemical modification of rat tail collagen by glucose, we have shown that chelators, sulfhydryl compounds and antioxidants have negligible effect on the glycation of protein, but inhibit glycoxidation and crosslinking of collagen by glucose (Fu et al., 1992, 1993). In effect, these agents, which inhibit autoxidative degradation of glucose or glucose adducts to protein, *uncouple* glycation from more permanent and damaging glycoxidation and crosslinking reactions.

There are also a number of reports on the effects of aspirin on glycation and/or crosslinking of proteins by glucose, both *in vitro* and *in vivo*. Studies *in vitro* have suggested that aspirin, by acetylating reactive lysine residues in target proteins, inhibits the glycation of albumin (Day et al., 1979; Rendell et al. 1986), hemoglobin (Rendell et al., 1986), lens proteins (Rao et al., 1985; Huby and Harding, 1988; Rao and Cotlier, 1988; Abraham et al., 1989) and retinal basement membrane proteins (Li et al., 1984). However, Blaktny and Harding (1992) showed that aspirin, as well as ibuprofen which does not have acetylating activity, provided protection against cataracts and inhibited glycation of lens proteins in diabetic rats, while van Boekel et al. (1992) showed that diclofenac, a non-acetylating, anti-inflammatory agent inhibited glycation of albumin *in vitro*, apparently by specific, non-covalent binding to the protein. Yue et al. (1984, 1985) showed that aspirin had no effect on glycation of tail tendon collagen in diabetic rats, but did inhibit the crosslinking of the collagen, as measured by thermal rupture time. Salicylate had similar effects, i.e. inhibition of crosslinking without an effect on glycation, both *in vitro* and *in vivo*, leading these investigators to conclude that acetylation was not involved in the mechanism of action of aspirin as an inhibitor of advanced glycation reactions.

In this study we have attempted to reconcile these various effects of aspirin by studying the effects of aspirin and the related compounds, benzoate and salicylate, on the glycation, glycoxidation, development of fluorescence and crosslinking of rat tail tendon collagen *in vitro*. Our results indicate that none of these compounds are effective as inhibitors of glycation of collagen, but that all inhibit subsequent glycoxidation and crosslinking reactions by virtue of their antioxidant activity.

Materials and Methods

Reagents were purchased from Sigma Chemical Co. (St. Louis, MO). The preparation of rat tail tendon collagen and reaction of glucose with collagen under air (oxidative conditions) were carried out as described previously (Fu et al., 1992, 1993). Briefly, collagen (250 mg wet weight) was incubated in 10 ml 0.2 M phosphate buffer, pH 7.4, in 20 mL scintillation vials under air in the dark at 37°C, in the presence or absence of glucose (250 mM) and various concentrations of benzoate, salicylate or aspirin. The latter compounds were dissolved in the incubation buffer at the start of the experiment. The pH of all samples was adjusted initially, then, as necessary, at least twice each day for the first 3 days, by addition of small volumes of 6 M NaOH. Procedures for measuring the extent of glycation and formation of glycoxidation products [N -(carboxymethyl)lysine (CML) and pentosidine] in collagen, generation of collagen-linked fluorescence, and crosslinking of collagen by glucose have been described previously (Fu et al., 1992, 1993; Dyer et al., 1993). Further experimental details are provided in legends to figures.

Results

The extent of chemical modification and crosslinking of collagen by glucose and the effects of benzoate, salicylate and aspirin were measured following incubation of collagen with glucose for 21 days under air. Incubation of collagen with glucose led to significant increases in glycation, measured as fructoselysine (FL), and formation of glycoxidation products (CML and pentosidine) in collagen (Figure 1, group 3 vs. group 1). There was also a corresponding increase in collagen-linked fluorescence and in crosslinking of collagen by glucose, as evidenced by increased resistance to solubilization with pepsin (Figure 2, group 3 vs. group 1). Notably, none of these chemical or physical changes were observed in collagen incubated without glucose (Figures 1 & 2, groups 2). Addition of benzoate, salicylate or aspirin to glucose-containing reaction mixtures had little effect on glycation of the collagen, but caused a significant, concentration-dependent inhibition of formation of CML and pentosidine (Figure 1, groups 4-12). Aspirin and salicylate were comparable in their inhibitory activity, while benzoate was much less potent. Aspirin and salicylate also inhibited the formation of collagen-linked fluorescence and crosslinking of collagen, as measured by resistance to limited digestion with pepsin, while benzoate was ineffective under the conditions tested (Figure 2, groups 4-12). In summary, aspirin and salicylate inhibited the advanced stages of the Maillard reaction, i.e. fluorescence and crosslinking (Figure 2), without a significant effect on glycation of the protein (Figure 1).

The experiments with aspirin were the most difficult to execute; they required frequent adjustment of pH because of gradual hydrolysis of aspirin to salicylate and acetic acid. The half-life of aspirin in these reactions was measured by HPLC (Figure 3A) and determined to be less than 1 day (Figure 3B). In unadjusted reaction mixtures the pH decreased to 7.3, 7.1 and 6.7 after 8 hours in the presence of 10, 25 and 100 mM aspirin,

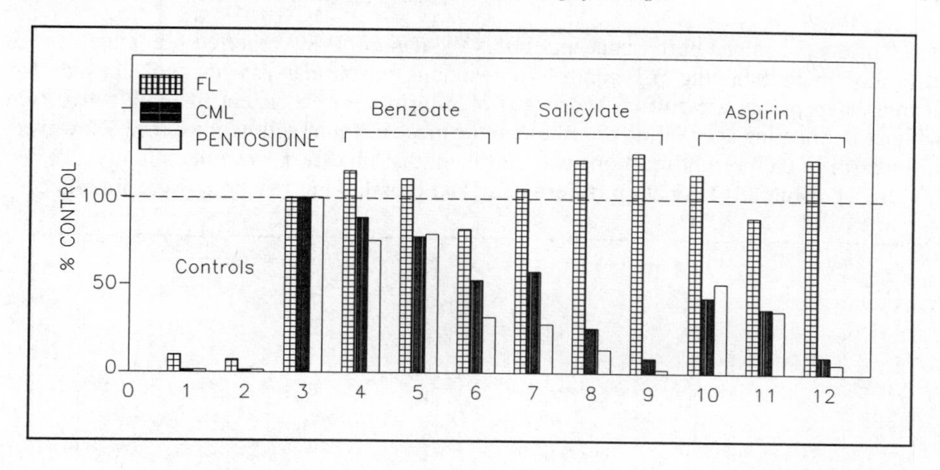

Figure 1. Effect of benzoate, salicylate and aspirin on glycation and glycoxidation of collagen. Reactions were conducted and assayed as described in *Materials and Methods*. 1) natural collagen at time zero; 2) collagen incubated for 21 days without glucose; 3) collagen incubated for 21 days with 250 mM glucose; 4-12) collagen incubated with glucose, but with 10, 25 or 100 mM benzoate (4-6), salicylate (7-9) or aspirin (10-12).

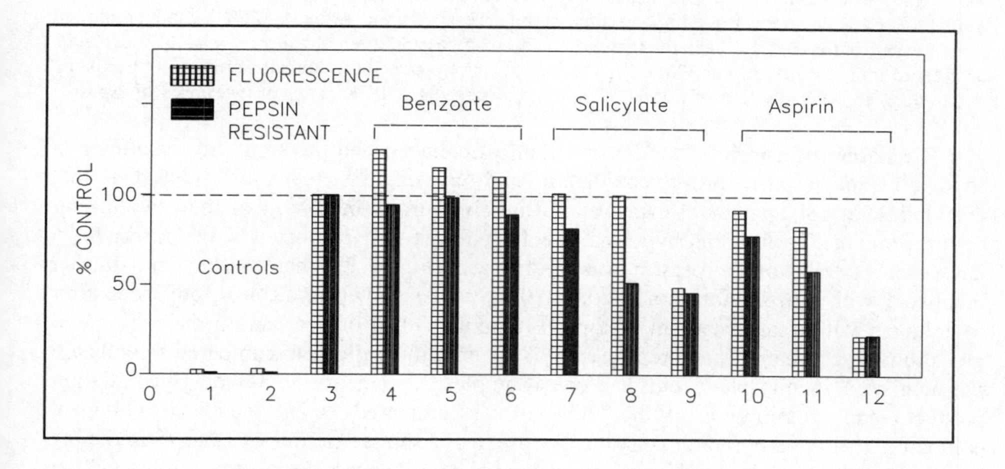

Figure 2. Effect of benzoate, salicylate and aspirin on formation of collagen-linked fluorescence and crosslinking of collagen. Samples described in Figure 1 were analyzed for collagen-linked fluorescence following complete digestion with pepsin (2% w/w, 24 hours in 0.5 M acetic acid) and for relative crosslinking by limited digestion by pepsin (2% w/w, 3 hours in 0.5 M acetic acid).

respectively. The effect of hydrolysis of aspirin on the pH of the reaction mixture was especially noticeable in the reaction at 100 mM aspirin, in which the aspirin concentration was half the phosphate buffer concentration. Vidal *et al.* (1989) reported previously on the difficulty in maintaining pH control in reactions containing aspirin and showed that decreases in pH as a result of hydrolysis of aspirin were sufficient to inhibit glycation without implicating acetylation of protein in the mechanism of action of aspirin. However, pH control in the present reactions was sufficient that glycation was only slightly affected (Figure 1), while glycoxidation, fluorogenic and crosslinking reactions were inhibited.

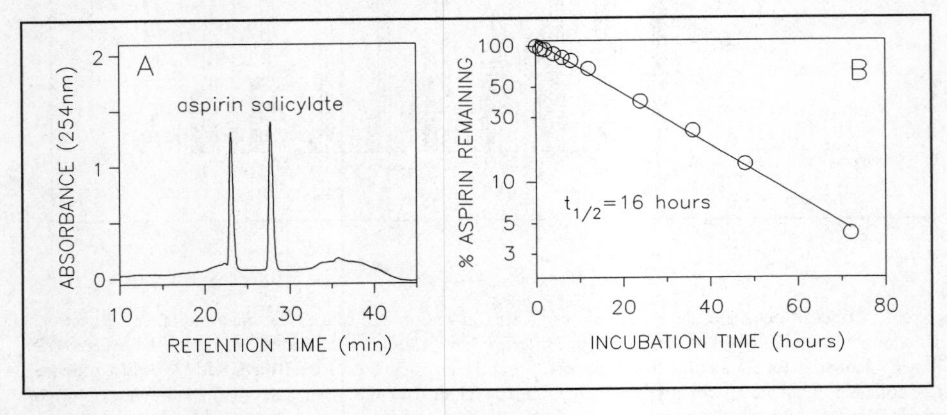

Figure 3. Kinetics of hydrolysis of aspirin (10 mM) in 0.2 M phosphate buffer, pH 7.4, at 37°C. Aliquots were removed at various times and analyzed by reversed-phase HPLC. Analytical conditions: column, Zorbax SB-C18; flow rate, 1 ml/min; detection, absorbance at 254 nm; buffer A, 0.1% heptafluorobutyric acid in water; buffer B, 0.1% heptafluorobutyric acid in 50% acetonitrile. Gradient: 0 min, 10% B; 5 min, 35% B; 25 min, 45% B; 30 min, 90% B. A) Chromatography of a 100 L aliquot of the reaction mixture at 24 hours, showing position of elution of aspirin and salicylate. B) Kinetics of hydrolysis of aspirin.

The failure of aspirin to inhibit glycation of collagen and the similarity of effects of salicylate and aspirin on glycoxidation and crosslinking reactions suggested that acetylation of collagen by aspirin was not involved in its inhibition of the crosslinking of the protein. To test this hypothesis, collagen was first incubated with aspirin for 6 days, when >95% of the aspirin was hydrolyzed (Figure 3), then washed and further incubated with glucose for an additional 10 days. As shown in Table I, pre-incubation of collagen with a large amount of aspirin (Sample 4) had little effect on the subsequent glycation, glycoxidation and crosslinking of the protein by glucose, compared to collagen pre-incubated in phosphate buffer alone (Sample 3). The formation of glycoxidation products and crosslinking of the collagen was increased in pre-incubated collagen, compared to that exposed directly to glucose for a similar period of time (Sample 2), possibly because of the binding of metal ions to the collagen during the pre-incubation stage. Overall, however, there was no evidence that acetylation of collagen by aspirin, while it probably occurs during exposure to aspirin, inhibits subsequent glycation, glycoxidation or crosslinking reactions.

Discussion

The above experiments indicate that aspirin and related antioxidant compounds, such as benzoate and salicylate, have little effect on glycation of collagen by glucose, and suggest that previous studies reporting that aspirin inhibits protein glycation by acetylation of

Table I: Effect of pre-incubation with aspirin on crosslinking of collagen by glucose

Sample	FL	CML	Pentosidine	Solubilization by pepsin (%)
	(mmol/mol lysine)		(mol/mol lysine)	
1	0.96	0.17	0.01	97
2	13.6	11.7	8.0	61
3	14.8	21.9	15.3	32
4	11.2	13.4	14.2	26

Sample 1 is a control, incubated for 10 days in phosphate buffer. Sample 2 was incubated for 10 days in the presence of 0.25 M glucose in phosphate buffer. Sample 3 was incubated for 6 days in phosphate buffer, then with glucose for an additional 10 days. Sample 4 was incubated in phosphate buffer with 100 mM aspirin for 6 days, then for an additional 10 days in the presence of glucose.

amino groups (Day et al., 1979; Li et al., 1984; Rendell et al., 1986; Huby and Harding, 1988; Rao and Cotlier, 1988; Abraham et al., 1989) were compromised by factors, such as: 1) use of very high concentrations of aspirin; 2) poor pH control; 3) measurements of glycation by radiochemical assays which do not distinguish glycation from glycoxidation products; 4) measurement of glycation by relatively non-specific assays, such as the thiobarbituric acid assay; or 5) specific binding of aspirin to the protein under study. Although benzoate, salicylate and aspirin had little effect on glycation, the latter two compounds were effective, albeit at high concentrations, as inhibitors of glycoxidation reactions and crosslinking of protein. These effects were graded, with salicylate being intermediate between benzoate and aspirin, and were concentration dependent. The inhibition of these advanced glycation reactions undoubtedly results from the oxygen radical scavenging activity of these compounds (Halliwell and Gutteridge, 1989) and possibly their ability to chelate metal ions (Woolard et al., 1990). Our observations support the work of Yue et al. (1984, 1985), indicating that both salicylate and aspirin inhibit collagen crosslinking by glucose *in vitro* and *in vivo*, but have no effect on glycation of the protein, and support earlier suggestions that aspirin and antioxidant therapy, in general, may be useful for inhibition of advanced glycation reactions in diabetes (Baynes, 1991).

Abbreviations

CML, N^{ε}-(carboxymethyl)lysine; FL, fructoselysine.

Acknowledgments

This work was supported by research grant, DK-19971, from the National Institute for Diabetes and Digestive and Kidney Diseases.

References

Abraham, E. C., Swamy, M. S., and Perry, R. E. (1989) Nonenzymatic glycosylation (glycation) of lens crystallins in diabetes and aging. *Prog. Clin. Biol. Res.* 304: 123-139.

Baynes, J. W. (1992) Role of oxidative stress in development of complications in diabetes. *Diabetes* 40: 405-412.

Day, J. F., Thorpe, S. R., and Baynes, J. W. (1979) Non-enzymatically glycosylated albumin. *J. Biol. Chem.* 254: 595-597.

Dyer, D. G., Dunn, J. A., Thorpe, S. R., Bailie, K. E., Lyons, T. J., McCance, D. R., and Baynes, J. W. (1993) Accumulation of Maillard reaction products in skin collagen in diabetes and aging. *J. Clin. Invest.* 91: 2463-2469.

Fu, M-X., Knecht, K.J., Thorpe, S. R., and Baynes, J. W. (1992) Role of oxygen in cross-linking and chemical modification of collagen by glucose. *Diabetes* 41 (Suppl. 2): 42-48.

Fu, M-X., Wells-Knecht, K. J., Blackledge, J. A., Lyons, T. J., Thorpe, S. R., and Baynes, J. W. (1993) Glycation, glycoxidation and crosslinking of collagen by glucose: kinetics, mechanisms and inhibition of late stages of the Maillard reaction. *Diabetes.* (in press).

Halliwell, B., and Gutteridge, J. M. C. (1989) *Free Radicals in Biology and Medicine*, 2nd Ed., Clarendon Press, Oxford.

Huby, R., and Harding, J. J. (1988) Non-enzymic glycosylation (glycation) of lens proteins by galactose and protection by aspirin and reduced glutathione. *Exp. Eye Res.* 47: 53-59.

Li, W., Khatami, M., Robertson, G. A., Shen, S., and Rockey, J. H. (1984) Nonenzymatic glycosylation of bovine retinal microvessel basement membrane *in vitro*: kinetic analysis and inhibition by aspirin. *Invest. Ophthalmol. Vis. Sci.* 25: 884-892.

Rao, G. N., and Cotlier, E. (1988) Aspirin prevents the nonenzymatic glycosylation and carbamylation of the human eye lens crystallins *in vitro*. *Biochem. Biophys. Res. Commun.* 151: 991-996.

Rao, G. N., Lardis, M. P., and Cotlier, E. (1985) Acetylation of lens crystallins: a possible mechanism by which aspirin could prevent cataract formation. *Biochem. Biophys. Res. Commun.* 128: 1125-1132.

Rendell, M., Nierenberg, J., Brannan, C., Valentine, J. L., Stephen, P. M., Dodds, S., Mercer, P., Smith, P. K., and Walder, J. (1986) Inhibition of glycation of albumin and hemoglobin by acetylation *in vitro* and *in vivo*. *J. Lab. Clin. Med.* 108: 826-893.

van Boekel, M. A. M., van den Bergh, P. J. P. C., and Hoenders, H. J. (1992) Glycation of human serum albumin: inhibition by diclofenac. *Biochim. Biophys. Acta* 1120: 201-204.

Vidal, P., Rodriguez-Requena, J., and Cabezas-Cerrato, J. (1989) Inhibition of protein glycation: a warning about the pH factor. *Med. Sci. Res.* 17: 21-22.

Woolard, A. C. S., Wolff, S. P., and Bascal, Z. A. (1990) Antioxidant characteristics of some potential anticataract agents. *Free Rad. Biol. Med.* 9: 299-395.

Yue, D. K., McLennan, S., Handelsman, D. J., Delbridge, L., Reeve, T., and Turtle, J. R. (1984.) The effect of salicylates on nonenzymatic glycosylation and thermal stability of collagen in diabetic rats. *Diabetes* 33: 745-751.

Yue, D. K., McLennan, S., Handelsman, D. J., Delbridge, L., Reeve, T., and Turtle, J. R. (1985) The effects of cyclooxygenase and lipoxygenase inhibitors on the collagen abnormalities in diabetic rats. *Diabetes* 34: 64-78.

An Investigation on *In Vivo* and *In Vitro* Glycated Proteins by Matrix Assisted Laser Desorption/Ionization Mass Spectrometry

A. Lapolla,[1] D. Fedele,[1] C. Gerhardinger,[1] L. Baldo,[1] G. Crepaldi,[1]
R. Seraglia,[2] S. Catinella,[2] and P. Traldi[2]

[1]INSTITUTE OF INTERNAL MEDICINE, UNIVERSITY OF PADOVA, VIA GIUSTINIANI 2, I-35100 PADOVA, ITALY; [2]CNR RESEARCH AREA, CORSO STATI UNITI 4, I-35020 PADOVA, ITALY

Summary
The reaction products of RNase with glucose and fructose, at different concentrations and incubation times were studied by Matrix Assisted Laser Desorption/Ionization Mass Spectrometry. Clear differences in the reactivity of the two sugars were found. Different spectra were obtained analyzing diabetic and healthy serum samples by the same analytical technique.

Introduction

Investigation on the structure of glycated proteins are usually carried out by means of spectroscopical methods or by other analytical techniques (mass spectrometry, ^{1}H and ^{13}C-nuclear magnetic resonance) on the products arising from chemical or enzymatic degradation of the substrate. More recently a new mass spectrometric technique has become available, leading to unequivocal data on molecular weight of large molecules, called Matrix Assisted Laser Desorption Ionization (MALDI)(Hillenkamp et al. 1991). It consists in the interaction of laser beams with the mixture of the compounds of interest with a suitable matrix, exhibiting an absorption maximum corresponding to the frequency of irradiating laser.

In a previous paper we investigated by MALDI/MS the *in vitro* glycation of Bovine Serum Albumin with glucose at different sugar concentrations and incubation times (Lapolla et al. 1993). An increase in molecular weight of glycated protein up to 70000 Da was found in the time range of 21-28 days, with the reaching of a steady state corresponding to the saturation of all the reactive sites present in the protein. In the present paper the data obtained by MALDI/MS on the products arising from incubation of RNase with two different sugars (glucose and fructose) are reported. Furthermore, the preliminary results achieved on plasma proteins of healthy and diabetic subjects are also described.

Materials and Methods

Samples Preparation

200 mg of bovine pancreatic RNase (Sigma, St. Louis, MO) were incubated with D-glucose and D-fructose at different concentrations (0.025 M and 0.25 M) in 2 ml of the sodium phosphate buffer (0.05 M) at pH 7.5. Following the addition of 0.5% of toluene, to avoid bacterial contamination, the solutions were subdivided in a series of samples which were incubated, in sterile conditions, at 37°C for different times (from 0 to 20 days). RNase in sodium phosphate buffer alone, at the same concentration and for the same incubation times, was used as the control. The samples were then dialyzed

extensively against distilled water, in order to eliminate free glucose, and lyophilized.

One mL of uranyl acetate was added to 50 μL of serum and the mixture was centrifuged for 4 min at 2000 x g at 4 °C. The precipitate was dissolved in 500 μL of 0.1% trifluoroacetic acid in water and directly analyzed.

Mass Measurements

MALDI experiments on RNase samples were performed on a Kompact MALDI III (Kratos Analytical, Manchester, U.K). Ions formed by a pulsed UV laser beam (nitrogen laser, λ=337 nm) were accelerated to 20 keV kinetic energy. The laser power density was about 10^6 W/cm^2. Sinapinic acid was used as a matrix.

Results

The MALDI spectra of samples of RNase incubated with 0.25 M glucose are shown in Figure 1.

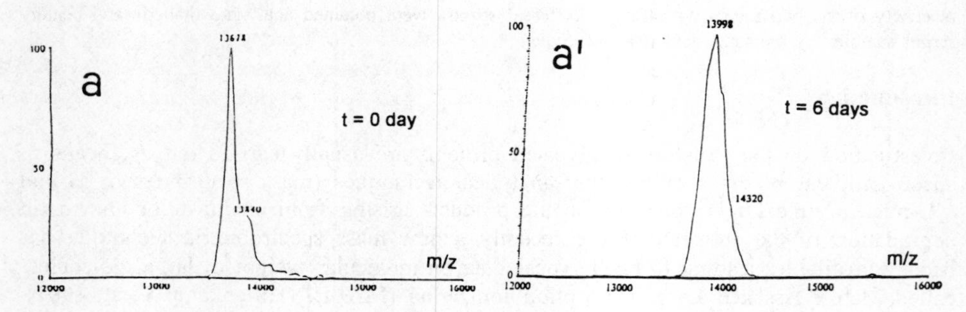

Figure 1. MALDI spectra of RNase incubated with 0.25M glucose at a) 0 day and a') 6 days.

A progressive increase in molecular weight with respect incubation time is observed: in fact the maximum of the peak due to RNase passes from 13678 Da at 0 days to 13998 Da at 6 days. The MALDI spectra of samples of RNase incubated with 0.25 M fructose are different from those obtained with glucose in the same concentration. A wide peak is always present and after 6 days of incubation a composite peak is observed, where compounds of molecular weight of 13729, 13790, 13842, 14001, 14165 and 14369 Da are present (Figure 2). The MALDI spectra of RNase incubated with glucose and fructose 0.025 M are reported in Figure 3.

With glucose, compounds of molecular weight of 13840, 13882, 13954 and 14040 Da are well detectable after 20 days of incubation whereas with fructose, only two peaks are observed, at 13730 and at 14006 Da.

In Figure 4 the MALDI spectra of serum samples of healthy subjects, diabetics and diabetics in bad metabolic control are reported.

Discussion

The results obtained by MALDI/MS on the products arising from the reaction of RNase with sugars clearly show that the molecular weight increase of the protein is strongly related to sugar concentration and incubation time. Clear differences in the reactivity of the two sugars have been observed. With 0.25 M glucose just after the addition of glucose (t = 0 days) a small peak, corresponding to the reaction of one molecule of sugar, is well detected at 13840 Da and it becomes the most abundant after 3 days. After 6 days the most abundant peak is at 13998 Da, corresponding to the condensation of two glucose units, but the presence of peaks at 14160 and 14320 Da indicates that three and four molecules of sugar respectively have been reacted in agreement with that found by Watkins et al. (1985). With 0.25 M fructose the spectra are different from those obtained with glucose: in fact at 0 day only the peak due to not-glycated RNase is detected, while that due to the reaction of one fructose unit become well detectable after 1 day. After 6 days the peak becomes larger and composite and species with different molecular weight are detectable, indicating the presence of different products that could reasonably have originated from the degradation reaction of the initial Amadori products. Leaving aside the species at 13842, 14001 and 14165 Da corresponding to the condensation of one, two and three fructose units on RNase respectively, the peaks at 13729, 13790 and 14369 Da could be reasonably attributed to the presence on the protein of pentosidine (Sell and Monnier, 1989) acetylamine and pyrrolaldehyde moieties (Ledl, 1990). The data obtained indicate that the rate of formation of the Amadori product is faster with glucose than that observed with fructose. Moreover, the rearrangement reactions described in literature as responsible for the propagation of non-enzymatic glycation, are favored using fructose as the reagent sugar.

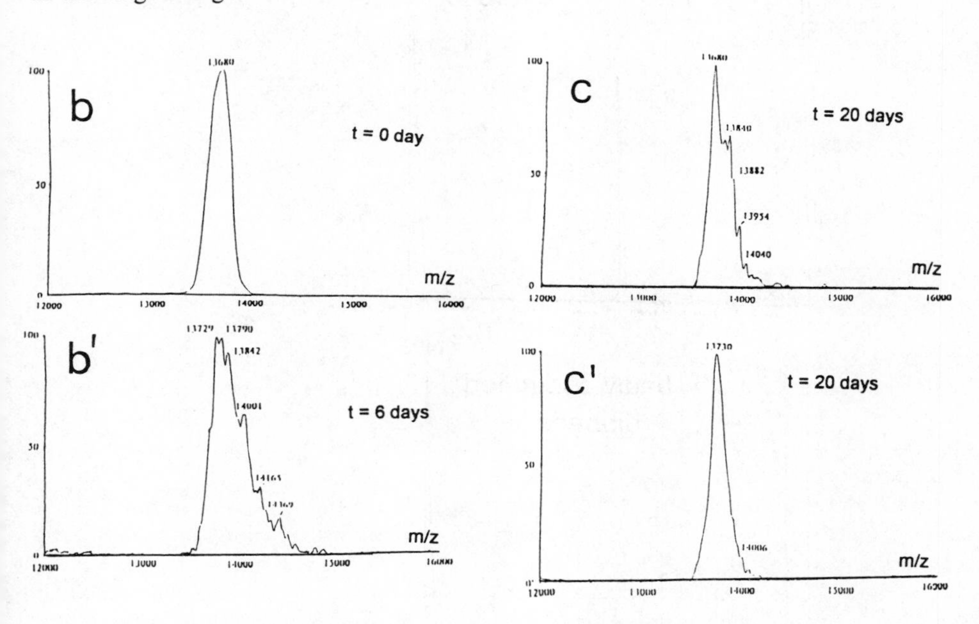

Figure 2. MALDI spectra of RNase incubated with 0.25M glucose b) 0 day and b') 6 days.

Figure 3. MALDI spectra of RNase incubated with c) 0.025M glucose and c') 0.025M fructose recorded at 20 days.

As expected the glycation rate is slower in presence of lower sugar concentration and only after 20 days peaks due to rerrangement products become detectable. With 0.025 M glucose the condensation and further rearrangements of three glucose units are

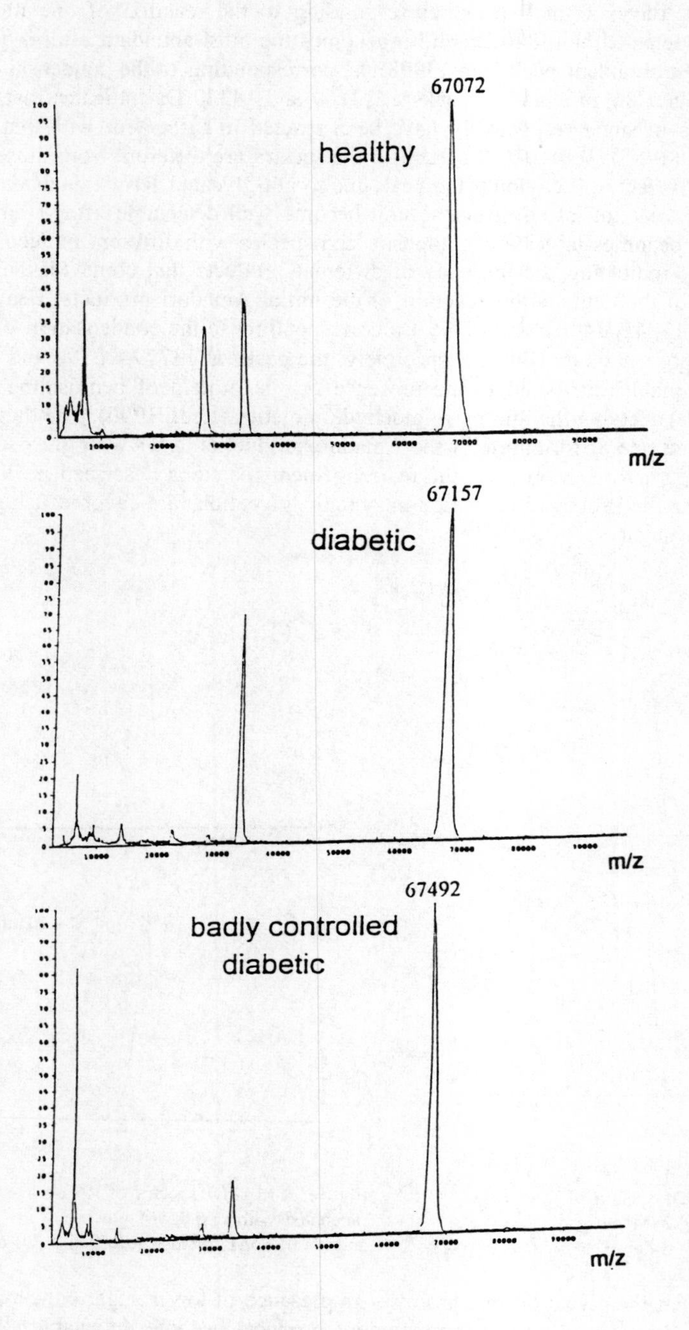

Figure 4. MALDI spectra of serum samples of healthy, diabetic and badly controlled diabetic subjects.

observed. The peak at 13840 Da corresponds to the condensation of one glucose unit, while for the other peaks the presence of acetylamine and pyrrolaldeyde could be tentatively proposed. With 0.025 M fructose the most abundant peak at 13730 could be explained by the presence of a pentosidine moiety, but the presence of a peak at 14006 Da indicates the reaction of two fructose units on RNase.

Unexpected results were obtained when analysing the control RNase. In fact the presence of a peak centered at about 11400 Da is observed just after 3 days of incubation. This peak is completely absent at 0 days, proving that it is not due to laser induced decomposition processes and its relative abundance increases with incubation time. Further works are in progress, in order to establish how this degradation reaction could affect the kinetics and yield of glycation. Interesting results were obtained analysing serum samples of diabetics with different degree of metabolic control and those of healthy subjects. The MALDI/MS spectra are different and in particular an increase of the molecular weight of Human Serum Albumin was found: in fact its value increases from 67072 Da in the healthy subject to 67492 Da in the badly controlled diabetic. Differences are also present in the pan of spectra ranging from 0 to 10000 Da. Further work is needed to confirm and to rationalize these findings.

Conclusions

The data so obtained indicate that MALDI/MS can be considered a powerful tool in the study of *in vivo* and *in vitro* glycation processes, due to its validity to gain information on the molecular weight increase of intact glycated proteins.

Acknowledgments:
Financial support by C.N.R., Roma, Progetto Finalizzato Invecchiamento co. No.00312PS40.

References

Watkins, N.G., Thorpe,S.R. and Baynes,J.W. (1985). Glycation of amino group in protein: Studies on the specificity of modification of RNase by glucose. J. Biol. Chem. 260:10628-10636.

Hillekamp, F., Karas, M., Beavis, R.C. and Chait, B.T. (1991). Matrix Assisted Laser Desorption/Ionization mass spectrometry of biopolymers. Anal. Chem. 63:1193A-1203A.

Lapolla, A., Gerhardinger,C., Baldo,L., Keane,A., Seraglia,R., Catinella,S. and Traldi,P. (1993). A Study on *in vitro* glycation processes by Matrix Assisted Laser Desorption/Ionization mass spectrometry. Biochim. Biophys. Acta (in press).

Ledl, F. (1990). In: The Maillard reaction in food processing, human nutrition and physiology (Finot,P.A., Aeschbacher, H.U., Hurrell, R.F., Liardon, R., Eds) Birkhauser Verlag, Basel pp l6-42.

Sell, D.R. and Monnier, V.M. (1989). Structure elucidation of a senescence cross-link from human extra cellular matrix: implication of pentose in the aging processes. J. Biol Chem. 264:21597-21602.

The Formation of Some Pyrido-[3,4-d]-imidazoles by Maillard Reaction

U. S. Gi and W. Baltes

INSTITUTE OF FOOD CHEMISTRY, TECHNICAL UNIVERSITY BERLIN,
GUSTAV-MEYER-ALLEE 25, D-13355 BERLIN, GERMANY

Summary
2-Acetyl- and 2-propionylpyrido-[3,4-d]-imidazole were formed by heating histidine with D-glucose to 120, 150 or 180°C in an aqueous buffer system or by roasting to 220°C. In the same way, the corresponding compounds methylated at the 1', 2' or 3'-position of the imidazole moiety were formed by heating of 1-, 2- or 3-methyl histidine with glucose to 150 °C. Tracer experiments with 1- or 6-^{13}C-D-glucose established that the terminal methyl group of the acyl moieties were formed from C-1 atom as well as from C-6 of glucose. Tentative reaction mechanisms of glucose degradation via formation of 3-^{13}C-2-propanone-1-al respectively 4-^{13}C-2-butanone-1-al are given.

Introduction

In the course of research into the reactions of histidine under the conditions of the Maillard reaction we have identified a new class of compounds with a pyridoimidazole ring structure that was previously unknown (Gi and Baltes, 1993). Histidine is relatively inert when it is heated with glucose in an aqueous phosphate buffer solution at pH 5.8 under atmospheric conditions (e.g. cooking a soup). But when this reaction mixture is heated in an autoclave to 150°C or when it is roasted like coffee beans by increased heating to 220°C during 10 minutes after mixing with sand numerous reaction products are formed. Most of them, e.g. furans, pyrroles, pyridines and pyrazines, were known as thermally generated aromas and will therefore not be further mentioned. On the other hand, our interest was focused on two peaks in the GC the mass spectra of which demonstrated molar masses of 161 and 165. Because of the very similar fragmentation mechanisms of their mass spectra, we were of the opinion that the one compound was the tetrahydro derivative of the other. The fragmentation pattern in the mass spectra demonstrated eliminations of acetyl, methyl and HCN. By exact mass determination of the compound with the molar mass peak of m/z 161, the sum formula of $C_8H_7N_3O$ was determined. From this composition the formula of an acetylpyrido-[3,4-d]-imidazole (**III**) was derived. Consequently, the second peak was assumed to be 2-acetylpyrido-[3,4-d]-imidazole.

In order to establish these structures we assumed a reaction of histamine with pyruvic aldehyde which is a well known sugar degradation product. It could be expected that by reaction of these two compounds, the tetrahydro derivative was formed which was dehydrated to yield the pseudoaromatic compound. Indeed, by reaction of these two compounds we obtained **III** in a sufficient amount to establish its structure by MS, IR and ^{1}H-NMR spectra. As by-products we identified hydroxyacetone and propanediol which makes clear that pyruvic aldehyde was not only active as a reagent causing ring closing but also as a dehydration agent. Concerning the reaction mechanism of formation of this compound from histidine and glucose, we assume a pathway which is demonstrated by Figure 1. This reaction pathway includes a condensation of histidine with pyruvic aldehyde yielding compound **IV** which is converted via the tautomeric formula **V** and loss of CO_2 to yield compound **VIa**. From **VIa** ring closing reaction leads by simultaneous dehydration to formation of compound **III**. On the other side, a cleavage of **VIa** to form the Strecker aldehyde seems not to occur by preference, because we could not establish any histamine which reacts with pyruvic aldehyde only slowly.

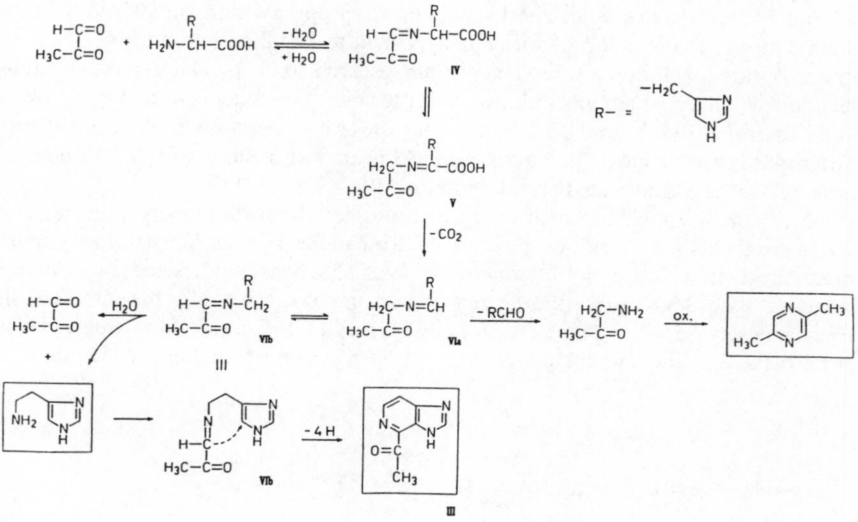

Figure 1. Formation of 2-acetylpyrido-[3,4-d]-imidazole via Strecker degradation and cyclization.

Materials and Methods

Histidine, histamine, methylhistidines, glucose, 1-^{13}C-D-glucose, 6-^{13}C-D-glucose, 2-butanone, pyruvic aldehyde: high-grade products of trade. 1,2-Butanedione was synthesized by the method of Riley et al. (1932) and of Organicum (1963). Solvents for extraction were freshly distilled before use. Sample preparation, GC/MS, ^{1}H-NMR, IR, preparation of compound **III**: See Gi and Baltes (1993). 2-Propionyl- pyrido-[3,4-d]-imidazole: 0,02 mol of histamine dihydrochloride was added to 4 mL of 2-butanone-1-al, freshly distilled; the mixture was dissolved in 60 mL of 1 mol/L phosphate buffer, pH 5.8 and heated at 150°C in a laboratory autoclave fitted with a PTFE insert for 1 h. After cooling to room temperature and extraction with 5 x 40 mL of ether, the combined ether fractions were treated with aqueous $NaHCO_3$ and the water was frozen out at -20°C. The carefully concentrated extract was injected into GC/MS Finnigan 4500.

Tracer experiments: 5 mmoles 1-^{13}C-D-glucose or 1.3 mmoles 6-^{13}C-D-glucose were added to the equimolar amount of histidine dissolved in 10 mL of 1 mol/L phosphate buffer, pH 5.8. The reaction mixtures were heated and extracted as was described above. The amount of ^{13}C was calculated via the mass spectra.

Results and Discussion

In the course of our investigations we identified a homologues compound of 2-acetylpyrido-[3,4-d]-imidazole, the mass spectrum of which showed as mole peak m/z 175. The base peak of this spectrum was m/z 119, meaning that there was no difference from the spectrum of 2-acetylpyrido-[3,4-d]-imidazole. Taking the radical cation of m/z 57 into account, we assumed the formula of a 2-propionyl derivative. On the other hand, methylation via a reaction of formaldehyde had to be taken into consideration. In order to check this assumption we extended our research to 1-,2-, and 3-methylhistidine. By reaction of those 3-methylhistidines with glucose in an autoclave at 150°C we obtained more than one pyridoimidazole each. But the acetyl derivatives were most prominent. Their mass spectra showed a strong m/z 133 peak each instead of m/z 119 being the base peak of the non methylated pyridoimidazoles.

In order to establish the structure of 2-propionyl-pyrido-[3,4-d]-imidazole, we have synthesized this compound by reaction of 2-butanone-1-al with histamine (Figure 2). 2-Butanone-1-al which is not commercially available was synthesized by oxidation of 2-butanone with SeO_2. Indeed, after reaction in an autoclave under the same conditions as already described, we obtained 2-propionylpyrido-[3,4-d]-imidazole which was identical in its behavior with the compound formed by reaction of histidine with glucose.

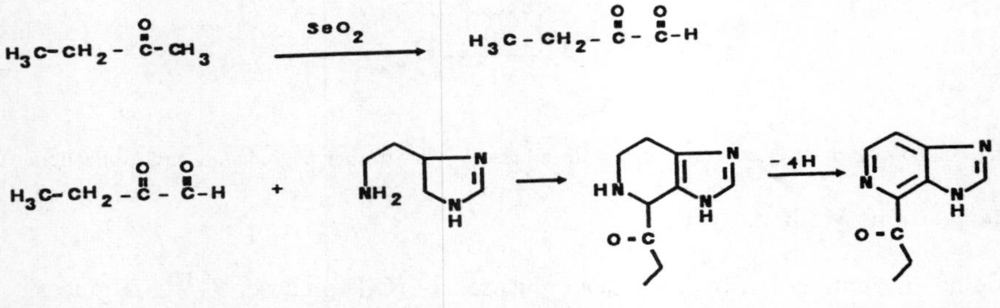

Figure 2. Synthesis of 2-butanone-1-al and its reaction with histamine to form 2-propionylpyrido-[3,4-d]-imidazole.

Figure 3 presents the pyridoimidazoles which were identified in our experiments. It can be recognized that the acetyl residue in the 2-position seems to be most prominent followed by the 2-propionyl compounds. In the case of 1-methylhistidine, we were successful in establishing the basic heterocyclic system methylated only at the imidazole residue. This compound might have been formed via ring closure by formaldehyde. Also the compound methylated in the 1-position by reaction of acetic aldehyde was identified. On the other side, the basic heterocycle bearing neither methyl nor acetyl groups is still unknown. We were not able to synthesize this compound because the tetrahydro derivative formed from a reaction of formaldehyde with histamine resisted our attempts of dehydration. But considerations made in correlation to these reactivities as well as to views of possible ring closures via a Pictet-Games reaction (1909) will not be discussed in this paper.

Glucose+histidine Glucose+1-methylhistidine Glucose+2-methylhistidine Glucose+3-methylhistidine

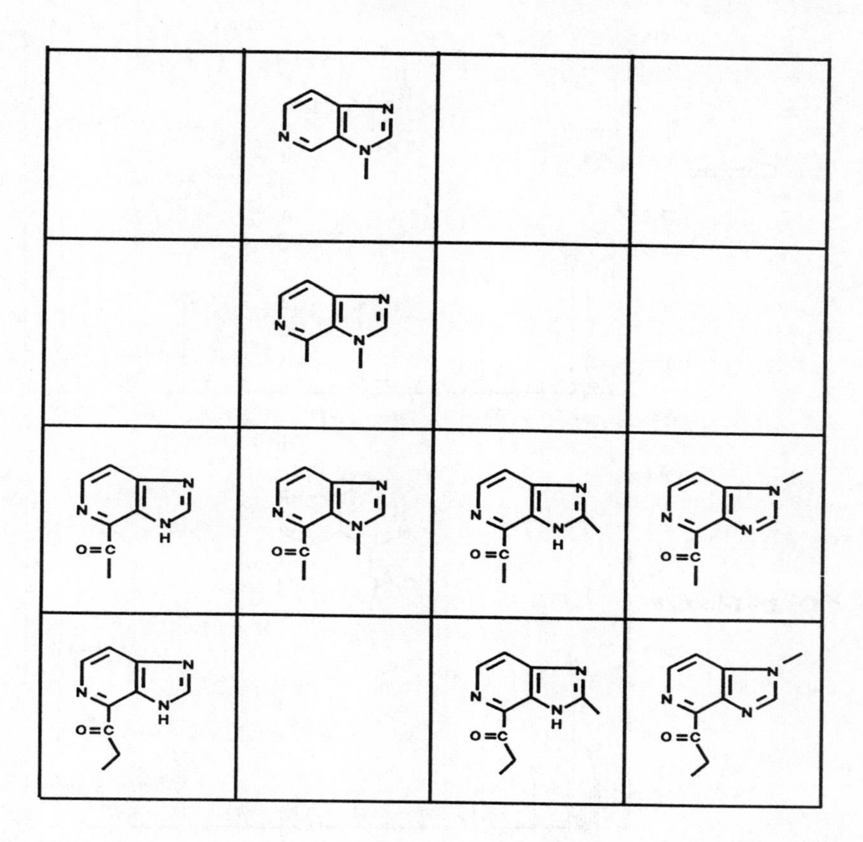

Figure 3. Formulas of pyrido-[3,4-d]-imidazoles formed by reaction of glucose with histidine and methylhistidines.

Origin of the Dicarbonyls

The starting point of our consideration was the fact that on the one hand pyruvic aldehyde is a well known sugar degradation product while 2-butanone-1-al is unknown. Its presence in volatiles after reaction of model mixtures of thermal aroma formation has been mentioned sometimes but detailed investigations about its formation were not carried out. Therefore we started some tracer experiments with glucose spiked with the isotope ^{13}C in the 1- or 6-position. This marked glucose was used in the same reaction with histidine. By analysis of the obtained mass spectra of acetylpyrido- or propionyl-pyrido-imidazole, the point of marking was determined.

Figure 4 shows the spectra of 2-acetylpyrido-[3,4-d]-imidazole which were obtained after reaction of histidine with unmarked glucose as well as glucose marked in the 1- or 6-position. It can be recognized that there are differences at m/z 43, 133 and 161. In the spectra m/z 162 (mole peak) appears after reaction of 1-^{13}C-glucose in an extent of about 45 % and from 6-^{13}C-glucose of about 55 %. The same $^{12}C/^{13}C$ ratio is shown at the fragment m/z 133. This fragment corresponds to the radical ion of 2-methyl-

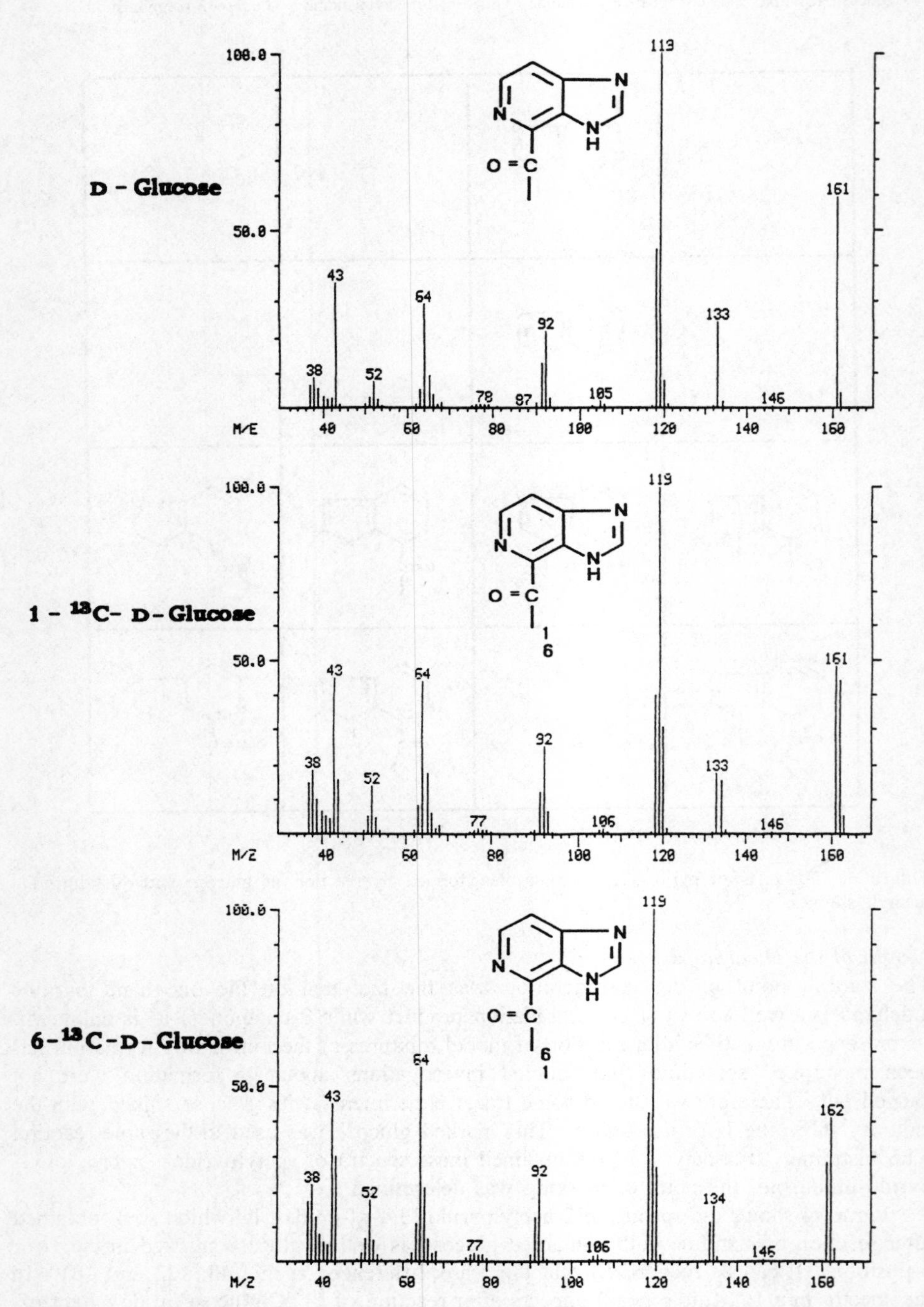

Figure 4. Mass spectra of 2-acetylpyrido-[3,4-d]-imidazole formed by reaction of histidine with D-glucose or 1-^{13}C- respectively 6-^{13} C-D-glucose

pyridoimidazole which was formed by CO-elimination. This type of CO-elimination has already described for 2-acetylpyridine (Tomer and Djerassi, 1973). Because of the identical $^{12}C/^{13}C$-ratio at m/z 133/134 compared with the mole peak, we assume that this methyl group is marked by the C-1 atom of glucose as well by the C-6 atom. Also m/z 44 is enhanced after use of 6-^{13}C glucose in a similar extent as demonstrated at the mole peak. Summing up these results the methyl residue of the acetyl group can be derived from the C-1 as well as from the C-6 atom of glucose.

This result required a correction to our hypothesis concerning the formation of pyruvic aldehyde from glucose during the Maillard reaction. The formation of pyruvic aldehyde from 1-desoxyglucosone by retro aldol scission is easily explainable. Pyruvic aldehyde is indeed one of the most prominent sugar degradation compounds. Also explainable are isomerizations of the 1-deoxyosone via tautomerism, which can be followed by water eliminations (Figure 5). By additional redox processes triosones can be formed. The formula of **VIII,** being a 1,6-dideoxy-2,4,5-trihexulose, was used to explain sugar degradations (e.g. to form acetic acid, Angrick and Rewicki, 1980). Up to now none of these derivatives has been isolated which are intermediates of sugar degradation. The inclusion of the C-6 atom of glucose as the methyl group of pyruvic aldehyde now enhances the probability of formation of compound **VIII** in Maillard reaction mixtures.

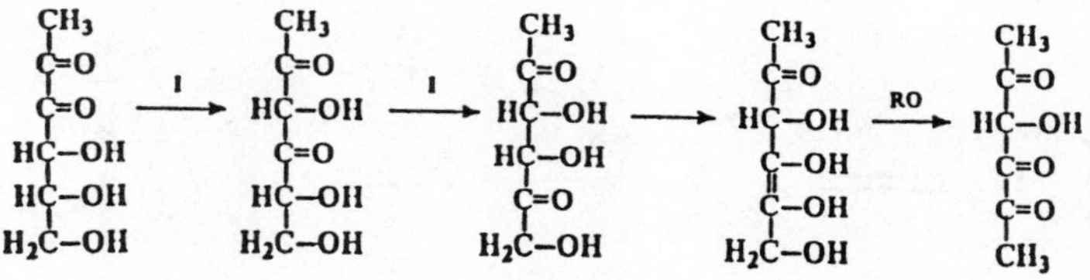

Figure 5. Mechanisms of formation of 1, 6-dideoxy-2,4,5-trihexulose.

However, this compound is not suitable to explain the formation of 2-butanone-1-al which then forms 2-propionylpyridoimidazole. Therefore, we compared the mass spectra which were obtained after reaction of 1-^{13}C- or 6-^{13}C-glucose with histidine. We found out, that the following mass units were changed:

m/z 175 to 175/176, ratio about 1:1 (molecular peak)
m/z 147 to 147/148, ratio about 2:1 (M-C0)
m/z 57 to 58 (C_2H_5 - $C0^+$)

The mass spectra obtained from reaction of both 1-^{13}C- and 6-^{13}C-glucose were identical. The base peak m/z 119 was not affected. Because of the similar fragmentation patterns of the 2-acetyl- and 2-propionyl derivatives we conclude that the terminal methyl group of the acyl moiety is spiked by ^{13}C from 1-^{13}C- as well as from 6-^{13}C-D-glucose.

In order to explain this behavior we postulate a mechanism which is based on the observation that 2-butanone-1-al can also be formed by heating of 2,5-dimethyl-4-

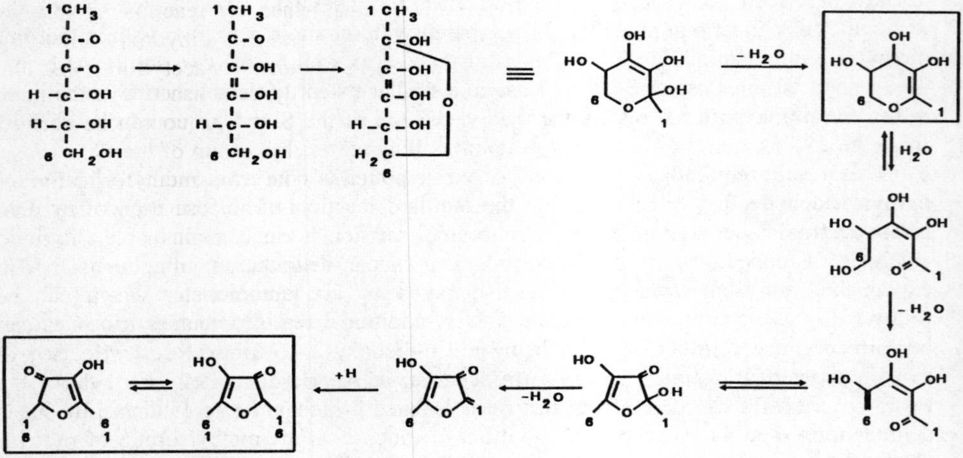

Figure 6. Thermal decomposition of 2,5-dimethyl-4-hydroxy-3(2H)-furanone.

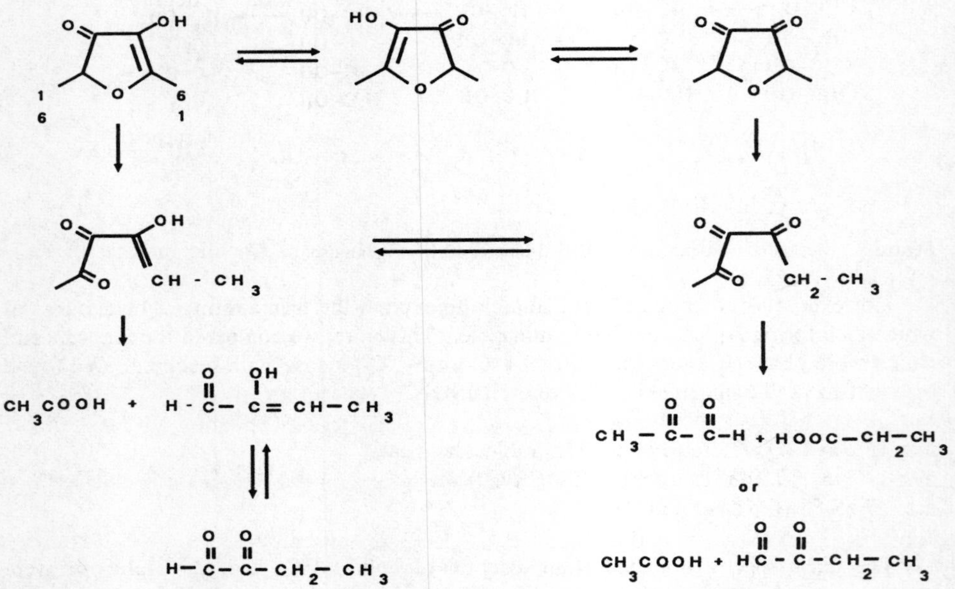

Figure 7. Reaction pathway of formation of 2,5-dimethyl-4-hydroxy-3(2H)-furanone via 3,5-dihydroxy-2-methyl-γ-pyranone.

hydroxy-3(2H)-furanone (Figure 6). This compound can also be formed by Maillard reaction of monosaccharides and was repeatedly established as reaction product of thermal aroma generation. But the amounts of the "γ-pyranone" (3,5-dihydroxy-2-methyl-γ-pyranone) were higher which can be converted to the furanone via acetylformoin. This observation originally proposed by Helak (1987) was shortly established by our group when we synthesized the γ-pyranone from 1-^{13}C-glucose. Since we could demonstrate that the C-1 of glucose was completely converted to the methyl group of the γ-pyranone. Thus, by ring opening the furanone can be formed. The two methyl groups are equivalent corresponding to a possible spiking of C-1 or C-6 of glucose because of keto-enol-tautomerism in an equilibrium shown in Figure 7. By ring opening a C-C-scission of the furanone can yield 2-butanone-1-al. This decomposition of 2,5-dimethyl-4-hydroxy-3(2H)-furanone has been suggested by the group of Ho et al. (1985). The equivalence of the two methyl groups makes clear the equal distribution of ^{13}C in the terminal methyl group of 2-butanone-1-al.

In order to test this result we reacted 2,5-dimethyl-4-hydroxy-3(2H)furanone with histidine. Indeed, we were successful in obtaining 2-propionylpyrido-[3,4-d]-imidazole.

Conclusions

The results of our tracer experiments made clear that 1) 1-deoxyglucosone can be converted to 1,6-dideoxy-2,4,5-trihexulose which can form pyruvic aldehyde by retro aldol scission between C-3 and C-4. This enables the carbon atoms 1 as well as 6 to become the methyl group in pyruvic aldehyde. 2) Another degradation of 1-deoxyglucosone takes place which forms 1,2-butanedione as well as pyruvic aldehyde and acetic acid via an intermediate cyclization.

Acknowledgment

The authors are grateful to the Deutsche Forschungsgemeinschaft for its financial support.

References

Angrick, M. and Rewicki, D. (1980) Die Maillard Reaktion, Chemie in Unserer. *Zeit* 14, 149-157.

Gi, U.S. and Baltes, W. (1993) Model Reactions on Roast Aroma Formation. 14. Formation of 2-Acetylpyrido-[3,4-d]-imidazole by Heating of Glucose with Histidine. *J. Agric. Food Chem.* 41:644-646.

Helak, B. (1987) Dissertation. Technical University, Berlin.

Organicum (1963) Organicum VEB Deutscher Verlag, Berlin, Oxidation mit Selendioxid, 336-337.

Pictet, A. and Games A (1910) A New Method for Synthesis of Isoquinoline Bases, Ber. Dtsch. Chem Ges. 43, 2384-2391.

Riley, H.L., Morley, J.F. and Friend, N.A.C. (1932) Selenium Dioxide, A New Oxidizing Agent. *J. of the Chem. Soc* (London), 1875-1883.

Shu, C-K., Mookherjee, B.D. and Ho, C-T. (1985) Volatile Compounds of the Thermal Degradation of 2,5-Dimethyl-4-hydroxy-3(2H)-furanone. *J. Agric. Food Chem.* 33:446-460.

Tomer, K.B. and Djerassi, C. (1973) *J. Organ. Chem.* 38:4152.

Alteration of Skin Surface Protein with Dihydroxyacetone: A Useful Application of the Maillard Browning Reaction

John A. Johnson[1,2] and Ramon M. Fusaro[1]

[1]SECTION OF DERMATOLOGY, DEPARTMENT OF INTERNAL MEDICINE; [2]DEPARTMENT OF BIOCHEMISTRY, UNIVERSITY OF NEBRASKA MEDICAL CENTER, 600 SOUTH 42ND ST., OMAHA, NEBRASKA, USA 68198-4360

Summary
Dihydroxyacetone (DHA) undergoes the Maillard reaction with skin surface protein to produce a durable brown color. This pigment protects persons sensitive to ultraviolet and visible radiation. Interest in photosensitivity encouraged us to examine browning activities of DHA and other hydroxycarbonyl compounds. Our observations provide clues of general interest in understanding the Maillard reaction. Although DHA has been used in the popular sunless tanning preparations for over 30 years, a more effective substitute has not been developed.

Introduction

Sunless tanning preparations contain dihydroxyacetone (DHA). In 1960, an estimated six million units were sold in the United States; unit sales now number in the tens or even hundreds of millions. Consequently, sunless tanning products represent a most positive aspect of the Maillard reaction. Because DHA-induced skin pigment (DISP) provides photo protection to persons sensitive to sunlight, we have investigated the browning activities of DHA and other hydroxycarbonyl compounds. This is a discussion of the Maillard reaction from a unique viewpoint, and reader inquiry and comment are invited.

Dihydroxyacetone-Induced Skin Pigment, DISP

Browning agents are colorless compounds that react with skin surface proteins to produce a durable brown color. Proprietary formulations contain 3 to 8 percent DHA and produce color within three hours; DHA solutions (50% aqueous isopropanol v/v) develop maximum color overnight. Early formation of blue-fluorescing intermediates may be monitored by exposure of DHA-treated skin to a black light fluorescent lamp (output 320 to 400 nm) or a Wood's lamp (line output at 365 nm). Fluorescence changes to orange-pink as visible color develops in the outer layers of the inert portion of skin, the stratum corneum. Pigment disappears within a few days due to natural sluffing of the skin. Color can be deepened by repeat applications, but final intensity depends on the concentration. Skin pigment withstands soap-and-water washing, and is conveniently maintained by application of DHA in the evening.

Photoprotection

The ultraviolet region of sunlight is divided into UVB (290-320 nm) and UVA (320-400 nm). Because UVB is the primary cause of sunlight-induced skin damage, commercial sunscreens were designed to protect against that radiation. Recent concern for harmful action of UVA radiation has encouraged development of broad spectrum sunscreens. *Photosensitivity* is characterized by exaggerated skin responses to sunlight, and often involves sensitivity to UVA and even visible radiation. Since commercial sunscreens afford protection to about 360 nm, they do not protect some photosensitive persons. In these instances, DHA often provides relief. The brown color of DISP of necessity absorbs in the shortwavelength (blue) end of the visible region, with overlap into long-wavelength UVA radiation (Johnson and Fusaro, 1987). Fluorescence of DISP under UVA radiation not only documents UVA screening ability, but also provides a means of dissipating absorbed energy harmlessly.

Sunscreens are rated by their Sun Protective Factor (SPF): Sunlight tolerance with sunscreen divided by tolerance without sunscreen. Although DISP provides a SPF of only 3 or 4, this continuous protection is more effective than episodic use of high SPF sunscreens. Furthermore, DISP has photoprotective potential in adjunctive use with sunscreens. Thus, with a baseline protection of SPF 3, a SPF 10 sunscreen should in theory have an effectiveness of SPF 30.

Dihydroxyacetone Analogs

We employed a rat model of UVA sensitivity to evaluate compounds similar to DHA (Follett *et al.*, 1987). Dihydroxyacetone browned the skin and provided a UVA protection factor of 4 (Figure 1). Monohydroxyacetone exhibited no browning activity and afforded little photo protection. The yellow color of 5-hydroxymethylfurfural stained the skin and provided a little photo protection, but this phenomenon did not reflect browning activity. Methylglyoxal was about as effective as DHA. Weak activity of glyceraldehyde was probably due to formation of a stable hydrate; enhanced activity after exposure to alkaline medium is consistent with base-catalyzed isomerization to DHA. Browning agents require two hydroxyl groups adjacent to the carbonyl, as in DHA and hydrated methylglyoxal.

Mechanism for Browning of Skin

This process is unique in several ways. Browning activity of DHA involves initial reaction with soluble amines in the stratum corneum, followed by incorporation of intermediates into skin proteins (Meybeck, 1977). Pigment formation proceeds over an interval of three or more hours, at a temperature of about 35°C, pH about 5, and in a moist environment. The ketotriose DHA bypasses many of the early steps of the Maillard reaction. Consequently, enolization, carbonyl migration and fission are severely restricted or absent. Likewise, rapid color development at low temperature suggests endothermic reactions such as Strecker degradation are of limited importance. Nevertheless, formation of melanoidin-like skin pigment suggests a high level of complexity.

Browning is inhibited by overly moist conditions and pretreatment of skin with formaldehyde. Color develops in aqueous solutions of DHA plus glycine, histidine, arginine or lysine; the ε-amino group of the latter is presumed to be a major reactant. Extraction of water-soluble skin components reduces browning activity (Maibach and Kligman, 1960). Since DHA browning is not compromised by prior soap-and-water washing, surface amines are not involved. Amino acids applied to the skin produce pigments that wash off readily (Goldman and Blaney, 1962), whereas soap-and-water washing of DHA-induced color does not reduce its intensity. These considerations

Potential Browning Agent		Browning Action	UVA[a] Protection Factor
Dihydroxyacetone	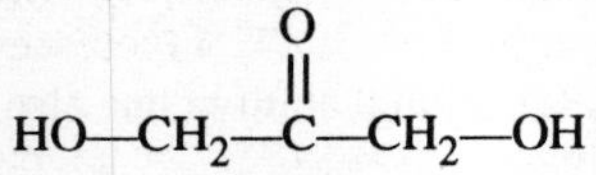 $HO{-}CH_2{-}C(=O){-}CH_2{-}OH$	+++	4
Monohydroxyacetone (acetol)	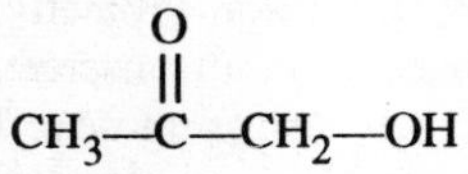 $CH_3{-}C(=O){-}CH_2{-}OH$	—	1-2
5-Hydroxymethylfurfural	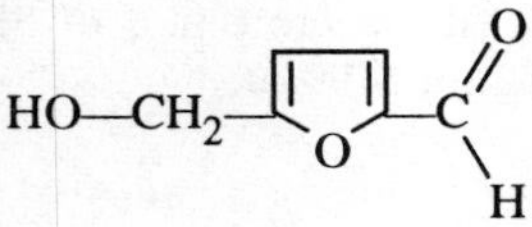 $HO{-}CH_2{-}C_4H_2O{-}CHO$ (furan ring)	—	2[b]
Methylglyoxal (pyruvaldehyde)	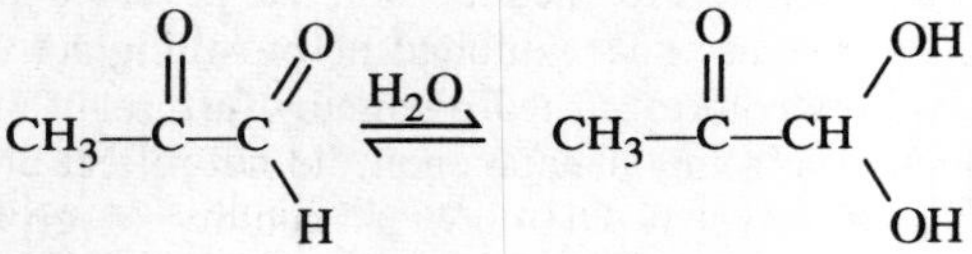 $CH_3{-}C(=O){-}CHO \overset{H_2O}{\rightleftharpoons} CH_3{-}C(=O){-}CH(OH)_2$	+++	3-4
Glyceraldehyde	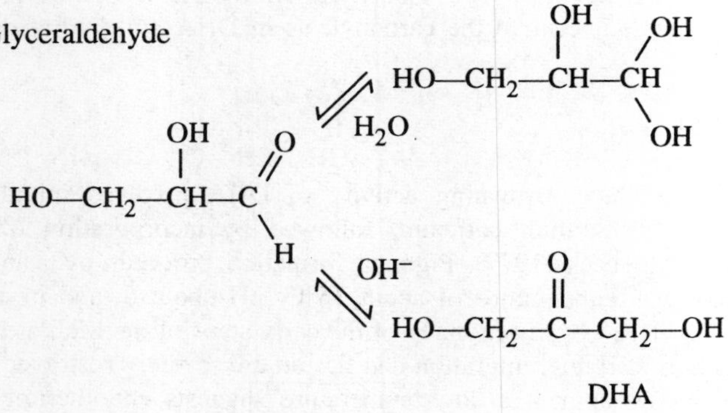 $HO{-}CH_2{-}CH(OH){-}CHO \overset{H_2O}{\rightleftharpoons} HO{-}CH_2{-}CH(OH){-}CH(OH)_2$	±	2-3
	$HO{-}CH_2{-}CH(OH){-}CHO \overset{OH^-}{\rightleftharpoons} HO{-}CH_2{-}C(=O){-}CH_2{-}OH$ (DHA)	+	3[c]

Figure 1. Structures and activities of dihydroxyacetone and analogs.

[a] UVA tolerance with agent divided by tolerance without agent.
[b] 5-HMF solution was yellow and stained the skin yellow.
[c] After exposure to pH 8.5 for 3 days.

suggest human skin contains just enough soluble amino compounds to trigger vigorous reaction with DHA, without blocking incorporation of intermediates into skin surface proteins. Variation in amount of soluble reactants may explain why the skin of some humans browns well with DHA, while that of others does not.

The action of hydroxycarbonyl compounds containing two to four carbons provides clues for understanding browning of skin. Glycolaldehyde reacts so rapidly that pigment is produced only in superficial skin layers, as revealed by rapid sluffing of the pigment. Methylglyoxal and DHA react at a rate that allows pigment to be distributed well into the stratum corneum. Reduced browning activity of glyceraldehyde probably relates to formation of a stable hydrate, which far exceeds the concentration of the carbonyl form in aqueous solution at 25°C (Rendina *et al.*, 1984). Monohydroxyacetone produces fluorescent materials that do not proceed to the pigment stage. Erythrulose, the largest monosaccharide that cannot form a stable furanose ring, exhibits weak browning activity.

The above considerations help us understand the browning of skin. The mechanism for conversion of DHA and methylglyoxal to nitrogen-containing ring structures characteristic of Maillard pigments is obscure. Oxidative deamination of the Heyns product from DHA would yield a reactive α-dicarbonyl structure. However, oxidative deamination of the Heyns product of monohydroxyacetone would yield methylglyoxal. Since the former has no browning activity, whereas the latter is very effective, it seems unlikely the pathway from DHA involves oxidative deamination. Calculations concerning skin surface amino groups suggest that tanning preparations deliver a several fold excess DHA. This excess would reduce chance interaction of more than one amino group with a DHA molecule. On the other hand, aldol condensation of DHA with a protein-bound Heyns product followed by cyclization and dehydration, would yield a multi-substituted pyrrole (Figure 2). Since this sequence does not accommodate prior reaction of DHA with soluble amines, it is tentatively presented as one of probably several pathways leading to skin melanoidins. The fact that maximum attainable skin color varies directly with DHA concentration supports a pathway involving unreacted DHA.

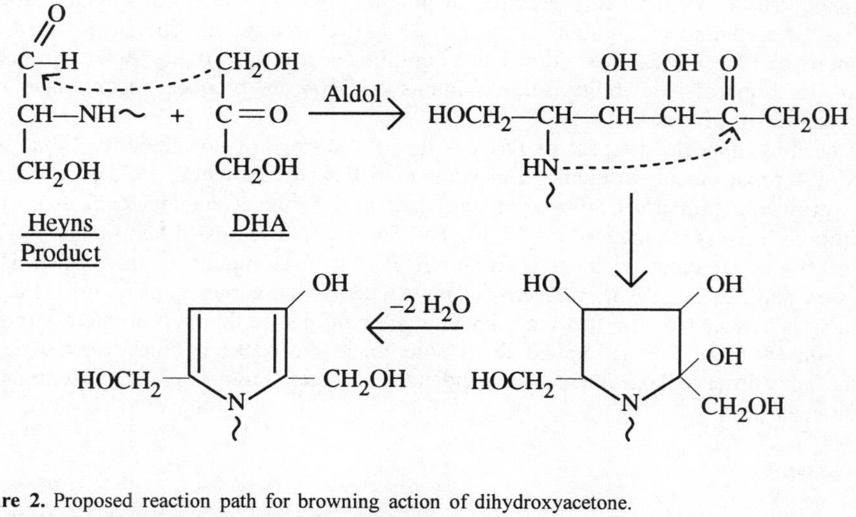

Figure 2. Proposed reaction path for browning action of dihydroxyacetone. NH~ and N~ represent bond to skin surface protein.

For a two-step process, one might postulate reaction of excess DHA with soluble amines and with ε-amino groups of peptide-bound lysine residues. Subsequent condensation of the Heyns products of soluble amines with those of peptide-bound amines would produce N,N' disubstituted dihydropyrazines, and possibly corresponding pyrazinium radical cations, as proposed by Narniki *et al.* (1977) for dimerization of the Amadori product of glycolaldehyde. As noted earlier, glycolaldehyde produces skin pigment rapidly. On the other hand, DHA would form structures with hydroxymethyl substituents on ring carbons 2 and 5. The need for elimination of these groups (Namiki, 1988) may explain why the DHA skin browning process is slower than that of glycolaldehyde.

Extensive review of the literature did not reveal a Maillard product with the structure presented in Figure 2. Strecker degradation yields 2,5-disubstituted pyrroles with appropriate functional groups (Olsson *et al.*, 1981), but requires α-dicarbonyl compounds with chain length sufficient to form the pyrrole ring. Methylglyoxal does not satisfy the chain-length requirement, and the hydrated form mimics the structure of DHA (Figure 1). The premise that browning action of methylglyoxal operates through the keto group is supported by formation of 3-hydroxy-1,2-dimethyl-5-formyl-pyrrole after heating methylammonium acetate with methylglyoxal (Wieland and Severin, 1973), and with DHA (Manninger and Severin, 1976). The pyrrole structure is consistent with condensation of methyl amine with the keto groups of two molecules of methylglyoxal or of DHA. Preferential reactivity of the keto group of methylglyoxal suggests Maillard reaction pathways evolving from other α-ketoaldehydes may proceed *via* this carbonyl group, rather than by involvement of the aldehyde function.

Why is Dihydroxyacetone Such a Good Browning Agent?

Failure to develop agents more efficient than DHA is due to rigorous requirements for a commercial browning agent: reactivity, chemical stability, availability and non toxicity. As noted, DHA is sufficiently reactive to produce, within hours, skin pigment that is distributed well into the stratum corneum. On the other hand, its chemical stability is enhanced by its occurrence as a dimer that equilibrates readily with monomer in aqueous media. In terms of availability, large amounts of DHA are produced economically by fermentation of glycerol.

The most appealing aspect of DHA is its solid history of non toxicity. It has been tested as a replacement for dietary glucose in diabetics (Rabinowitch, 1925), and interest in its browning properties arose from the studies of Wittgenstein and colleagues with diabetic children (Goldman *et al.*, 1960). The few reported cases of skin irritation from use of sunless tanning products were probably due to components other than DHA (Johnson and Fusaro, 1993). Concern for mutagenicity and carcinogenicity of browning products is lessened by the fact that pigments are confined to the inert stratum corneum. Akin and Marlowe (1984) applied DHA solution to the backs of mice weekly for 80 weeks and observed no increased incidence of solid tumors or of leukemias or lymphomas.

Conclusions

Our experience with the Maillard browning reaction in skin suggests: 1) dihydroxyacetone is the browning agent of choice; 2) dihydroxyacetone-induced skin pigment affords all day protection against UVA and visible radiation, and potentiates the action of sunscreen; 3) browning by trioses bypasses early steps of the Maillard reaction, but is sufficiently complex to produce melanoidin-like pigment; and 4) dihydroxyacetone may produce a multi-substituted pyrrole by aldol condensation of free triose with the Heyns product.

References

Akin, F. J., and Marlowe, E. (1984) Non-carcinogenicity of dihydroxyacetone by skin painting. J. Environ. Pathol. Toxicol. Oncol. 5(4/5): 349-351.

Follett, K.A., Johnson, J.A., and Fusaro, R.M. (1987) Protection of photosensitized rats against long ultraviolet radiation by topical application of compounds with structures similar to that of dihydroxyacetone. Dermatologica. 175:58-63.

Goldman, L., and Blaney, D.J. (1962) Dihydroxyacetone. Arch. Dermatol. 85:86-90.

Goldman, L., Barkoff, J., Blaney, D., Nakai, T., and Suskind, R. (1960) Investigative studies with the skin coloring agents dihydroxyacetone and glyoxal. J. Invest. Dermatol. 35:161-164.

Johnson, J.A., and Fusaro, R.M. (1987) Protection against long ultraviolet radiation: Topical browning agents and a new outlook. Dermatologica. 175:53-57.

Johnson, J.A., and Fusaro R.M. (1993) Therapeutic potential of dihydroxyacetone. J. Am. Acad. Dermatol. 29:284-285.

Maibach, H.I., and Kligman, A.M. (1960) Dihydroxyacetone: A suntan-simulating agent. Arch. Dermatol. 82:505-507.

Manninger, G., and Severin, T. (1976) Reaction of dihydroxyacetone with methylammoniumacetate. Studies on the Maillard-reaction, X. Z. Lebensm. Unters.-Forsch. 161:45-51.

Meybeck, A. (1977) A spectroscopic study of the reaction products of dihydroxyacetone with amino acids. J. Soc. Cosmet. Chem. 28:25-35.

Namiki, M., Hayashi, T., and Ohta, Y. (1977) Novel free radicals formed by the amino-carbonyl reactions of sugars with amino acids, amines, and proteins. Adv. Exp. Med. Biol. 86B:471-501.

Namiki, M. (1988) Chemistry of Maillard reactions: Recent studies on the browning reaction mechanism and the development of anti oxidants and mutagens. Adv. Food Res. 32:115-184.

Olsson, K., Pernemalm P.A., and Theander, O. (1981). Reaction products and mechanism in some simple model systems. Prog. Food Nutr. Sci. 5:47-55.

Rabinowitch, I.M. (1925). Observations on the use of dihydroxyacetone in the treatment of diabetes mellitus. Can. Med. Assoc. J. 15:374-381.

Rendina, A.R., Hermes, J.D., and Cleland, W.W. (1984). A novel method for determining rate constants for dehydration of aldehyde hydrates. Biochemistry. 23:5148-5156.

Wieland, T., and Severin, T. (1973) Formation of a hydroxypyrrole from methylglyoxal and methylammoniumacetate. VIII. Studies on the Maillard reaction. Z. Lebensm. tinters.-Forsch. 153:201-203.

Colour Development in an Intermediate Moisture Maillard Model System

Jennifer M. Ames, Lisa Bates, and Douglas B. MacDougall

DEPARTMENT OF FOOD SCIENCE AND TECHNOLOGY, UNIVERSITY OF READING, WHITEKNIGHTS, PO BOX 226, READING RG6 2AP, UK

Summary

A Maillard model system comprising starch, glucose and lysine monohydrochloride (96:3:1, m:m:m) was heated in a reaction cell designed to simulate the temperature, residence time, pressure and rapid heating and cooling encountered during extrusion cooking. Different temperatures (130-150°C), times (32 and 45 s), pHs (3.1-6.8) and moisture contents (13-18%) were examined. The model system was also cooked in a twin-screw corotating extruder using similar sets of conditions. CIE L*a*b* values for mixtures cooked in the extruder and the reaction cell were compared and showed similar trends. In general, the L* values decreased and the a* and b* values increased with increasing temperature, time and pH. The reaction cell has the potential to study low moisture heat processing systems (such as extrusion cooking) quickly and on a laboratory scale.

Introduction

The development of colour is an extremely important feature of the Maillard reaction but relatively little is known about the chemical nature of the compounds responsible. It is, however, possible to group the coloured compounds into two general classes: the low molecular weight colour compounds which typically comprise two to four linked rings, and the melanoidins which are brown polymers and possess molecular weights of several thousand daltons (Ames and Nursten, 1989). Due to the complexity of the chemistry underlying the Maillard reaction and the difficulties associated with the analysis of the compounds responsible for colour, studies have been largely confined to model systems. These systems normally comprise a single reducing sugar and a single amino compound heated in a solvent (usually water, methanol or ethanol) and have resulted in much useful information regarding the nature of the compounds concerned. We are particularly indebted to the late Professor Ledl and his groups at Stuttgart and Munich for their outstanding contributions in this area (e.g. Ledl, 1990; Ledl and Schleicher, 1990). The structures of only five coloured compounds formed in aqueous sugar-amino acid model systems have been elucidated (Ames, 1992; Ames et al., 1993). In addition, a great deal of spectral data has been gathered for a sixth compound which possesses four rings (including two terminal furan rings, each substituted in the 2-position), a molecular weight of 445 daltons and an empirical formula of $C_{25}H_{19}O_7N$ (Ames et al., 1993).

Several factors influence the Maillard reaction and hence colour development, including temperature and duration of heating, pH and moisture content (Ames, 1990). It is generally accepted that colour development is increased by increases in temperature and time of heating, increases in pH and by intermediate moisture contents (a_w = 0.3-0.7). The effect of pH on the profile of ethyl acetate extractable coloured compounds formed in an aqueous xylose-lysine system has been studied using HPLC with diode array detection (Ames et al., 1993; Apriyantono and Ames, 1994; Ames and Apriyantono,

1994). There are few studies reporting colour development in Maillard model systems of reduced moisture content at temperatures relevant to food processing operations, e.g., extrusion cooking (Berset, 1989) although such studies would be more relevant to most foods than investigations of aqueous solutions. There are no reports of the chemistry underlying the formation of colour compounds in reduced moisture systems.

Extrusion cooking is a high temperature, short time, high mechanical shear operation which has been applied to foods for more than fifty years. The equipment available, its use and applications have been described (Harper, 1979; 1989). The process is ideal for the manufacture of ready-to-eat cereals and expanded snack foods where it involves processing ingredients at e.g., 120-180°C at typical moisture contents of 12-18%. These conditions favour the Maillard reaction. In addition, extrusion cooking can bring about changes in the conformation of starches and proteins as well as their partial degradation (Harper, 1979). These changes can increase the availability of reactive groups which subsequently take part in the reaction. Many previous investigations have shown the effect of extrusion cooking on lysine availability but only a few have examined the effect on colour (Berset, 1989). The most useful study is that of Noguchi et al. (1982) who extrusion cooked a mixture of wheat flour, corn starch, sucrose, soya protein isolate, sodium caseinate and sodium chloride and studied the effects of various processing parameters on colour development. Decreases in moisture content and increases in temperature both resulted in a more intense colour. A previous study in our laboratory using a starch-glucose-lysine model system showed that increases in the extruder die temperature and decreases in moisture content increased colour development in the extrudates (Sgaramella and Ames, 1993). The data reported here extends that study to the effect of pH on colour development and also illustrates the use of a reaction cell to model colour development in an extruder.

Materials and Methods

Homogeneous mixtures of wheat starch type A, D-(+)-glucose and L-lysine monohydrochloride (96:3:1, m:m:m) were prepared and cooked in both an MPF50 twin screw corotating extruder (APV Baker, Peterborough, UK) and in a reaction cell. Temperature, residence time, pH and moisture content were all varied for both processes. Residence time was altered during extrusion cooking by changing the screw speed and the feed rate but keeping the specific mechanical energy constant at 0.25-0.28 kWh/kg. Variations in pH were achieved by adding citric acid or sodium bicarbonate to the dry ingredients before mixing. The design of the reaction cell and its use have been described by Bates et al. (1994). Samples from both the reaction cell and the extruder were ground and sieved, and the 250-500 μm fractions were used for colour measurements. CIE L*a*b* values were obtained in a HunterLab ColorQuest spectrophotometer. Hue angle values, h*, $\tan^{-1}(b^*/a^*)$ and chroma values, C*, $(a^{*2} + b^{*2})^{1/2}$ were calculated (Little, 1976; Hunt, 1987). SAS software was used for the statistical evaluation of the data including analysis of variance and regression analysis. Full experimental details have been reported previously (Sgaramella and Ames, 1993; Bates et al., 1994).

Results and Discussion

Initially, extrusion experiments were performed to establish which parameter (temperature, residence time, moisture content or pH) had the greatest effect on colour development. Two temperature ranges for the product in the final barrel zone (136-145°C and 146-155°C), two residence times (32 and 45 s), three total moisture levels (13, 15 and 18%) and three pH values (3.1, 5.0 and 6.8) were used. The values of the processing parameters were chosen for their relevance to extrusion cooking. The most significant effect on colour development was brought about by pH (p = 0.01 for L*). Colour development increased with increasing pH of the feed material, and the extrudate ranged in colour from pale yellow (pH = 3.1) to dark orange-brown (pH = 6.8). Starch, extruded in the absence of glucose and lysine, was white in colour. Increasing temperature, increasing residence time and decreasing moisture also increased browning during extrusion. Thus, over the ranges of values tested, increases in temperature, time and pH each increase colour development due to the Maillard reaction in a starch-glucose-lysine system of intermediate moisture contents as they do for aqueous reducing sugar-amino acid model systems. Results obtained for temperature are in line with those reported in the literature (Berset, 1989). The moisture content of the starch-glucose-lysine mixture before processing was 11.4 ± 0.1% (a_w = 0.43 ± 0.01) (Sgaramella and Ames, 1993). Since it was not possible to extrude the starch-glucose-lysine mixture at a total moisture content of less than 13%, it was not possible to establish the moisture level corresponding to maximum browning.

The reaction cell was designed to examine the colour development which takes place during extrusion cooking on a small scale. It simulated the rapid heating and cooling and the pressure encountered during extrusion but it was not able to mimic mechanical shear. Extruder trials typically required 50 kg of raw material and one day to run. In contrast, the reaction cell required 20 g of material and samples for analysis could be obtained in five minutes. Samples were heated in the reaction cell using temperatures of 130, 140 and 150°C and the same values for time, moisture and pH as for the extrusion experiments. Again, pH resulted in the most significant effect on colour development, followed by temperature, moisture and, lastly, by the time of heating (see Table I). The b* coordinate was more significantly affected by modifications to the processing conditions than the a* value. Thus, as browning developed, lightness decreased as chroma increased but not with constant hue angle because of the different rates of increase in a* and b*. The hue became more orange-brown than yellow-brown as shown by the decrease in h* as the material darkened.

Table I. Statistical data for L*a*b* data for reaction cell samples.

VARIABLES & DEGREES OF FREEDOM		L*		a*		b*		h*	
		R-Square: 0.93		R-Square: 0.95		R-Square: 0.98		R-Square: 0.96	
		F Value	P[a]	F Value	P[a]	F Value	P[a]	F Value	P[a]
pH	2	107.49	***	191.71	***	419.93	***	241.68	***
Temperature	2	39.37	***	37.13	***	48.29	***	31.40	***
Moisture	2	10.27	***	0.13	NS	10.00	***	7.41	**
Time	1	3.60	NS	5.24	*	9.69	**	14.38	***

P[a] (significance level) = *** at $p<0.001$, ** at $p<0.01$, * at $p<0.05$, NS not significant.

Of all the variables examined, pH clearly had the greatest influence on colour development in the starch-glucose-lysine system heated in both the extruder and the reaction cell. Therefore, further reaction cell experiments were performed over a greater number of pH values, i.e., 3.1, 4.5, 5.0, 6.3, 6.7 and 6.8, but keeping the other parameters constant at 140°C, 32 s and 15% moisture. The L*, a*, b*, h* and C* values are shown in Table II. For both methods of heating chroma increased with increasing pH but more colour developed in the extruded samples than in those heated in the reaction cell (see Figure 1). This may be due to high shear mixing in the extruder. Mechanical shear can result in starch degradation, open up the starch structure and give an increase in reducing sugars. The slope of the plot of C* against pH is greater for the reaction cell samples than for the extrudates. This is mainly due to a greater rate of increase in b* for the reaction cell samples. The slopes of the plots of L* and a* against pH were approximately the same for the extrudates and the reaction cell samples. Table II also shows that the variation in L*a*b* responses between replicates was greater at low pH for the extruded samples and at higher pH for the reaction cell samples. As a result of the increased colour development in the extruded samples, the L*a*b* values obtained for the extruded samples at low pH were similar to those obtained for the reaction cell samples at high pH and it is suggested that the relatively large variation in L*a*b* responses at these conditions may correspond to a critical stage of colour development during the Maillard reaction.

In spite of some anomalies between the L*a*b* data obtained for the extrudates and the reaction cell samples, linear regression of the L*and a* responses for the reaction cell samples shown in Table II was used to predict the L* and a* values for extrudates prepared at pH 6.3. The predicted values of 63.9 for L* and 5.6 for a* lie within the experimental ranges of 60.5-70.2 (mean 65.4) for L* and 5.2-7.3 (mean 6.3) for a*.

Table II. Mean response values for extrudates and reaction cell samples.

pH	L*		a*		b*		h*		C*	
	Mean	S.D.	Mean	S.D.	Mean	S.D.	Mean	S.D.	Mean	S.D.
EXTRUSION										
3.1	70.31	6.84	3.14	2.23	16.47	1.63	79.61	6.47	16.81	2.01
5.0	65.08	1.68	5.52	0.64	19.53	0.56	74.21	2.15	20.31	0.38
6.8	57.14	1.55	8.21	0.35	18.37	2.04	65.82	1.84	20.12	1.97
REACTION CELL										
3.1	85.11	0.38	-0.29	0.05	4.11	0.23	94.04	0.51	4.12	0.23
4.5	85.95	0.40	-0.15	0.06	5.53	0.48	91.52	0.76	5.54	0.48
5.0	84.24	0.98	0.48	0.10	8.88	0.80	86.89	0.57	8.89	0.80
6.3	80.28	0.99	1.40	0.51	13.26	1.33	84.11	1.66	13.34	1.37
6.7	77.89	2.56	2.64	0.81	15.31	1.99	80.42	2.05	15.54	2.09
6.8	74.82	2.66	3.66	0.92	16.93	1.42	77.93	2.12	17.33	1.56

Conclusions

The colour of a starch-glucose-lysine mixture subjected to extrusion cooking and heating in a reaction cell changes from pale yellow to dark orange-brown on increasing the pH of the mixture from 3.1 to 6.8 and this is reflected by the L*a*b* responses. The results obtained may be due to different profiles of colour compounds formed or to differences in intensity of the same profile of compounds at the different pH values. This will be established by isolating and analysing the colour compounds which is the next stage of the project.

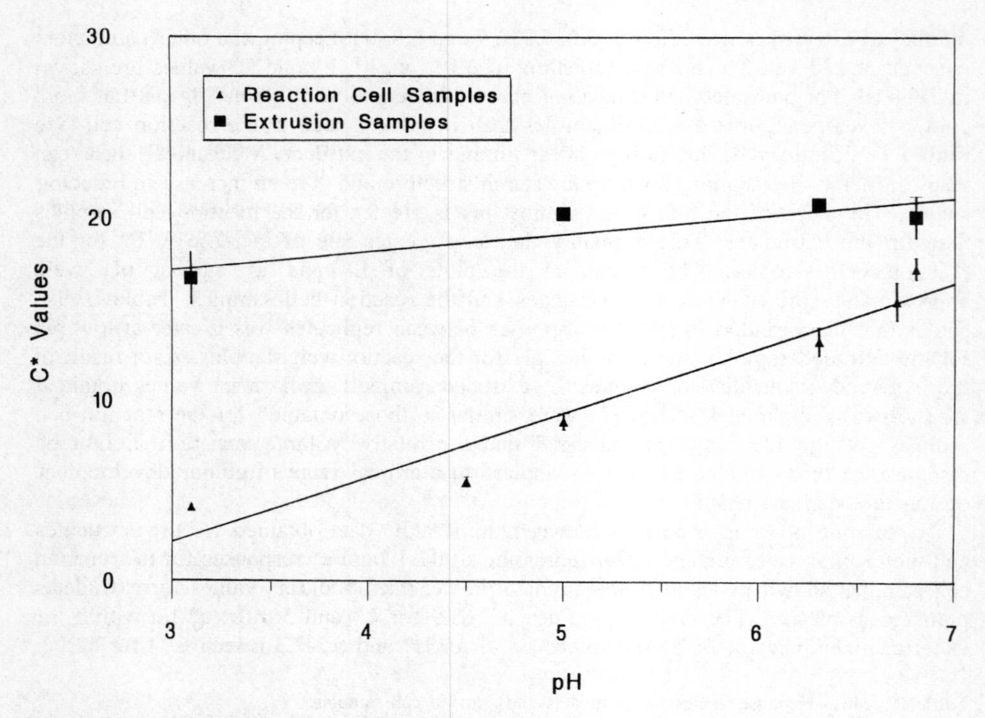

Figure 1. Changes in C* values with increasing pH.

Acknowledgements
The AFRC, RHM Research and Engineering Ltd., Leatherhead Food RA and APV Baker are thanked for financial support. The expertise of Mr G. Adcock, APV Baker; Mr A. Brooks, RHM Research and Engineering Ltd, Dr. S. Jones, Leatherhead Food RA and Professor C.R. Chaplin and Mr. N.D. Cowell, University of Reading, is gratefully acknowledged. The Royal Society and the Research Board of the University of Reading are thanked for travel assistance.

References

Ames, J.M. (1990) Control of the Maillard reaction in food systems. Trends in Food Sci. Technol. 1: 150-154.

Ames, J.M. (1992) The Maillard Reaction. In Biochemistry of Food Proteins B.J.F. Hudson (ed.) Elsevier Applied Science, London. pp. 99-153.

Ames, J.M., and Apriyantono, A. (1994) Comparison of the non-volatile ethyl acetate extractable reaction products formed in a xylose-lysine model system heated with and without pH control. Food Chem. (in press).

Ames, J.M., Apriyantono, A., and Arnoldi, A. (1993) Low molecular weight coloured compounds formed in xylose-lysine model systems. Food Chem. 46: 121-127.

Apriyantono, A., and Ames, J.M. (1994) Low molecular weight coloured compounds formed in xylose-lysine model systems heated with and without pH control. In Proceedings of this symposium, p 410.

Ames, J.M., and Nursten, H.E. (1989) Recent advances in the chemistry of coloured compounds formed during the Maillard reaction. In Trends in Food Science W.S. Lien and C.W. Foo (eds.) Singapore Institute of Food Science and Technology, pp. 8-14.

Bates, L., Ames, J.M., and MacDougall, D.B. (1994) The use of a reaction cell to model the development and control of colour in extrusion cooked foods. Lebensm. Wiss. u. Technol. (in press).

Berset, C. (1989) Colour. In Extrusion Cooking C. Mercier, P. Linko and J.M. Harper (eds.) Amer. Assoc. Cereal Chem., St Paul, MN, USA, pp. 371-85.

Harper, J.M. (1979) Food extrusion. CRC Crit. Rev. Food Sci. Nutr. 11: 155-215.

Harper, J.M. (1989) Food extruders and their applications. In Extrusion Cooking Mercier, P. Linko and J.M. Harper (eds.) Amer. Assoc. Cereal Chem., St Paul, MN, USA, pp. 1-15.

Hunt, R.W.G. (1987) Measuring Colour. Ellis Horwood Ltd, Chichester, UK.

Ledl, F. (1990) Chemical pathways of the Maillard reaction. In The Maillard Reaction in Food Processing, Human Nutrition and Physiology P.A. Finot, H.U. Aeschbacher, R.F. Hurrell and R. Liardon (eds.) Birkhauser, Basel, pp. 19-42.

Ledl, F., and Schleicher, E. (1990) New aspects of the Maillard reaction in foods and in the human body. Angew. Chem. Int. Engl. 29: 565-594.

Little, A.C. (1976) Physical measurements as predictors of physical appearance. Food Technol., October: 74-82.

Noguchi, A., Mosso, K., Aymanrod, C., Jeunink, J., and Cheftel, J.C. (1982) Maillard reactions during extrusion cooking of protein-enriched biscuits. Lebensm. Wiss. u. Technol. 15: 105-110.

Sgaramella, S., and Ames, J.M. (1993) The development and control of colour in extrusion cooked foods. Food Chem. 46: 129-132.

Release of Ammonia from Peptides and Proteins and Their Effects on Maillard Flavor Generation

Chi-Tang Ho, Jie Zhang, Hui-Ing Hwang, and William E. Riha, III

DEPARTMENT OF FOOD SCIENCE AND CENTER FOR ADVANCED FOOD TECHNOLOGY, COOK COLLEGE, NEW JERSEY AGRICULTURAL EXPERIMENT STATION, RUTGERS, THE STATE UNIVERSITY OF NEW JERSEY, NEW BRUNSWICK, NJ 08903, USA

Summary
Ammonia has been recognized as an important reactant for food flavor development. Ammonia could be produced by thermal and hydrolytic deamidation of asparagine and glutamine residues of proteins. The contribution of amide nitrogen to pyrazine formation was confirmed by the use of glutamine labeled with ^{15}N at the amide side chain. The comparison of volatile compounds produced from the reaction of glucose with either dipeptide Ala-Asn or Ala-Asp also indicated the contribution of deamidation to the Maillard reaction.

Introduction

The nonenzymatic deamidation of proteins is a reaction which has recently captured the interest of food processors trying to improve the functionality of food proteins and understand the changes which occur during processing. However, this reaction has been recognized by biochemists for many years (Wright, 1991). Deamidation is described as the loss of the amide function of a glutamine (Gln) or asparagine (Asn) side chain resulting in the formation of glutamic and aspartic acid. As a consequence of this reaction, free ammonia is liberated from the protein.

Deamidation may take place via one of two mechanisms. The first mechanism is the direct hydrolysis of the amide bond in which an acid or base acts as a catalyst. The second mechanism, known as β-shift involves the formation of a succinimate intermediate. There are a number of factors which may influence the reaction rate and mechanism such as pH, water activity, amino acid sequence of the protein, available ions, and nonionic catalysts.

Since the deamidation of an amide residue in proteins results in the release of a molecule of ammonia, one can speculate that ammonia would react with reducing sugars in much the same role as free amino acids. In recent studies, we have shown that ammonia can contribute to the Maillard browning reactions in complex systems (Izzo and Ho, 1992).

Contribution of Deamidation Reaction to Pyrazine Formation in a Glutamine Model System

Pyrazines are nitrogen-containing heterocyclic compounds that are potent characteristic flavorants found in a wide range of raw and processed foods (Maga, 1992). Pyrazines are usually associated with the generation of roasted and burnt flavor notes. These unique and desirable sensory properties make pyrazines essential to the food industry.

As shown in the past, glutamine and asparagine have been shown to produce considerably more pyrazines than do their corresponding acids when heated with reducing sugars (Koehler et al., 1969). This seems to suggest that the amide nitrogen, possibly through deamidation, is available to contribute to amino/carbonyl interactions leaking to pyrazine generation. In order to clarify the participation of the deamidation reaction in pyrazine formation, we used glutamine with a labeled ^{15}N isotope at the amide side chain and a ^{14}N at the α-amino group to investigate the relative contribution of the α-amino nitrogen and the amide nitrogen to pyrazine formation.

Fifty g of wheat starch, 500 mg of glucose, and 100 mg of L-glutamine/or L-glutamine-^{15}N *(amide-*^{15}N*)* were mixed with 500 mL deionized water and then freeze-dried. The moisture content of the solid mixture was adjusted to 12-14%. The samples were then transferred into a reaction vessel and heated at 180°C for one hour. The volatile compounds generated from the reaction were analyzed by purge-and-trap short path thermal desorption GC-MS (Hartman et al., 1993).

The pyrazines that were identified from heating isotope-labeled glutamine with glucose in the dry system included pyrazine, methylpyrazine, ethylpyrazine, 2,5-dimethylpyrazine, 2,6-dimethylpyrazine, 2,3-dimethylpyrazine, vinylpyrazine, and 2-ethyl-5-methylpyrazine. The relative contributions of amide nitrogen to pyrazine formation are shown in Figure 1. From this figure it is obviously revealed that more than half of the pyrazines consisted of ^{15}N nitrogen atoms that came from the amide chains of glutamine. These data demonstrated that deamidation did happen and could participate in the pyrazine formation. Similar results were also investigated by Bohnenstengel and Baltes (1992). They found that in the asparagine and glucose mixtures, predominantly nitrogen-containing heterocycles (pyrazines, pyridines, pyrroles) were formed, whereas the mixtures of aspartic acid/ glucose yielded furans and aliphatic carbonyls. These results

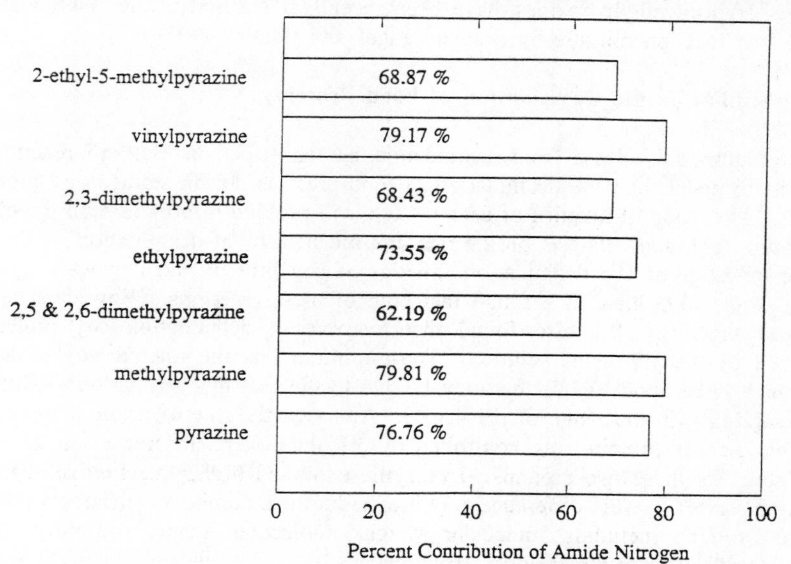

Figure 1. The relative contribution of amide nitrogen to pyrazine formation in the reaction of labeled glutamine with glucose.

further implied that ammonia from deamidating the side chains of glutamine more easily reacted with dicarbonyls than the α-amino groups of glutamine. This was reasonable because the α-amino groups could not directly react with dicarbonyls like ammonia and had to proceed through Strecker degradation to generate pyrazines. These results supported the hypothesis that deamidation has a profound effect on pyrazine generation.

Effect of Deamidation Reaction on Pyrazine Formation in Peptide Model System

Maillard reactions of dipeptides, Ala-Gln and Ala-Glu, and glucose were studied. 0.005 mol of each peptide was dissolved with an equal mole of glucose in 100 mL deionized water. The pH of each solution mixture was adjusted to 8 and heated at 180°C for two hours. These reaction mixtures were further adjusted back to alkaline pH and extracted with methylene chloride. After drying over anhydrous sodium sulfate, the methylene chloride extracts were concentrated down to 0.2 mL by using a stream of nitrogen gas. These volatiles were analyzed by GC and GC/MS and quantified by the internal standard p-cymene. Both dipeptides generated significant amounts of pyrazines. Table I lists pyrazines identified in these two systems. It is interesting to observe that Ala-Asn generated higher amounts of unsubstituted pyrazine and methylpyrazine, and Ala-Asp produced greater amounts of higher substituted pyrazines such as 2,5-dimethyl-3-ethylpyrazine. Shibamoto (1977) has demonstrated that the reaction between ammonia and sugar produced simply substituted pyrazines such as methylpyrazine and ethylpyrazine. It is, therefore, reasonable to explain that Ala-Asn can release ammonia upon heating which in turn will lead to the formation of more unsubstituted pyrazine and methylpyrazine.

It is also interesting to observe the presence of hydroxyethylpyrazines in these model systems. Hydroxyethylpyrazines have been recently identified in the Asp/glucose and Asn/glucose reaction mixture by Bohnenstengel and Baltes (1992).

Kinetic Studies on the Deamidation of Food Proteins

Since the ammonia release from deamidation has an effect on Maillard reaction, it is important to understand the deamidation of food proteins during storage and processing of foods. The study of kinetics of deamidation will provide information on the effect of temperature, pH, and catalysts on the rate and mechanism of deamidation.

We investigated the deamidation kinetics of soy protein and egg white lysozyme (Zhang et al., 1993 a,b). We found that both of these reactions followed an apparent first-order kinetics. We also found that the rate of deamidation soy protein was influenced by the pH of the solution. The minimum reaction rate for soy protein was determined to be about pH 5, this may be due to the fact that soy protein exhibits the lowest solubility in the range of pH 4 to 5. Although the rate of deamidation for both lysozyme and soy protein was controlled by pH, there were differences in the rates of deamidation for these two proteins. Lysozyme exhibited higher reaction rates than soy protein above pH 5. This difference may also bedue to a number of differences between the two proteins including molecular weight, molecular shape, solubility, primary sequence and the three-dimensional structure.

Table I. The concentration of pyrazines identified in the Ala-Asn/glucose and Ala-Asp/glucose model systems.

Pyrazine identified	ppm	
	Ala-Asn/glucose	Ala-Asp/glucose
pyrazine	168	21
methylpyrazine	2241	878
2,5-dimethylpyrazine	1189	1180
ethylpyrazine	36	23
2,3-dimethylpyrazine	36	34
vinylpyrazine	2	3
2-ethyl-6-methylpyrazine	35	25
2-ethyl-5-methylpyrazine	14	9
trimethylpyrazine	30	36
2-vinyl-6-methylpyrazine	45	36
3-ethyl-2,5-dimethylpyrazine	23	1747
2-ethyl-3,5-dimethylpyrazine	14	39
2-methyl-6-(1-propenyl)pyrazine	30	60
5,6-dimethyl-2-vinylpyrazine	18	54
5-n-butyl-2,3-dimethylpyrazine	9	68
3-(1-methylpropyl)-2,5-dimethylpyrazine	7	442
3-(2-methylpropyl)-2,5-dimethylpyrazine	57	362
2-hydroxyethyl-5-methylpyrazine	128	117
2-hydroxyethyl-5,6-dimethylpyrazine	10	11
2-hydroxyethyl-3,5,6-trimethylpyrazine	11	58
Total	4103	5203

Acknowledgements
This is publication No. F-10535-1-93 of the New Jersey Agricultural Experiment Station supported by the State Funds and the Center for Advanced Food Technology (CAFT). The Center for Advanced Technology is a New Jersey Commission on Science and Technology Center. This work was also supported in part by the U.S. Army Research Office.

References

Hartman, T.G., Lech, J., Karmas, K., Salinas, J., Rosen, R.T., and Ho, C.-T. (1993) Flavor characterization using adsorbent trapping-thermal desorption or direct thermal desorption-gas chromatography and gas chromatography-mass spectrometry. In *Flavor measurement*, Edited by C.-T. Ho and C.H. Manley, Marcel Dekker Inc.: New York, pp 37-60.

Bohnenstengel, C., and Baltes, W. (1992) Model reactions on roast aroma formation. XII. Reaction of glucose with aspartic acid or asparagine at three different temperatures. *Z. Lebensm. Unters. Forsch.* 194: 366-371.

Izzo, H.V., and Ho, C.-T. (1992) Ammonia affects Maillard chemistry of an extruded autolyzed yeast extract: Pyrazine aroma generation and brown color formation. *J. Food Sci.* 57: 657-659 & 674.

Koehler, P.E., Mason, M.E., and Newell, J.A. (1969) Formation of pyrazine compounds in sugar-amino acid model systems *J. Agric. Food Chem.* 17: 393-396.

Maga, J.A. (1992) Pyrazine update. *Food Rev. Internat.* 8: 479-558.

Shibamoto, T., Benhard, R.A. (1977) Investigation of pyrazine formation pathways in glucose-ammonia model systems *J. Agric. Food Chem.* 25; 609-614.

Wright, H.T. (1991) Nonenzymatic deamidation of asparaginyl and glutaminyl residues in proteins. *Crit. Rev. Biochem. Mol. Biol.* 26: 1-51.

Zhang, J., Lee, T.C.and Ho, C.-T. (1993) Kinetics and mechanism of nonenzymatic deamidation of soy protein. *J. Food Processing Preservation.* (in press).

Zhang, J., Lee, T.C., and Ho, C.-T. (1993) Comparative study on kinetics of nonenzymatic deamidation of soy protein and egg white lysozyme. *J. Agric. Food Chem.* (in press).

The Reaction Kinetics for the Formation of Isovaleraldehyde, 2-Acetyl-1-pyrroline, di(H)di(OH)-6-Methylpyranone, Phenylacetaldehyde, 5-Methyl-2-phenyl-2-hexenal, and 2-Acetylfuran in Model Systems

F. Chan and G. A. Reineccius

DEPARTMENT OF FOOD SCIENCE AND NUTRITION, UNIVERSITY OF MINNESOTA, 1334 ECKLES AVENUE, ST. PAUL, MN 55108, USA

Summary
A kinetic study on the formation of the Strecker aldehydes, isovaleraldehyde and phenylacetaldehyde, plus four additional, "secondary" aroma compounds was conducted. A gas chromatograph atomic emission detector was used to collect kinetic data on the formation of these volatiles using a glucose (0.5 mole) and amino acid (0.075 mole) model system dissolved in 400 mL phosphate buffer (pH 6, 7 & 8) that had been heated at 75 to 115°C from 5 min. to 7.5 h. The formation of these volatiles followed zero order reaction kinetics and activation energies ranged from 14.3 to 23.1 kcal/mole.

Introduction

The Maillard reaction is one of the most important reactions providing flavor to heated foods (Maarse and Visscher 1989a,b,c; Maarse 1991; Parliment et al., 1989, 1994; Finot et al., 1989). Unfortunately, it is also responsible for some undesirable flavors occurring during storage or overheating of food (Heath and Reineccius, 1986). The Strecker degradation is a minor pathway of the Maillard reaction but it is very important in the formation of many volatile organic flavor compounds. The Strecker degradation involves the oxidative deamination and decarboxylation of an α-amino acid in the presence of a dicarbonyl compound. The products formed from this reaction are an aldehyde containing one less carbon than it's original amino acid and an α-aminoketone. The interaction of Strecker degradation products with other Maillard reaction products leads to the formation of many important classes of flavor compounds (e.g. aldehydes, pyrazines, oxazoles and thiophenes).

In the course of our work in this laboratory, we have noticed that everytime we have done flavor isolation and aromagram work from products that have been even minimally heated, we have always found caramel notes (isovaleraldehyde), baked potato aroma (methional) and floral notes (phenylacetaldehyde). This gave rise to the question why were they always smelled. It could be that they have a low sensory threshold, are easily formed or a combination of both. A literature search revealed that there has been little kinetic work done on the Maillard reaction and, specifically, no kinetic work on the Strecker degradation.

This kinetic study focuses on the reaction kinetics of formation for two Strecker aldehydes and four related Maillard reaction products. Even though numerous other volatiles were detected, this paper will present information only on those that yielded sufficient data to permit kinetic analysis.

Materials and Methods

Sample Preparation

Glucose (0.5 mole = 90 g; Aldrich Chemical, Milwaukee, WI), methionine (0.075 mole = 11.19 g; Sigma Chemical, St. Louis, MO), phenylalanine (0.075 mole = 12.39 g; Sigma Chemical, St. Louis, MO), proline (0.075 mole = 8.64 g; Aldrich Chemical, Milwaukee,

WI) and leucine (0.075 mole = 9.84 g; Aldrich Chemical, Milwaukee, WI) were dissolved in 400 mL of distilled water (Glenwood Inglewood, Minneapolis, MN) and pH adjusted to 6, 7 or 8 with a 0.1M phosphate buffer and the appropriate amount of NaOH. A 600 mL pressure reactor with a temperature controller (Parr 4563, Parr 4842; Parr Instrument Co., Moline, IL) was filled with the entire solution, sealed and heating and stirring started. Initial temperature was noted and start time determined when the solutions reached reaction temperatures. Sample (50 mL) was withdrawn from the reactor via a sampling port at the following schedule:

75°C	1.5, 3.0, 4.5, 6.0, 7.5 h
85°C	0.5, 1.0, 1.5, 2.0, 2.5 h
95°C	20, 40, 60, 100, 120 min.
105°C	10, 20, 30, 40, 50 min.
115°C	5, 10, 15, 20, 25 min.

Ultra high purity N_2 (ca. 50 mL) was added after each sampling to restore pressure to the reactor. Each sample was cooled to room temperature in an ice bath, the pH measured and the sample extracted with 3 x 5 mL of dichloromethane (EM Science, Gibbstown, NJ) containing 500 ppm of 4-methylthiazole (Aldrich Chemical, Milwaukee, WI) as an internal standard in a 150 mL separatory funnel. The CH_2Cl_2 fraction was dried with anhydrous magnesium sulfate (Fisher Scientific, Fairlawn, NJ), filtered and evaporated to 0.5 to 1.0 mL volume under a stream of high purity nitrogen. Duplicates at each temperatue and pH series was done.

Gas Chromatography/Mass Spectrometry
A Hewlett Packard (HP) Model 5890 gas chromatograph (GC) (Hewlett Packard, Avondale, PA) equipped with a mass selective detector (MSD) HP Model 5970 was used to identify compounds in this study. The capillary column was interfaced directly into the MSD operating at 70 eV ionization potential, with an ion source temperature of 220°C and a scan threshold of 750, scanning from m/z 29 to 400 at 0.86 s/cycle. GC separation was achieved on a 30m x 0.32mm i.d. x 1μm film thickness DB-5 fused silica capillary column (J & W Scientific, Folsom, CA). All injections were performed under the following conditions: injection port temperature 230°C, initial column temperature 40°C, initial time 3 min., final temperature 250°C, final time 10 min. and an oven ramp rate of 5°C/min. Helium was used as a carrier gas at 7 psi of head pressure, a 20:1 inlet split and 1.0 μL of the extract was injected. The compounds studied were identified by comparing the mass spectrum obtained with the National Bureau of Standards Mass Spectra library, published retention indices and cochromatography with authentic compounds.

Gas Chromatography/Atomic Emission Detection
A HP Model 5890 Series II GC equipped with an Atomic Emission Detector (AED) HP Model 5921A (sulfur, carbon and nitrogen signal) was used to collect kinetic data. Compounds were separated using the same column and GC operating conditions as noted above for the GC/MS. An injection port temperature of 250°C and AED transfer line temperature 275°C were used. Helium (ca. 2 mL/min.) was used as a carrier gas in the system at 15 psi of head pressure. For the AED detector, the flow rates of the reagent gases are based on Fox and Wylie (1989). For the carbon (193 nm), nitrogen (174 nm) and sulfur signal (181 nm), O_2 and H_2 were used as scavenger gases and the makeup flow was 30 mL/min. A split ratio of 45:1 was used and injection size was 1.0 μL.

Results and Discussion

General Observations
The amount of volatile compounds formed increased both in number and quantity as either time or temperature of heating increased. This result is expected based on previous studies too numerous to cite (e.g. Leahy and Reineccius 1989a,b; Shu and Ho, 1989; Shaw and Ho 1989; Baltes et al., 1989).

The quantity of volatiles produced generally increased as pH increased from 6 to 8. One would expect this relationship since the Maillard reaction occurs between the uncharged amine and a carbonyl. The pH influences the amount of uncharged amine and thus the rate of Maillard browning. While generalizations on pH effects can be made for total volatiles, the effect of pH on individual volatile compounds cannot readily be made. The formation of some volatiles may be enhanced at low pH's while the formation of other volatiles may be retarded at low pH's (e.g. Shu and Ho, 1989).

Individual Compounds
The concentration of the Strecker aldehyde, isovaleraldehyde, found in the model system during heating was higher at pH 7 than pH 8. While one would expect the rate of formation to increase with pH, one must note that isovaleraldehyde is not an end product of the reaction but an intermediate. The amounts determined in this study at any given time are the net result of formation and consumption through further reaction. Thus the greater amount of isovaleraldehyde found during heating at pH 7 may be the result of less losses of isovaleraldehyde through secondary reactions.

The concentration of the Strecker aldehyde, phenylacetaldehyde, found in the model system during heating was highest at pH 6. This would again appear contrary to expectations. However, Lindsay (1985) has shown that phenylacetaldehyde can react with isovaleraldehyde in an aldol condensation reaction to form 5-methyl-2-phenyl-2-hexenal. It is likely that pH 6 does not favor the aldol condensation reaction but pH 7 and 8 do. This hypothesis is supported by the data which show that the concentration of 5-methyl-2-phenyl-2-hexenal was found to be higher at pH 7 and 8.

The formation of the secondary reaction product of proline, 2-acetyl-1-pyrroline, (Schieberle, 1989) followed the general trend of increasing with time and pH but it appeared that there was little difference between pHs 6 and 7. The kinetics of the Maillard products in a proline-glucose model system have also been investigated by Stahl and Parliment (1994). Studies by previous workers (Leahy and Reineccius, 1989a,b) on the formation of pyrazines found that rate of formation of pyrazine, 2-methylpyrazine and dimethylpyrazine increased with either an increase in temperature or pH. These compounds are secondary products of the Maillard reaction and their rate of formation would be strongly influenced by the overall rate of the Maillard reaction.

The formation of the sugar fragmentation products (furans and pyranones) followed the general trend of increasing with time and pH. This was especially true for the formation of 2-acetylfuran which increased with temperature, pH and time. The formation of di(H)di(OH)-6-methylpyranone only began to occur at temperatures of 95°C or higher and it's formation increased as temperature, time and pH increased. It should be noted that the kinetics of the sulfur-containing compounds in this study and the GC/AED have been discussed in a separate paper (Chan and Reineccius, 1994).

Kinetic Analysis
Before discussing the kinetic data, it is worthwhile to make the point that the kinetic data obtained in this study are from a complex model system (i.e., there was more than one amino acid present in the model system). This is likely less significant for the data on Strecker aldehydes (i.e., phenylacetaldehyde and isovaleraldehyde) than it would be for secondary volatiles (2-acetyl-1-pyrroline) and sugar fragmentation products (2-acetylfuran and di(H)di(OH)-6-methylpyranone). The secondary volatiles are those which are formed from the reaction of two or more Maillard products. In our system, one reactant may come from the degradation of one amino acid while another may come from a different amino acid. Thus including a very reactive amino acid in a model system will rapidly give many reactive fragments that will greatly change the kinetics of the formation of a secondary volatile versus having the system composed of a single amino acid or amino acids which are very slow in reaction. One can readily see the same dependence for sugar fragmentation products since the amino acids act as catalysts for sugar degradation. Therefore the kinetics observed for secondary reaction products or sugar fragmentation products in any study are

very system dependent. This will become evident when the current data are compared to previous studies.

An additional kinetic consideration has already been noted. The amount of a volatile found in the system at any given time is the net result of formation and further reaction. Thus the kinetics observed are the dynamic result of formation vs consumption via further reaction. Since we know so little about the reactions occurring in the Maillard reaction, we can do little other than accept an "observed" reaction rate and the resultant kinetics.

The Macintosh program "Water Analyzer Series - Reaction Kinetics Program Version 2.09" (Labuza et al., 1991) was used to do the kinetic analysis of the data. The reaction order for the formation of volatiles was determined by plotting quantity formed vs time and then using linear regression to determine how well the data fit the straight line. The r^2 for all the compounds except 5-methyl-2-phenyl-2-hexenal at zero order were consistently better than any other order. The rate constants were obtained by calculating the slope of the quantity formed vs. time plots. The rate constant is a characteristic of each particular reaction, it's value depends on the conditions of the reaction, especially the reaction temperature. This temperature dependence can be expressed with the Arrhenius equation and the Arrhenius equation was used to determine the activation energy (Table I).

Table I. Average activation energies for the Strecker aldehydes and Maillard reaction products

Compound	Activation Energy (kcal/mole)
2-Acetyl-1-pyrroline	14.4
di(H)di(OH)-6-Methylpyranone	16.1
2-Acetylfuran	17.7
Isovaleraldehyde	19.2
Phenylacetaldehyde	21.5
5-Methyl-2-phenyl-2-hexenal	23.1

There was good reproducibility between the first and second replicate but no overall trend between activation energy and pH was observed for any of the compounds. When the 95% confidence limits are taken into account, there is essentially no difference between pH 6, 7 and 8. The data were also analyzed by an analysis of covariance method using "MacAnova 3.1" (Oehlert, 1993) to determine if replicates, pH and temperature:pH had a significant effect ($p \leq 0.05$) on the rate of formation. It was found that there was a significant difference between the 2-acetylfuran and 5-methyl-2-phenyl-2-hexenal replicates and the temperature:pH effect was not significant. It showed that pH had a significant effect on the rate of formation for all compounds except 2-acetylfuran, phenylacetaldehyde and 5-methyl-2-phenyl-2-hexenal. This indicates that pH does have an effect on formation of most of the volatiles studied and the contradiction with the relationship between E_a and pH was probably caused by the small degrees of freedom associated with using the 95% confidence intervals.

Leahy's (1985) study on the formation of pyrazine and 2-methylpyrazine at pH 5, 7 and 9 found that the activation energy was highest at pH 5 and lowest at pH 7. The study had rate constant data at three temperatures and three data points at each temperature which are inadequate for statistical calculations (i.e., confidence interval calculations). The activation energies found in the current study ranged from 14 to 23 kcal/mole. There is little data in the literature for comparison. Schirlé-Keller and Reineccius (1990) studied the formation of oxygen containing heterocyclic compounds and found that their activation energies ranged from 28 to 33 kcal/mole. When these numbers are compared with ours (Table II), there are substantial differences. The model system used by Schirlé-Keller and Reineccius (1990) had no buffer [phosphate has been reported by Potman and van Wijk (1989) to increase the rate of the Maillard reaction], used a single and different amino acid and sugar/amino acid molar ratio. Oxygen-containing heterocyclic compounds such as furans and pyranones are considered to be sugar fragmentation products and not products of amino acid degradation. As noted earlier, their formation would be catalyzed by amino acids and their rate of formation would be related to the reaction rate of sugar with the most reactive amino acid. One would expect to see major variations in the kinetics of the formation of these compounds because of their extreme system dependency. This is illustrated by the comparison of our data with that of Schirlé-Keller and Reineccius (1990).

Table II. Comparison of our activation energy data with Schirlé-Keller and Reineccius (1990)

	Published[1]	**Current study**
2-Acetylfuran	36.2 kcal/mole	17.7 kcal/mole
di(H)di(OH)-6-Methylpyranone	30.7 kcal/mole	16.0 kcal/mole

[1] - data of Schirlé-Keller and Reineccius (1990)

Leahy and Reineccius (1989a, b) study on the formation of pyrazines such as pyrazine, 2-methylpyrazine and dimethylpyrazine found activation energies to range from 27 to 45 kcal/mole. These values are much higher than were found by either us or Schirlé-Keller and Reineccius (1990) for nonpyrazine volatiles. While the high values are expected based on sensory data which has shown that the roasted, toasted notes associated with pyrazines are generally found only at high cooking temperatures, they also are system dependent since Rizzi (1987), Biede and Hammond (1979) and Griffith and Hammond (1989) have found pyrazines to be formed at low temperatures.

Formation of 5-methyl-2-phenyl-2-hexenal

It was observed that pH had a strong influence on the formation of 5-methyl-2-phenyl-2-hexenal which is the aldol condensation product of isovaleraldehyde and phenylacetaldehyde. There was substantially more 5-methyl-2-phenyl-2-hexenal formed at pH 7 and 8 than pH 6. It should also be noted that the concentration of 5-methyl-2-phenyl-2-hexenal reached a maximum at 20 min. heating and decreased with continued heating demonstrating that 5-methyl-2-phenyl-2-hexenal is not an end product of the reaction but yet another intermediate in the overall Maillard reaction. The results obtained showed that while the formation of 5-methyl-2-phenyl-2-hexenal was favored at pH 7 and 8 as compared to pH 6, the fate of the initial reactants was strikingly dissimilar. It was found that the reaction/degradation of phenylacetaldehyde was lowest at pH 6 and increased with increasing pH. This was similar to our observation in our initial experiment where we found that the accumulation/formation of phenylacetaldehyde was highest at pH 6 and decreased with increasing pH. It was found that the concentration of isovaleraldehyde in the reaction mixture did not vary with pH's. This was unexpected because 5-methyl-2-phenyl-2-hexenal is formed from the condensation reaction between isovaleraldehyde and phenylacetaldehyde and while the levels of phenylacetaldehyde went down, the levels of isovaleraldehyde did not change significantly (Figure 1-3). This suggests that once the initial products from the Maillard reaction are formed, it doesn't take much energy (i.e. E_a) for the formation of subsequent intermediates or end products. Another observation that bolstered this hypothesis was that a zero time extract had been taken for the pH 7 sample to serve as an anchor point and it revealed that the condensation between isovaleraldehyde and phenylacetaldehyde was extremely reactive because at room temperature and zero time, already 151 ppm of 5-methyl-2-phenyl-2-hexenal was found.

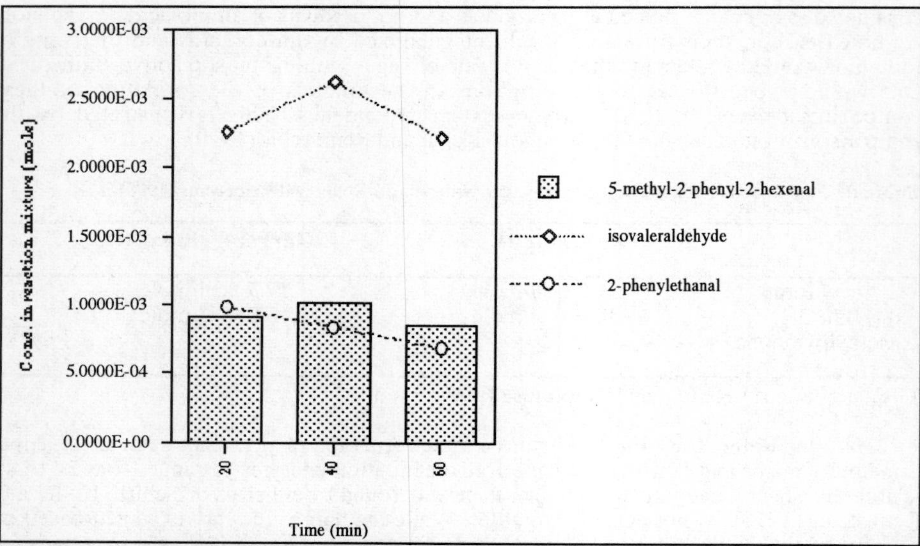

Fig. 1. Reaction of isovaleraldehyde and 2-phenylethanal at 95°C-pH 6.

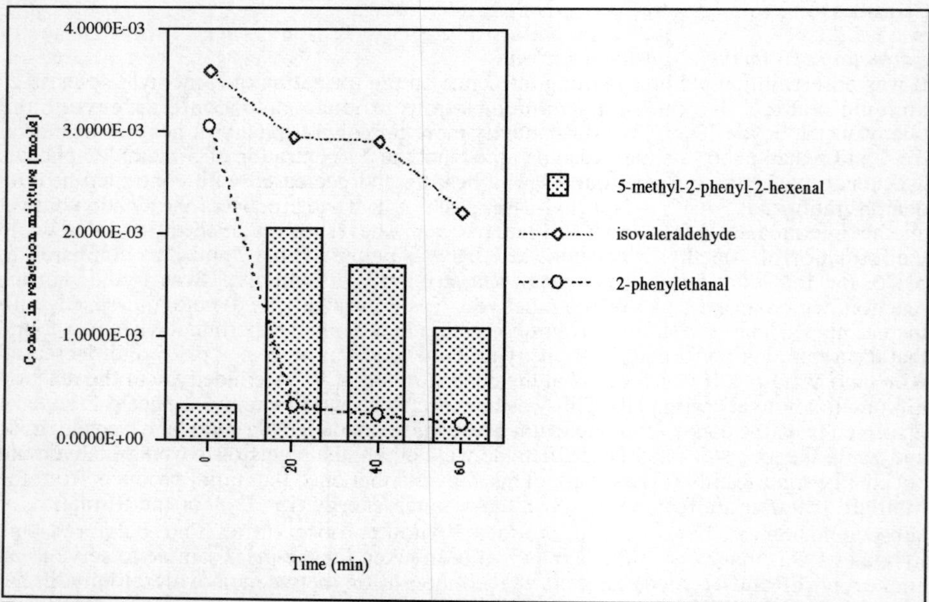

Fig. 2. Reaction of isovaleraldehyde and 2-phenylethanal at 95°C-pH 7.

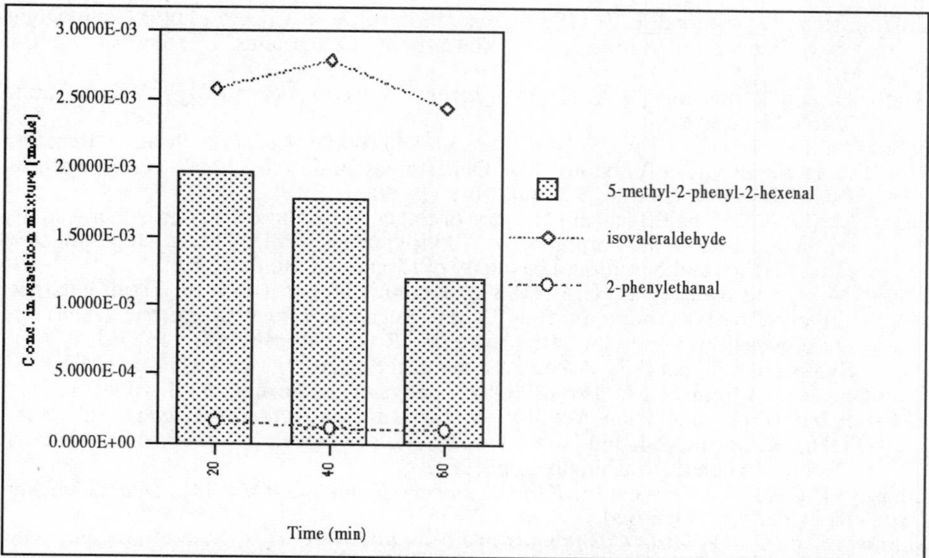

Fig. 3. Reaction of isovaleraldehyde and 2-phenylethanal at 95°C-pH 8.

Conclusions

The quantity of Strecker aldehydes and other Maillard reaction products formed generally increased with time and temperature of heating. Effects of pH were not as predictable but much more compound dependent. The formation of the Strecker aldehydes and other Maillard reaction products were found to follow zero order reaction kinetics with the exception of 5-methyl-2-phenyl-2-hexenal which followed first order. The activation energies for these compounds were relatively low ranging from 14 to 23 kcal/mole. Their ubiquitous presence in foods appears to be related to their ease of formation and low sensory thresholds.

CAS Registry No. Isovaleraldehyde, 590-86-3; 2-acetyl-1-pyrroline, 85213-22-5; phenylacetaldehyde, 122-78-1; 4-methylthiazole, 693-95-8; 5-methyl-2-phenyl-2-hexenal, 21834-92-4; 2-acetylfuran, 1192-62-7.

References

Baltes, W., Kunert-Kirchhoff, J. and Reese, G. (1990) Model Reactions on Generation of Thermal Aroma Compounds. In: *Thermal Generation of Aromas*, Parliment, T.H., McGorrin, R.J. and Ho, C.T. Eds., ACS Symposium Series #409, American Chemical Society, Washington, pp. 143-155.

Biede, S.L. and Hammond, E.G. (1979) Swiss Cheese Flavor: I. Chemical Analysis. J. Dairy Sci. 62:227-237.

Chan, F. and Reineccius, G.A. (1994) Kinetics of the Formation of Methional, Dimethyl Disulfide and 2-Acetylthiophene via the Maillard Reaction. In: *Sulfur Compounds in Foods*, Mussinan, C.J. and Keelan, M.E. Eds., in press.

Finot, P.A., Aeschbacher, H.U., Hurrell, R.F. and Liardon, R. (1989) *The Maillard Reaction in Food Processing, Human Nutrition and Physiology*. Birkhäuser Verlag, Berlin.

Fox, L. and Wylie, P. (1989) Qualitative Analysis of Extracted Cooked Meat Flavors by Gas Chromatography with the HP 5921A Atomic Emission Detector. Hewlett Packard Appl. Note 228-275: 1-3.

Griffith, R. and Hammond, E.G. (1989) Generation of Swiss Cheese Flavor Components by the Reaction of Amino Acids with Carbonyl Compounds. J. Dairy Sci. 72:604-613.

Heath, H. and Reineccius, G.A. (1986) *Flavor Science and Technology.* Van Nostrand Reinhold, New York.

Labuza, T.P., Nelson, K.A., Nelson, G.J. (1991) Water Analyzer Series - Reaction Kinetics Program Version 2.09. Department of Food Science and Nutrition, University of Minnesota, St. Paul, MN.

Leahy, M.M. (1985) The Effects of pH, types of Sugar and Amino Acid and Water Activity on the Kinetics of the Formation of Alkylpyrazines. Ph.D. Thesis. Department of Food Science and Nutrition, University of Minnesota, St. Paul, MN.

Leahy, M.M. and Reineccius, G.A. (1989a) Kinetics of Formation of Alkylpyrazines: Effect of Type of Amino Acid and Type of Sugar. In: *Flavor Chemistry: Trends and Developments*, Teranishi, R., Buttery, R.G. and Shahidi, F. Eds., ACS Symposium Series #388, American Chemical Society, Washington, pp. 76-91.

Leahy, M.M. and Reineccius, G.A. (1989b) Kinetics of the Formation of Alkylpyrazines: Effect of pH and Water Activity. In: *Thermal Generation of Aromas*, Parliment, T.H., McGorrin, R.J. and Ho, C.T. Eds., ACS Symposium Series #409, American Chemical Society, Washington, pp. 198-208.

Lindsay, R.C. (1985) Flavors. In: *Food Chemistry*, Fennema, O.R. Ed., Marcel Dekker, New York, pp. 615-620.

Marrse, H. (1991) *Volatile Compounds in Foods and Beverages,* Marcel Dekker, New York.

Marrse, H. and Visscher, C.A. (1989a) *Volatile Compounds in Food. Qualitative and Quantitative Data,* Vol. I, TNO-CIVO, Netherlands.

Marrse, H. and Visscher, C.A. (1989b) *Volatile Compounds in Food. Qualitative and Quantitative Data,* Vol. II, TNO-CIVO, Netherlands.

Marrse, H. and Visscher, C.A. (1989c) *Volatile Compounds in Food. Qualitative and Quantitative Data,* Vol. III, TNO-CIVO, Netherlands.

Oehlert, G.W. (1993) MacAnova Version 3.1. Department of Applied Statistics, University of Minnesota, St. Paul, MN.

Parliment, T.H., McGorrin, R.J. and Ho, C.T. (1989) *Thermal Generation of Aromas.* ACS Symposium Series #409, American Chemical Society, Washington.

Parliment, T.H., Morello, M.J. and McGorrin, R.J. (1994) *Thermally Generated Flavors - Maillard, Microwave, and Extrusion Processes*, ACS Symposium Series #543, American Chemical Society, Washington.

Potman, R.P. and van Wijk, T. A. 1989. Mechanistic Studies of the Maillard Reaction with Emphasis on Phosphate-Mediated Catalysis. In: *Thermal Generation of Aromas*, Parliment, T.H., McGorrin, R.J. and Ho, C.T. Eds., ACS Symposium Series #409, American Chemical Society, Washington, pp. 182-195.

Rizzi, G.P. (1987). New Aspects on the Mechanism of Pyrazine Formation in the Strecker Degradation of Amino Acids. In: *Flavor Science and Technology*, Martens, M., Dalen, G.A. and Russwurm Jr., H. Eds., J. Wiley & Sons, Chichester, pp. 23-28.

Schieberle, P. 1989. Formation of 2-acetyl-1-pyrroline and other important flavor compounds in wheat bread crust. In: *Thermal Generation of Aromas*, Parliment, T.H., McGorrin, R.J. and Ho, C.T. Eds., ACS Symposium Series #409, American Chemical Society, Washington, pp. 268-275.

Schirlé-Keller, J.P. and Reineccius, G.A. (1990) Reaction Kinetics for the Formation of Oxygen Containing Heterocyclic Compounds in Model Systems. In: *Flavor Precursors*, Teranishi, R., Takeoka, G.R. and Guntert, M. Eds., ACS Symposium Series #490, American Chemical Society, Washington, pp. 244-258.

Shaw, J.J. and Ho, C.T. (1989) Effects of Temperature, pH and Relative Concentration on the Reaction of Rhamnose and Proline. In: *Thermal Generation of Aromas*, Parliment, T.H., McGorrin, R.J. and Ho, C.T. Eds., ACS Symposium Series #409, American Chemical Society, Washington, pp. 217-228.

Shu, C.K. and Ho, C.T. (1989) Parameter Effects on the Thermal Reaction of Cystine and 2,5-Dimethyl-4-hydroxy-3-(2H)-furanone. In: *Thermal Generation of Aromas*, Parliment, T.H., McGorrin, R.J. and Ho, C.T. Eds., ACS Symposium Series #409, American Chemical Society, Washington, pp. 229-241.

Stahl, H.D. and Parliment,T.H. (1994) Formation of Maillard Products in the Proline-Glucose Model System : High Temperature Short-Time Kinetics. In: *Thermally Generated Flavors - Maillard, Microwave, and Extrusion Processes*, Parliment, T.H., Morello, M.J. and McGorrin, R.J. Eds., ACS Symposium Series #543, American Chemical Society, Washington, pp. 251-262.

Temperature Effect on the Volatiles Formed from the Reaction of Glucose and Ammonium Hydroxide: A Model System Study

Chi-Kuen Shu and Brian M. Lawrence

R. J. REYNOLDS TOBACCO COMPANY, PRODUCT DEVELOPMENT DEPARTMENT, BOWMAN GRAY TECHNICAL CENTER, 950 REYNOLDS BOULEVARD, WINSTON-SALEM, NC 27105, USA

Summary

A study of the reaction between glucose and ammonium hydroxide revealed that in general, the number and amount of volatiles increased as the reaction temperature increased from 75 to 150°C. At low temperatures, imidazole formation was predominant, while at higher temperatures the predominant compounds were alkyl pyrazines and hydroxyalkyl pyrazines. Positive identification of the hydroxymethyl pyrazines was confirmed by synthesis. As the products obtained from the reactions between glucose/NH_4OH and glyceraldehyde/NH_4OH were very similar, this supports the postulate that the initial step in glucose degradation is glyceraldehyde formation. It is proposed that hydroxymethyl dihydropyrazine generated from glyceraldehyde may follow oxidation and dehydration pathways to form 2-hydroxymethyl pyrazine and methyl pyrazine, respectively.

Introduction

Pyrazines are generally considered as important flavor components in foods and as such have been identified as components of various food products (Maga 1992). Numerous model system studies have suggested that pyrazines are derived from the reaction between a reducing sugar and an amino acid (Vernin and Parkanyi 1982). Basically, the α-dicarbonyl compound, which is generated from the reducing sugar, reacts with an amino acid, via the Strecker degradation, to produce a Strecker aldehyde and an α-amino carbonyl compound. Condensation of the α-amino carbonyls followed by oxidation yields pyrazines.

Among the model systems studied, the reaction between glucose and ammonium hydroxide (as the nitrogen source) has been the subject of greatest interest. Kort et al. (1952), Hough et al. (1953), and van Praag et al. (1968) identified several alkyl pyrazines and alkyl imidazoles as products of this reaction. Shibamoto and Bernhard (1976) studied the influence of reaction parameters on pyrazine formation using a combination of an open system and an enclosed system, and concluded that during the 2-hour heating period from -5 to 160°C, the optimal temperature to generate the highest yield of pyrazine was found to be 120°C. Shibamoto et al. (1979) further studied the reaction and identified the principal components as imidazoles, pyrazines and pyrroles, and also proposed the formation mechanism of cyclopentapyrazines. The objective of this study was to examine the temperature effect on the title reaction in an enclosed system and to determine whether other types of pyrazines in addition to alkyl pyrazines could be formed.

Materials and Methods

Preparation of Reaction Mixtures of Glucose and Ammonium

In an enclosed reaction vessel (Parr Instrument Company, Model 4563) a solution of 0.1 mole of glucose (18 g), 0.1 mole of ammonia (6.8 ml of 28% NH_4OH) and 100 mL of water was heated at 75, 100, 125 and 150°C for 5 minutes, after which the reaction mixture was cooled to room temperature.

Preparation of the Reaction Mixture of Glyceraldehyde and Ammonium Hydroxide

In an enclosed reaction vessel, a solution of 0.0125 mole of glyceraldehyde (1.13 g), 0.0125 mole of ammonia (0.9 mL of 28% NH_4OH) and 50 mL of water was heated at 150°C for 5 minutes, after which it was cooled to room temperature.

Solvent Extraction

An aliquot (20 mL) of each reaction mixture was extracted in a separatory funnel with ethyl acetate (3 x 30 mL), the extracts were combined and concentrated using a rotary evaporator to about 1 mL.

GC/MS Analysis

Each concentrate was analyzed by GC/MS with a mass selective detector (MSD, HP 5970) on a DBWAX fused silica column (30 m x 0.32 mm, 0.15 μm film thickness). The oven temperature was programmed from 50 to 190°C at 6°C/min., then held for 35 minutes. The mass spectrometer was set at 70 eV, a source temperature of 220°C, and a filament emission current of 1 mA.

Preparative GC

The peaks of interest were collected from a preparative glass capillary column (Supelcowax 10, 30 m x 0.75 mm, 1.0 μm film thickness) using a thermal conductivity detector (TCD) under the same chromatographic conditions.

IR Analysis

IR spectrum of the material prepared or trapped from GC was analyzed on a Mattson Polaris FTIR microscope.

Semi-quantitation of the Volatile Compounds

Octadecane was added as an internal standard to the reaction mixture prepared from glucose and ammonium hydroxide. The mixture was analyzed by GC with a flame ionization detector (FID), under similar chromatographic conditions as described above. Based on the response factor of octadecane, the quantities of the volatile compounds were estimated by assuming that all of the volatiles had the same response factors as octadecane. The yield of the volatile compounds formed was calculated as the total weight of volatiles divided by the total weight of glucose (18 g) and ammonia (1.7 g).

Synthesis of 2-Hydroxymethyl Pyrazines

In order to confirm the structures of the components identified, three 2-hydroxymethyl pyrazines (2-hydroxymethyl-3-methyl pyrazine, 2-hydroxymethyl-5-methyl pyrazine and 2-hydroxy-methyl pyrazine) were synthesized using methods described in the literature (Koelsch and Gumprecht, 1958; Klein et al., 1961) with some modifications. In general, a corresponding methyl pyrazine was oxidized into methyl pyrazine oxide, which was then acetylated/rearranged to an acetoxymethyl pyrazine, which was in turn hydrolyzed to hydroxymethyl pyrazine. The reactions were performed at 0.1 mole level.

In this modified procedure, methylene chloride was used to extract the crude pyrazine oxide, after which it was removed by a rotary evaporator. During concentration solid material was deposited. By washing the solid material with hexane, pure 2,5-dimethyl pyrazine 1-oxide and pure 2,3-dimethyl pyrazine 1-oxide were obtained. However, 2-methyl pyrazine 1-oxide and 2-methyl pyrazine 4-oxide were obtained as a mixture. Because only 2-methyl pyrazine 1-oxide could be used for the further preparation, one g of this oxide mixture was loaded on a silica gel column (5% H_2O deactivated, 20 g). The column was packed and eluted with methylene chloride/hexane (1/1 ratio). From this chromatographic separation, 400 mg of the 1-oxide was collected. The crude acetoxymethyl pyrazines obtained were not further purified. Ethyl acetate was used to extract the crude final products, after which it was removed by a rotary evaporator. The crude final products obtained were not further purified either.

All chemicals used in this study were purchased from standard laboratory chemical suppliers.

Results

From this title reaction, most components were initially identified by GC/MS and retention times as alkyl pyrazines and imidazoles except three unknown peaks. However, the mass spectra of these peaks indicated that they were possibly pyrazine related compounds. Furthermore, the IR spectra of these peaks trapped from GC showed a strong absorption at 3000-3700 cm^{-1} which is characteristic for alcohols. Combination of both MS and IR information suggested that these three components could be homologs, possibly as pyrazines with a hydroxy group attached to a methyl side chain. Therefore, synthesis of the three potential compounds namely 2-hydroxymethyl pyrazine, 2-hydroxymethyl-3-methyl pyrazine and 2-hydroxymethyl-5-methyl pyrazine was undertaken.

Table I compiles the mass spectra data of pyrazine oxides, 2-acetoxy pyrazines and 2-hydroxymethyl pyrazines obtained from the syntheses. The mass spectra of 2-hydroxymethyl-3-methyl pyrazine, 2-hydroxymethyl pyrazine and 2-hydroxymethyl-5-methyl pyrazine synthesized are identical to those obtained from the reaction mixtures corresponding to the three unknown peaks. The retention times of the authentic samples and the peaks from the reaction mixture also matched; therefore, the three 2-hydroxymethyl pyrazines have been positively identified.

Table II shows the volatile components identified from the reactions performed along with the semi-quantitative data. Three 2-hydroxyethyl pyrazines (2-hydroxyethyl, 2-hydroxyethyl-5-methyl, and 2-hydroxyethyl-6-methyl) were also identified based on the mass spectral data recently published (Bohnenstengel and Baltes, 1992).

Discussion

Table II reveals that in general the number and amount of the volatile components produced increased as the temperature of the reaction increased The yield of the volatile compounds generated was found to be 0.03% at 75°C, and 1.39% at 150°C. At 75°C, only 4-methyl imidazole and 4-methyl-2-acetyl imidazole, 2-hydroxymethyl pyrazines and

Table I. Mass spectral data of the pyrazines synthesized.

Pyrazine	Mass Fragment of Mass Spectrum, m/e
2,3-dimethyl pyrazine 1-oxide	124(M^+,100),107(45),80(34),67(35),66(28), 53(33),42(31),39(43)
2,5-dimethyl pyrazine 1-oxide	124(M^+,100),107(50),80(76),66(27),42(34), 39(96)
2-methyl pyrazine 1-oxide	110(M^+,100)93(35),66(58),42(26),39(52)
2-methyl pyrazine 4-oxide	110(M^+,100),55(13),54(20),53(14),42(13), 40(16),39(21)
2-acetoxymethyl-3-methyl pyrazine	166(M^+,3),124(100),123(43),106(29),93(21), 43(74),42(23)
2-acetoxymethyl-5-methyl pyrazine	166(M^+,2)124(100),123(50),95(15),93(14), 43(88),42(17),39(48)
2-acetoxymethyl pyrazine	152(M^+,3),110(100),109(31),81(12),52(12), 43(68),39(16)
2-hydroxmethyl-3-methyl pyrazine	124(M^+,91),106(63),95(100),93(40),67(34), 53(33),52(32),42(50),39(35)
2-hydroxymethyl-5-methyl pyrazine	124(M^+,88),123(61),95(100),55(25),42(34), 40(25),39(53)
2-hydroxymethyl pyrazine	110(M^+,100),109(59),81(89),80(24),55(21), 54(20),53(53),52(41)

Table II. Volatile compounds identified from the reaction of glucose and ammonium hydroxide (5 min at different temperatures).

	75°C mg	100°C mg	125°C mg	150°C mg
pyrazine		1.8	6.0	9.9
methyl pyrazine	0.9	24.3	88.5	141.6
acetol			1.5	7.8
2,5-dimethyl pyrazine		1.5	3.6	5.7
2,6-dimethyl pyrazine		1.5	8.7	16.8
2,3-dimethyl pyrazine		0.3	1.8	3.3
ethyl methyl pyrazine			t	t
trimethyl pyrazine			1.2	2.1
furfuryl alcohol			t	2.4
cyclotene			t	0.3
2-hydroxymethyl-3-methyl pyrazine			1.8	3.3
2-hydroxymethyl pyrazine	0.3	2.4	5.4	9.9
2-hydroxymethyl-5-methyl pyrazine	1.0	15.0	36.6	54.3
2-hydroxyethyl pyrazine		1.2	5.7	11.7
2-hydroxyethyl-5(6)-methyl pyrazines			0.9 1.2	0.9 1.5
4-methyl imidazole	2.5	9.6	9.0	
2-acetyl-4-methyl imidazole	0.1	0.6	0.6	
hydroxy maltol				0.8
yield, mg	4.8	58.2	172.5	272.3
yield, %	0.03	0.29	0.88	1.39

t = trace amount

methyl pyrazine were formed with the predominant compounds being imidazoles. On increasing the temperature to 100°C, alkyl pyrazines, 2-hydroxymethyl pyrazines and 2-hydroxyethyl pyrazines were formed in amounts greater than the imidazoles, consequently at this reaction temperature the imidazole reaction products no longer predominated. At 125°C, the volatile profile continued to show a trend similar to that experienced at 100°C, except that the formation of acetol, furfuryl alcohol and cyclotene commenced. The appearance of these compounds indicates that glucose degradation/rearrangement was readily initiated.

There was a considerable increase in acetol, furfuryl alcohol, cyclotene and 2,3-dihydro-3,5-dihydroxy-6-methyl-4-pyranone (hydroxy maltol) when the temperature was increased to 150°C; in the meanwhile, imidazoles were no longer reaction products. The appearance of hydroxy maltol as a reaction product infers that glucose degradation/rearrangement had become the major reaction. At this temperature, alkyl pyrazines, 2-hydroxymethyl pyrazines, and 2-hydroxyethyl pyrazines continued to be predominant. Also the yield of the volatiles from the reaction continued to increase.

During the increase of reaction temperature, both alkyl pyrazines and hydroxyalkyl (methyl or ethyl) pyrazines increased as major compounds. These results do not completely agree with the previous findings (Shibamoto and Bernhard, 1976). They found that the maximal yield of alkyl pyrazines was at 120°C.

Literature information regarding the formation of the substituted 2-hydroxyalkyl pyrazines is limited (Bohnenstengel and Baltes, 1992). Based on the results obtained from this study, the formation mechanism of such compounds is proposed using 2-hydroxymethyl-5(6)-methyl pyrazines as an example as shown in Figure 1. Under basic conditions, glucose was partially converted into fructose by an enediol rearrangement. Both glucose and fructose were degraded/reacted with ammonia to form 1-amino-2-ketone (I) and 2-amino-1-aldehyde (II) respectively. Self-condensation of (I) or (II) leads to the formation of the corresponding dihydropyrazines, (III) and (IV) respectively. Dehydration of (III) and (IV) generated the same compound, 2-hydroxymethyl-5-methyl pyrazine as one of the major components in the reaction mixture. Cross-condensation of (I) and (II), instead of self condensation, is expected to generate 2-hydroxymethyl-6-methyl pyrazine.

Substituted hydroxyalkyl dihydropyrazine may lead to two pathways: oxidation and dehydration. For instance, hydroxy acetaldehyde, a product from retro-aldol condensation of glucose, reacts with ammonia to form amino acetaldehyde, which condenses with intermediate compounds (I) or (II), to form the dihydropyrazine intermediate (V). This intermediate may be converted into 2-hydroxymethyl pyrazine by oxidation, or into methyl pyrazine by dehydration (Figure 2). The traditional mechanism for methyl pyrazine formation is the condensation of acetol (which was identified from this study) and hydroxy acetaldehyde in the presence of ammonia, followed by oxidation.

On consideration of the chemical structures of the reaction products of this model system, it is interesting to note that most of the heterocyclic compounds formed contain one or two three-carbon moieties. This finding implies that glucose is readily degraded into a three-carbon moiety. In order to prove this phenomenon, a reaction between glyceraldehyde and ammonium hydroxide was performed under the similar conditions, and the reaction products identified are compiled in Table III.

A comparison between results obtained from the reaction of glucose/NH_4OH with those obtained from glyceraldehdye/NH_4OH shows that both reactions generate the same major groups of compounds, alkyl pyrazines, 2-hydroxymethyl pyrazines, 2-hydroxyethyl pyrazines and imidazoles, and the GC profiles are very similar to each other. We believe that glucose is first degraded into glyceraldehyde and other reactive carbonyl compounds via a retroaldol condensation during the sugar ammonia reactions.

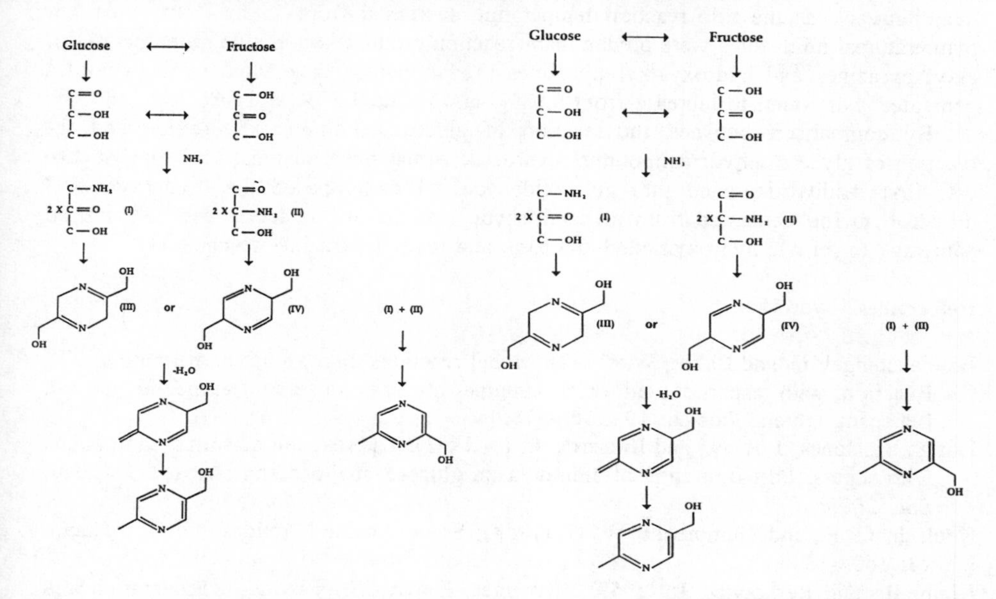

Figure 1. Proposed formation mechanism of 2-hydroxymethyl-5(6)-methyl pyrazine.

Figure 2. Proposed formation mechanism of 2-hydroxymethyl pyrazine & 2-methyl pyrazine.

Table III. Volatile components identified from the reaction of glyceraldehyde and ammonium hydroxide.

Component Identified	Area %
pyrazine	1.49
methyl pyrazine	21.47
acetol	4.03
2,5 & 2,6-dimethyl pyrazine	9.35
2,3-dimethyl pyrazine	1.51
trimethyl pyrazine	1.68
2-methyl pyrrole	0.98
3-methyl pyrrole	0.74
cyclotene	0.42
2-hydroxymethyl-3-methyl pyrazine	0.81
2-hydroxymethyl pyrazine	5.44
2-hydroxymethyl-5-methyl pyrazine	23.00
2-carboxaldehyde-1(or 5)-methyl pyrrole	0.74
2-hydroxyethyl pyrazine	1.57
2-hydroxyethyl-5&6-methyl pyrazine	1.58
2,4-dimethyl imidazole	0.87
4-methyl imidazole	10.65
2-acetyl-4-methyl imidazole	2.23

Conclusions

This study reveals that in general, the volatiles formed increased, qualitatively and quantitatively, as the title reaction temperature increased from 75 to 150°C. At low temperatures, imidazoles were predominant reaction products while at high temperatures, alkyl pyrazines and hydroxyalkyl pyrazines predominated. The yield of the volatiles generated was found to increase from 0.03% at 75°C to 1.39% at 150°C.

By comparison between the reaction of glucose/ammonium hydroxide and the reaction of glyceraldehyde/ammonium hydroxide, it has been demonstrated that glucose was first readily degraded into glyceraldehyde. It is proposed that hydroxymethyl dihydropyrazine generated from glyceraldehyde may follow oxidation and dehydration pathways to form 2-hydroxymethyl pyrazine and methyl pyrazine, respectively.

References

Bohnenstengel, C. and Baltes, W. (1992); Model reactions on roast aroma formation, XII. Reaction with aspartic acid or asparagine at three different temperatures. Z. Lebensm.-Unters. Forsch. 194: 366-371.

Hough, L., Jones, J. K. N., and Richards, E. L. (1953); The reaction of amino compounds with sugars, Part II, Action of ammonia on glucose, maltose and lactose. J. Chem. Soc. 2005.

Koelsch, C. F., and Gumprecht, W. H. (1958); Some diazine-N-oxides. J. Org. Chem. 23: 1603-1606.

Klein, B., and Berkowitz, J., (1959); Pyrazines, I. pyrazine-N-oxides. Preparation and spectral characteristics. J. Am. Chem. Soc. 81: 5160-5166.

Klein, B., Berkowitz, J., and Hetman, N. E. (1961); Pyrazines II, the rearrangement of pyrazine-N-oxides. J. Org. Chem. 26: 126-131.

Kort, M. J., (1952); Reaction of free sugars and aqueous ammonia. J Chem. Soc. 3954.

Maga, J. A. (1992); Pyrazine update. Food Rev. Int. 8(4): 479-558.

van Praag, M., Stein, H. S., and Tibbetts, M. S. (1968); Steam volatile aroma constituents of roasted cocoa beans. J. Agric. Food Chem. 16: 1005-1008.

Shibamoto, T., and Bernhard, R. A. (1976); Effect of time, temperature, and reactant ratio on pyrazine formation in model systems. J. Agric. Food Chem. 24: 847-852.

Shibamoto, T., Akiyama, T., Sakaguchi, M., Enomoto, Y., and Masuda, H. (1979); A study of pyrazine formation. J. Agric. Food Chem. 27: 1027-1031.

Vernin, G., and Parkanyi, C. (1982); Occurrence and formation of heterocyclic compounds. In The Chemistry of Heterocyclic Flavouring and Aroma Compounds; Vernin, G. Ed.; Ellis Horwood: New York, pp 192-198.

Incorporation of ^{14}C-Glucose into Mutagenic Heterocyclic Amines

K. Skog and M. Jägerstad

DEPARTMENT OF APPLIED NUTRITION AND FOOD CHEMISTRY, CHEMICAL CENTER, PO BOX 124, S-221 00 LUND, SWEDEN

Summary

Mixtures of creatinine, glucose, and threonine with the addition of a small amount, 250 μCi (0.6 mCi/mmol) of each [U-^{14}C]-glucose, [1-^{14}C]-glucose or [6-^{14}C]-glucose were heated at 180°C for 30 min in an aqueous model system. The mixtures were purified and analyzed using HPLC, a scintillation counter and the Ames test. MeIQx (2-amino-3,8-dimethyl-imidazo-[4,5-f] quinoxaline) and 4,8-DiMeIQx (2-amino-3,4,8-trimethylimidazo-[4,5-f] quinoxaline) were detected as the main radioactive mutagens. The amount of MeIQx and 4,8-DiMeIQx produced from threonine was estimated to be 18 and 60 nmol/mmol glucose, respectively. Radioactive carbon atoms originating from glucose were also shown to be incorporated into IQx (2-amino-3-methylimidazo-[4,5-f]-quinoxaline). The specific activity was calculated to be 0.6, 0.3, and 0.1-0.3 mCi/mmol for MeIQx, 4,8-DiMeIQx, and IQx, respectively, for all three labeled forms of glucose. By establishing the incorporation of carbon atoms originating from glucose into the imidazoquinoxaline mutagens it was clearly demonstrated that glucose is a precursor in the formation of these food mutagens.

Introduction

A tentative pathway for the formation of a group of mutagenic heterocyclic amines, e.g. imidazoquinolines (IQ) and imidazoquinoxalines (IQx) via the Maillard reaction was suggested by Jägerstad et al. (1983a) as illustrated in Figure 1. It was postulated that creatine formed the amino-imidazo part by cyclization and water elimination. The remaining parts of the IQ compounds were assumed to arise from Strecker degradation products, e.g. pyridines or pyrazines, formed in the Maillard reaction between the hexose and the amino acid. Aldol condensation was thought to link the two parts together via a Strecker aldehyde (or a related Schiff base). The hypothesis has been more thoroughly outlined by Nyhammar in his thesis (1986). One of the tentative reaction routes, including the same precursors, where creatinine was assumed first to condense with an aldehyde before reacting with a pyridine or a pyrazine, was also proposed by Jones and Weisburger (1989). Further support for the formation of IQ and IQx compounds via pyrazines and pyridines has recently been demonstrated by Milic et al. (1993).

This hypothesis has so far been verified concerning the proposed precursors using a model system by heating creatin(in)e, amino acids, and glucose for various times and temperatures in the range of 130-180°C. Not only IQ and IQx, but also methyl homologues of these compounds and imidazopyridines (PhIP) have been isolated from this model system, as recently reviewed by Skog and Jägerstad (1993). In order to further study the role of sugars in the formation of IQ and IQx compounds, ^{14}C-labeled glucose was added as a precursor in the model system. Three different labeling positions of glucose were compared - 1, 6, or all positions (U). The mutagens produced by heating creatine, threonine, and ^{14}C-labeled glucose were isolated, identified and analyzed for radioactivity.

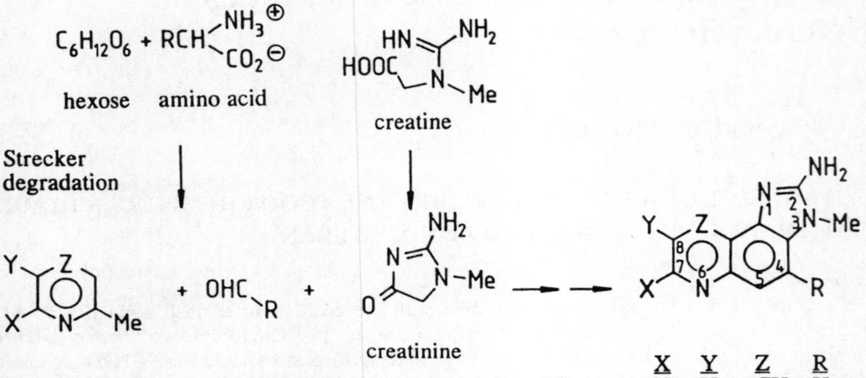

$C_6H_{12}O_6$ + $RCH(NH_3^{\oplus})CO_2^{\ominus}$

hexose amino acid

creatine

Strecker degradation

creatinine

Pyridine (Z = CH)

Pyrazine (Z = N)

aldehyde

	X	Y	Z	R
IQ:	H	H	CH	H
MeIQ:	H	H	CH	Me
MeIQx:	H	Me	N	H
7,8-DiMeIQx:	Me	Me	N	H
4,8-DiMeIQx:	H	Me	N	Me

Figure 1. Suggested reaction routes for IQ and IQx homologues (Jägerstad et al, 1983).

Materials and Methods

Sample preparation

Samples were prepared by heating threonine (900 μmol), creatinine (900 μmol), glucose (450 μmol) and ^{14}C-labeled glucose (250 μCi; [U-^{14}C]-Glucose, [1-^{14}C]-glucose, or [6-^{14}C]-glucose) in 3 mL distilled water in closed metal tubes at 180°C for 30 min. After heating, the mixtures were purified with the solid-phase extraction method of Gross (1990), with minor modifications as described by Johansson et al. (1993).

Analysis

After purification, the samples were injected into a Varian 9010 liquid chromatograph equipped with an ODS column (ToyoSoda TSK gel ODS80TM, 250 x 4.6 mm, 5 μm particle size, Varian, Sweden) and a precolumn (Supelguard LC-18-DB, Varian, Sweden) as earlier described (Johansson et al., 1993). The effluent was monitored at 263 nm with a photodiode array UV detector (Varian 9065 Poly View). Fractions were collected every minute throughout the runs and aliquots of each were tested for mutagenic activity. The amounts of IQx, MeIQx, and 4,8-DiMeIQx were estimated by peak area measurements and comparison with synthetic references.

Following UV detection, the effluent from HPLC was automatically mixed with scintillation liquid (Flow Scint V, Packard, CIAB, Stockholm) and passed through an on-line radiochromatography detector (Radiomatic Flo-One Beta, A 500). The detector was equipped with a lead shielding to reduce the background counts.

Mutagenic activity in the fractions was determined with the Salmonella test using the tester strain TA 98 with the addition of 2 mL S9-mix containing 5% Chlorophene-induced rat liver protein per plate, as described earlier (Ames et al., 1975).

Results

The UV-chromatogram after fractionating a heated mixture of threonine, creatinine, and ^{14}C-labeled glucose showed three major UV peaks; their retention times corresponding to IQx, MeIQx, and 4,8-DiMeIQx (Figure 2). The Ames test showed one minor and two major mutagenic fractions, also corresponding to IQx, MeIQx and 4,8-DiMeIQx. It was clearly seen that that the UV-spectra were practically identical to their corresponding synthetic references (Figure 2). The yields of IQx, MeIQx, and 4,8-DiMeIQx were estimated to be 14, 18, and 60 nmol/mmol glucose, respectively. The simultaneous on-line radioactivity monitoring of purified mutagenic fractions corresponding to IQx, MeIQx, and 4,8-DiMeIQx all indicated one single radioactive and one single UV-peak at the same retention time (Figure 2). Similar results were obtained using glucose labeled with ^{14}C-carbon atoms at different positions. The estimated radioactivity, expressed in specific activity, was 0.1-0.3 mCi/mmol for IQx, 0.6-0.8 mCi/mmol for MeIQx, and 0.3 mCi/mmol for 4,8-DiMeIQx. These data clearly demonstrate that carbon atoms from glucose are incorporated in the IQx, MeIQx, and 4,8-DiMeIQx molecules. In a similar experiment, using glycine instead of threonine, carbon atoms from [U-^{14}C]-glucose and [1-^{14}C]-glucose have been shown to be incorporated into the MeIQx molecules (Skog, 1992).

Discussion

The proposed pathway outlined in Figure 1, presented at a Maillard Meeting held in Las Vegas 1982, suggested the imidazole moiety to be derived from creatinine and the rest of the molecule from Strecker degradation products formed in the Maillard reaction between sugars and α-amino acids (Jägerstad et al., 1983). Thus, one benzene carbon (C-4) was believed to originate from the α-carbon of an amino acid (after oxidative decarboxylation), while a sugar was thought to provide another benzene carbon (C-5) and the pyridine or pyrazine carbons (via a methylpyridine or a methylpyrazine). Methylpyrazines are very common Maillard reaction products (Bemis-Young et al., 1993), and their carbon atoms are derived almost entirely from a sugar, while amino acids mostly supply only nitrogen, as demonstrated in model experiments using ^{14}C-labeled glucose and ^{14}C-labeled amino acids (Koehler et al., 1969). "The Las Vegas Hypothesis" also included a pathway for the formation of IQ and its methyl homologues based on methylpyridines instead of pyrazines. Methylpyridines are, however, a minor and fairly unusual product making the formation of IQ-compounds less likely in this model system. In fact only IQx and its methyl homologues have been more frequently identified in our model systems than IQ and MeIQ. It should also be clearly stated that "the Las Vegas Hypothesis" does not apply to the formation of imidazopyridines, like PhIP, a major principle among the food mutagens.

According to the hypothesis, MeIQx and 4,8-DiMeIQx are formed from dimethylpyrazines (DMP), while IQx is formed from methylpyrazines (MP). Our experiments demonstrated that U-, 1-, and 6-^{14}C-glucose produced MeIQx and 4,8-DiMeIQx, each with the same radioactivity with respect to the labeling position of glucose. These results are in agreement with works by Koehler et al. (1969) showing that glycine and ^{14}C-glucose labeled in positions 1 or 6 or positions 3 and 4 produced DMP labeled to the same extent. These data indicate that the glucose molecule is split into two three-carbon fragments both contributing equally to the formation of DMP.

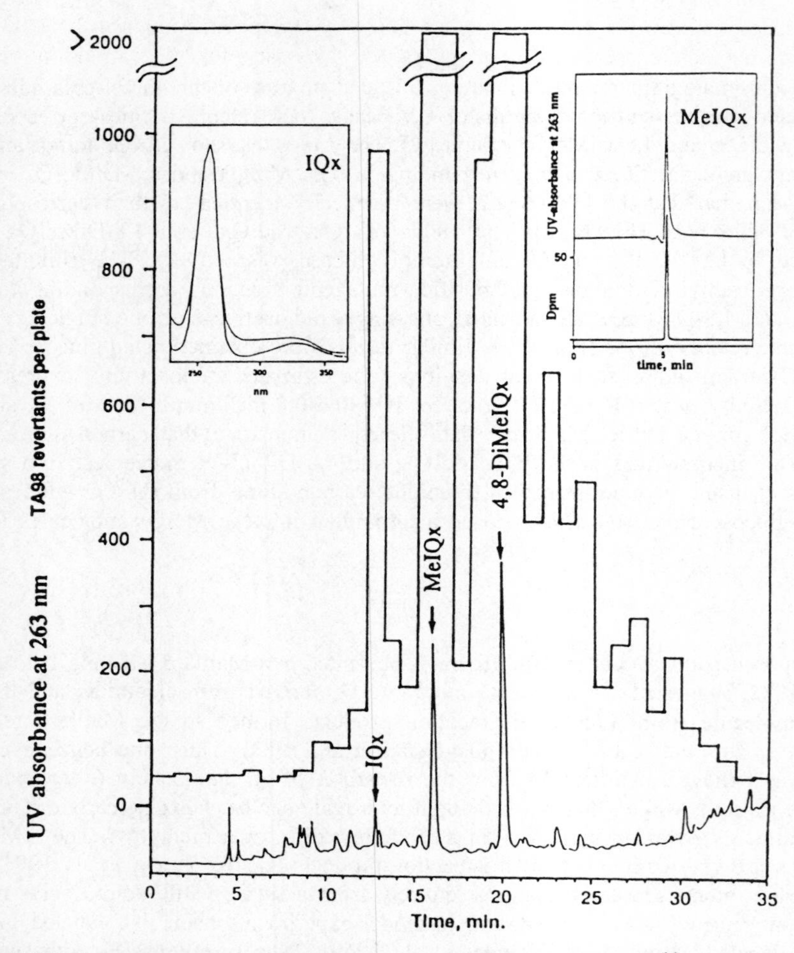

Figure 2. HPLC chromatogram of a heated reaction mixture containing creatinine ^{14}C-labeled glucose, and threonine heated at 180°C for 30 min. The upper plot shows the mutagenic activity of the collected fractions and the lower one shows the UV absorption at 263 nm. The left-hand scale shows the number of revertants. In the top left corner a UV spectrum from IQx is shown in comparison with synthetic IQx. The upper right inset shows the UV absorption at 263 nm (upper plot) and below it a chromatogram from on-line analysis of the radioactivity (dpm) in MeIQx from the sample originally containing [6-^{14}C]-glucose.

The chemical yield of 4,8-DiMeIQx (about 60 nmol/mmol glucose) was three times higher than for MeIQx (about 18 nmol/mmol glucose) in all three experiments, but the radioactivity of 4,8-DiMeIQx (0.3 mCi/mmol) was only about half that of MeIQx (0.6-0.8 mCi/mmol). This means that in comparison with MeIQx, a considerably lower proportion of 4,8-DiMeIQx molecules is formed from glucose. A possible explanation for this unexpected result may be in terms of reaction kinetics. Thus, threonine, which is known to produce pyrazines after heating, also in the absence of glucose (Wang and Odell, 1973), might produce pyrazines that are more prone to produce 4,8-DiMeIQx than MeIQx or IQx.

The role of sugars for the formation of IQ or IQx compounds has been unclear, and conflicting data exist about whether sugars are necessary for the formation or not. However, in spite of the very low yields, this study clearly demonstrates that carbon atoms from glucose are incorporated into the IQx, MeIQx, and 4,8-DiMeIQx, showing glucose to be an active precursor in the formation of such food mutagens as the IQx compounds and their homologues.

Acknowledgements
The gifts of synthetic compounds from Professor Kjell Olsson and Dr Spiros Grivas, Uppsala, Sweden, and from Mark G. Knize, Livermore, CA, USA are gratefully acknowledged. This study was supported by the Swedish Cancer Foundation (1824-B92-11XAC) and the Swedish Council for Forestry and Agricultural Research (50. 0440/91).

Abbreviations

HPLC	high performance liquid chromatography
MeIQx	(2-amino-3,8-dimethylimidazo[4,5-f]quinoxaline)
4,8-DiMeIQx	(2-amino-3,4,8-trimethylimidazo[4,5-f]quinoxaline)
PhIP	(2-amino-1-methyl-6-phenylimidazo[4,5-b]pyridine)
IQ	(2-amino-3-methylimidazo[4,5-f]quinoline)
IQx	(2-amino-3-methylimidazo[4,5-f]quinoxaline)

References

Ames, B. N., Mc Cann, J., and Yamasaki, E. (1975) Methods for detecting carcinogens with the Salmonella/mammalian-microsomal·mutagenicity test. Mutat. Res. 31: 347-364.

Bemis-Young, G. L., Huang, J., and Bernhard, R. A. (1993) Effect of pH on pyrazine formation in glucose-glycine model systems. Food Chem. 46: 383-387.

Gross, G. A. (1990) Simple methods for quantifying mutagenic heterocyclic aromatic amines in food products. Carcinogenesis 11: 1597-1603.

Jägerstad, M., Laser Reutersward, A., Öste, R., Dahlqvist, A., Grivas, S., Olsson, K., and Nyhammar, T. (1983) Creatinine and Maillard reaction products as precursors of mutagenic compounds formed in fried beef. In Waller, G.R. and Feather, M.S. (Eds). The Maillard Reaction in Foods and Nutrition. ACS Symposium Series 215, Washington DC, pp. 507-519.

Johansson, M., Skog, K., and Jägerstad, M. (1993) Effects of edible oils and fatty acids on the formation of mutagenic heterocyclic amine in a model system. Carcinogenesis, 14, 89-94.

Jones, R. C. and Weisburger, H. H. (1989) Characterization of aminoalkylimidazol-4-one mutagens from liquid-reflux models. Mutat. Res. 222:43-51.

Koehler, P. E., Mason, M. E., and Newell, J. A. (1969) Formation of pyrazine compounds in sugar-amino acid model systems. J. Agric. Food Chem. 17: 393-396.

Milic, B.Lj., Djilas, S.M. and Canadanovic-Brunet, J.M. (1993) Synthesis of some heterocyclic aminoimidazoazaarenes. Food Chem. 46: 273-276.

Nyhammar, T. (1986) Studies on the Maillard reaction and its role in the formation of food mutagens. Doctoral thesis, Swedish University of Agricultural Sciences, ISBN 91-576-2658-8.

Skog, K. (1992) Formation of thermic mutagens in cooked meat products and model systems. Doctoral Thesis, University of Lund, Sweden. ISBN 91-87818-04-3.

Skog, K., and Jägerstad, M. (1993) Frying and food mutagens - a literature review. Fd. Chem. Tox. 31, 665-675, 1993.

Wang, P-S. and Odell, G. V. (1973) Formation of pyrazines from thermal treatment of some aminohydroxy compounds. J. Agric. Food Chem. 21, 868-870.

The Influence of Temperature and Water Activity on the Formation of Aminoimidazoquinolines and -quinoxalines (IQ Compounds) in Meat Products

K. Eichner and E. Schuirmann

INSTITUTE FOR FOOD CHEMISTRY, UNIVERSITY OF MÜNSTER, PIUSALLEE 7, D-48147 MÜNSTER, GERMANY

Summary

During the processing of meat mutagenic imidazoquinolines and -quinoxalines ("IQ compounds") may be formed by the Maillard reaction of reducing sugars, amino acids, and creatine/creatinine. For analytical determination, the IQ compounds were extracted with methanol, fractionated, separated by reversed phase HPLC and identified by gas chromatography/mass spectrometry. Although IQ compounds are already formed at moderate temperatures, increasing temperature causes a large increase in the rate of formation; a decrease of water activity further promotes the formation of IQ compounds.

Introduction

In 1977 Japanese scientists (Sugimura et al., 1977) reported that polycyclic aromatic hydrocarbons account only for a small part of the total mutagenic activity of the charred surfaces of meat and fish. The main mutagenic compounds were identified as amino acid pyrolysates and aminoimidazoquinolines, -quinoxalines and -pyridines (IQ compounds, cf. Figure 1). IQ compounds were also detected in meat products heated at moderate temperatures (Uhde and Macholz, 1986; Sugimura, 1986). These mutagens may be formed by reaction of methylpyridines and -pyrazines, which are typical products of the Maillard reaction, with a Strecker aldehyde and creatine/creatinine (Nyhammar et al., 1986).

Our goal was to investigate the technological conditions favouring the formation of IQ compounds during processing of meat products such as meat extract and low-moisture products containing meat extract.

Materials and Methods

For our heating experiments we used low-moisture model systems consisting of glucose or fructose, glycine and creatinine (molar ratio: 0.5:1:1) on microcrystalline cellulose (Avicel, 0.49 g/mmol glycine;a_w=0.75) as carrier material (Schuirmann and Eichner, 1991a,b) and meat extract of different origins.

The IQ compounds formed by heating were extracted from 3 g of the model system or 20 g of the meat product with methanol (3 x 50 mL). After addition of 100 mL water to the evaporated extracts, a liquid-liquid partition between methylene chloride (3 x 100 mL) and water was performed at pH 12. Following evaporation of the solvent, the extracts were dissolved in 500 µL methanol and fractionated by HPLC on a sulfopropyl Si 100 cation exchange column (125 x 4.6 mm) using 0.15 M ammonium acetate (pH 4.5) and acetonitrile (7:3 v/v) as the mobile phase (Schuirmann and Eichner, 1991a) at

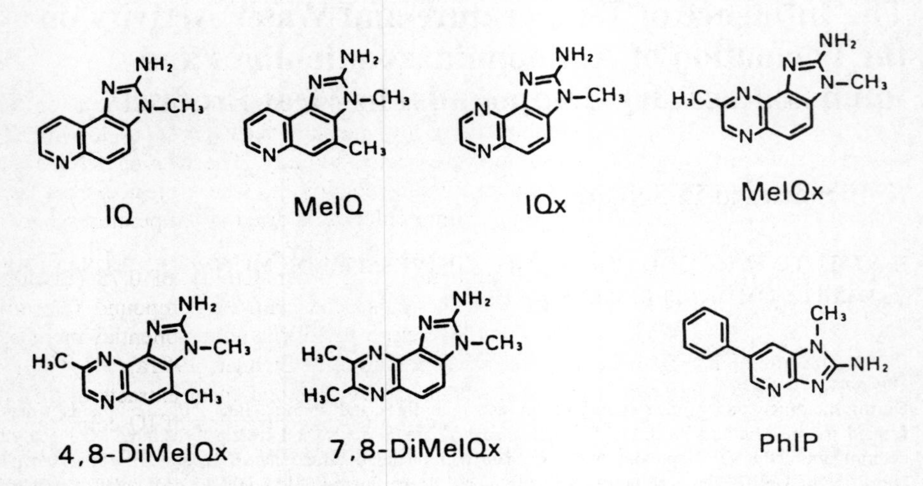

Figure 1. Heterocyclic amines formed during heating of meat and meat products.

a flow rate of 1 mL/min (UV-detection at 265 nm). The fraction containing the IQ compounds (determined by compison with a standard mixture) was dissolved in 300-500 μL methanol and separated on a LiChrospher 60 RP-select B column (250 x 4.0 mm; Merck, Darmstadt) using a triethylammonium phosphate buffer (pH 3.0) and acetonitrile gradient (Schuirmann and Eichner, 1991a). The retention times and concentrations of the individual IQ compounds were determined by using standard compounds as references.

An electrochemical detector (dual-electrode cells) can also be used for the detection of IQ compounds. The cells give signals at +0.65 V on the first electrode and at -0.40 V on the second electrode (redox mode). Each IQ compound shows a characteristic ratio between the intensities (peak areas) of both signals supporting the identification.

For identification the fractionated IQ compounds were collected from the mobile phase on a RP 18 cartridge, acetylated with acetic anhydride/pyridine, and separated by gas chromatography on a DB 5.625 capillary column (length 30 m, diameter 0.32 mm, film thickness 0.25 μm). The temperature program used was as follows: 80°C (1 min); 30°C/min to 230°C; 10°C/min to 290°C (hold for 15 min). The individual compounds were transferred to an ion trap detector (Finnigan Mat Ltd.). For comparison, reference compounds were treated in the same way. Electron-impact ionization of the acetylated IQ compounds resulted in the following characteristic fragment ions (arranged according to decreasing intensities):

2-Amino-3-methylimidazo[4,5f]quinoline (IQ):	198/225/240
2-Amino-3,4-dimethylimidazo[4,5f]quinoline (MeIQ):	212/254/239
2-Amino-3,8-dimethylimidazo[4,5f]quinoxaline (MeIQx):	240/255/213
2-Amino-3,4,8-trimethylimidazo[4,5f]quinoxaline (4,8-DiMeIQx):	254/269

Results and Discussion

Since our investigations were aimed at optimizing the processing conditions so as to minimize the formation of IQ compounds in low-moisture meat products, low-moisture model systems of known composition were heated initially. The heating temperatures were choosen according to industrial processing conditions, where meat extract (water content about 22 %) after mixing with sodium chloride is dried at temperatures between 95 and 120°C for about 5 hours.

The model systems were first equilibrated to a water activity of 0.75 (Rockland, 1960) and then heated at 100°C for 5 or 10 hours. The main IQ compound formed by heating was MeIQx. By increasing the temperature to 150°C, an exponential increase in the formation of MeIQx was observed (Schuirmann and Eichner, 1991a).

In a second series of experiments, we heated commercial meat extracts at different temperatures and water activities. Table I shows the concentrations of IQ compounds in different meat extracts (MeIQx being the main component) which were used for our heating experiments. The lowest concentration of MeIQx was found in a sample (number 3) which was produced under recently optimized conditions. The analytical data of other authors (Nyhammar et al., 1986; Takahashi et al., 1985; Wakabayashi et al., 1986; Gross et al., 1989) show concentrations up to 52 ng/g for IQ, MeIQx and 4,8-DiMeIQx in meat extract. Bacteriological grade meat extracts generally contain higher concentrations of IQ compounds. The influence of reaction conditions on the formation of IQ compounds in meat extracts shows the same trends as their formation in model systems (Schuirmann and Eichner, 1991a).

Figure 2 demonstrates the increase in the concentration of IQ compounds in meat extract at 100°C as a function of heating time; the increase in MeIQx shows an almost linear dependence on time.

Table I. Concentrations of IQ compounds in meat extracts in ng/g (related to wet matter).

	IQ	MeIQ	MeIQx	4,8-DiMeIQx
Meat extract 1 (food grade)	1.9	n.d.	28.3	3.6
Meat extract 2 (food grade)	2.1	n.d.	24.4	3.4
Meat extract 3 (food grade)	4.8	n.d.	6.2	2.9
Difco meat extract (bact.grade)	52	n.d.	79	10

n.d. = not detectable

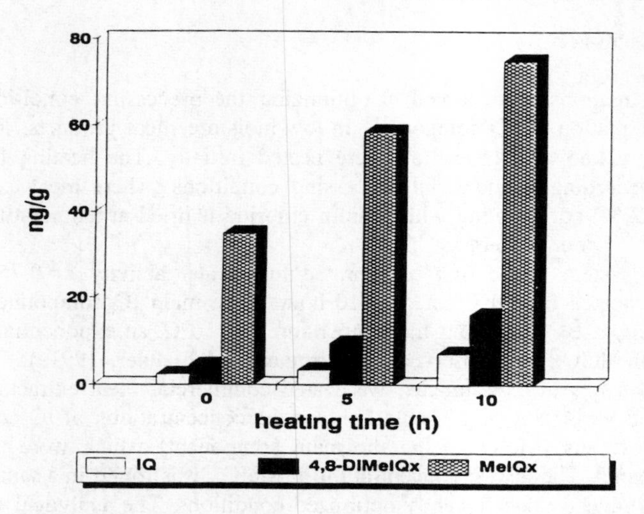

Figure 2. Formation of IQ compounds (related to dry matter) in meat extract (1 in Table I) at 100°C (a_w = 0.42; water content = 19.1 %).

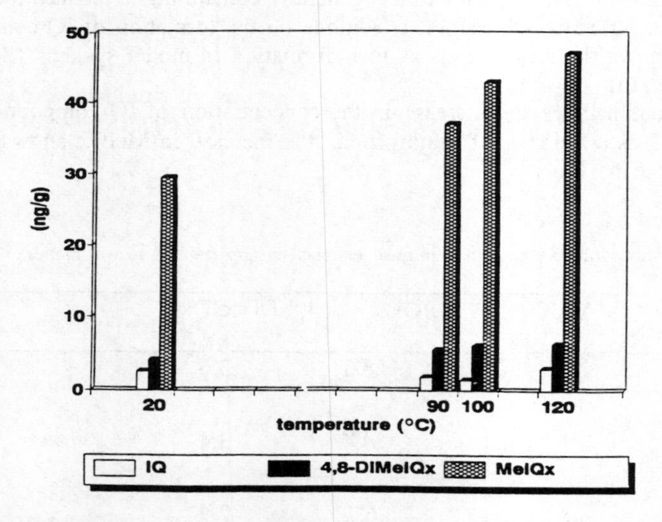

Figure 3. Formation of IQ compounds (related to dry matter) in meat extract (2 in Table I) at different temperatures (reaction time: 1 h; a_w = 0.42; water content = 17.4 %) (The values for 20°C correspond to the initial concentrations).

Figure 3 shows the influence of heating temperature on the formation of IQ compounds. Prediction of the amount of IQ compounds formed during heat processing, especially drying, also requires information on the influence of water content on their rate of formation. Figure 4 shows the influence of water activity (determined at room temperature) on the increase of MeIQx during heating of meat extract (initial concentration of MeIQx: 30 ng/g dry matter) for 4 h at 100°C and 2 h at 120°C, respectively.

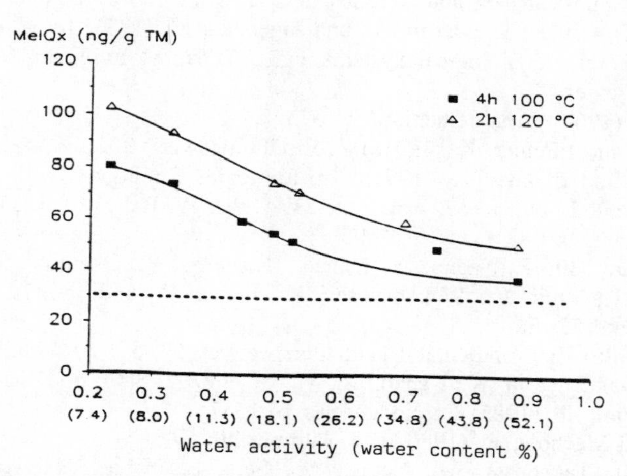

Figure 4. Formation of MeIQx (related to dry matter) in meat extract dependent on temperature and water activity/water content (dotted line: initial concentration of MeIQx).

The formation of MeIQx is increased by lowering the water content. Thus it can be concluded that especially during the last drying section of products containing meat extract, the temperature should be lowered by applying a vacuum in order to minimize the formation of IQ compounds.

Conclusions

The experimental results obtained with meat extract prove that IQ compounds may be formed at relatively moderate temperatures applied during processing of meat and meat products. Also the water activity of the food plays an important role in the formation of IQ compounds. The present results must be taken into consideration in optimizing the processing conditions for minimizing the formation of IQ compounds in meat and meat products.

Acknowledgements

These investigations were supported from industrial joint research funds granted by the German Federal Ministry of Economy (BMWi) via AiF and Forschungskreis der Ernährungsindustrie e.V. (FEI), for which we are very grateful.

References

Gross, G.A., Philippossian, G., and Aeschbacher, H. U. (1989) Carcinogenesis 10: 1175
Nyhammar, T., Grivas, S., Olsson, K., and Jägerstad, M. (1986) In: Amino-Carbonyl Reactions in Food and Biological Systems, p. 323 (Eds.: M. Fujimaki, M. Namiki and H. Kato), Elsevier

Rockland, L. B. (1960) Anal. Chem. 32: 1375

Schuirmann, E., and Eichner, K. (1991a) Z. Ernährungswiss. 30: 36

Schuirmann, E., and Eichner, K. (1991b) In: Strategies for Food Quality Control and Analytical Methods in Europe, Vol. 2, p. 739 (Eds.: W. Baltes et al.), Behr's Verlag Hamburg

Sugimura, T., Kawachi, T., Nagao, M., Yahagi, T., Seino, Y., Okamoto, T., Shudo, K., Kosuge, T., Tsuji, K., Wakabayashi, K., Iitaka, Y., and Itai, A. (1977) Proc. Japan Acad. 53: 58

Sugimura, T. (1986) Environmental Health Perspectives 67: 5

Takahashi, M. Wakabayashi, K., Nagao, M., Yamaizumi, Z., Sato, S., Kinae, N., Tomita, I., and Sugimura, T. (1985) Carcinogenesis 6: 1537

Uhde, W.-J., and Macholz, R. (1986) Die Nahrung 30: 59

Wakabayashi, K., Takahashi, M., Nagao, M., Sato, S., Kinae, N., Tomita, I., and Sugimura, T. (1986) In: Amino-Carbonyl Reactions in Food and Biological Systems, p. 363 (Eds.: M. Fujimaki, M. Namiki and H. Kato), Elsevier

Enhancement of the Gelation of Food Macromolecules Using the Maillard Reaction and Elevated Temperatures

Helen J. Armstrong, Sandra E. Hill, and John R. Mitchell

DEPARTMENT OF APPLIED BIOCHEMISTRY AND FOOD SCIENCE, UNIVERSITY OF NOTTINGHAM, SUTTON BONINGTON, LOUGHBOROUGH, LEICESTERSHIRE LE12 5RD, UK

Summary
The use of the Maillard reaction to enhance the gelling properties of globular proteins has been investigated. The incorporation of reducing sugars with proteins, such as bovine serum albumin and soy isolate,enhanced the break strength of the gels formed on heating. Increase in gel strength and color development plus the decline in pH followed the reducing sugar reactivity series typical for the Maillard reaction. Solubility studies and stress relaxation tests indicate that the gels formed in the presence of reducing sugars have additional permanent crosslinks.

Introduction

The Maillard reaction in foods is normally associated with a decrease in nutritional quality along with color and flavor development. However, the commercial value of a food protein is often determined by its functionality, particularly by its ability to form gels, rather than its nutritional value. If the concentration is sufficiently high, most globular proteins will form gels when heated to temperatures in excess of 80°C. Although extensively studied (Ziegler and Foegeding, 1990), the relationship between the structure and gelling properties is not fully understood, but factors such as the final pH of the gel, net charge on the protein and the covalent and non-covalent crosslinking within the gel network are all known to be important in gel formation. Chemical modification is often used to alter proteins so enhancing their ability to form gels (Kinsella, 1976). This type of approach has two major drawbacks; it is expensive and the legislative status of the resultant product's use in food is dubious. It is reported that the Maillard reaction will; a) alter the charge on the protein (for example by modifying the ε-amino group of lysine), b) cause a pH decline in the gelling system (breakdown of sugars to organic acids) and c) polymerize the proteins (Watanabe *et al.*, 1980). It was, therefore, decided to investigate the gelation properties of some globular proteins after heat treatment in the presence of reducing sugars and to try and ascertain which, if any, of these phenomena altered the gelling behavior of the protein.

Materials and Methods

Materials

Bovine serum albumin (BSA), egg white, D (+) xylose and glucono-δ-lactone (GDL) were obtained from Sigma Chemical Co. Ltd. Soya isolate (Purina 500E) and pea protein were obtained from Ralston Purina and Grinstead Products, respectively.

Methods

Gel Preparation - Solutions of the proteins were prepared in deionized distilled water to a range of concentrations. 'Maillard gels' were prepared by addition of 2% or 3% D (+) xylose. The necessary amount of GDL was added to the solutions in order to keep the final pH of the GDL gels approximately constant and like those in the 'Maillard' gels. Gelation was initiated by heating 20 mL of the solutions in tubing (21 mm diam) in an autoclave for 30/60 min at 121°C.

Gel Evaluation

The final protein content of the gel was determined using the Kjeldahl method of analysis. Color of the gels was assessed using the HunterLab Colorquest spectrocolorimeter.

The rheological properties of the gels were evaluated by compressing the cylindrical gel samples (2.0 x 1.5 cm) using a TA.XT-2 Texture Analyzer (Stable Microsystems, UK.). The plunger speed was 2mm/sec and the measurements performed at 15±2°C. The gel was deformed until rupture (break strength) as evidenced by a peak in the force-distance curve. Stress relaxation was followed for 240 seconds after compression to a constant strain level (25% deformation). The data obtained was linearized using the method proposed by Peleg (1979). Parameters K_1 and K_2 were determined as described by Nussinovitch *et al.*(1990) employing the following equation;

$$\frac{[F_o t]}{[F_o - F(t)]} = K_1 + K_2 t$$

where t is time (sec), F_o and F(t) are the maximum and momentary force and K_1 and K_2 are constants.

Gels were homogenized by pressing through a copper sieve (200 μm aperture). A given weight of particles was stirred overnight in 1% β-mercaptoethanol (β-ME) and 1% sodium dodecyl sulfate (SDS). Any undissolved material was removed by centrifugation (1200g for 10 min) and total protein determination using a modified Lowry method carried out on the supernatant.

Molecular Weight Studies - Sample Preparation

Aqueous solutions of 3% BSA (w/v) and 3% BSA + 2% xylose (w/v) were prepared. Aliquots were heated in a water bath at 95°C for a range of times. Diffusion coefficient values were obtained by dynamic light scattering using a Malvern Instruments System 4700C (Malvern, UK.). Sedimentation velocity experiments were performed using an XLA analytical ultracentrifuge (Beckman, USA) equipped with scanning absorption optics. The methodologies used for these two techniques and subsequent analysis were those of Errington *et al.* (1993).

Results and Discussion

Table I illustrates that the inclusion of different sugars has a marked effect on the gelling properties of BSA when heated at 121°C for 1 hour. The presence of reducing sugars promotes gelation and there is a correlation between the extent of the Maillard reaction, as indicated by a darkening of the gel (low L* values represent dark color), and the gel strength (Pearson's correlation = -0.87). It was also noticeable that the gels with the lowest pH were the most dark. The effectiveness of the sugars to promote gelation followed the reactivity series reported by Hurrell *et al.* (1979).

Protein gels consist of a network of aggregated partially unfolded protein molecules. To monitor the aggregation and change in molecular properties of BSA the molecular weight was determined as a function of time of heat treatment. Samples consisted of 3% BSA and 3% BSA + 2% xylose heated at 95°C for a range of times. For the first 30 minutes the increase in molecular weights for the two systems was very similar to about 0.6 MDalton. However, further heat treatment produced no increase for the BSA while the BSA/ xylose continued to show an increase to approximately 2 MDaltons. Further heating of this system gave a weak gel network.

As well as molecular weight being important in gelation the concentration of protein is also vital. The critical concentration of protein in order for a gel matrix to be formed is highly pH dependent with far lower protein concentrations being required at the protein's isoelectric point (Stading and Hermansson, 1990). The incorporation of 3% xylose to a range of globular proteins demonstrated that the threshold level of protein required for gelation could be substantially lowered. Soya and pea protein isolates needed a level of 15-20% to form a gel, but the inclusion of 3% xylose reduced the critical value to less than 8%. To further investigate whether this phenomena was due to the decline in pH of the system brought about by the formation of acidic byproducts of the Maillard reaction, another set of samples were prepared with glucono-δ-lactone (GDL). Over time, GDL hydrolyses to gluconic acid, the greater the amount of GDL used the greater the pH fall. Gels were constructed so that the pH fall of the Maillard reaction was mimicked by the fall induced by the GDL. It was therefore possible to compare gels of similar protein concentrations and final hydrogen ion concentrations as well as gels formed without incorporation of any low molecular weight compounds (neutral gels).

Table I. A comparison of the pH, color and gel strengths of samples of 2% BSA + 2% sugars after 1 hour at 121°C

Sugar	pH	Gel Strength (g)	Color (L*)
Sucrose	7.10	No Gel	No measurement
Lactose	6.81	186 ± 6	69.76 ± 1.12
Fructose	6.24	253 ± 10	63.56 ± 0.20
Glucose	6.06	277 ± 3	61.44 ± 0.93
Mannose	5.84	285 ± 5	60.05 ± 0.15
Xylose	5.02	322 ± 18	42.59 ± 0.20

Figure 1 demonstrates the breakstrengths of these three types of gel. The low critical protein concentration and the high breakstrength can be seen for the gels formed by the incorporation of xylose. Visually the Maillard gels are very different, being brown in colour compared to the white neutral and GDL gels. At similar protein concentrations it was noticed that the GDL gels showed far greater levels of syneresis for both the BSA and soya samples. The superior water holding ability of the Maillard gels probably reflects the higher net charge that the network has at pH 5.0 since there would have been a reduction in the number of positively charged amino groups present, resulting in a lowering of the isoelectric point of the protein. Gels are known to have reduced water holding capacity as the isoelectric pH is approached (Van Kleef, 1986).

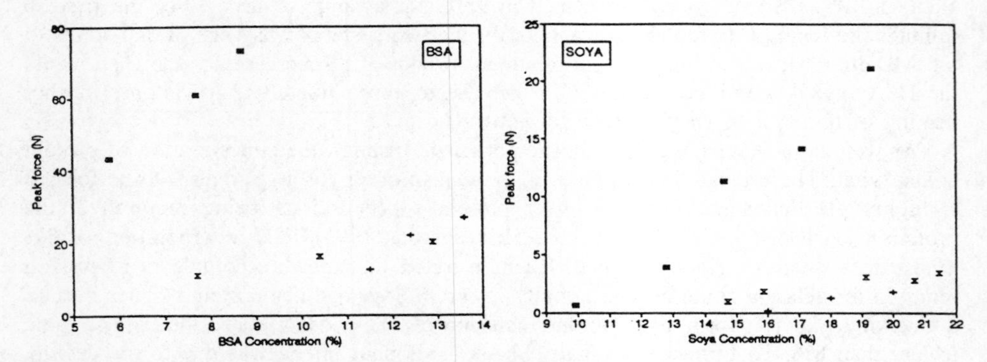

Figure 1. Peak force (break strength) of neutral pH (+), Maillard (■) and GDL (⌶) gels expressed as a function of protein concentration. Heating regime 121°C for 30 min and 60 min for BSA and Soya respectively.

It would be expected that the formation of additional crosslinks in a gel network would effect the rupture and viscoelastic properties of a gel. To investigate this, gel cylinders were compressed and the values of K_1 obtained after linearisation of the stress relaxation data (Fig. 2). K_1 can be considered as a measure of gel elasticity (Peleg, 1979). For an ideal elastic solid $K_1 \rightarrow \infty$, whereas for a non elastic liquid $K_1 = 1$. At similar concentrations the Maillard gels have higher K_1 values than the GDL gels. The decay in stress with time reflects the rearrangement and/or rupture of crosslinks within the gel structure. Increasing the degree of covalent bonding within a system would reduce this tendency. Thus, it could be thought that the higher K_1 values for the Maillard gel may represent additional crosslinking of the protein network.

Supporting evidence for additional crosslinking comes from differences in the solubilities of the gels in a mixture of SDS and β-ME. All the GDL gels are completely soluble in the solvent mix, while the Maillard gels exhibit much reduced solubility. Material not solubilized can be thought of as having crosslinks additional to hydrogen bonding, electrostatic interaction and disulphide bridging. For the BSA gels less than 6% of the protein is solubilized from the Maillard gel. Soya Maillard gels, in comparison, are still approximately 40% soluble. This quantity of soya protein is solubilized by SDS alone and it is interesting to reflect if the difference between the two gels indicates the difference in the disulphide bridging in the proteins. While the BSA is highly disulphide bridged the 7S soya globulin fraction contains no such crosslinks (Van Kleef, 1986).

Hence, it can be concluded that foods containing reducing sugars and globular proteins, subjected to high temperatures, will have gelation properties dependant on the Maillard reaction. The changes in gelation are a consequence of both the pH fall and the formation of additional covalent crosslinks.

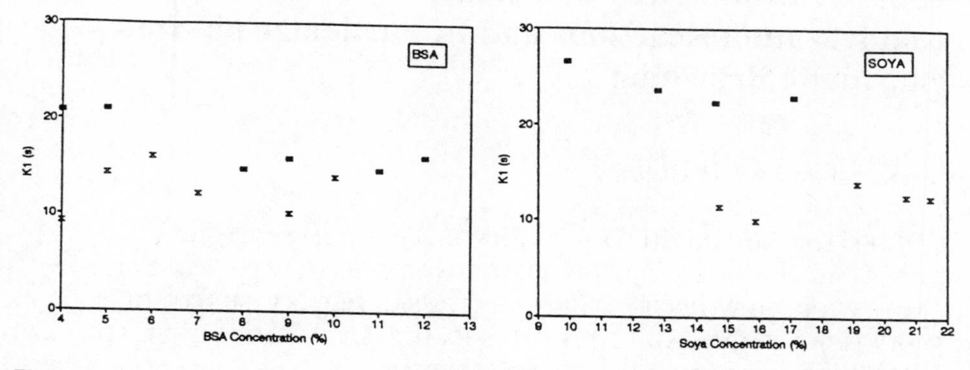

Figure 2. Values of K_1 (intercept) from stress relaxation data of Maillard (■) and GDL (X) gels prepared as in Figure 1 expressed as a function of BSA/Soya concentration .

References

Errington, N., Harding, S., Värum, K.M. and Illum, L. (1993) Hydrodynamic characterization of chitosans varying in degree of acetylation. *Int. J. Biol. Macromol.* **15**: 113 - 117.

Hurrell, R.F., Lerman, P. and Carpenter, K.J. (1979) Reactive lysine in foodstuffs as measured by a rapid dye-binding procedure. *J. Food Sci.* **44**: 1221-1227.

Kinsella, J.E. (1976) Functional properties of proteins in foods: A survey. *Crit. Rev. Fd. Sci. Nutr.* **7**: 219-280.

Nussinovitch,A., Kalentunc, G., Normand, M.D. and Peleg,M. (1990) Recoverable work versus asymptotic residual modulus in agar, carrageenan and gellan gels. *J. Texture Studies* **21**: 427-438.

O'Brien, J.O. and Morrissey, P.A. (1989) Nutritional and toxicological aspects of the Maillard reaction in foods. *Crit. Rev. Fd. Sci. Nutr.* **28**: 211-248.

Peleg, M. (1979) Characterization of the stress relaxation curves of solid foods. *J. Food Sci.* **44**: 277-281.

Stading, M. and Hermannson, A-M. (1990) Viscoelastic behaviour of β-lactoglobulin gel structures. *Food Hydrocolloids* **4**: 121-135.

Van Kleef, F.S.M. (1986) Thermally induced protein gelation: Gelation and rheological characterization of highly conc ovalbumin and soybean protein gels. *Biopolymers* **25**: 31- 59.

Watanabe, K., Kato, Y. and Sato, Y. (1980) Chemical and conformational changes of ovalbumin due to the Maillard reaction. *J. Food Proc. Preserv.* **3**: 263-274.

Ziegler, G.R. and Foegeding, E.A. (1990) The gelation of proteins. In "Advances in Food and Nutrition Research" (Ed. Kinsella, J.E.) **34** Academic Press Inc., New York, 203-298.

Glass Transition and its Potential Effects on Kinetics of Condensation Reactions and In Particular on Non-enzymatic Browning

M. Karel and M. P. Buera[1]

CENTER FOR ADVANCED FOOD TECHNOLOGY, FOOD SCIENCE DEPARTMENT, COOK COLLEGE, RUTGERS UNIVERSITY, PO BOX 231, NEW BRUNSWICK, NEW JERSEY 08903-0231, USA; [1]DEPARTAMENTO DE INDUSTRIAS, FACULTAD DE CIENCIAS EXACTAS, CIUDAD UNIVERSITANA, (1428) NUNEZ, BUENOS AIRES, ARGENTINA

Summary
Non-enzymatic browning rates decrease with decreasing water contents below a critical water activity. Recent results indicate that this effect may be related to the temperature above the glass transition temperature, and may be due to diffusion-limited reactions.

Introduction

Effects of water activity on reaction rates in foods
Food deterioration reactions depend in a complex manner on the water content of foods. The components of the food may have different water contents at equilibrium, but they have the same water activity (a_w). For this reason, many workers have attempted to correlate relative reaction rates to a_w. Such a map of relative rates as a function of a_w was presented by Labuza et al. (1970) and is shown in Figure 1.

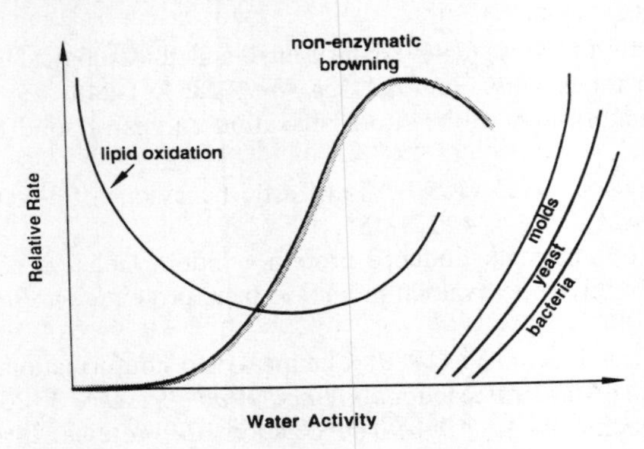

Figure 1. Dependence of rates of food deterioration on water activity. (Labuza et al. 1970).

The rate dependence of non enzymatic browning (NEB) at high a_w has been attributed to dilution effects and to "product inhibition" effect of water that is produced by the several condensation steps in the reaction. The decrease in rates below the maximum is generally attributed to decreasing availability of water, which is necessary for mobility of the reactants. Two approaches to description of this mobility are possible.

In the "solution" scheme (Duckworth, 1981), it may be assumed that the reactions occur only in solution. At the lowest a_w values, no reactants are in solution, and there is no reaction. Above this point, the concentration of reactants remains constant (because excess of solute maintains a saturated solution level), but the total volume in which the reaction takes place increases. The apparent reaction rate constant calculated for the total system increases, because the "reactive volume fraction" increases. At the a_w corresponding to maximum reaction rate (a_{wm}) all the reactants are in solution and further water additions result in dilution.

The "diffusion hypothesis" avoids the consideration of "reactive solution volume" and considers the food as a continuous and homogeneous system. The reduction in rate is attributed to diffusional limitations. The observed rate constant k' is then lower than the true constant k of equation (1) (observable in well-stirred solutions).

$$k' = \frac{k}{1 + \alpha/D} \qquad (1)$$

where α = a constant
D = diffusivity of the critical reactant.

As a result, at low values of D, the reaction may be controlled by diffusion, and it is well known that diffusivity of sugars, various small organic molecules, and water itself decreases exponentially as water contents approach zero. The range in which diffusivity shows the steepest decrease does in fact correspond to water activities below a_{wm} (King, 1988).

Glass transition temperature and mobility

It has been proposed that the mobility effects on rates of changes are related to the physical state of the food, and specifically to the temperature above the glass transition temperature (*Tg*). The development of a stability map based on the *Tg* curve in a temperature-concentration or temperature water content plot has been undertaken and reported in a very extensive series of papers reviewed by Slade and Levine, 1991 and by Roos and Karel, 1991. The empirical equation proposed by Williams et al. (1955) (WLF equation) is useful in describing several time-dependent processes. (equation 2):

$$\log \frac{\eta}{\eta_r} = \frac{C_1 (T - T_r)}{C_2 - (T - T_r)} \qquad (2)$$

where T_r is a reference temperature, η is viscosity, and η_r is viscosity corresponding T_r.

Williams et al. (1955) found good correlation for mechanical properties of polymers, using $T_r = T_g$, $C_1 = -17.44$ and $C_2 = 51.6$. Ferry (1980) noted that the best viscosity correlations in polymers above T_g were in many cases obtained with T_r different from T_g and C_1 and C_2 different from the original constants.

In solution, D is inversely proportional to η, and it is not unreasonable to assume this relationship in more complex systems. The effect of T_g on D has been studied in polymers, and it has been observed that the temperature dependence of diffusion in amorphous systems is often well represented by one of the following models (Vrentas and Duda, 1978; Karel and Saguy, 1991):

1) Arrhenius equation with two different activation energies, one valid below Tg and

the other above. Typically, the activation energy is much higher above Tg. (Vrentas and Duda, 1978)
2) WLF equation (equation 2), assuming D inversely proportional to η.

Recent studies at Rutgers and at the University of Minnesota have shown that diffusion in food-related systems above T_g is strongly dependent on $(T-T_g)$ and can often be represented by the WLF equation, as suggested by Slade and Levine (1992). In an experiment conducted in our laboratory, the diffusivity of propanol in amorphous sugars was shown to be negligibly small below T_g, and increased by a factor of 10 to 20 when $(T-T_g)$ increased from 15°C to 25°C. The results were consistent with the WLF equation. Nelson (1993) analyzed literature data on diffusion of spin probes in maltodextrin and reported a break in the Arrhenius plots at $1/T_g$. Phillies and Quinlan (1992) observed that diffusion of optical probes in dextran solutions followed an equation very similar in form to the WLF equation.

Effect of glass transition on rates of NEB

NEB involves a number of condensation steps involving pairs of reactants, and it is therefore at least potentially a set of reactions subject to diffusional limitations. If it can be assumed that the diffusion coefficient is inversely proportional to viscosity, and if the viscosity of the systems in which NEB is studied decreases drastically above T_g, then the rates of browning should show a strong dependence on $(T-T_g)$. If in addition, in the systems studied, viscosity decreases in accordance with the WLF equation, then the rates may also follow the WLF equation, and the Arrhenius plot may show a break in the vicinity of $1/T_g$.

We analyzed both data from the literature and our own experimental data. In the case of dehydrated foods for which browning data were available in the literature, this literature did not provide information on T_g. We obtained these glass transition temperatures experimentally on comparable materials. The details of the experimental procedures and results have been reported by Buera and Karel (1993).

We used the WLF equation as modified by Ferry (1980), to relate literature-derived and experimental NEB rates to $(T-T_g)$. The reference temperature of the Ferry method was located at the midpoint of the experimental temperature range. $T\infty$, the fixed temperature at which the rate constant k is assumed to approximate zero, was proposed by Ferry (1980) to be T_g-50°C. However, a better fit was obtained by setting $T\infty = T_g - T_{go}$, where T_{go} is the T_g of the anhydrous material. A plot for browning rates is shown in Figure 2.

The analysis of experimental and literature data on foods and related models led Buera and Karel (1993) to the conclusion that a good correlation was obtained using equation 3.

$$\log \frac{k_r}{k} = \frac{-C_{1r}(T - T_r)}{T - T_g + T_{g0}} \quad (3)$$

In this equation, the subscript r refers to the reference temperature. C_{1r} is a constant that varied with the system, but was independent of water content, and therefore of T_g. In fact, for all of the systems studied, including milk, vegetables, fruit and model systems, the average C_{1r} was 7.9, and this value can be used as a good first approximation. The

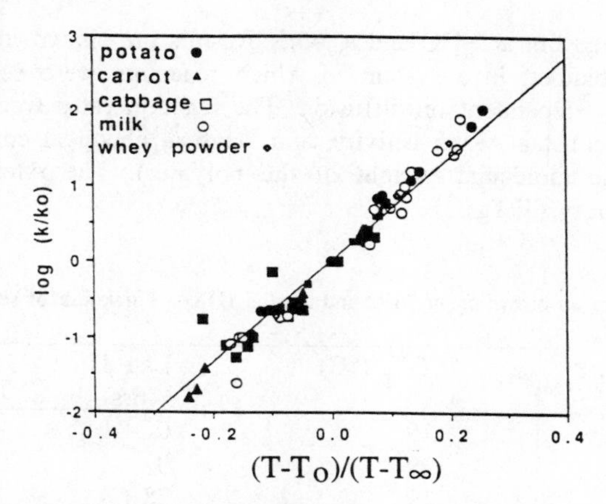

Figure 2. Linearized WLF plot for browning rates in several food materials.

coefficients can be estimated easily, providing the T_g values are known, and kinetic data at several temperatures are available. The dependence of the NEB rate constants on moisture content is implied by the inclusion of T_g in the equation.

The occurrence of breaks in an Arrhenius plot of NEB reaction rate constants has been observed in our laboratory. Figure 3 shows this phenomenon for a model system of lactose:carboxymethyl cellulose:xylose:lysine (88:10:1:1) reacted at an a_w of 0.12. More recent experimental studies are presented elsewhere in this volume by Karmas and Karel.

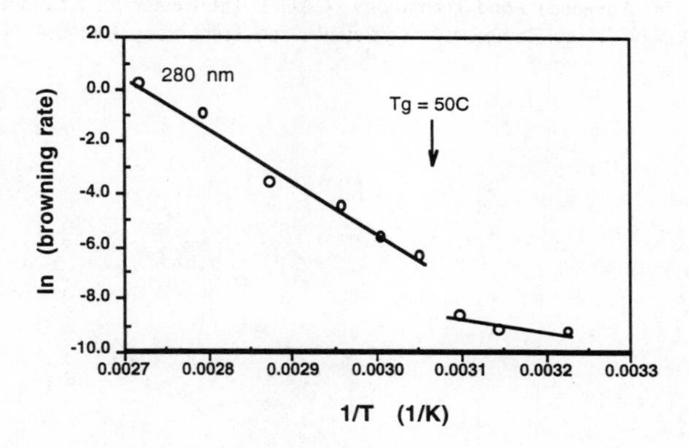

Figure 3. Arrhenius plot for browning in a model system.

We have also demonstrated the effect of $(T\text{-}T_g)$ on diffusion of reactants in NEB, by preparing samples of polyvinylpyrrolidone (PVP), of two different molecular weights, with reactants xylose and lysine dispersed within them, as follows:

a) PVP with both reactants premixed in ratio 99:0.5:0.5
b) PVP with xylose (99:0.5), either separated from PVP with lysine (99:0.5) by a layer of PVP 2 or 8 mm thick, or brought into direct contact (0 mm).

The reaction was carried out at 74°C and a_w=0.5. Results are shown in Table I.
These results were obtained in a system in which reactants were separated and the reaction was obviously dependent on diffusion. The reaction rates were measured at a constant temperature, constant water activity, and identical chemical composition of the matrix (except for the molecular weight of the polymer). The rates were strongly dependent on the quantity (T-Tg).

Table I. Effect of diffusion on browning at 74°C and a_w=0.5 (DX = Thickness of separating layer).

Matrix	DX (mm)	T-T_g (°C)	Lag Time (hrs)	Browning Rate
PVP24	premixed	19	0	1.00
PVP24	0	19	0	0.25
PVP24	2	19	28	0.18
PVP24	8	19	43	0.13
PVP40	premixed	5	0	0.14
PVP40	0	5	0	0.03
PVP40	2	5	110	0.02
PVP40	8	5	165	0.01

Acknowledgements
This is publication No. F10535-2-93 of the New Jersey Agricultural Experiment Station supported by State Funds and the Center for Advanced Food Technology (CAFT). The Center for Advanced Food Technology is a New Jersey Commission on Science and Technology Center.

References

Buera, M.P. and M. Karel (1993) Application of the WLF equation to describe the combined effects of moisture and temperature on non enzymatic browning rates in food systems. J. Food Process. Preser. 17:31-45.

Duckworth, R.B. (1975) Water Relations in Foods. Academic Press, N.Y.

Ferry, J.D. (1980) Dependence of viscoelastic behavior on temperature and pressure. In: Viscoelastic properties of polymers, pp. 264-272. John Wiley and Sons, N.Y.

Karel, M. and I. Saguy (1991) Effects of water on diffusion in food systems. In: H. Levine and L. Slade, eds. Water Relationships in Foods. Plenum Press, N.Y., pp. 157-174.

King, C.J. (1988) Spray drying of lipids, and volatiles retention. In: Bruin, S., ed. Preconcentration and drying of foods. Elsevier, Amsterdam pp. 147-162.

Labuza, T.P., Tannenbaum, S.R. and Karel, M. (1970) Water content and stability of low-moisture and intermediate-moisture foods. Food Technol. 24:543-550.

Nelson, K.A. (1993) Application of WLF and Arrhenius models to chemical reaction and diffusion rates in polymeric model food systems. Ph.D. Thesis University of Minnesota, St. Paul, MN

Phillies, G.D.J. and Quinlan, C.A. (1992) Glass temperature effects on probe diffusion in dextran solutions. Macromolecules 1993:25.

Roos, Y. and Karel, M. (1991) Applying state diagrams to food processing and development. Food Technology, December 66, 68-71, 107.

Slade L. and Levine H. (1991) Beyond Water Activity. Recent Advances Based on an Alternative Approach to the Assessment of Food Quality and Safety. CRC Crit. Revs. Food Sci. and Nutrition, 30(2,3) 115.

Vrentas, J.S. and Duda, J.L. (1978) A free volume interpretation of the influence of the glass transition on diffusion in amorphous polymers. J. Appl. Polymer Sci. 22:2325-2339.

Williams, M.L., Landel, R.F. and Ferry, J.D. (1955) The temperature dependence of relaxation mechanisms in amorphous polymers and other glass-forming liquids. J. Am. Chem. Soc.

Kinetics of the Early Maillard Reaction During Heating of Milk

M. A. J. S. Van Boekel and H. E. Berg

DEPARTMENT OF FOOD SCIENCE, WAGENINGEN AGRICULTURAL UNIVERSITY, PO BOX 8129, 6700 EV WAGENINGEN, THE NETHERLANDS

Summary
Study of the kinetics of the heat-induced Maillard reaction in milk is difficult due to competing reactions of lactose: it is also involved in isomerization and degradation reactions yielding lactulose, galactose and formic acid as the main products. To unravel the reaction network in which lactose is involved in heated milk, the protein-bound Amadori product was isolated and heated: the main reaction products were lactose, galactose and formic acid. A kinetic model was established that takes into account both isomerization/ degradation and the Maillard reaction. This model consisted of coupled ordinary differential equations which were solved by numerical integration and fitted to experimental data to obtain reaction rate constants. The model was able to describe the course of the reaction in heated milk for 0-15 min at 110-150 °C, but not at longer heating times, especially not at 150 °C, probably due to interference of advanced Maillard reactions. Overall, the degradation of lactose via isomerization/degradation was quantitatively of more importance than via the Maillard reaction.

Introduction

Very often, milk is heat-treated which may affect its quality due to heat-induced chemical reactions. Reactions of lactose are important in this respect. Reaction products studied are lactulose and hydroxymethylfurfural (Burton, 1984), available lysine (e.g. Horak & Kessler, 1981) and furosine (e.g. Finot et al., 1981). The kinetics of formation of such components is studied using a zero-, first- or second-order approach, whichever fits the data best. However, in our view the kinetics of heat-induced breakdown of lactose is too complicated to be treated in such a relatively simple way because of possible parallel and consecutive reactions. Therefore, we had two objectives in this study: 1) to identify reaction products and reaction paths of lactose and its breakdown products, and 2) to establish kinetic parameters for the complete reaction network by kinetic modelling. Here, we concentrate on the behaviour of lactose in the Maillard reaction, which in milk is largely confined to lactose and lysine-residues of milk protein.

Materials and methods

All reagents used were of analytical grade. Raw skim milk was heated in an oil bath in stainless steel tubes (contents 12 mL) and after heating immediately cooled in ice. Heat treatments ranged from 0-30 min at 110-150 °C. The reported heating times include the 2 min heating-up time.

Lactose, lactulose and galactose were determined by HPLC (IDF, 1992). Formic acid was determined by making up 15 g milk with 0.5 M perchloric acid; the filtrate was analyzed by HPLC (Marsili *et al.*, 1981). Furosine was determined by HPLC (Resmini *et al.*, 1990). Hydroxymethylfurfural, furfural and furfurylalcohol were measured by HPLC (Van Boekel & Zia-ur-Rehman, 1987). Lysine was determined by a dye-binding procedure using Crocein Orange G (Hurrell *et al.,* 1979).

Gear's algorithm was used for numerical integration (Chesick, 1988), and data fitting was by non-linear least-squares using the direct search algorithm described by Lobo & Lobo (1991).

Results

The main reaction products of lactose in heated milk appeared to be lactulose, galactose, formic acid and protein-bound lactulosyllysine in quantities ranging from 0-20 mmol/L. HMF and furfurylalcohol were only formed in small quantities (0-400 μmol/L), furfural was produced in even smaller amounts (0-10 μmol/L). The early Maillard reaction was studied by following the changes in available lysine and furosine. Furosine was recalculated to the Amadori compound lactulosyllysine as 2.25xfurosine. (The factor 2.25 accounts for the yield of furosine from the Amadori product (Molnár *et al.*, 1987).) Figure 1 shows some of the results.

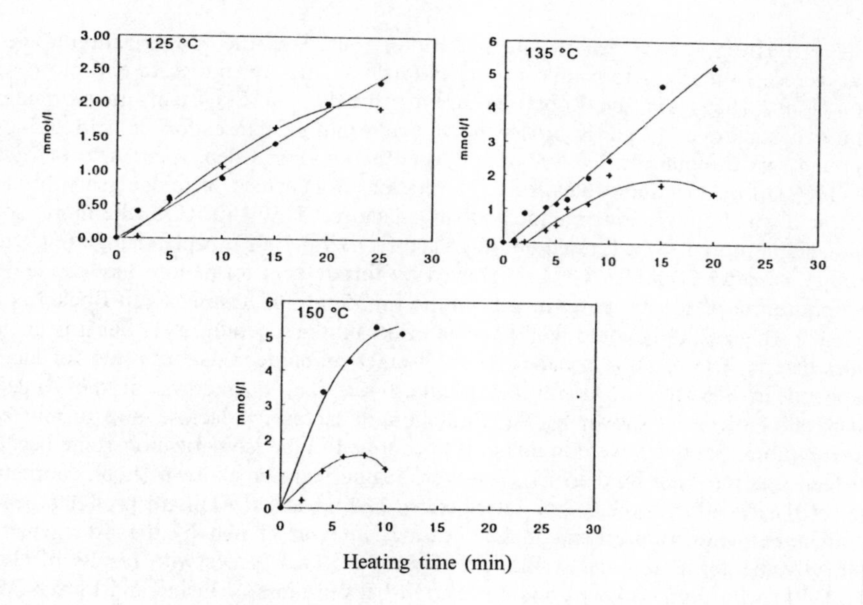

Figure 1. Lysine loss (•) and lactulosyllysine formation (+) as a function of heating time at 125, 135 and 150 °C.

It is not well known how the protein-bound Amadori product behaves in heated milk; to our knowledge no quantitative data are available. Therefore, we isolated it by mildly heating of a lactose/casein/ milk salts solution (resembling the composition of milk), followed by dialysis until all lactose was removed; a yield of about 600 μmol/L protein-bound lactulosyllysine was obtained in this experiment. This dialysed solution was heated at 120 and 140 °C and analysed for reaction products (Figure 2).

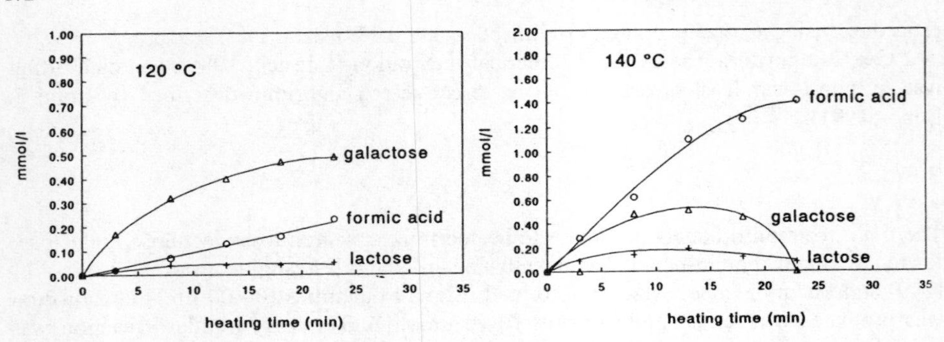

Figure 2. Reaction products formed from protein-bound Amadori-product during heating at 120 and 140 °C.

Discussion

It appeared (Berg, 1993) that the main reaction route was the isomerization route in which lactose isomerizes to lactulose (and in much smaller quantities to epilactose, e.g. Olano et al., 1989) via the Lobry de Bruyn-Alberda van Ekenstein transformation; lactulose is subsequently partly broken down again into galactose, formic acid and a C5 compound, which could be deoxyribose (Berg, 1993) and/or 3-deoxypentulose (Troyano et al., 1992). Two reaction routes for the formation of galactose were necessary because formic acid was found in lower amounts than galactose below 140 °C; furthermore, some formic acid appeared to be formed out of galactose (Berg, 1993). A pH change (e.g. from pH 6.6 to 6.4 after 20 min 120°C) is induced by formic acid formation, though changes in salt-equilibria also contribute, mainly in the first 5 min of heating (Van Boekel et al., 1989). The change in pH could well have an effect on the reaction rates, but it is unclear to what extent. The reaction scheme for the isomerization/degradation route for lactose is depicted in Figure 3A. As for the Maillard reaction, breakdown of the Amadori product takes place as shown by the formation of lactose, galactose and formic acid (Figure 2); these products were in this case not formed in the isomerization route because all lactose was removed by dialysis. However, a significant breakdown (in a quantitative sense) of the Amadori product into intermediate and advanced Maillard products occurs only at higher temperatures and longer heating times as shown by the divergence of lactulosyllysine formation and lysine loss (Figure 1). This agrees with results of Henle et al. (1991) who found a very good correlation between loss of lysine and formation of lactulosyllysine in milk heated at 100-120 °C. It is noteworthy that no lactulose was formed from the Amadori product; this rules out the possibility that lactulose is formed in the Maillard reaction, as has been suggested (Adachi & Patton, 1961). The small amounts of lactose formed suggest that perhaps not all of the Schiff base was transformed into the Amadori product. At any rate, the early Maillard reaction should be described as a reversible reaction. The formation of galactose and formic acid from the Amadori product indicates the degradation of its carbohydrate moiety and resembles very much the degradation of lactulose shown in Figure 3A. Formation of galactose due to hydrolysis of the carbohydrate moiety is not very likely as Henle (1991) was not able to identify any fructoselysine in heated milk. The Maillard reaction scheme based on these results is shown in Figure 3B.

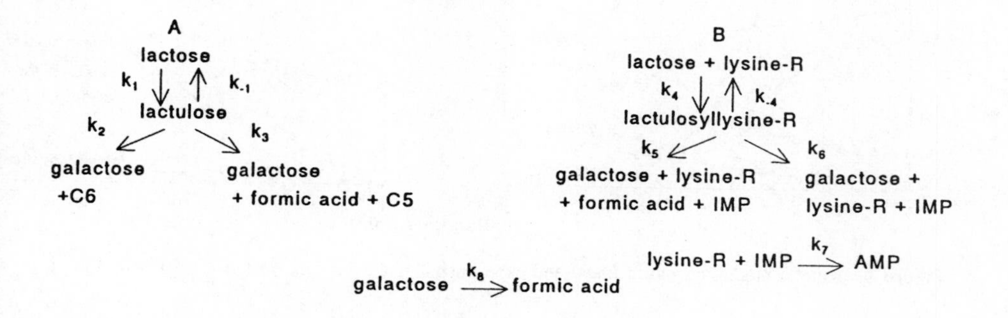

Figure 3. Reaction scheme for the isomerization/degradation (A) and Maillard reaction (B) route of lactose in heated milk. IMP = Intermediate Maillard Products, AMP = Advanced Maillard Products.

Lactulose and galactose are probably also reacting with lysine in the Maillard reaction, but the quantitative contribution of this was found to be negligible (Berg, 1993). The much more pronounced formation of formic acid at the higher temperatures may indicate a change in reaction mechanism at these higher temperatures; we have accounted for this by inclusion of the two reaction rate constants k_3 and k_5 (cf. Figure 3).

Differential equations were derived based on the reaction scheme of Figure 3. These were solved numerically and fitted to experimental results obtained for milks heated between 110 and 150 °C. Two examples are given here (Figure 4). It appeared that the kinetic model was able to describe the formation of reaction products quite well, except for the longer heating times at the higher temperatures (≥ 140 °C) as shown in Figure 4B. This is very likely due to the interference of intermediate and advanced Maillard reactions, and further reactions of both lactulose and galactose. Especially the formation of galactose and formic acid is not correctly predicted. This shows the power of computer modelling to test proposed kinetic models; in fact we have tried numerous kinetic models, of which the one presented in Figure 3 appeared to be the best. The kinetic analysis showed that k_1, k_2 and k_4 were quite critical while k_3 and k_6 were necessary to account for the formation of formic acid at the higher temperatures. The numerical values of the other reaction rate constants were not so critical; however, additional experiments described by Berg (1993) showed that they were necessary to include in the kinetic model and their order of magnitude was derived from experiments with model solutions of casein and various sugars.

The temperature dependence of the reaction rate constants was derived from Eyring's absolute rate theory (Berg, 1993). The results for k_4 (formation of Amadori product) and, for comparison, for k_1 (lactulose formation) are given in Table 1. The activation enthalpies and entropies found are quite normal for chemical reactions; the confidence interval for k_4 is unfortunately too wide to draw conclusions about a possible difference with k_1.

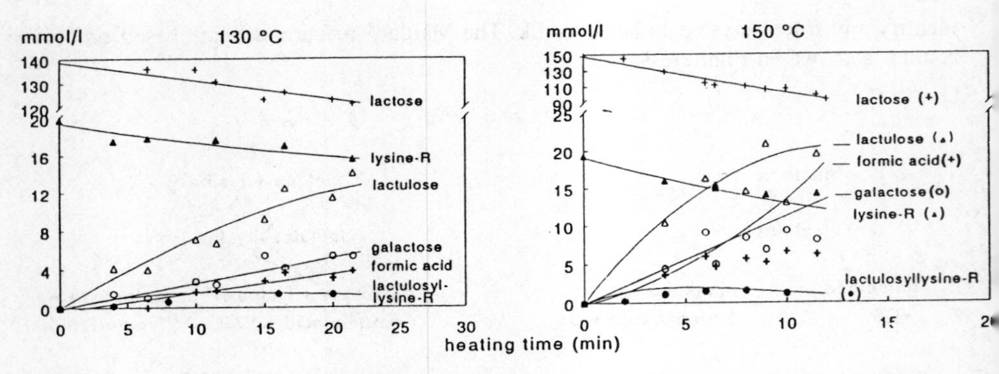

Figure 4. Kinetic modelling (solid lines) and experimental data for milk heated at 130 and 150 °C.

Table 1. Activation enthalpy $\Delta H^{\mp}$, activation entropy $\Delta S^{\mp}$ ($\pm$ 95% confidence intervals) and Q_{10} for reaction rate constants k_1 and k_4.

	$\Delta H^{\mp}$ (kJ.mol^{-1})	$\Delta S^{\mp}$ (J.mol^{-1}K^{-1})	Q_{10} (at 130 °C)
k_1	101 ± 6	-73 ± 5	2.2
k_4	94 ± 27	-121 ± 41	2.1

Conclusions

Degradation of lactose via the Maillard reaction in heated milk is quantitatively of less importance than the isomerization/degradation route. The early Maillard reaction as well as the isomerization route can well be described by the proposed kinetic model. Only at prolonged heating times and temperatures ≥ 140 °C become advanced Maillard reactions of quantitative significance. Kinetic modelling appears to be a powerful tool to analyse complicated reaction networks such as the Maillard reaction.

References

Adachi, S. and Patton, S. (1961) Presence and significance of lactulose in milk products: a review. *J. Dairy Sci.* 44: 1375-1393

Berg, H.E. (1993). Reactions of lactose during heat treatment of milk: a quantitative study. PhD thesis, Wageningen Agricultural University, Wageningen, The Netherlands.

Burton, H. (1984) Reviews of the progress of Dairy Science: the bacteriological, chemical and physical changes that occur in milk at temperatures of 100-150 °C. *J. Dairy Res.* 51: 341-363

Chesick, J.P. (1988) Interactive program system for integration of reaction rate equations. *J. Chem. Ed.* 65: 599-602

Finot, P.A., Deutsch, R and Bujard, E. (1981) The extent of the Maillard reaction during processing of milk. *Progr. Food Nutr. Sci.* 5: 345-355

Henle, Th., Walter, H. and Klostermeyer, H. (1991) Evaluation of the extent of the early Maillard reaction in milk products by direct measurement of the Amadori-product lactuloselysine. *Z. Lebensm. Unters. Forsch.* 193:119-122

Henle, Th. (1991) [Studies on the reactive behaviour of milk proteins in the Maillard reaction] (in German). PhD thesis, Technical University Munich, Germany.

Horak, F.P and Kessler, H.G. (1981) The influence of UHT heating and sterilisation on lysine in milk. *Milchwissenschaft* 36: 543-547

Hurrell, R.F., Lerman, P. and Carpenter, K.J. (1979) Reactive lysine in foodstuffs as measured by a dye-binding procedure. *J. Food Sci.* 44: 1221-1227

IDF (1991) Heat-treated milk-Determination of lactulose content. HPLC method (reference method). Provisional standard 147. International Dairy Federation, Brussels.

Lobo, L.S. and Lobo, M.S. (1991) Robust and efficient nonlinear regression of kinetic systems using a direct search method. *Comp. Chem. Eng.* 15: 141-144

Marsili, R.T., Ostapenko, H., Simmons, R.E. and Green, D.E. (1981) HPLC determination of organic acids in dairy products. *J. Food Sci.* 46: 52-57

Molnár-Perl, J., Pintér-Szakács, J.,Wittmann, R., Reuter, M., and Eichner, K. (1986) Optimum yield of pyridosine and furosine originating from Maillard reactions monitored by ion-exchange chromatography. *J. Chromatogr.* 361: 311-320

Olano, A., Calvo, M.M. and Corzo, N. (1989) Changes in the carbohydrate fraction of milk during heating processes. *Food Chem.* 31: 259-265

Resmini, P., Pellegrino, L. and Battelli, G. (1990) Accurate quantification of furosine in milk and dairy products by a direct HPLC method. *Ital. J. Food Sci.* 3: 173-183

Troyano, E. I., Olano, A., Jimeno, M.L. and Sanz, I. (1992) Isolation and characterization of 3-deoxypentulose and its determination in heated milk *J. Dairy Res.* 59: 507-515

Van Boekel, M.A.J.S. & Zia-ur-Rehman, (1987) Determination of hydroxymethylfurfural in heated milk by HPLC. *Neth. Milk Dairy J.* 41: 297-306

Van Boekel, M.A.J.S., Nieuwenhuijse, J.A. & P. Walstra, P. (1989). The heat coagulation of milk. 1. Mechanisms. *Neth. Milk Dairy J.* 43: 97-127

Interpreting the Complexity of the Kinetics of the Maillard Reaction

T. P. Labuza

DEPARTMENT OF FOOD SCIENCE AND NUTRITION, UNIVERSITY OF MINNESOTA, ST. PAUL, MINNESOTA, USA, 55112

Summary
The Maillard reaction in foods is a very complex reaction comprising some well studied initial steps, much data on intermediate product formation (flavors) and the overall production of brown color, but little information exists that can be used to model the extent of the reaction based on basic physical chemical principles. It is known that the reaction has a high activation energy which is reactant, moisture and pH dependent but the exact influence is unclear. This paper takes available data and tries to put some basic physical chemical sense into them so that future studies can be better designed.

Introduction

It is well known that the Maillard, or non-enzymatic browning, reaction in foods and living biological systems (NEB), is much more complex than the simple glucose-glycine condensation reaction studied by Louis Maillard in 1912 (Baynes and Monnier, 1989; Finot et al., 1990). By 1953 the classic Hodge scheme (Hodge 1953) indicated the very complex nature of the reaction with the myriad of intermediate steps. Yet in these past 80+ years most food researchers have been involved with finding out what intermediate compounds are produced (literally thousands) rather than developing the mathematical equations based on sound physical chemical principles that could be used to make valid predictions of color, flavor etc. under various formulation, processing and storage conditions as has been done in part for lipid oxidation in foods (Labuza 1971). For example, there is little data available to help a ketchup processor choose the sugar level and time temperature sequence in cooking to optimize the desirable flavors and color of the product. In fact many of the NEB studies in model systems have been carried out under conditions which have little extrapolation to reality, such as high temperatures and high pH values (Kaanane and Labuza 1989). In 1992 we thoroughly reviewed the kinetics of NEB and established the guidelines of what should be done if future studies are to contribute to both the food technology and biomedical arena (Labuza and Baisier 1992). This short paper highlights some of those criteria and poses questions for future researchers.

Discussion

Although the Hodge scheme is very complex, the NEB reaction can be kinetically described as a reaction between two reactants. One reactant, the amine, after being cleaved from an important first intermediate, can react with more reducing compounds and can continue on to eventually condense into brown pigment. It must be understood that the reactants are not just reducing sugars and amino acids in proteins. In foods (which are very complex) the reducing moiety in the reaction can be represented as the various reducing sugars (ribose, xylose, glucose, fructose, lactose, maltose), ascorbic acid which is the main form of reactant in orange juice, aldehydes and ketones which are critical odors in foods (cinnamic aldehyde), orthophenolics (present in abundance in coffee, tea and chocolate) and sucrose when reacted in acid conditions. The reacting amines can comprise the free amino acids naturally present in many fresh fruits and vegetables or added to make TPN (total parenteral nutrition) solutions, the N-terminal amine group on a protein, the ε-amino group on lysine in a protein, aspartame which is used as an artificial sweetener and MSG and disodium guanylate used to create meat-like (umami) flavors. Each of these react at different rates depending on pH and water activity (related to water content), which makes prediction of the extent of the reaction in any food difficult unless specific data are available for the particular system. Thus the variety of flavors developed in cereals, coffee, and chocolates, are a result of this complexity and lead to a vast variety of odors and color.

The overall scheme of product flow is seen in Figure 1 where, as would be expected for a multi-step reaction, the level of concentration of intermediates increases with time, reaches a maximum and then decreases again, as occurs with hydro-peroxides in lipid oxidation (Labuza 1971). Recently, Namiki et al. (1993), using chemiluminescence, has confirmed that indeed the concentration of intermediates can be described by the rise and fall shown in this scheme.

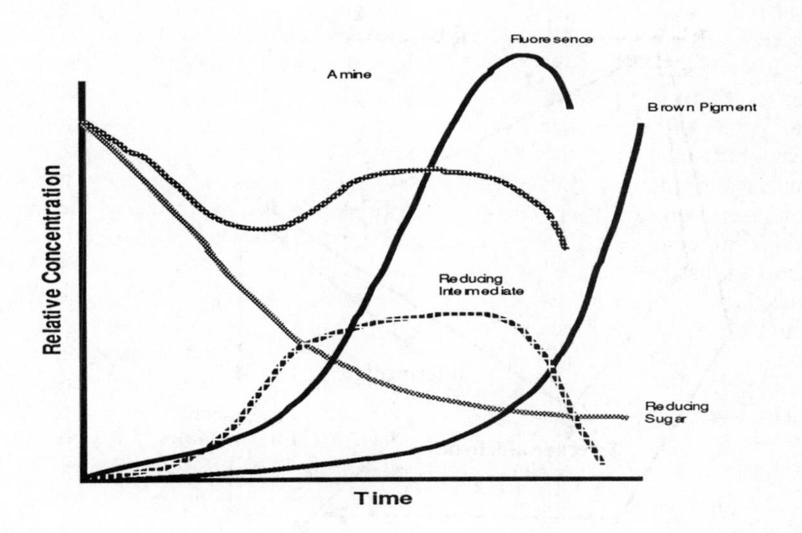

Figure 1. Concentration of reacting and formed species as a function of time during the Maillard reaction.

This scheme can be represented kinetically by the following series of reactions as represented in Figure 2 (Labuza, 1984). Based on this scheme one can see that the reaction is first order with respect to reducing sugar (compound) consumption as is generally found, initially first order for amine loss until the recycle step becomes of significance which then leads to an increased concentration reaching a no loss period (Massaro and Labuza 1990), second order overall for sugar/amine loss (Baisier and Labuza (1992) and very complex for all intermediates and end products (see other kinetics papers in this volume).

As found by Baisier and Labuza (1992), the absolute rate of the loss of glucose in a sugar/amine solution increases as the amine concentration increases since the amine is limiting. For a 0.4 M glucose system the rate increased in about the expected ratio (e.g. k = 3.62 hr^{-1} at 0.2 M glycine and 7.78 hr^{-1} at 0.4 M glycine, i.e. a doubling of the rate for a doubling of the amino acid concentration. Thus the observed first order rate constant initially is a product of k_1 times the amine concentration until the reaction begins to go further. In many cases in the literature, this concentration dependence of the rate constant is not well understood and thus reported rate constants are compared between studies at different concentrations. Baisier and Labuza (1992) found a similar pattern for the effect on increasing sugar concentration at constant amine concentration. Thus the observed first order rate constant reported in most studies is not an absolute value but depends on the pH and concentration of the other reactant which might not be reported. More importantly, if browning extent is measured, the rate increase with concentration follows a different pattern, the reasons for which are not well studied or understood. For example in the same system, a four fold increase in the glycine content at constant glucose level only increased the browning rate (color formation) by 50%, while increasing the glucose content at constant glycine increased the rate by about 66%.

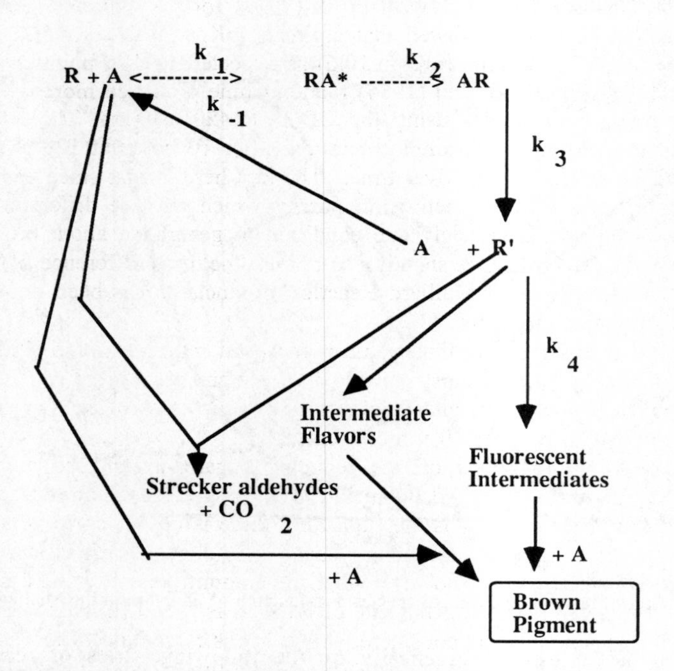

Figure 2. Kinetic scheme for the general steps in the Maillard reaction.

The same problem of reporting absolute rates exists for the pH effect. In many studies reviewed by Labuza and Baisier (1992), the pH was not controlled although some reported that the pH generally increased with reaction extent. From physical chemistry, the pH decreases as a function of increased temperature, e.g. the pH of a neutral solution at 25°C is about 6.0 at 100°C (Bell and Labuza 1992). Thus comparing studies of systems of similar initial pH at room temperature but different study temperatures (e.g. 65 and 120°C) may give false conclusions. More importantly the pH is a major factor for base catalyzed reactions such as NEB. For a pure system with no interference, the rate constant should increase by 10 fold for a unit pH increase. Thus most reactions studied are done in basic solution to accelerate it, (e.g. Dworschak and Orsi, 1977) but back extrapolation to normal pH requires at least 3 or 4 pH values since there are many complications. For example, Pilkova et al. (1990) found that as pH was increased from 4 to about 9.5, tryptophan (pK_a 9.39) reacted faster (disappearance) with fructose than methionine ($pK_a = 9.2$), the opposite of that expected since more tryptophan would be unprotonated. In addition, tryptophan showed only a 1.2 fold increase for a unit pH increase while methionine was almost 8 fold. At pH 9.3 both amino acids disappeared at the same rate.

The major problem with this study and many others like it is that only one data point was taken to make a rate comparison, i.e. the reaction mixture was heated for 8 hours at 110°C and then sampled. Not only is this meaningless in terms of kinetics, it ignores the potential recycling of the amino acid which would show a decrease followed by an increase to a steady state and then a decrease again as was found by Massaro and Labuza (1990) for model TPN solutions and Wolf et al. (1977) for soy proteins. As another example, Wolfram et al. (1974) showed that arginine ($pK_a = 9.1$) browned 20 times faster with glucose than did alanine (pKa = 10.6) as expected (6 data points over 10-80 hours at 65°C), while Ashoor and Zent (1984) found alanine produced more color (about 4 fold than arginine), both studies using the OD near 420-490 nm. The later study, however, was done with only one sample measured after 10 hours at 120°C while the former was based on several points over time. The key here is that since each step in the overall scheme has a different activation energy which may be different for each amino acid and sugar used, using only one condition to generalize about browning or amino acid reactivity is in error. One should also not use the mass difference of the sugar and amine to calculate a mass of Maillard Reaction products as has been done since it should be based on molar quantities.

Lastly, the well publicized fact that lysine is the most reactive amino acid, in fact, has no foundation. Certainly it is most reactive in a protein because of the availability of the ε-amino group, but both Fry and Stegink (1982) and Massaro and Labuza (1990) have shown tryptophan to be over 30 x more reactive.

The pH also has an influence on the reducing sugar through the mutarotation reaction. As reviewed by Kaanane and Labuza (1989), the greater the amount of acyclic form in solution, the faster reacting is the sugar, but there is no 1:1 correspondence. In fact the actual amount of acyclic state of any sugar is in dispute so it is hard to make theoretical judgments. It is known, however, that the amount generally increases as the pH increases, theoretically increasing the rate of NEB as is found (Cantor and Peniston 1940), however this is generally not considered when one is evaluating different systems. Higgins and Bunn (1981) found that fructose reacted 7.5 times faster than glucose with hemoglobin (pH 7.3 @ 37°C), while theoretically fructose has 350 fold more acyclic state present in solution. Kaanane and Labuza (1989) also noted, in an evaluation of this data, that the aldehydes had a 3 log cycle faster reaction rate with the protein at similar % acyclic state.

The equation below represents a theoretical depiction of the extent of brown pigment formation as a function of time. Interestingly, formation of both fluorescence (Hildago

$$B = R_0 + B_0 + [F_0][1 - e^{-k_B t}] - k_3 k_A k_B [R_0] x$$
$$\left\{ \frac{e^{-k_3 t}}{k_3(k_A - k_3)(k_B - k_3)} + \frac{e^{-k_A t}}{k_A(k_3 - k_A)(k_B - k_A)} + \frac{e^{-k_B t}}{k_B(k_3 - k_B)(k_A - k_B)} \right\}$$

and Zamora, 1993) and brown pigment (Warmbier et al., 1976; Labuza and Saltmarch, 1981; Labuza and Baisier, 1992) can generally be represented as a pseudo-zero order reaction after a variable but sometimes significant induction period despite this complex equation. Unfortunately, for most systems we do not have the magnitude of the individual rate constants so we cannot solve this equation and must revert to the simplistic plot as a zero order reaction, which it truly is not.

Generally, fluorescence has a lower activation energy than browning, with that of browning varying from 84 to 170 Kjoules/mole. In addition, because of the complex reaction scheme, with numerous pathways all of which are both pH and temperature dependent, it is clear that results at high temperature for loss of or production of compound X can not be extrapolated by any simple means to other conditions.

Because of limited space, no discussion has been made in this paper about the influence of water content or water activity on the Maillard reaction. This topic is of great importance and is covered elsewhere in this book.

References

Ashoor, S.H. and Zent. 1984. Maillard browning of common amino acids and sugars. *J. Fd. Sci.* 49:1206-1207.

Baisier W. and Labuza, T.P. 1992. Maillard Browning Kinetics in a liquid model system. *J. Agric. Fd. Chem.* 40(5): 707-712.

Baynes J.W. and Monnier V. (editors) 1989 *Maillard Reaction in Aging.*, Alan R. Liss, Inc.NY.

Bell, L.N. and Labuza, T.P. 1991 Potential pH implications in the freeze-dried state. *Cryoletters* 13:235-244.

Cantor, S.M. and Penniston, Q.P. The reduction of aldoses at the dropping mercury cathode: estimation of aldehydo structure in aqueous solutions. 1940. *JAOC* 62:211-2121.

Dworschak, E. and Orsi, F. (1977) Study into the Maillard reaction occurring between methionine and tryptophan on the one hand and glucose on the other *Acta Aliment.*, 6:59.

Finot, P.A., Aeschbacher, H.U., Hurrell, R.F. and Liardon, R., eds. (1990) *The Maillard Reaction in Food Processing, Human Nutrition and Physiology.* Birkhauser Verlag.Basel.

Fry, L.K. and Stegink, L.D. 1982. Formation of Maillard reaction products in parenteral alimentation solutions. *J. Nutrition* 112:1631-1637.

Kaanane, A. and Labuza, T.P. 1989. The Maillard Reaction in Foods in *The Maillard Reaction in Aging Diabetes and Nutrition* J. Baynes and V. Monnier (eds) A.R. Liss Press N.Y. pp 301-328

Higgins, P.J. and Bunn, H.F. (1981) Kinetic Analysis of the Nonenzymatic glycosylation of hemoglobin *J. Biol. Chem.* 256(10) 5204-5208.

Hildago, F. and Zamora, R. 1993. Non-enzymatic browning and fluorescence development in a (E)-4,5- epoxy-(E)-2-heptenal/lysine model system. *J. Fd. Sci.* 58(3) 667-670.

Hodge, J. 1953. Chemistry of browning reactions in model systems. *J. Agric. Fd. Chem.* 15:928-943

Labuza, T.P. and Baisier, W.M. (1992) "Kinetics of non-enzymatic browning" in *Physical Chemistry of Foods.* H. Schwartzberg, ed. Marcel Dekker, New York. pp 595-649.

Labuza, T.P. 1971. Kinetics of lipid oxidation of foods. *Crit. Rev. Food Tech.* 2:355-405

Labuza, T.P. (1984) Applications of chemical kinetics to the deterioration of foods *J. Chem. Ed.* 61(4):348-388.

Labuza, T.P. and Saltmarch, M. (1981)Kinetics of browning and quality loss in whey powders during steady state and non-steady state storage conditions *J. Food Sci.* 47:92-96.

Lerici, C.R., Barbanti, D. and Cherubin, S. (1990) Early indicators of chemical changes in foods due to enzymatic or nonenzymatic browning reactions *Lebensm.-Wiss. u.Tech.,* 23: 289-294.

Massaro, S.A. and Labuza, T.P. (1990) Browning and amino acid losses in model Total Parenteral Nutrition Systems with specific reference to cysteine *J. Food Sci.* 55(3) 821-826.

Namiki, M. Oka,M. Otsuka, M. Miyazawa, T. Fujimoto, K. and Namiki, K. 1993. Weak chemiluminescence at an early stage of the Maillard reaction. *J. Agric. Fd. Chem.* 41: 1704-1709.

Pilkova, L. Pokorny, J. and Davidek, J. 1990. Browning reactions of Heyns rearrangement products. *Die Nahrung* 34(8):759-764

Warmbier, H.C., Schnickels, R.A. and Labuza, T.P. (1976) Nonenzymatic browning kinetics in an intermediate moisture model system: effect of glucose to lysine ratio *J. Food Sci.* 41: 981-983.

Wolf, J.C., Thompson, D.R. and Reineccius, G.A. (1977) Initial losses of available lysine in model systems *J. Food Sci.* 42:1540.

Wolfrom, M.L., Kashimura, N., and Horton, D. (1974) Factors affecting the Maillard browning reaction between sugars and amino acids *J. Agr. Food Chem.* 792:796.

The Effect of Glass Transition on Maillard Browning in Food Models

R. Karmas and M. Karel

CENTER FOR ADVANCED FOOD TECHNOLOGY, FOOD SCIENCE DEPARTMENT, COOK COLLEGE, RUTGERS UNIVERSITY, PO BOX 231, NEW BRUNSWICK, NEW JERSEY 08903-0231, USA

Summary
Maillard browning is influenced by the glass transition. In systems where the matrix crystallizes, an increase in browning is noticed as a consequence of crystallization, which in turn is a result of storing the system above the glass transition temperature. In other systems the browning kinetics is best described by including moisture and temperature considerations in addition to the glass transition temperature.

Introduction

Non-enzymatic browning (NEB) is an important cause of food quality loss. Maximum browning rates typically occur in the water activity (a_w) range of 0.50 to 0.75 (Labuza and Saltmarch, 1981). At lower a_ws, the reaction proceeds more slowly, and this has been attributed to diffusional limitations (Eichner and Karel, 1972). The glass transition may be an important factor in limiting diffusion in low moisture amorphous food systems (Karel and Saguy, 1991; Slade and Levine, 1991). Karel and Buera, elsewhere in these proceedings, discuss the relevance of Tg as a critical point for food stability. White and Cakebread (1966), Roos and Karel (1991), and Slade and Levine (1991) have discussed the effect of glass transition on physical changes in amorphous foods. Recent work has studied the influence of glass transition temperature (Tg) on chemical changes and NEB (Buera and Karel, 1993; Karmas et al., 1992; Lim and Reid, 1991; Nelson, 1993). The following research further addresses the importance of the glass transition, as well as temperature, moisture content and composition on the browning reaction.

Materials and Methods

The effect of glass transition on browning of xylose and lysine in various matrices was studied. The reactants were chosen because of their high reactivity. A variety of matrix materials were used: simple sugars, polysaccharides, crystallizing and non-crystallizing systems, other polymeric compounds and mixtures. Samples were prepared from 20% (w/w) solutions that were freeze dried (0.5 mL in 2 mL vials) and equilibrated in desiccators to various moisture contents. Three types of experiments were done: 1) extent of browning at room temperature (20°C) and various moisture contents: T=20°C, Tg decreases with increasing moisture; 2) browning rates at different temperatures and a fixed moisture content; T is variable, Tg is fixed; and 3) browning rates in systems in which both temperature and moisture content were changed; both T and Tg change.

Browning measurements, DSC scans, and isotherms were done as described by Karmas et al., 1992.

Results and Discussion

Figure 1 shows browning results for the matrix/xylose/lysine system (94:5:1) at 20°C after 4 weeks of storage in the desiccators of 10 different water activities. For the lactose, raffinose, and trehalose matrices a typical browning curve is obtained (including diffusion limited, maximum browning, and dilution regions) . The arrows show the onset of browning or the a_w at which a noticeable increase in browning is observed. For each of these systems browning starts when the matrix material crystallizes, resulting in phase separation and subsequent concentration of the reactants in the non-crystalline volume. This behavior has been noticed in other systems and has been reported in the literature (Vautaz, 1988; Saltmarch et al., 1981; Kim et al., 1981). Crystallization is initiated at the a_w where Tg drops below the storage temperature (20°C). Crystallization in these systems was confirmed using DSC scans and moisture sorption measurements. In the matrix containing both maltodextrin (MD150) and trehalose, crystallization is inhibited and browning is retarded.

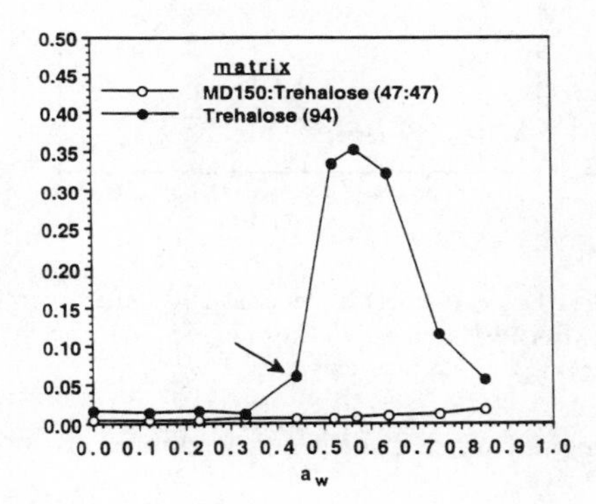

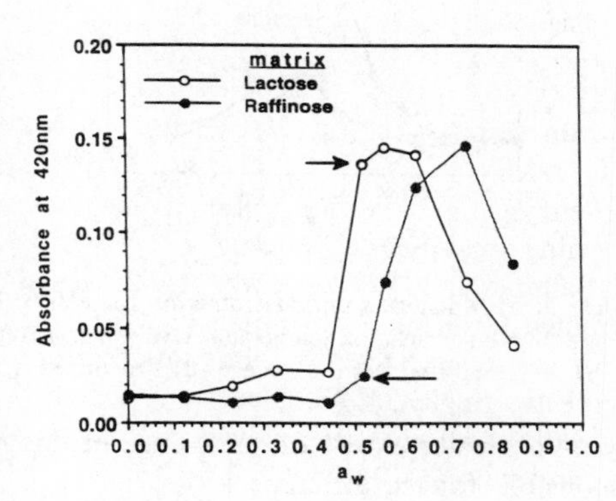

Figure 1. Browning of xylose and lysine in crystallizing and non-crystallizing matrices after storage at 20°C in desiccators. Lactose and raffinose systems were stored for 2 weeks. The trehalose containing systems were held for 4 weeks.

Figure 2 shows a similar type of experiment except in non-crystallizing polymeric matrices. Polyvinylpyrrolidone (PVP) of 3 different molecular weights (MW) were used. These systems had similar sorption characteristics but different Tgs. PVP/xylose/lysine (94:5:1) samples were held for 1 week in desiccators. Figure 2a shows the isotherms for the three systems. Figure 2b shows the Tgs for the three systems over the range of a_ws. For all systems, Tg decreases with increasing a_w, as expected. Higher Tgs were observed in the higher MW samples. The horizontal line represents the 20°C experimental storage temperature. If the Tg can serve as a "critical point" for reactivity, the onset of browning would be expected at the a_ws indicated by the arrows. This is what is in fact observed (Figure 2c). Browning starts between an a_w of 0.23 and 0.33 in the PVP10 system. In the other two systems browning occurs at about a_w=0.50. Another way to see the effect of Tg on browning is to consider the same results plotted as a function of (T-Tg). The storage temperature, as stated before, is 20°C for all the samples; however, Tg is unique to the MW and moisture content. Figure 2d shows that the onset of browning occurs at a (T-Tg) of -25. All the curves superimpose below (T-Tg)=0, when the systems are in a glassy state. Above (T-Tg)=0 browning curves diverge with the lower MW system

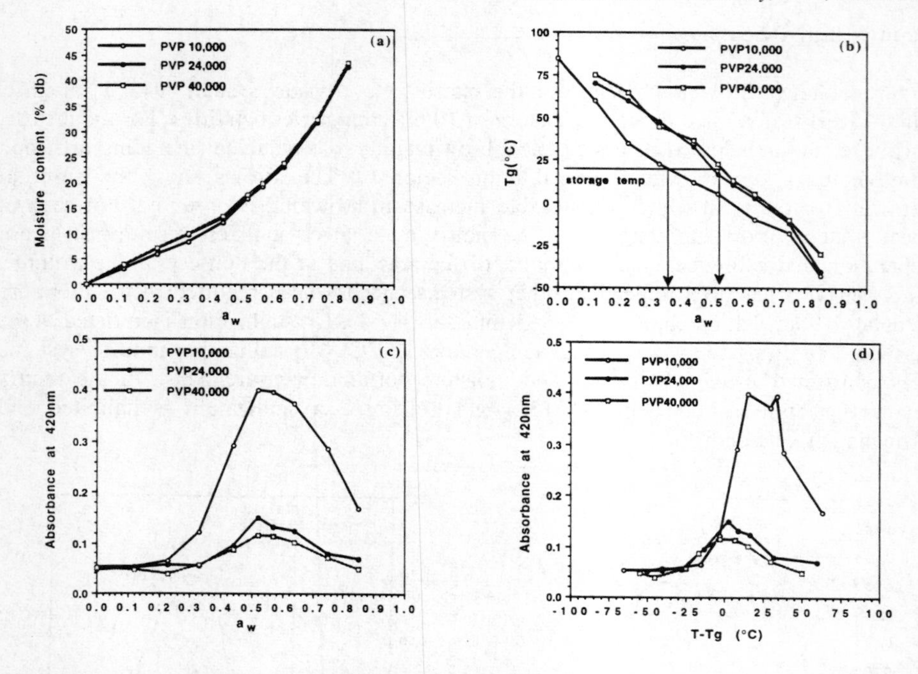

Figure 2. a) Moisture sorption isotherms for PVP/xylose/lysine (94:5:1) systems at 20C° after 1 week storage. b) Glass transition temperatures vs. water activity for PVP systems. c) Browning vs. water activity in PVP systems after 1 week storage. d) Browning vs. (T-Tg) in PVP systems after 1 week.

having a greater extent of browning. In this region, other effects not accounted for are responsible for the differences.

In another type of experiment, the effect of Tg on browning in anhydrous systems was evaluated. Maltodextrins (MD) of molecular weight 1,200 and 920 were used as matrices. In these experiments, only temperature and Tg should influence browning since there is no moisture. After freeze-drying, samples were held over P_2O_5 for 1 week, to remove any residual moisture. Vials were then sealed and heated in dry baths after which browning was determined. Zero-order rate constants were calculated and used to construct Arrhenius plots.

Figure 3 shows the results for the MD/xylose/lysine (94:5:1) systems. Curves for the two different MW maltodextrins are shown together. The arrow indicates the Tg for the lower MW system. For the higher MW maltodextrin, matrix all experimental temperatures were below the Tg. When comparing the browning rates of the two systems, we see that they are similar at the lower temperatures, when both systems are in the glassy state. As the temperature is increased beyond the Tg of the MW920 system, the curves diverge. Since both systems are at the same water content and temperature, we attribute the difference between browning rates for the two systems above Tg to the different values of (T-Tg). Similar results were observed for the other concentrations of MD/xylose/lysine.

In Figure 4, browning results for PVP(MW=40,000/ xylose/lysine (98:1:1) are shown. In these experiments, (T-Tg) was changed by using different experimental temperatures (changing T), as well as by having systems of different moisture contents

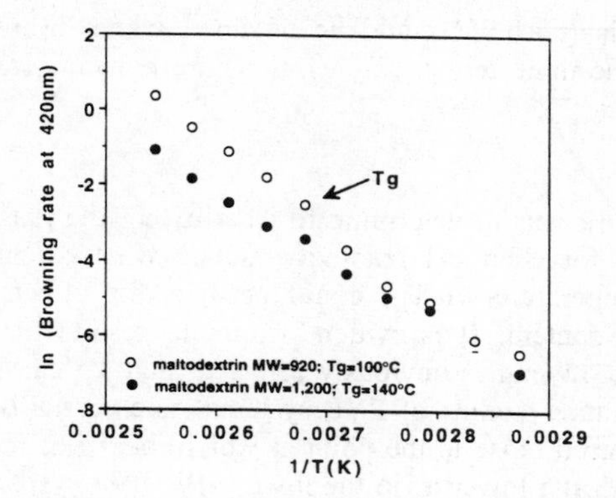

Figure 3. Arrhenius plot of browning in anhydrous maltodextrin/xylose/lysine (94:5:1).

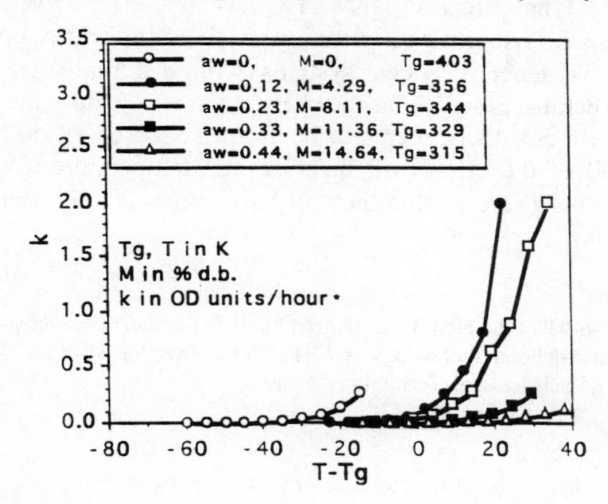

Figure 4. Browning vs. (T-Tg) in PVP/xylose/lysine (98:1:1) at various temperatures and moisture contents.

(having different Tg). The browning results, moisture contents and Tgs for the samples are shown in Figure 4. Data are plotted as a function of T-Tg, with like symbols representing samples of the same moisture content.

From the figure, two conclusions can be made. First, browning rates below (T-Tg)=0 are relatively low for most of the moisture contents and temperatures. Second, separate curves are obtained for browning at the different moisture contents. This indicates that plotting the browning rates vs. T-Tg does not account for all the temperature and moisture effects. Statistical analysis has indicated that the variables temperature, moisture content and glass transition temperature are all influential.

Modeling the results in Figure 4 with JMP statistical software resulted in the following equation, in which the temperature dependence does have the Arrhenius -equation form, but additional effects of (T-Tg) and moisture are discerned (r^2=0.98):

$$\ln k = 18.42 - 7111(^1/_T) - 0.15M + 0.07(T - Tg)$$

Adequate empirical models (r =0.95) could be obtained when ln(browning rates) are described by any two of the three terms: (T-Tg), moisture content, 1/T(K).

Conclusions

The initial intent of the work was to determine to what extent the glass transition could serve as a "critical point" for chemical reactivity and to what extent the temperature above glass transition temperature (T-Tg) could account for effects of composition, temperature and moisture content. It is evident from our results that in systems that crystallize above the Tg, browning is obviously affected, and Tg can serve as a critical point. In the room-temperature studies of PVP systems (non-crystallizing), an increase in extent of browning occurred close to the point at which the Tg of the system dropped below 20°C. This happened at a lower a_w in the lower MW PVP system. Plotting results as a function of (T-Tg) caused the curves for the three different MW PVP systems to superimpose below Tg. This suggests that Tg has an effect on browning in non-crystallizing systems. In experiments done at a fixed moisture content and at different temperatures, divergence between systems with different Tgs occur in the vicinity of the Tg in Arrhenius plots. Finally, when moisture, temperature and Tg are changed in the same experiment, there seems to be an influence of Tg on browning, but it does not account for all of the effects of moisture and temperature. While moisture content, temperature and (T-Tg) are related they all have some degree of an independent effect on browning.

Acknowledgments
This is publication No. F10544-6-93 of the New Jersey Agricultural Experiment Station supported by State Funds and the Center for Advanced Food Technology (CAFT). The Center for Advance Food Technology is a New Jersey Commission of Science and Technology Center.

References

Buera, M.P. and Karel, M. (1993) Application of the WLF equation to describe the combined effects of moisture and temperature on nonenzymatic browning rates in food systems. J. Food Processing and Preservation 7:31-45.

Eichner, K. and Karel, M. (1972) The influence of water content and water activity on the sugar-amino browning reaction in model systems under various conditions. J. Agric. Food Chem. 20:218-223.

Karel, M. and Saguy, I. (1991) Effect of glass transition on diffusion in food systems. In Water Relationships in Foods, (H. Levine and L. Slade, eds.) pp.157-174, Plenum Press, New York.

Karmas, R., Buera, M.P., and Karel, M. (1992) Effect of glass transition on rates of nonenzymatic browning in food systems. J. Agric. Food Chem. 40:873-879.

Kim, M.N., Saltmarch, M., and Labuza, T.P. (1981) Nonenzymatic browning of hygroscopic whey powders in open versus sealed pouchs. J. Food Processing and Preservation 5:49-57.

Labuza, T.P. and Saltmarch, M. (1981) The nonenzymatic browning reaction as affected by water in foods. In Water Activity: Influences on Food Quality, pp. 605-649. Academic Press, San Francisco.

Lim, M.H., and Reid, D. S. (1991) Studies of reaction kinetics in relation to the Tg of polymers in frozen model systems. In Water Relationships in Foods, (H. Levine and L. Slade, eds.), pp. 103-122. Plenum Press, New York.

Roos, Y. and Karel, M. (1991) Applying state diagrams to food processing and development. Food Technol. 45(12): 66-71, 107.

Slade, L. and Levine, H. (1991) Beyond Water Activity: Recent advances based on an alterative approach to the assessment of food quality and safety. Crit. Rev. Food Sci. Nutr. 30:115-360.

Vautaz, G. (1988) Preservation of skim milk powders: role of water activity and temperature in lactose crystallization and lysine loss. In Food Preservation by Water Activity Control, (C.C. Seow, ed.) pp. 73-101, Elsevier, Amsterdam.

White, M.L. and Cakebread, S.H. (1966) The glassy state in certain sugar containing food products. J. Food Technol. 1:73-82.

Analysis of Lactose–Protein Maillard Complexes in Commercial Milk Products by Using Specific Monoclonal Antibodies

Y. Kato,[1] T. Matsuda,[2] N. Kato,[3] and R. Nakamura[2]

[1]TOKAIGAKUEN WOMEN'S COLLEGE, TENPAKU-KU, NAGOYA 468;
[2]DEPARTMENT OF APPLIED BIOLOGICAL SCIENCES, SCHOOL OF AGRICULTURE, NAGOYA UNIVERSITY, CHIKUSA-KU, NAGOYA 464-01;
[3]CHUKYO WOMEN'S UNIVERSITY, OHBU 474, JAPAN

Summary
Lactose-protein Maillard complexes were immunochemically analyzed in various commercial milk products by ELISA and Immunoblotting using a specific monoclonal antibody. The Maillard complexes were detected in all samples analyzed, i.e., modified milk powder, skim milk powder, market pasteurized milk, milk beverages and concentrated milk. The apparent contents of Maillard complexes did not necessarily correlate to the loss of free amino groups, and the contents were generally higher in powdered milk products and milk beverages than in the market pasteurized milk. There appeared to be some relationship between the content of Maillard complexes and the time and temperature for pasteurization. Caseins were the major proteins detected by the antibody as lactose-protein Maillard complexes in various commercial milk products, though several whey proteins and unidentified polymerized proteins were also detected in some of the milk products. Thus, the monoclonal antibody was useful for *in situ* detection of lactose-protein Maillard adducts in milk and milk products.

Introduction

The polyclonal and monoclonal antibodies specific for lactose-protein Maillard adducts are now available (Matsuda et al., 1985, 1992). These antibodies reacted with the Maillard adducts between lactose and various proteins, but not with the adducts of several other reducing sugars. The production of lactose-protein Maillard complexes were successfully monitored by these antibodies in model systems of β-lactoglobulin and bovine serum albumin with lactose (Matsuda et al., 1985, 1992). In the present study, to estimate lactose-protein Maillard complexes in complex food systems, we applied a monoclonal antibody for the specific detection of the Maillard complexes in various commercial milk products such as modified milk powder, skim milk powder, pasteurized milk, milk beverages and concentrated milk.

Materials and Methods

A total of 44 samples of commercial milk and milk products were purchased from local markets. The samples used for the analyses are listed with some properties in Tables I and II. The monoclonal antibody (L101) to the lactose-protein Maillard complex was prepared as ascitic fluid by injecting the hybridoma L101 into CDF1 mice (Matsuda et al., 1992).

The lactose-protein Maillard complexes in the commercial milks and milk products were semi-quantitatively measured by ELISA (Engvall and Perlmann, 1975) using the monoclonal antibody, alkaline phosphatase-coupled anti-mouse IgG and *p*-nitrophenylphosphate/disodium as described previously (Matsuda et. al., 1992). SDS-polyacrylamide gel electrophoresis (PAGE) using 12% acrylamide and protein blotting were done by the methods of Laemmli (1970) and Towbin et al. (1979), respectively. Nitrocellulose sheets were incubated with the monoclonal antibody, and the protein bands reactive to the specific antibody were visualized with peroxidase-coupled anti-mouse IgG and 4-chloro-1-naphthol (Matsuda et al., 1992).

Free amino groups of milk proteins were determined by the fluorometric method using fluorescamine according to the procedure of Böhlen et al. (1973) as described previously (Kato et al., 1988; Matsuda et al., 1991). The fluorescence was measured by a Jasco FP-550A spectrofluorometer with excitation at 390 nm and emission at 475 nm. Protein concentration of the diluted milk samples was determined by the method of Lowry et al. (1951).

Results

The results of the amino group analysis and immunoassay on the modified milk powder and skim milk powder are shown in Table I. Free amino groups per mg protein of the milk powder samples were compared with those of fresh raw milk. Decreases in free amino group probably due to Maillard reaction was observed for all samples tested. The free amino groups varied among the modified milk samples; they decreased to about 30-75% of that of raw milk. Six skim milk powder items tested showed similar amino group contents, which were about 30 to 45% of that of raw milk.

Table I. Commercial modified milk powders and skim milk powder used in the analysis, and the relative amount of free amino groups and lactose-protein Maillard complexes as measured by the fluorometric method and ELISA using the specific monoclonal antibody, respectively.

Modified milk powder			
Code	Additives	Maillard complexes*	Free amino group (%)
---	Raw milk	0.00	100
A	Glucose, Soluble starch, (Lactose-free)	0.00	73
B	Starch sugar	0.59	29
C	None	0.59	41
D	Lactulose	0.73	54
E	Lactulose	0.81	51
F	Galactosyl-lactose, Dextrin	0.80	56
G	Dextrin	0.72	37
H	Corn syrup	0.66	49
Skim milk powder			
Code	Additives	Maillard complexes*	Free amino group (%)
a	None	0.52	44
b	None	0.50	39
c	Maltose, Fe	0.56	37
d	Glucose, Maltose, Isomaltose, Dextrin	0.57	40
e	None, (Filled with nitrogen gas)	0.32	38
f	Lactulose, Corn syrup, Dextrin, Fe	0.80	32

* The relative amount of lactose-protein Maillard complexes are shown as ELISA value (A492).

Lactose-protein Maillard complexes in various modified milk powders and skim milk powders were estimated by ELISA using the specific antibody, and the results are summarized in Table I. All milk powder samples but one reacted well with the monoclonal antibody. The exceptional one was sample A, from which lactose had industrially been eliminated, showing no immunoreactivity to the antibody and relatively high content of residual amino groups. The ELISA value of sample e, which had been filled with nitrogen gas, was a little lower than that of the others.

Table II summarizes contents of lactose-protein Maillard complexes in various liquid milk products as measured by ELISA. Among 19 commercial market milk, two which had been pasteurized at 140 °C (samples 3 and 4) showed higher ELISA values than those pasteurized at 130° or 120 °C, though milks containing Fe (samples 18 and 19) also showed high ELISA values. Milk pasteurized below 100 °C did not react with the monoclonal antibody except for two samples pasteurized at 85 °C for 15 sec. Lactic fermented and canned milk beverages generally showed high reactivity to the antibody, especially one lactic fermented beverage heated at 112 °C for 10 sec (sample 22) and one canned milk beverage pasteurized at 117 °C for 20 min. (sample 27). Non-sweetened evaporated milk heated at 116°C for 15-20 min. (samples 33 and 34) strongly reacted with the antibody, whereas sweetened condensed milks (samples 31 and 32) did not.

Table II. Commercial market milk, milk beverages and concentrated milk samples used in the analysis, and the relative amount of lactose-protein Maillard complexes as measured by ELISA using a specific monoclonal antibody.

Code No.	Market milks	Pasteurization	Maillard complexes*
1, 2	UHT-A, B	140 °C, 2 sec	0.09, 0.09
3,4	UHT-C, D	140 °C, 2 sec	0.14, 0.18
5-8	UHT-E, F, G, H	130 °C, 2 sec	0.03, 0.03, 0.04, 0.04
9, 10	UHT-I, J	120 °C, 2-3 sec	0.02, 0.03
11, 12	HTST-A, B	85 °C, 15 sec	0.10, 0.11
13, 14	HTST-C, D	75 °C, 15 sec	0.01, 0.02
15 - 17	LTLT-A, B, C	65 °C, 30 min.	0.01, 0.02, 0.01
18, 19	UHT-Fe-A, B	130 °C, 2 sec	0.15, 0.22
Code No.	**Milk beverages**	**Heat-treatment**	**Maillard complexes***
21	Lactic fermented-A	130 °C, 2 sec	0.11
22	Lactic fermented-B	112 °C, 10 sec	1.09
23	Lactic fermented-C	90 °C, 15 sec	0.34
24	Lactic fermented-D	90 °C, 15 sec	0.03
25	Lactic fermented-E	128 °C, 2 sec	0.23
26	Lactic fermented-F	140 °C, 2 sec	0.11
27	Canned milk beverage	117 °C, 20 min.	0.91
Code No.	**Concentrated milk**	**Heat-treatment**	**Maillard complexes**
31	Condensed milk-A	100 °C, 15 sec	0.07
32	Condensed milk-B	100-120 °C, 1-3 min.	0.06
33	Evaporated milk-A	116 °C, 15-20 min.	0.96
34	Evaporated milk-B	116 °C, 16 min.	0.96

* The relative amount of lactose-protein Maillard complexes are shown as ELISA value (A492).

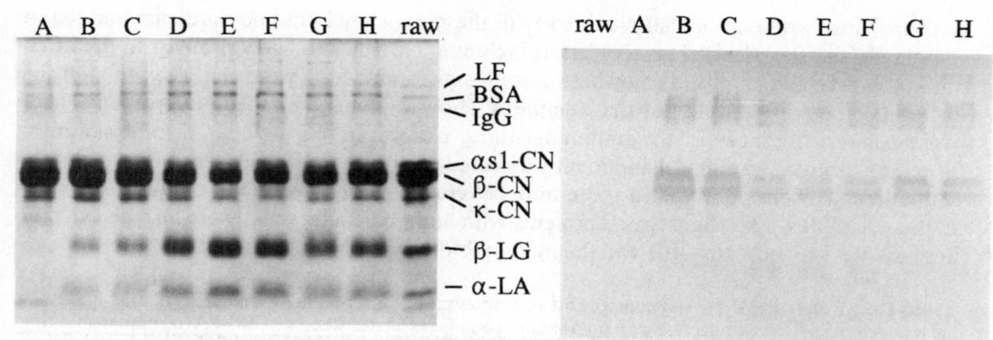

Figure 1. SDS-PAGE of proteins in various modified milk powders and detection of lactose-protein Maillard complexes by immunoblotting. A sheet of acrylamide gel was stained with CBB (left), and protein blot was immunologically stained with the monoclonal antibody (L101). Some properties of samples A to H are shown in Table I. LF; lactoferrin; BSA: bovine serum albumin; CN: casein; LG: lactoglobulin; LA: lactalbumin.

Table III. Reactivity of protein components in modified milk and skim milk powders with the monoclonal antibody to lactose-protein Maillard complexes as measured by SDS-PAGE/immunoblotting.

Milk proteins	Modified milk powder							
	A	B	C	D	E	F	G	H
Lactoferrin	-	++	++	+	-	±	+	+
Serum albumin	-	+++	+++	+++	+	+	++	++
Immunoglobulins	-	+	+	++	+	+	+	+
αs1-Casein	-	+++	+++	++	++	++	+++	++
β-Casein	-	+++	+++	++	++	++	+++	++
κ-Casein	-	+	+	-	-	-	-	-
β-lactoglobulin	-	+	+	++	++	++	+	+
α-lactalbumin	-	±	±	+	+	+	±	±
	Skim milk powder							
	a	b	c	d	e	f		
Lactoferrin	-	-	-	+	+	-		
Serum albumin	-	-	-	±	±	+		
Immunoglobulins	-	-	-	+	+	++		
αs1-Casein	++	+++	+++	+++	+++	+++		
β-Casein	++	++	++	++	++	+		
κ-Casein	-	-	-	-	-	-		
β-lactoglobulin	+++	+++	+++	+++	+++	++		
α-lactalbumin	+++	+++	+++	+++	+++	++		

* The reactivity was represented by arbitrary intensities of each protein band stained with the monoclonal antibody to lactose-protein Maillard complexes.
-: not stained, ±: faintly stained, +: weakly stained, ++: stained, +++: strongly stained

Protein components with lactose-protein Maillard complexes were analyzed by SDS-PAGE using the specific monoclonal antibody. A typical electrophoretogram and an immunoblot of the modified milk powder are shown in Figure 1. The relative intensities of protein bands immunologically stained with the monoclonal antibody are summarized in Tables III and IV. Several protein bands including caseins and whey proteins in modified milk powders and skim milk powders were clearly stained with the antibody, though the lactose-free modified milk powder, sample A, showed no band on the immunoblot. There were some differences in the immunostaining pattern among powdered milk samples. Weak or almost unstained bands were detected in market milk, whereas many clearly stained bands were observed in some milk beverages and unsweetened evaporated milks. In these samples, some aggregated proteins with high molecular mass were also detected in both the gel stained with CBB and the immunoblot stained with the antibody.

Table IV. Reactivity of protein components in market milk, milk beverages and concentrated milk with the monoclonal antibody to lactose-protein Maillard complexes as measured by SDS-PAGE/immunoblotting

Milk proteins	Market milk							
	1	2,3	4-7	8	9,10	11,12	13-17	18,19
Lactoferrin	-	-	-	-	-	-	-	-
Serum albumin	-	-	-	-	-	-	-	±
Immunoglobulins	-	-	-	-	-	-	-	±
αs1-Casein	-	+	+	-	-	++	+	++
β-Casein	-	+	+	-	-	++	+	++
κ-Casein	-	±	-	-	±	±	±	±
β-lactoglobulin	-	-	-	-	-	-	-	-
α-lactalbumin	-	-	-	-	-	-	-	-
	Milk beverages					Concentrated milk		
	21,24	22	23,25	26	27	31,32	33	34
Polymers	-	+++	-	-	+++	-	+++	++
Lactoferrin	-	++	-	-	+	-	+	+
Serum albumin	-	-	-	-	+	-	+	+
Immunoglobulins	-	-	-	-	+	-	+	+
αs1-Casein	+	++	++	++	++	+	++	++
β-Casein	+	++	++	++	++	+	+	+
κ-Casein	-	++	±	+	++	-	++	++
β-lactoglobulin	-	+	+	+	++	+	++	++
α-lactalbumin	-	-	++	+	++	+	++	+

* The reactivity was represented by arbitrary intensities of each protein band stained with the monoclonal antibody to lactose-protein Maillard complexes.
-: not stained, ±: faintly stained, +: weakly stained, ++: stained, +++: strongly stained

Discussion

Lactose-protein Maillard complexes in milk and milk products have been evaluated by various chemical analyses (Finot et al., 1981). In the present study, the Maillard products were detected by ELISA in various commercial milks and milk products, and milk protein components with the Maillard complexes were identified by immunoblotting in these milk samples. Among various milk products, powdered milk products generally showed higher reactivity to the antibody than market milk samples. This might indicate that Maillard reaction occurred in the dried milk products during storage, since the storage periods after the manufactured production of the market milks and the dried samples were a few days and two to three months, respectively.

Production of the lactose-protein Maillard complexes in milks and milk beverages appeared to be affected by both temperature and time of pasteurization. In some cases, there are certain differences in reactivity to antibody among samples pasteurized under the same or similar conditions (Table II). The immunoblotting patterns were not necessarily the same among samples, i.e., serum albumin was well stained in some modified milk powders, whereas α-lactalbumin was one of the major stained bands in skim milk powders (Table III). These results suggest that the chemical properties of each protein also affect the production of lactose-protein Maillard complexes during food processing. The results of immunoblot analysis did not completely agree with those of ELISA in some cases (Table I and Figure 1). This might be due to the difference in samples analyzed; whole protein mixtures were used for ELISA, whereas individual proteins were analyzed by the immunoblot after electrophoretic separation.

Conclusions

Lactose-protein Maillard complexes were detected in various commercial milk and milk products by ELISA and immunoblotting. This immunochemical analysis using the monoclonal antibody would be quite useful for the analysis of Maillard reaction between proteins and lactose, since many samples could be analyzed simultaneously and with high sensitivity; no special pre-treatment, e.g. acid hydrolysis, is required for the analysis. This analytical method may also be useful for quality control of commercial pasteurized milks and milk products.

References

Böhlen, P., Stein, S., Dairman, W. and Udenfriend, S. (1973) Fluorometric assay of proteins in the nanogram range. *Arch. Biochem. Biophys.*, 155: 213-220.

Finot, P. A., Deutsch, R. and Bujard, E. (1981) The extent of Maillard reaction during the processing of milk. *In Maillard reaction of food*, Eriksson, C., Ed., Pergamon Press, Oxford, U.K., pp345-355.

Engvall, E. and Perlmann, P. (1971) Enzyme-linked immunosorbent assay (ELISA) - Quantitative assay of immunoglobulin G. *Immunochemistry* 8: 871-874.

Kato, Y., Matsuda, T., Kato, N. and Nakamura, R. (1988) Browning and protein polymerization induced by amino-carbonyl reaction of ovalbumin with glucose and lactose. *J. Agric. Food Chem.* 36: 806-809.

Laemmli, U. K. (1970) Cleavage of structural proteins during the assembly of the head of bacteriophage T4.*Nature (London)* 227: 680-685.

Lowry, O. H., Rosebrough, N. J., Farr, A. L. and Randal, R. J. (1951) Protein measurement with the folin phenol reagent. *J. Biol.Chem.* 193: 265-275.

Matsuda, T., Kato, Y. and Nakamura, R. (1991) Lysine loss and polymerization of bovine β-lactoglobulin by amino carbonyl reaction with lactulose (4-*o*-β-D-galactopyranosyl-D-fructose). *J.Agric. Food Chem.* 39: 1201-1204.

Matsuda, T., Kato, Y., Watanabe, K. and Nakamura, R. (1985) Direct evaluation of β-lactoglobulin lactosylation in early Maillard reaction using an antibody specific to protein-bound lactose. *J. Agric. Food Chem.* 33: 1193-1196.

Matsuda, T., Ishiguro, H., Ohkubo, I., Sasaki, M. and Nakamura, R. (1992) Carbohydrate binding specificity of monoclonal antibodies raised against lactose-protein Maillard adducts. *J. Biochem.* 111:383-387.

Towbin, H., Staehelin, T. and Gordon, J. (1979) Electrophoretic transfer of proteins from polyacrylamide gels to nitrocellulose sheets: Procedure and some applications. *Proc. Natl. Acad.Sci. U.S.A.* 76: 4350-4353.

Simultaneous Determination of Protein-bound Maillard Products by Ion Exchange Chromatography and Photodiode Array Detection

T. Henle, A. W. Walter, and H. Klostermeyer

LEHRSTUHL FÜR MILCHWISSENSCHAFT, TECHNISCHE UNIVERSITÄT MÜNCHEN, VÖTTINGER STR. 45, D-85350 FREISING-WEIHENSTEPHAN, GERMANY

Summary
Following complete enzymic hydrolysis, simultaneous determination of protein-bound advanced-stage Maillard compounds in addition to the Amadori products lactuloselysine and fructoselysine as well as the common amino acids including tryptophan was achieved by ion-exchange chromatography with photodiode array measurement and subsequent ninhydrin detection. Direct ultraviolet detection at the absorption maximum allowed the quantification of pyrraline (λ_{max} = 297 nm), an acid labile lysine derivative with a pyrrole structure, at levels lower than 2 mg/kg protein. Values for pyrraline in a number of commercially available foods (milk products, bakery goods, pasta products) ranged from not detectable up to 3500 mg/kg protein. Pyrraline proved to be a suitable indicator for the advanced stages of the Maillard reaction in foods. In samples of severely heated whey powder, four previously unknown UV-active compounds were detected. Their formation correlated with heating time and temperature. One of these compounds could be identified as maltosine, a pyridone derivative of lysine of the 1-deoxyosone pathway, which up to now has not been described in protein hydrolysates.

Introduction

For decades, the Maillard reaction has been intensively studied with the attention focused mainly on the reactivity of the carbonyl compounds and the formation of sugar degradation products (Ledl & Schleicher, 1990). However, little information is available concerning the course of the complex reaction for proteins. To date, only the Amadori products, lactuloselysine and fructoselysine (Möller et al., 1977) and their oxidative degradation product carboxymethyllysine (Büser & Erbersdobler, 1986), have been identified as protein-bound Maillard compounds in heated foods. The pentose mediated lysine-arginine crosslink pentosidine was isolated from collagen (Sell & Monnier, 1989). Pyrraline, an acid labile pyrrole formed from lysine via the 3-deoxyglucosone pathway, was first identified in heated reaction mixtures of lysine and glucose (Nakayama et al., 1980; Miller et al., 1984) and a few years later in alkaline hydrolysates of proteins that had been reacted with glucose (Sengl et al., 1989). The formation of small amounts of free pyrraline during heating of food samples was demonstrated by Chiang (1988). Hayase et al. (1989) as well as Miyata and Monnier (1992) succeeded in the immunological detection of the lysine derivative in serum proteins. It was the purpose of our study, to investigate the formation of pyrraline, which is reported to show mutagenic and antiproteolytic properties (Omura et al., 1983; Öste et al., 1987), as well as of other protein-bound Maillard products in various foods.

Materials and Methods

Food Samples

A total of 56 samples were investigated (Table I). All samples were obtained from local retail stores.

Acid and Enzymic Hydrolysis

Defatted samples were hydrolyzed directly without further preparation in the presence of 6 N HCl as well as with a combination of four proteinases as described recently (Henle et al., 1991).

Amino Acid Analysis

This was performed with an Alpha Plus amino acid analyzer (LKB Biochrom, Cambridge, UK) using cation-exchange resin DC4A-spec 8 µm, sodium form (Benson, Reno, NE, USA). Chromatographic conditions are described elsewhere (Henle & Klostermeyer, 1993). Pyrraline and other aromatic amino acids were detected using a photodiode array detector PDA 991 (Waters, Milford, MS, USA) within a wavelength range from 240 to 350 nm. After flowing through the PDA detector, common amino acids were detected with ninhydrin at 570 and 440 nm. Data acquisition and peak integration were evaluated using Millennium and Maxima chromatography software (Waters, Milford, MS, USA), respectively, on 486 and 386 personal computers.

Results and Discussion

Under the conditions used for amino acid analysis, pyrraline eluted with an identical retention time as phenylalanine (Figure 1 a). However, UV-detection at the pyrrolic absorption maximum (λ_{max} = 297 nm) in combination with subsequent ninhydrin reaction enabled the unambiguous quantification of pyrraline along with all common amino acids including tryptophan as well as the acid-labile Amadori products lactuloselysine and fructoselysine (Henle & Klostermeyer, 1993). The enzymic hydrolysis procedure, previously established for the complete and careful digestion of milk proteins (Schmitz et al., 1976; Henle et al., 1991), proved suitable also for sufficient hydrolysis of proteins from cereals and other foods. Thus, pyrraline was detectable in protein hydrolysates of various heat treated foods at levels lower than 2 mg/kg protein. Pyrraline eluted as a well resolved peak after tyrosine in the chromatograms obtained after detection at 297 nm (Figure 1 b). At this wavelength, phenylalanine exhibited no detectable absorption. Quantification was achieved using a reference standard sample of pyrraline isolated via a modified procedure according to Schüßler & Ledl (1989). As reported previously, pyrraline increased linearly with heating time during storage of milk powder samples at 100 and 110° C for up to 6 hours, followed by a remarkable decrease after long-term heating. The formation of pyrraline proved to be independent of a degradation of the corresponding Amadori product lactuloselysine (Henle & Klostermeyer, 1993).

Great differences in the pyrraline contents of commercially available foods could be observed (Table I). For milk products, the amounts of pyrraline ranged between not detectable and 150 mg/kg protein. Increased values found for individual samples of sterilized milk (260 mg/kg protein), skim milk powder (1450 and 3150 mg/kg protein) and whey powder (850 and 1150 mg/kg protein) indicated elevated temperatures during heating or high relative humidity during storage. Bakery products usually are considered as foods processed under heating conditions which promote "advanced" Maillard reactions. Considerable higher pyrraline contents could therefore be found in bread crust as well as in special products prepared for long-term storage (crisp bread, rusk, pumpernickel, crackers).

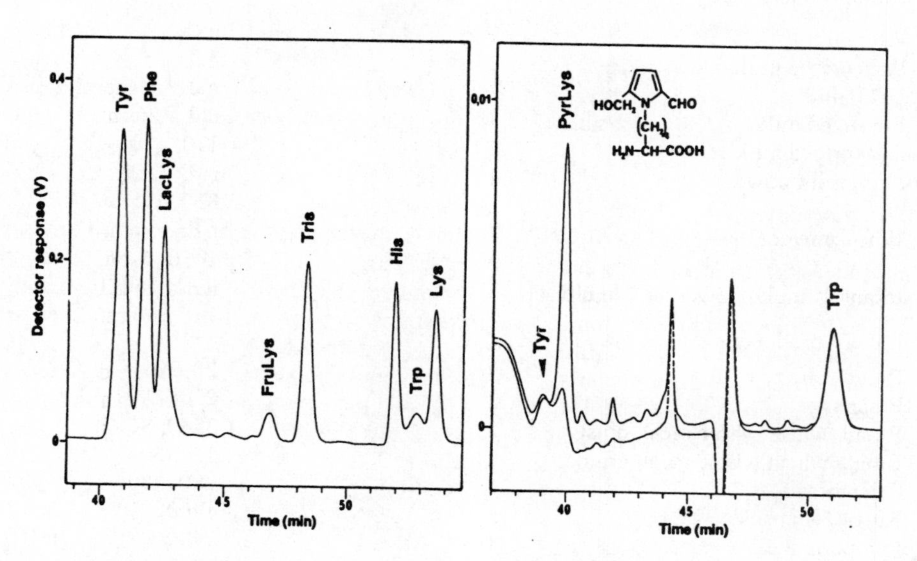

Figure 1. Chromatograms (detail) of enzymic hydrolysates (PyrLys = pyrraline; LacLys = lactuloselysine; FruLys = fructoselysine). 1 a: Skim milk powder (pyrraline 1150 mg/kg protein), detection at 570 nm after reaction with ninhydrin. 1 b: Skim milk powder samples (pyrraline 150 and 1150 mg/kg protein), PDA detection at 297 nm.

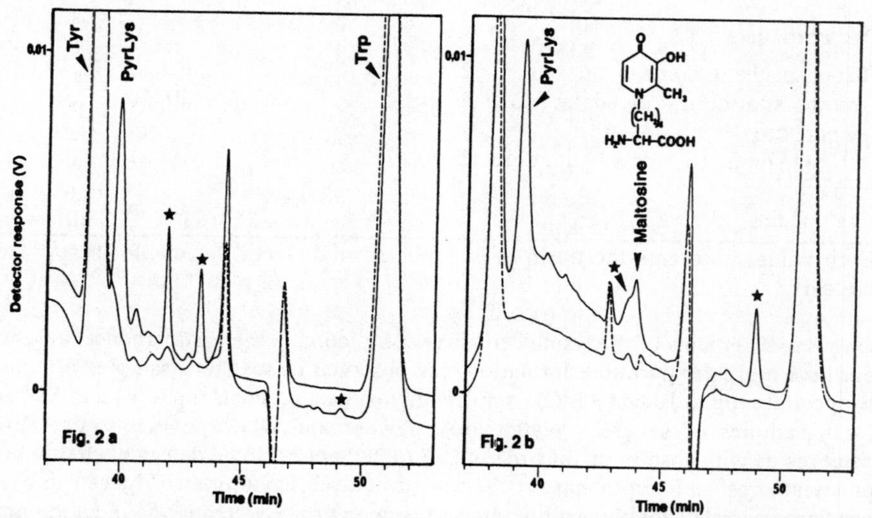

Figure 2. Overlay chromatograms (detail) of enzymic hydrolysates, PDA-detection at 280 nm (PyrLys = pyrraline; unknown compounds are marked with an asterisk). 2 a: Skim milk powder (pyrraline 150 and 1150 mg/kg protein). 2 b: Whey powder (pyrraline 20 and 3150 mg/kg protein).

Table I. Pyrraline contents of commercially available foods.

Sample	Number of samples	Pyrraline (mg/kg protein)
Milk products		
Raw milk	2	n.d.
Pasteurized milk	2	n.d.
UHT-milk	2	n.d.
Sterilized milk	3	n.d. - 260
Evaporated milk	2	110; 130
Skim milk powder	4	n.d. - 150
	2	850; 1150
Whey powder	8	n.d.-90
	2	1450; 3150
Infant formula, powdered / liquid	5	n.d. - 160
Bakery products		
Bread rolls	2	20; 30
Pretzels	2	230; 240
Bread, wheat whole meal, crust	2	3250; 3680
Bread, wheat whole meal, crumb	2	25; 35
Crisp bread	2	280; 480
Rusk (zwieback)	1	1400
Pumpernickel	1	1130
Pizza	2	80; 110
Butter cookies	1	120
Wafers	1	570
Crackers	1	1320
Pasta products		
Italian spaghetti, home made	2	n.d.
Italian spaghetti, dried at low temperature	2	20; 40
Italian spaghetti, dried at high temperature	2	100; 125
Egg noodles	2	30; 50

Each value represents the mean of a duplicate; n.d = not detectable (below 2 mg/kg protein).

Only small amounts of the lysine derivative were found in the pasta products investigated, but here a greater pyrraline formation was observed in spaghetti samples dried at high temperature (up to 10 h at 85 °C) compared to low-heat products (up to 14 h at 55 °C).

In samples of severely heated milk powder and whey powder, four UV-active compounds with absorption maxima of 275 to 280 nm could be detected, eluting between phenylalanine and tryptophan in the amino acid chromatogram (Figure 2 a, b). Their formation correlated with heating time and temperature. By comparing retention time and UV-spectra with a standard sample, one of these compounds could be identified as maltosine, a pyridone derivative of lysine, which up to now has not been described in protein hydrolysates. In contrast to pyridosine, another pyridone formed together with furosine from lysine Amadori products (Finot et al., 1981), protein-bound maltosine is a natural advanced Maillard reaction product of the 1-deoxy-glucosone pathway (Ledl et al., 1989). It is well known, that this route is favored at higher pH-values, whereas the degradation of the Amadori products via 3-deoxyglucosones and thus the formation of pyrraline proceeds mainly at pH-values between 5 and 7 (Ledl & Schleicher, 1990). This fact might explain the observation that rather small amounts of maltosine (about 100 mg/kg protein) compared to the corresponding pyrraline values (3150 mg/kg protein) were found only in a few browned samples.

Conclusion

With the proposed method it was possible to obtain information about the formation of protein bound pyrraline in foods. The lysine derivative may serve as a suitable indicator for the advanced stages of the Maillard reaction. As its content in some foods processed under the severe heating conditions of baking proved to be remarkably high, more information is necessary about the physiological properties of pyrraline as well as concerning the parameters influencing its formation. In contrast to pyrraline, protein bound maltosine represents only a minor compound of lysine derivatization during advanced browning. In order to clarify pathways of the Maillard reaction for proteins, isolation and structural identification of other amino acid derivatives is now in progress.

Acknowledgments
We thank Mrs. E. Hofmair for skillful technical assistance and Dr. H. Osiander, formerly at the Institut für Pharmazie und Lebensmittelchemie, Universität München, who kindly provided a pure standard sample of maltosine. This work was supported by a research grant from Deutsche Forschungsgemeinschaft.

References

Büser, W., and Erbersdobler, H.F. (1986). Carboxymethyllysine, a new compound of heat damage in milk products. *Milchwiss.* 41: 363-368.

Chiang, G.H. (1988). High-performance liquid chromatographic determination of ε-pyrrole-lysine in processed foods. *J. Agric. Food Chem.* 36: 506-509.

Finot, P.A., Deutsch, R., and Bujard, E. (1981). The extent of the Maillard reaction during the processing of milk. *Prog. Food Sci. Nutr.* 5: 345-355.

Hayase, F., Nagaraj, R.H., Miyata, S., Njoroge, F.G., and Monnier, V.M. (1989). Aging of proteins: Immunological detection of a glucose-derived pyrrole formed during Maillard reaction in vivo. *J. Biol. Chem.* 363: 3758-3764.

Henle, T., Walter, H., and Klostermeyer, H. (1991). Evaluation of the extent of the early Maillard reaction in milk products by direct measurement of the Amadori product lactuloselysine. *Z. Lebensm. Unters. Forsch.* 193: 119-122.

Henle, T., and Klostermeyer, H. (1993). Determination of protein-bound 2-amino-6-(2-formyl-5-hydroxymethyl-1-pyrrolyl)-hexanoic acid ("pyrraline") by ion-exchange chromatography and photodiode array detection. *Z. Lebensm. Unters. Forsch.* 196: 1-4.

Ledl, F., and Schleicher, E. (1990). New aspects of the Maillard reaction in food and in the human body. *Angew. Chem. Int. Ed. Engl.* 29: 565-594.

Ledl, F., Osiander, H., Pachmayr, O., Severin, Th. (1989). Formation of maltosine, a product of the Maillard reaction with a pyridone structure. *Z. Lebensm. Unters. Forsch.* 188: 207-211.

Miller, R., Olsson, K., and Pernemalm, P.A. (1984). Formation of aromatic compounds from carbohydrates. IX. Reaction of D-glucose and L-lysine in slightly acidic, aqueous solution. *Acta Chem. Scand.* 38B: 689-694.

Miyata, S., and Monnier, V. (1992). Immunohistochemical detection of advanced glycosylation end products in diabetic tissue using monoclonal antibody to pyrraline. *J. Clin. Invest.* 89: 1102-1112.

Möller, A.B., Andrews, A.T., and Cheeseman, G.C. (1977). Chemical changes in ultra-heat treated milk during storage. II. Lactuloselysine and fructoselysine formation by the Maillard reaction. *J. Dairy Res.* 44: 267-275.

Nakayama, T., Hayase, F., and Kato, H. (1980). Formation of ε-(2-formyl-5-hydroxymethyl-pyrrol-1-yl)-L-norleucine in the Maillard reaction between D-glucose and L-lysine. *Agric. Biol. Chem.* 44: 1201-1202.

Öste, R.E., Miller, R., Sjöström, H., and Noren, O. (1987). Effect of Maillard reaction products on protein digestion. Studies on pure components. *J. Agric. Food Chem.* 35: 938-942.

Omura, H., Jahan, N., Shinohara, K., and Murakami, H. (1983). Formation of mutagens by the Maillard reaction. in: Waller, G.R., and Feather, M.S. (eds): *The Maillard reaction in foods and nutrition.* ACS Symp. Ser. 5: 537-563.

Schüßler, U., and Ledl, F. (1989). Synthesis of 2-amino-6-(2-formyl-5-hydroxymethyl-1-pyrrolyl-)-hexanoic acid. *Z. Lebensm. Unters. Forsch.* 189: 138-140.

Schmitz, I., Zahn, H., Klostermeyer, H., Rabbel, K., Watanabe, K. (1976). Zum Vorkommen von Isopeptid-bindungen in erhitztem Milcheiweiß. *Z. Lebensm. Unters. Forsch.* 160: 377-381.

Sell, D.R., and Monnier, V.M. (1989). Stucture elucidation of a senescence cross-link from human extracellular matrix. *J. Biol. Chem.* 264: 21597-21602.

Sengl, M., Ledl, F., and Severin, Th. (1989). Maillard-Reaktion von Rinderserumalbumin mit Glucose. Hochleistungsflüssigkeitschromatographischer Nachweis des 2-Formyl-5-(hydroxymethyl)-pyrrol-1-norleucins nach alkalischer Hydrolyse. *J. Chromatogr.* 463: 119-125.

Catalytic Aspects of the Glycation Hot Spots of Proteins

Seetharama A. Acharya and Parimala Nacharaju

DIVISION OF HEMATOLOGY, ALBERT EINSTEIN COLLEGE OF MEDICINE, 1300 MORRIS PARK AVENUE, BRONX, NEW YORK 10461, USA

Summary
A systematic investigation of the protein structural factors that could determine the chemical reactivity of the amino groups of proteins for nonenzymic glycation (glycation hot spots) has been carried out. The studies suggest that the microdomains of the glycation hot spots of proteins are Amadori rearrangement catalytic centers. The isomerization reaction catalysed by these microdomains is comparable to the reactions catalyzed by glucose isomerase and triose phosphate isomerase. The protein microdomains containing a constellation of positively charged functional groups thereby generating a proton rich microenvironment seem to act as the Amadori rearrangement catalytic centers. The α and/or the ε-amino groups located in such microdomains and accessible to aldoses to form the aldimine adducts are the glycation hot spots of proteins. It is suggested that the catalytic power of the glycation hot spots of proteins may have some role in the post glycation reactions leading to the generation of the advanced glycation end products and thus in the pathophysiology of diabetes.

Introduction

The nonenzymic glycation of proteins belongs to the class of *in vivo* post translational protein modification reactions that proceed spontaneously. The chemistry of the reaction involves the addition of glucose to the α or ε-amino groups of the protein through ketoamine linkages (Bunn et al., 1975). The glycation of proteins is initiated by the formation of a reversible Schiff base adduct of (aldimine) aldose with the amino group of protein. The aldimine adducts of simple aldehydes can be stabilized by reducing them using sodium cyanoborohydride or borohydride (Acharya et al., 1983). This results in the formation of an alkylated protein. Nature has developed an ingenious approach to stabilize the aldimine adducts of aldoses generated under

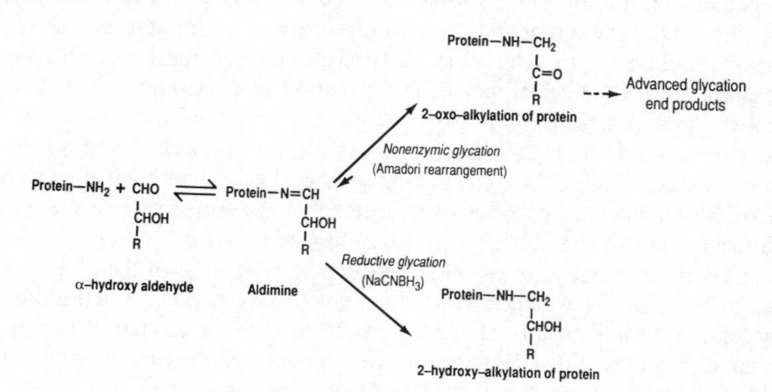

Figure 1. Schematic representation of reductive and nonreductive alkylation of proteins by aldose.

physiological conditions. This involves an intra molecular rearrangement of aldimine to a more stable ketoamine adduct (Amadori rearrangement). This is an alkylation reaction and the ketoamine adduct is a 2-oxo alkylated protein (Figure 1).

There is a high degree of site selectivity in the nonenzymic glycation of proteins; only a limited number of amino groups of a given protein undergo this modification reaction. These reactive amino groups are referred to as the glycation hot spots of proteins. The protein bound ketoamine adduct is the substrate for the post glycation reactions to generate the advanced glycation end products. Delineation of the unique structural aspects of the glycation hot spots of proteins has been the subject of considerable interest in recent years (Watkins et al., 1987; Iberg & Flukinger, 1986; Shilton & Walton, 1991; Nacharaju & Acharya, 1992).

Results and Discussion

The organic chemical aspects of the nonenzymic glycation reaction suggests that the reaction of α-hydroxyaldehydes with proteins should serve as a simple model for the nonenzymic glycation reaction. Consistent with this it has been now demonstrated that the aldimine adducts of other aldoses like glycolaldehyde, glyceraldehyde and erythrose also undergo the Amadori rearrangement. The open chain aldehydic form of glucose is the reactive species in the nonenzymic glycation reaction. However, the concentration of the reactive species of glucose under the physiological conditions is very low and accordingly the glycation reaction proceeds very slowly as compared with the lower homologues of aldoses, like glyceraldehyde. Aldodiose, aldotriose and aldotetrose are present as the reactive aldehydic form and react with proteins at a faster rate. Accordingly, we have used glyceraldehyde as the model aldose to delineate the molecular aspects of the initial phases of the nonenzymic glycation reaction. The site selectivity of the nonenzymic glycation of HbA and RNAse A with glyceraldehyde is comparable to that of glucose (Acharya & Manning, 1980), suggesting that the structure of the protein is the determinant of the site selectivity of the glycation reaction.

The nonenzymic addition of glucose to proteins will be facilitated at least by two structural aspects of the protein. First, a lowering of the pK_a of the amino group facilitates the glycation reaction as the unprotonated form of the amino group is the reactive species for the aldimine formation. The reaction could also be facilitated by endowing the microenvironment of an amino group a propensity to catalyze the isomerization of aldimine to ketoamine. A synergy between these two structural aspects is also possible. If the site selectivity of the glycation is a consequence of the pK_a of the amino group, the site selectivity of nonenzymic glycation and reductive glycation of a protein will be same as both of these reactions proceed through the same intermediate, the aldimine adduct (Figure 1). A comparative study of the reductive and nonenzymic glycation of HbA and RNAse A has established that the Amadori rearrangement step is the rate limiting step in the nonenzymic glycation of proteins.

An immediate question is to establish whether the potential of the glycation hot spots of proteins to catalyze the Amadori rearrangement (catalytic power) is the same. The site selectivity of the amino groups of HbA and RNAse A for reductive and nonenzymic glycation are different and this suggests that the Amadori rearrangement catalytic potential of glycation sites are distinct. An estimate of the catalytic power of the glycation sites has been obtained by comparing the extent of the chemical reaction at the glycation sites under the conditions of reductive glycation with that of nonenzymic glycation (Acharya et al., 1991). In HbA the microenvironment of Lys-16(α) appears to be the most efficient in isomerizing the aldimine adduct to the corresponding ketoamine. The isomerization potential (catalytic power) of Lys-16(α) is about 650 times higher than that of Val-1(α), and 50 times higher than that of Val-1(β). Similar results have also been obtained for the glycation hot spots of RNAse A. The catalytic power of Lys 7, 37, and 41, are 750, 1300, and 1000 times higher than that at the α-amino group of Lys 1 of RNAse A.

The catalytic power concept invokes that a generality exists in the structural and/or conformational aspects of the glycation hot spots. The ultimate goal of all the studies on the mechanistic aspects of the site selectivity of the glycation reactions is to identify the unique aspects in the conformational design of the glycation prone amino groups of proteins and establish the commonality between them (nonenzymic glycation motif). The catalytic power at the glycation hot spots could be determined either by the amino acid sequence of the protein around the glycation site (nearest neighbor linear effect) or by the conformational aspects of the microenvironment of the glycation hot spots that bring unique functional groups of the protein to proximity (nearest neighbor three dimensional effect).

Nonenzymic glycation of performic acid oxidized RNAse A (Hirs, 1956) has given new insight as to which one of these two mechanisms best explains the site selectivity of the glycation reactions. The performic acid oxidized RNAse A lacks the tertiary interactions of the native protein, but retains the linear amino acid sequence of the polypeptide chain. Nonetheless, nearly 70% decrease was observed in the over all chemical reactivity of the amino groups of the protein for nonenzymic glycation as a result of performic acid oxidation. Interestingly, the influence of cleavage of the disulfide bonds of RNAse A on the chemical reactivity of all the glycation hot spots of the protein is not the same. Thus, the chemical reactivity of the glycation hot spots of RNAse A is not a consequence of the nearest neighbor linear effect.

On the other hand, nearest neighbor three dimensional effect appears to be the primary determinant of the site selectivity in the nonenzymic glycation of proteins. A comparison of the amino acid residues located in the microenvironment of the glycation hot spots of RNAse A and HbA suggested that the microdomains of proteins containing a constellation of positively charged functional groups, thereby a proton rich microenvironment, is the first approximate description of the Amadori rearrangement catalytic centers (Acharya et al., 1989). Amino groups (α or ε) located in such a region and accessible to form aldimine adducts with aldoses are the glycation hot spots of proteins.

A corollary to the nearest neighbor three dimensional effect concept is that the site selectivity depends heavily on the structural integrity of the native three dimensional structure of the proteins. A comparison of the nonenzymic glycation potential of Lys-16(α) of HbA in three different forms of the substrate has provided further support to the nearest neighbor three dimensional effect concept. When α-chain is used as the substrate for nonenzymic glycation, the influence of the quaternary structural aspects of the protein at Lys-16(α) will be absent. Similarly, the contribution of the tertiary and quaternary structural aspects of the protein to the microenvironment of Lys-16(α) will be absent when the segment α_{1-30} is used as the substrate. The microenvironment of Lys-16(α) exhibited a lower catalytic power in the α-chain, but had hardly any Amadori rearrangement activity in the segment α_{1-30}. This establishes that the high catalytic power of the microenvironment of Lys-16(α) in the tetramer is not a direct consequence of the nearest neighbor linear effect (Nacharaju & Acharya, 1992).

Conceptually, the nearest neighbor three dimensional effect discussed above could be a direct consequence of a unique functional group (catalytic residue) located in the microenvironment of the glycation site or the catalytic power is an integrated property of a number of functional groups present in the microdomain. The unique functional group mechanism invokes an 'enzyme active site' concept to the catalytic centers. Mechanistically, the Amadori rearrangement reaction could be compared to the isomerization of the glyceraldehyde-3-phosphate to dihydroxy acetone phosphate that is catalyzed by triose phosphate isomerase or the isomerization of glucose to fructose catalyzed by glucose isomerase. The major difference is that the substrate is bound non covalently at the catalytic center in the enzyme systems where as in the protein glycation reactions the substrate (aldose) is covalently bound as the aldimine adduct at the catalytic centers. Therefore, it is tempting to speculate that the glycation hot spots of proteins represent 'degenerate isomerase-like active centers'. A corollary of the 'enzyme active site mimicry' concept is the implication of catalytic residues in the microdomains of the glycation hot spots. If this indeed is the case, the chemical modification or the site specific mutagenesis of such residues of the region will abolish the catalytic power of the microenvironment of glycation hot spots in much the same way as the chemical modification or mutation of the catalytic residues of the active sites does to the catalytic activity of enzymes.

A comparison of the general structural features of the microdomains of glycation hot spots of HbA and RNAse A revealed the presence of one or more His residues in the vicinity of the glycation hot spots. The His residues of the catalytic center could contribute to the catalytic power of the glycation hot spots by being only a contributor to the charge density of the domain. Alternatively the His residues could act as the catalytic residues of the Amadori rearrangement catalytic centers. Nonenzymic glycation of RNAse A carboxymethylated at its active site His residues has been undertaken to gain an insight into the possible role of the His residues in facilitating the glycation of Lys 7, 37, and 41. All of these three Lys residues are near the active site of RNAse A. His 12 and His 119 are the catalytic residues of the active site of RNAse A. Carboxymethylation of either one of these two His residues inactivates the enzyme. The carboxymethylation of one of these His residues of RNAse A could similarly influence the nonenzymic glycation catalytic power of this domain. The carboxymethylation of His-119 lowered the extent of nonenzymic glycation at Lys-7 and Lys-41, by nearly 70%, and that at Lys-37 by about 20%. The differential influence of the modification of His-119 of the domain on the chemical reactivities of three Lys residues of the region suggests the facilitation of the Amadori rearrangement by multiple structural factors (Nacharaju & Acharya, 1993).

The presence of residual amount of Amadori rearrangement catalytic power in the carboxymethylated RNAse A implies that His-119 by itself is not the sole contributor to the catalytic power of the glycation hot spots. The partial loss of the catalytic power of the glycation hot spots on carboxymethylation is comparable to the situations that generate partially active enzymes as a result of chemical modification. The interpretation put forward in such cases is that the modification introduces significant amount of conformational changes to the active site region of the enzyme, thereby decreasing the catalytic efficiency of the active site. If such conformational changes are invoked to explain the reduction in the chemical reactivity of glycation hot spots, it is also conceivable that these conformational changes could have influenced the pK_a of the ε-amino groups of this microdomain (Lys 7, 37, and 41). The observed loss in the chemical reactivity of Lys residues may be an indirect consequence of changes in the propensity of the Lys residues to form aldimine adducts as a result of the chemical modification rather than being a reflection of the direct participation of the His residue in the catalytic process. A comparison of reductive glycation of carboxymethylated RNAse A with that of the unmodified RNAse A demonstrated that the propensity of the ε-amino groups of Lys 7, 37, and 41 to form the aldimine adducts is not significantly perturbed by the carboxymethylation of His-119. Thus, a direct role is implicated for His-119 in the catalytic power of the microenvironment.

In summary, these structural studies have established the basic concept, namely the role of the protein structure in the generation of protein bound ketoamine adducts. Further modification of these ketoamine adducts to generate the so called 'advanced glycation end products' (Figure 1) on one hand and the efficient removal of these advanced glycation end products by specific receptors has been suggested to determine the severity in the pathophysiology of diabetes. In view of the studies described here, it seems worthwhile to investigate the contribution of the structural elements of the microdomains of glycation hot spots on the formation of the advanced glycation end products.

Acknowledgements
This research is supported by a grant-in-aid from Juvenile Diabetes Foundation and the National Sickle Center award from NIH. We thank Professor R.L. Nagel for the facilities extended to us.

References

Acharya, A.S., & Manning, J.M., (1980) Reactivity of the amino groups of carbonmonoxy Hemoglobin S with glyceraldehyde J. Biol. Chem. **255,** 1406-1412.

Acharya, A.S., Cho, Y.J., & Dorai, B. (1989) Nonreductive modification of proteins by glyceraldehyde Ann. NY. Acad. Sci., **565,** 349-350.

Acharya, A.S., Roy, R.P., & Dorai, B., (1991) Aldimine to ketoamine isomerization (Amadori rearrangement) potential at the individual glycation sites of Hemoglobin A: Preferential Inhibition of glycation by nucleophiles at sites of low isomerization potential. J. Prot. Chem. **10,** 345-358.

Acharya, A.S., Sussman, L.G., & Manning, J.M. (1983) Schiff base adducts of glyceraldehyde with hemoglobin: Differences in the Amadori rearrangement at the α-amino groups. J. Biol. Chem. **258,** 2296-2302.

Bunn, H.F., Haney, D.N., Gabbay, K.H., & Gallop, P.M., (1975) Further identification of the nature and linkage of the carbohydrate in Hemoglobin A_{1C}. Biochem. Biophys. Res. Commun. **67,** 103-118.

Hirs, C.H. (1956) The oxidation of ribonuclease with performic acid. J. Biol. Chem. **219,** 611-621.

Iberg, N., & Fluckiger, R. (1986) Nonenzymic glycation of albumin **in vivo**: Identification of multiple glycosylated sites J. Biol. Chem. **261,** 13542-13545.

Nacharaju, P., & Acharya, A.S. (1992) Amadori rearrangement potential of Hemoglobin A at its glycation sites is dependent on the three-dimensional structure of protein. Biochemistry, **31,** 12673-12679.

Nacharaju, P. & Acharya, A.S. (1993) Catalytic role of His-119 of RNAse A in the nonenzymic glycation of its active site Lys residues. Protein Science **2,** Suppl. 1. 135.

Shilton, B.H., & Walton, D.J. (1991) Sites of glycation of human and horse liver alcohol dehydrogenase **in vivo.** J. Biol. Chem., **266,** 5587-5592.

Watkins, N.G., Neglia-Fisher, C. I., Dyer, D.G., Thorpe, S.P., & Baynes, J. W. (1987) Effect of phosphate on the kinetics & specificity of glycation of protein J. Biol.Chem. **262,** 7207-7212.

Examination of Site Specificity of Glycation of Alcohol Dehydrogenase by Computer Modeling

D. J. Walton,[1] R. L. Campbell,[2] and B. H. Shilton[1]

[1]DEPARTMENT OF BIOCHEMISTRY, QUEEN'S UNIVERSITY, KINGSTON, ONTARIO, CANADA K7L 3N6; [2]INSTITUTE FOR BIOLOGICAL SCIENCES, NATIONAL RESEARCH COUNCIL, OTTAWA, ONTARIO, CANADA K1A 0R6

Summary

Molecular modeling was used to examine the site specificity of glycation of horse liver alcohol dehydrogenase in terms of structural features of the enzyme molecule. The results indicated that the catalyst for glycation of Lys-231 was likely to be the imidazole group of His-348, exerting its effect through the hydroxyl of Thr-347. For glycation of Lys-228, which requires phosphate, the base catalyst could be a phosphate ion, bound to the enzyme at a positive region of the coenzyme binding site. NAD^+ inhibited glycation of Lys-228 by binding to the enzyme and restricting access to glucose. Molecular modeling appears to be a useful tool for gaining and understanding the influence of protein groups, bound ions and bound coenzymes upon the site specificity of glycation.

Introduction

The site specificity of glycation of proteins has been ascribed to variations in the rates of Amadori rearrangement of the Schiff bases formed at different amino groups. The rate of rearrangement of a particular Schiff base is thought to be governed by the "isomerization potential" (Nacharaju and Acharya, 1992) of groups that are close to the glycation site.

Sequencing methods have been used to identify the sites of glycation of liver alcohol dehydrogenase (ADH) *in vivo* (Shilton and Walton, 1991) and *in vitro* (Walton and Shilton, 1991). The most significantly glycated amino groups are those of lysines 231 and 228, which are within the coenzyme binding domain. We now describe computer modeling studies that were conducted in an attempt to account for this site specificity.

Methods

Models of the Schiff bases formed by reaction of glucose with Lys-231 or Lys-228 of horse liver ADH were constructed with the Quanta program (Molecular Simulations Inc.). Torsional angles of (a) lysine side-chains, (b) side-chains of nearby amino acid residues, and (c) the sugar residue, were adjusted manually to optimize the relative positions of potential base catalysts and H-2 of the sugar. Atomic overlaps were avoided; energy minimization methods were not used.

Atomic coordinates were taken from Protein Data Bank file 6ADH (Eklund et al., 1981) for the holoenzyme containing NADH, and file 8ADH (Jones and Eklund, personal communication) for the apoenzyme.

Results and Discussion

Glycation of Lys-231

This reaction occurs *in vivo* (Shilton and Walton, 1991) and *in vitro* (Walton and Shilton, 1991), is relatively independent of the type of anion present in the solution, is unaffected by NAD^+, and is inhibited by guanidine hydrochloride (Walton and Shilton, 1991). It therefore seemed likely that rearrangement of the Schiff base formed at Lys-231 was promoted by nearby protein groups.

A preliminary examination of a model of holo-ADH showed that the amino group of Lys-231 is unique, as it is the only one of the thirty such groups that is within 8 A of a histidine (residue 348) imidazole group. It seemed possible that the latter promotes the Amadori arrangement of the carbohydrate of the Schiff base of Lys-231 by removing its C-2 proton. However, adjustments to the conformation of the peptide backbone would be needed to bring the imidazole and the attached sugar closer together. A more satisfactory rationalization of glycation of Lys-231 was obtained by the molecular modeling which is described next.

In a model of the holoenzyme containing a hexose residue attached to Lys-231, changes were made to the conformations of the sugar moiety, and the side-chains of Lys-231, His-348 and Thr-347. In the resulting structure (Figure 1, upper panel) the Thr-347 hydroxyl proton can form a hydrogen bond to one of the His-348 imidazole nitrogen atoms. The Thr-347 hydroxyl oxygen atom is in van der Waals contact with H-2 of the attached sugar. It was therefore possible to create an alignment of groups that would permit H-2 abstraction, without changing the conformation of the protein backbone. In this arrangement the basicity of the His-348 imidazole could be relayed to the sugar C-2 proton via the hydroxyl group of Thr-347. This would result in removal of the C-2 proton of the hexose to afford an enol, as shown in Figure 2. It therefore seems possible that the relatively high rate of glycation of Lys-231 is caused by catalysis of the Amadori rearrangement by side-chain groups of His-348 and Thr-347.

Modeling demonstrated that a bound coenzyme molecule, or a bound phosphate ion, would be too distant from Lys-231, Thr-347 or His-348 to affect the rate of glycation of Lys-231. Modeling also showed that glycation of Lys-231 would not be expected to be affected by a phosphate ion or a bound coenzyme molecule. The lack of glycation of Lys-231 occurring in the presence of guanidine hydrochloride was probably due to the distancing of this residue from Thr-347 and His-348 caused by denaturation.

Glycation of Lys-228

Glycation of the amino group of Lys-228, was interesting because if affected enzymatic activity by interfering with coenzyme binding (Walton and Shilton, 1993). Glycation at this site occurred only *in vitro*, and only if phosphate was included in the medium. The reaction was inhibited by NAD^+.

A phosphate ion can occupy the site of ADH that normally accommodates the coenzyme (Dahl and McKinley-McKee, 1980). Therefore the bound ion might be responsible for catalyzing glycation of Lys-228.

This hypothesis was tested by molecular modeling, as described next. The following procedure was adopted to produce a model (Figure 1; lower panel) of apo-ADH with an aldimine at Lys-228, containing bound phosphate. A phosphate ion was added to a model of the holoenzyme, in a position close to that normally occupied by the adenyl phosphate of NADH. A model of apo-ADH, with an aldimine at Lys-228, was added,

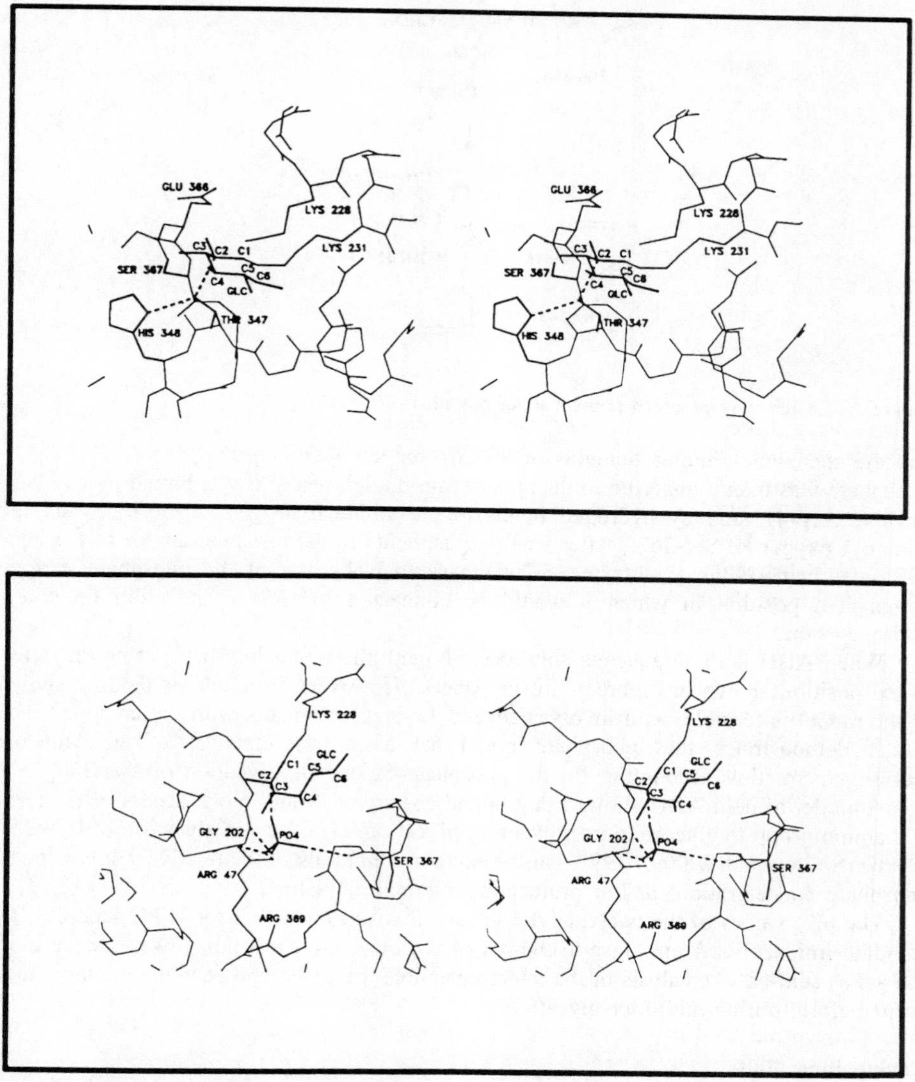

Figure 1. Views of portions of models of ADH derivatives. Upper panel: Holo-ADH, in which an aldimine has been formed by reaction of Lys-231 with glucose. The resulting model was modified as described in the text. Residues 228 and 231 are part of the coenzyme-binding domain. Other numbered residues are in the catalytic domain. Atoms that are likely to be involved in an Amadori rearrangement are joined by dotted lines. Lower panel: Apo-ADH, in which Lys-228 has formed an aldimine by reaction with glucose. A phosphate ion was added, and the model was adjusted as described in the text. Residues 47 and 367 are in the catalytic domain. Other numbered residues are in the coenzyme-binding domain.

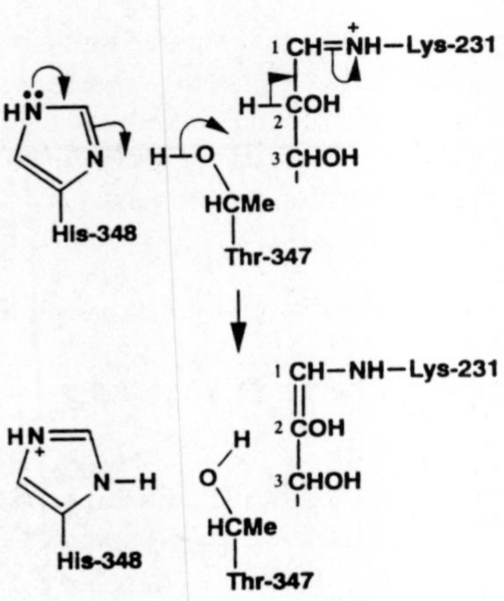

Figure 2. Charge-relay proposed to account for glycation of Lys-231.

and the coenzyme-binding domains of the two models were superimposed. The phosphate was then transferred to the apoenzyme model, where it was bound by two salt-links to Arg-47, and by hydrogen bonds to the amide hydrogen of Gly-202 and the carbonyl oxygen of Ser-367. After small adjustments to the torsional angles of the Lys-228 side-chain of the apoenzyme, it was easy to place one of the phosphate oxygen atoms at a position in which it would be connected to H-2 of the aldimine, via a hydrogen bond.

When ADH is in an aqueous solution, a bound phosphate ion may not be restricted to the position shown in Figure 1 (lower panel). However, the mode of binding shown in the model is feasible, as it involves several favorable contacts with the enzyme. The model demonstrates that phosphate could act as a base catalyst of an Amadori rearrangement, thus accounting for the phosphate-dependent glycation of Lys-228.

A model of holo-ADH showed that bound coenzyme would block access of the Lys-228 amino group to glucose, thus preventing glycation. The lack of glycation of Lys-228 *in vivo* (Shilton and Walton, 1991) can be explained in terms of the relatively low hepatic phosphate concentration, and/or protection by bound cofactors.

The pK_A values of the two catalytic groups invoked here, (7.2 for H_2PO_4 and 6.0 for histidine imidazolium) are close to the pH of the incubation medium (7.4). Hence they can act as acid-base catalysts of the interconversions of cyclic and acyclic tautomers that occur before, during and after glycation.

Conclusions

Previous attempts to account for the site-selectivity of glycation of a protein were based upon a consideration of the position of a reactive amino group in the primary (Arai et al., 1987; Baynes et al., 1989) or tertiary (Baynes et al., 1989; Nacharaju and Acharya, 1992; Shilton and Walton, 1991) structure. Since the critical step in glycation is the Amadori rearrangement, it is important to examine spatial relationships between potentially catalytic groups and a covalently-attached hexose, in the Schiff base form.

In the present study this was achieved by computer modeling. The method was used to demonstrate that interactions between a covalently-bound hexose, and side-chain groups of nearby amino acid residues (or a bound anion), could account for relatively rapid glycation at two locations in ADH. This new approach may prove to be useful in further studies of the site-specificity of glycation of proteins.

Acknowledgments
This work was supported by the Canadian Diabetes Association and the National Sciences and Engineering Research Council of Canada.

References

Arai, K., Maguchi, S., Fujii, S., Ishibashi, H., Oikawa, K., and Taniguchi, N. (1987) Glycation and Inactivation of Human Cu-Zn-Superoxide Dismutase. *J. Biol. Chem.* 262:16969-16972.

Baynes, J.W., Watkins, N.G., Fisher, C.I., Hull, C.J., Patrick, J.S., Dahl, K.H., Ahmed, M.U., Dunn, J.A., and Thorpe, S.R. (1989) The Amadori Product on Protein: Structure and Reactions. In *The Maillard Reaction In Aging, Diabetes, and Nutrition*. Baynes, J.W. and Monnier, V.M. (eds). Alan R. Liss, New York. pp. 43-67.

Dahl, K.H., and McKinley-McKee, J.S. (1990) Phosphate Binding to Liver Alcohol Dehydrogenase Studied by the Rate of Reaction with Affinity Labels. *Eur. J. Biochem.* 103:47-51.

Eklund, H., Samara, J.-P., Wallén, L., Bränden, C.-I., Akeson, A., and Jones, A.L. (1981) Structure of a Triclinic of Horse Liver Alcohol Dehydrogenase at 2.9 A Resolution. *J. Mol. Biol.* 146:561-587.

Nacharaju, P., and Acharya, S. (1992) Amadori Rearrangement Potential of Hemoglobin at its Glycation Sites is Dependent on the Three-Dimensional Structure of Protein. *Biochemistry* 31:12673-12679.

Shilton, B.H., and Walton, D.J. (1991) Sites of Glycation of Human and Horse Liver Alcohol Dehydrogenase *In Vivo*. *J. Biol. Chem.* 266:5587-5592.

Walton, D.J. and Shilton, B.H. (1991) Site Specificity of Protein Glycation. *Amino Acids* 1:199-203.

Walton, D.J. and Shilton, B.H. (1993) Effect of Glycation Upon Activity of Liver Alcohol Dehydrogenase. In *Enzymology and Molecular Biology of Carbonyl Metabolism 4* Weiner, H., Crabb, D.W., and Flynn, T.G. (eds.), Plenum Press, New York, pp. 493-499.

Advanced Glycation Dependent Formation of Modified α-Crystallin and High Molecular Weight Aggregates

M. Cherian, A. Abraham, S. Swamy-Mruthinti, and E. C. Abraham

DEPARTMENT OF BIOCHEMISTRY AND MOLECULAR BIOLOGY, MEDICAL COLLEGE OF GEORGIA, AUGUSTA, GA 30912, USA

Summary
Modified α-crystallin (peak 3) fraction from 70-88 year old human lens protein was isolated by RP-HPLC and further analyzed by Sephacryl/guanidine HCl chromatography and SDS-PAGE and advanced glycated products were estimated by analyzing fluorescence spectra. Likewise, *in vitro* generated peak 3 from calf α-crystallin incubated with various sugars was isolated and studied. Highly fluorescent and cross-linked aggregates were present in both types of peak 3 preparations. Almost complete inhibition of peak 3 formation by aminoguanidine, an inhibitor of the advanced glycation process, suggested a direct involvement of advanced glycation in the formation of the peak 3 fraction.

Introduction

Human α-crystallin normally exists as a polymer having molecular weight of approximately 800,000 daltons. It undergoes yet unidentified modification becoming an aggregate of > 1 million daltons (Spector *et al.*, 1971). By applying reverse-phase (RP) HPLC we have shown earlier that the high molecular weight (HMW) α-crystallin fraction contains a predominantly modified α-crystallin peak (peak 3) with a longer retention time than either αB- or αA-crystallin (Swamy and Abraham, 1991). The same study also showed that the peak 3 component increases with lens age. In this report we show that the modified α-crystallin fraction is heterogeneous and contains cross-linked HMW aggregates enriched with advanced glycated products. *In vitro* glycation studies in the presence and absence of aminoguanidine (AG), an inhibitor of advanced glycation, suggested that the formation of peak 3 is advanced glycation dependent.

Materials and Methods

Calf lenses and 70-88 year old human lenses were decapsulated, homogenized in 50 mM phosphate buffer, pH 7.4 containing 50 mM NaCl (for calf lenses) or 50 mM Tris-HCl, pH 7.4 containing 0.2 M NaCl and 1 mM EDTA (for human lenses) and centrifuged for 1 hr at 10,000 g at 4°C and the supernatants were used for further studies. α-Crystallin fraction was purified by chromatography on a Sephacryl-S-300-HR column. About 30 mg of the protein was applied to a 80 x 1.6 cm column and developed isocratically with the respective homogenizing buffer. Calf α-crystallin was incubated with different glycating agents for up to 20 days in the presence or absence of 5 mM AG. Incubation mixture (0.5 mL) contained 2.5 mg α-crystallin in 150 mM phosphate buffered saline, pH 7.0 containing 0.5 mM EDTA 0.02% sodium azide and 1.5 mM phenylmethylsulfonyl fluoride. For the separation of αB- and αA-crystallin and the modified α or peak 3 fraction, RP-HPLC was carried out with a vydac C_4 column as decribed before (Swamy and Abraham, 1991). To separate the monomeric proteins from the cross-linked aggregated proteins in the peak 3 fraction, chromatography on a Sephacryl-S-300-HR column was performed in the presence of 6 M guanidine HCl and SDS-PAGE was done on a slab gel system (15% gels and 4 hr run) according to Laemmli (1990).

Results

Analysis of the *in vivo* formed modified α-crystallin or peak 3 fraction from human lenses for advanced glycated products and *in vitro* generation of peak 3 by glycation in the presence and absence of AG were the two lines of approaches followed for showing that peak 3 formation is advanced glycation dependent. Fluorescence scan at two different sets of excitation and emission wavelengths, one for the age related blue fluorescence (Figure 1A) and the other for pentosidine fluorescence (Figure 1B), showed significant level of fluorescence in the peak 3 component, as compared to αB or αA-crystallin, from aged human lenses. Moreover, peak 3 constituted about 30% of the total α-crystallin. *In vitro* glycation of calf α-crystallin with various sugars resulted in different levels of peak 3, fast reacting sugars producing the highest levels (Figure 2 and Table I). As in the case of human lenses, large amounts of fluorescent products were present in the peak 3 formed *in vitro*. The presence of AG showed substantial inhibition of the production of peak 3 and the fluorescence associated with it (Figure 2 and Table I). Gel electrophoresis of the peak 3 proteins showed considerable heterogeneity by having monomeric α-crystallin as well as oligomeric protein bands formed by non-disulfide cross-links (Figure 3). Heterogeneity was further evident when RP-HPLC purified peak 3 was chromatographed on a Sephacryl-S-300-HR column in the presence of guanidine HCl. About 25% of the peak 3 proteins existed as monomers (peak c) and the remaining (peaks a & b) were present as aggregates of varying sizes (Figure 4). Examination of the aggregate and monomer fractions from this column by SDS-PAGE confirmed that the monomer fraction is indeed monomeric α-crystallin and the aggregate fractions contain mostly cross-linked proteins (data not shown).

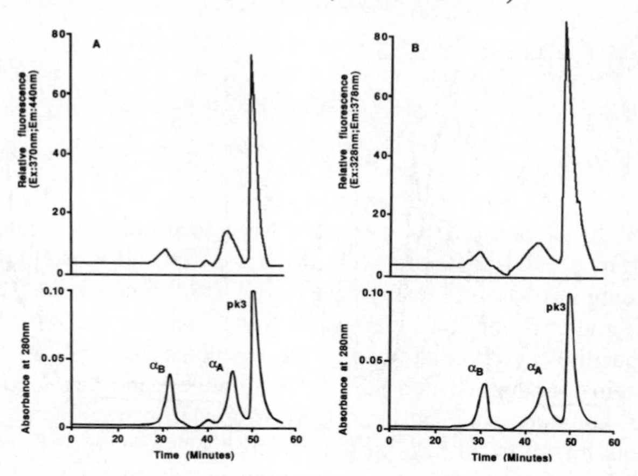

Figure 1. RP-HPLC separation of αB- and αA- crystallin and the modified α-crystallin or the peak 3 component from the total α-crystallin fraction isolated from 70-88 year old human lenses. Absorbance at 280 nm and fluorescence at two different sets (A and B) of fixed excitation and emission wavelengths were recorded.

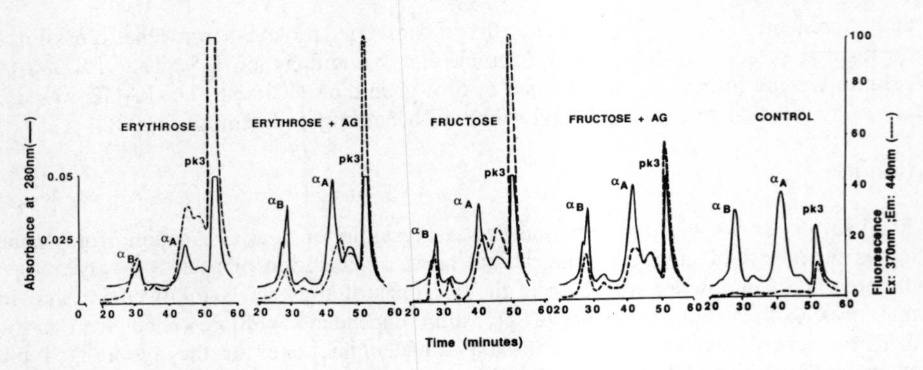

Figure 2. Separation of αB- and αA-crystallin and the peak 3 component after incubation of calf α-crystallin fraction with 5 mM erythrose or 1 M fructose in the presence or absence of 5 mM aminoguanidine (AG). A chromatogram of a control incubation of α-crystallin without any sugar is also provided.

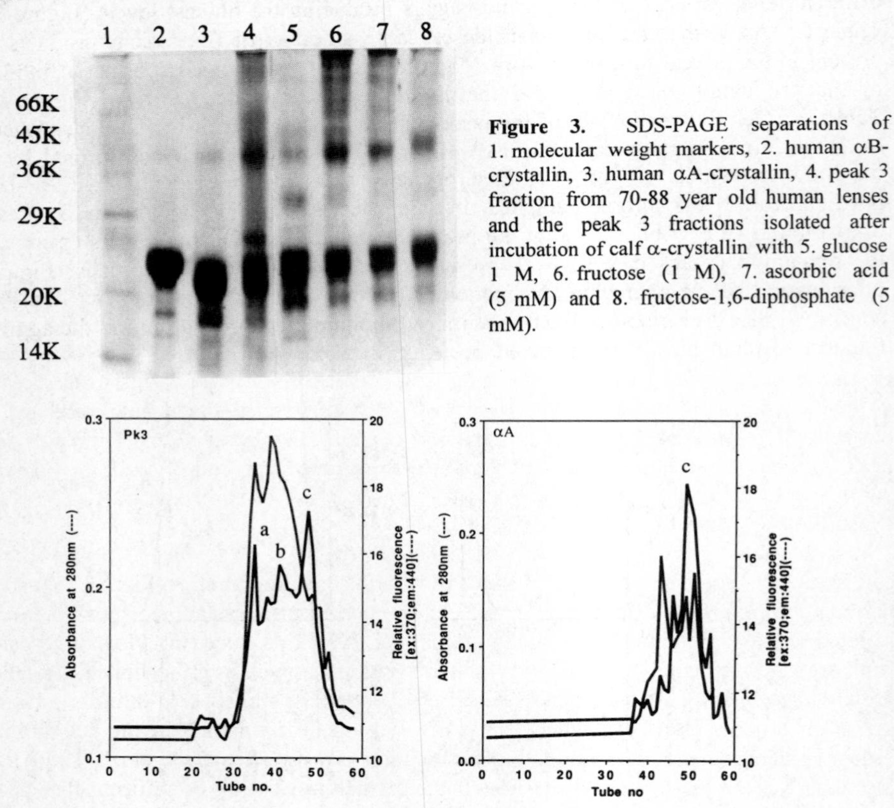

Figure 3. SDS-PAGE separations of 1. molecular weight markers, 2. human αB-crystallin, 3. human αA-crystallin, 4. peak 3 fraction from 70-88 year old human lenses and the peak 3 fractions isolated after incubation of calf α-crystallin with 5. glucose 1 M, 6. fructose (1 M), 7. ascorbic acid (5 mM) and 8. fructose-1,6-diphosphate (5 mM).

Figure 4. Further fractionation of the peak 3 proteins isolated from senile lenses on a Sephacryl-S-300-HR column in the presence of 6 M guanidine HCl. A control αA-crystallin run under the same condition is also included.

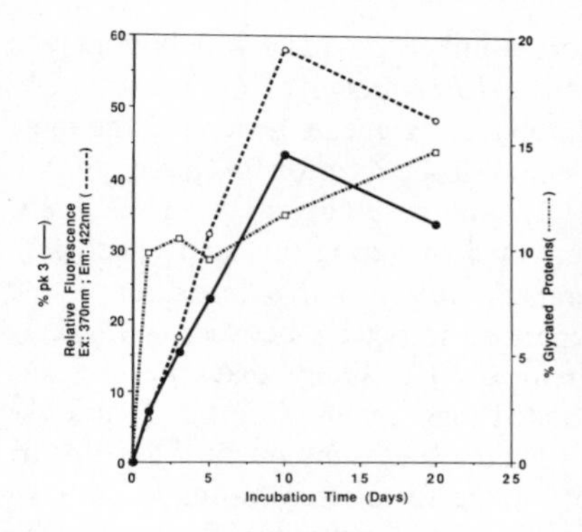

Figure 5. Time-course of the formation of the peak 3 component, advanced glycated products and Amadori products during incubation of 5 mM erythrose with calf α-crystallin. Amadori products were quantitated by a previously established affinity chromatographic method (Perry *et al.*, 1987).

Table I. Influence of advanced glycation on levels of peak 3*.

Glycating agent	Peak 3 (% of total α)		Glycating Agent	Peak 3 (% of total α)	
	-AG	+AG		-AG	+AG
1 M Glucose	20.7	16.9	5 mM Ascorbic acid	38.3	31.9
1 M Fructose	38.0	19.8	5 mM Glucose-6-phosphate	24.0	19.3
5 mM Ribose	26.9	17.6	5 mM Fructose-1,6-diphosph.	30.0	21.7
5 mM Erythrose	54.1	32.5	Control (α-crystallin alone)	17.6	

*Levels of peak 3 after *in vitro* glycation of calf α-crystallin with various glycating agents for 20 days in the presence and absence of 5 mM aminoguanidine (AG). Protein fractions were estimated by Lowry method (1951).

To find out whether peak 3 formation is closely related to ketoamine formation (early glycation) or advanced glycation, time-course of the formation of the various products formed during glycation of calf α-crystallin with erythrose was determined. Erythrose being a fast glycating agent, the ketoamine formation nearly plateaued within 24 hours whereas the formation of advanced glycation related fluorescent products and the peak 3 component proceeded in a parallel fashion at a slower rate (Figure 5). Thus, peak 3 generation seems to be closely related to advanced glycation.

Discussion

The present study shows that, on the basis of the electrophoretic (Figure 3) and the chromatographic behavior (Figure 1 and 2), the modified α-crystallin or peak 3 fraction isolated from senile lenses has striking similarities to that produced by *in vitro* glycation. Both are heterogeneous containing monomeric proteins as well as cross-linked oligomeric proteins presumably formed by advanced glycation cross-links. AG inhibits advanced glycation process by a mechanism which involves specific trapping of the ketoamine or 3-deoxyglucosone product through a derivatization process (Brownlee *et al.*, 1986). The fact that AG almost completely inhibited the formation of peak 3 confirms that peak 3 formation is advanced glycation dependent (Figure 2 and Table I). The time-course studies also support this conclusion (Figure 5). It should be emphasised however, that glycation may not be the sole cause for *in vivo* generation of the modified α-crystallin. All the sugars that were used in the present *in vitro* study were tested before for their ability to form ketoamine products and fluorescent cross-linked products with bovine lens proteins (Abraham *et al.*, 1990; Swamy *et al.*, 1993). Comparison of the present data with the earlier data indicated that the most powerful glycating agents generated the most amounts of peak 3 (Table I). Ascorbic acid was shown to be a powerful glycating and cross-linking agent (Prabhakaram and Ortwerth, 1991) and its ability to form relatively large quantities of heavily fluorescent peak 3 became evident during this study (Table I).

As shown in an earlier study (Swamy and Abraham, 1991) *in vivo* formation of the peak 3 component is age dependent. A two-fold increase in peak 3 between a 17-year-old and an 82-year-old human lens was noted before. It is likely that most of this increase is due to the formation of the cross-linked aggregates (Figure 3 and 4). Sephacryl/guanidine HCl purification of peak 3 proteins and subsequent SDS-PAGE characterization of each fraction showed almost 75% in the form of cross-linked proteins and the remaining as monomers. Interestingly, even the monomer fraction when reinjected into an RP-HPLC column, behaved as peak 3 and not as αB- or αA-crystallin (data not given) indicating the presence of some unidentified modifications causing this change in the chromatographic behavior.

Acknowledgements
This work was supported by NIH Grant EY 07394.

References

Abraham, E.C., Tsai, C., Abraham, A., and Swamy, M.S. (1990) Formation of early and advanced glycation products of lens crystallins with erythrose, ribose and glucose. In *The Maillard Reaction in Food Processing, Human Nutrition and Physiology* (Eds Finot, P.A., Aeschbacher, H.U., Hurrell, R.F., and Liardon, R.) Pp. 437-442. Birkhäuser Verlag: Basel, Switzerland.

Brownlee, M., Vlassara, H., Kooney, A., Ulrich, P., and Cerami, A. (1986) Aminoguanidine prevents diabetes-induced arterial protein cross-linking. *Science* 232: 1629-1632.

Laemmli, U.K. (1970) Clevage of structural proteins during the assembly of bacteriophage T_4. *Nature* 227: 680-685.

Lowry, O.H., Rosebrough, N.J., Farr, A.L., and Randall, R.J. (1951) Protein measurement with the Folin Phenol reagent. 193: 265-275.

Perry, R.E., Swamy, M.S., and Abraham, E.C. (1987) Progressive changes in lens crystallin glycation and high molecular weight aggregate formation leading to cataract development in streptozotocin-diabetic rats. *Exp. Eye Res.* 44: 269-282.

Prabhakaram, M., and Ortwerth, B.J. (1991) Glycation associated cross-linking of lens proteins by ascorbic acid is not mediated by oxygen free radicals. *Exp. Eye Res.* 53: 261-268.

Spector, A., Li, L.K., Augusteyn, R.C., Schneider, A., and Freund, T. (1971) α-Crystallin: Isolation and characterization of distinct macromolecular fractions. *Biochem. J.* 124: 337-343.

Swamy, M.S., and Abraham, E.C. (1991) Reverse-phase HPLC analysis of human α-crystallin. *Current Eye Res.* 10: 213-220.

Swamy, M.S., Tsai, C., Abraham, A., and Abraham, E.C. (1993) Glycation mediated lens crystallin aggregation and cross-linking by various sugars and sugar phosphates *in vitro*. *Exp. Eye Res.* 56:177-185.

Site-specific and Random Fragmentation of Cu,Zn-Superoxide Dismutase (Cu,Zn-SOD) by Glycation Reaction: Implication of Reactive Oxygen Species

N. Taniguchi,[1] T. Ookawara, [1,2] and H. Ohno[2]

[1]DEPARTMENT OF BIOCHEMISTRY, OSAKA UNIVERSITY MEDICAL SCHOOL, 2-2 YAMADAOKA, SUITA, OSAKA 565, JAPAN; [2]DEPARTMENT HYGIENE, NATIONAL DEFENSE MEDICAL COLLEGE, 3-2 NAMIKI, TOKOROZAWA, SAITAMA 359, JAPAN

Summary

Site-specific and random fragmentation of human Cu,Zn-superoxide dismutase (Cu,Zn-SOD) was observed following the glycation reaction (the early stage of the Maillard reaction). The fragmentation occurred at a peptide bond between Pro 62 and His 63, as judged by amino acid analysis and sequencing of fragment peptides, yielding a large (15 kD) and a small (5 kD) fragment. In the second step, random fragmentation occurred. The ESR spectrum of the glycated Cu,Zn-SOD suggested that reactive oxygen species (ROS) were implicated in both the steps of fragmentation. Incubation with glucose resulted in a time-dependent release of Cu^{2+} from the Cu,Zn-SOD molecule. The released Cu^{2+} then likely participated in a Fenton's type of reaction to produce hydroxyl radical, which may cause the non-specific fragmentation. This is the first report that a site-specific fragmentation of a protein is caused by reactive oxygen species formed by the Maillard reaction.

Introduction

Our previous studies indicated that human Cu,Zn-SOD undergoes glycation (Maillard reaction) at specific lysine residues and that the enzyme is inactivated by glycation *in vitro* (Arai et al., 1987a) as well as *in vivo* (Arai et al., 1987b) The level of glycated Cu,Zn-SOD is increased in the erythrocytes of patients with diabetes mellitus (Kawamura et al., 1992) as well as patients with Werner's syndrome, an age-accelerated disease (Taniguchi et al., 1989). Involvement of the glycation reaction in the pathogenesis of diabetic complications has also been suggested. Sakurai et al. (1988) reported that the Schiff base and/or Amadori products generate ROS. Implication of divalent metals for the production of ROS also has been suggested. In the present study we found that glycation of Cu,Zn-SOD initially brought about a site-specific fragmentation of the molecule, and further incubation of the enzyme with glucose brought about complete fragmentation. The role of reactive oxygen species in the fragmentation reaction is discussed.

Materials and Methods

Recombinant human Cu,Zn-SOD was a kind gift from Ube Industries, Japan. The amino acid sequence of the recombinant enzyme is the same as that of the human erythrocyte enzyme except that the N-terminal Ala is not acetylated.

Glycation of Cu,Zn-SOD in vitro
The standard reaction mixture contained 50 mM potassium/sodium phosphate buffer, pH 7.4, 150 mM NaCl. 0.025% NaN_3, 0.1 M reducing sugar and 1-2 mg/ml Cu,Zn-SOD (Ookawara et al., 1992). After incubation at 37°C for the indicated duration with gentle shaking in a water bath, the reaction was stopped by freezing the mixture. Each sample was stored at -35°C until use. All solutions were sterilized by filtration through a disposable cellulose nitrate filter. Glycated Cu,Zn-SOD was observed by SDS-PAGE using 15% slab gels (90 x 55-mm).

Amino acid sequence analysis
For the amino acid sequence analysis, recombinant Cu,Zn-SOD which had been incubated with 0.1 M glucose for 7 days was subjected to SDS-PAGE. The proteins were then transferred onto PVDF membrane using a semi-dry electroblotting system. The blotted spots were subjected to a gas phase sequencer. Amino acid sequencing was carried out with an Applied Biosystems 477A Protein Sequencer equipped with an on-line analyzer for phenylthiohydantoin-derivatized amino acids.

ESR spectra measurements
ESR spectra were measured at room temperature using an E-12 EPR spectrometer system (Varian). After a 3-day incubation of Cu,Zn-SOD in the presence of 0.1 M glucose at 37°C, DMPO was added to the incubation mixture to a final concentration of 0.1 M as the hydroxyl radical trapping reagent and ESR signals were measured.

Results

Cleavage of Cu,Zn-SOD by glycation reaction
Incubation of human Cu,Zn-SOD with 0.1 M glucose for 0 to 14 days at 37°C resulted in a time-dependent decrease in the amount of intact enzyme. SDS-PAGE showed a gradual decrease in the intensity of the original band with an approximate molecular weight of 20 kD and the simultaneous appearance of a large molecular weight fragment with a distinct 15 kD band and a small molecular weight fragment of about 5 kD (Figure 1). The site-specific fragmentation was apparent at 3 days and increased gradually until 7 days. After 7 days, the fragments gradually decreased. These results indicate that the site-specific fragmentation occurred at first, and after 7 days, random fragmentation occurred predominantly.

The fragmentation of Cu,Zn-SOD depended on the glucose concentration. Fructose also induced a rapid and marked fragmentation of the protein. This suggests that the fragmentation is related to interaction of the Cu,Zn-SOD with reducing sugar and that the Maillard reaction (glycation reaction) may play a role in the fragmentation. Under anaerobic conditions, the fragmentation was not observed, indicating that the dissolved oxygen molecules are requisite for the fragmentation reaction.

Effect of H_2O_2 on the fragmentation reaction
A bolus of H_2O_2 (5 mM) was directly added to the Cu,Zn-SOD solution (2 mg/ml) to determine whether a similar fragmentation reaction would occur. The site-specific fragmentation occurred rapidly. The extent of fragmentation increased up to 60 min., after which time random fragmentation occurred. In the presence of 5 mM EDTA, the site-specific fragmentation was not inhibited, but most of the random, non-specific

fragmentation was inhibited. This indicates that trace metal ion(s) are involved in the 2nd step of the fragmentation process, but metal ion is not required for the site-specific 1st step of fragmentation due to H_2O_2.

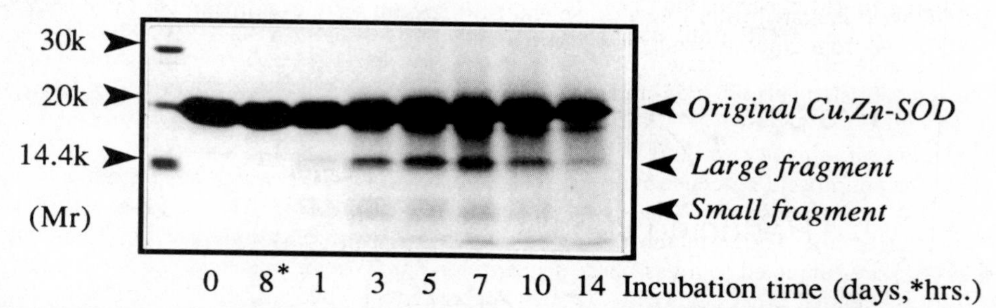

Figure 1. Fragmentation of Cu,Zn-SOD after incubation with glucose. After incubation with 0.1 M glucose as described under Materials and Methods, an aliquot of the incubation mixture containing 30 μg of Cu,Zn-SOD was directly loaded onto an SDS-PAGE.

Identification of the site of Cu,Zn-SOD cleavage by glycation

Amino acid analysis of the 5 kD fragment indicated that the fragment contained 3 prolines. The value was consistent with the reported amino acid residues, suggesting that the Pro 62 remained intact. This suggests that the cleavage site was a peptide bond between Pro 62 and His 63 of the human Cu,Zn-SOD. His 63 is one of four histidines coordinated with the Cu ion of Cu,Zn-SOD, and His 63 is also coordinated with the Zn ion. Consequently, cleavage at this position results in structural and functional damage to the enzyme.

Most mammalian Cu,Zn-SODs have the Pro 62- His 63 sequence, while the enzyme from horse erythrocytes has Ala 62 - His 63. In order to determine whether Pro 62 is requisite for the cleavage site, horse Cu,Zn-SOD was incubated with glucose and SDS-PAGE was carried out. The same fragmentation pattern was seen, indicating that the Pro residue is not essential for this fragmentation. Addition of H_2O_2 also resulted in fragmentation of the horse Cu,Zn-SOD.

Copper ion release from Cu,Zn-SOD during glycation reaction

Hydroxyl radical can be formed from H_2O_2 through a Fenton-type reaction in the presence of metals. As described above, metal ion(s) are implicated in the second step, random fragmentation of Cu,Zn-SOD. Glycation lead to release of Cu ions from the Cu,Zn-SOD molecule. The released Cu ions could then enhance the Fenton-type reaction to produce hydroxyl radicals from the H_2O_2 generated by the glycation reaction and play a major role in the random fragmentation.

Electron spin resonance spectra of DMPO-OH adduct by glycation and fragmentation of reaction mixture

In order to identify radical species, ESR spectra were measured after incubation of Cu,Zn-SOD with 0.1 M glucose at 37°C for 3 days using DMPO as a trapping agent. In the presence of DMPO, glycation of Cu,Zn-SOD generated OH• which was detected by ESR as a DMPO-OH adduct. The adduct spectrum consisted of a quartet (1;2;2;1) with hyperfine splitting constants of $a^N = a^H =$ 15.0 G. No signals were observed after addition of EDTA, which suggests involvement of metal ions in this reaction as suggested by several groups (Sakurai et al., 1988; Kawakishi et al., 1991). Addition of catalase also prevented the appearance of the DMPO-OH signals. This suggests that superoxide anion is generated at first from glycation reaction and then H_2O_2 is formed by dismutation reaction of Cu,Zn-SOD. In order to eliminate autoxidation of glucose as a source of ROS, glycated Cu,Zn-SOD was separated from free glucose and ESR spectra were examined. Glycated Cu,Zn-SOD free from glucose also gave a DMPO-OH adduct.

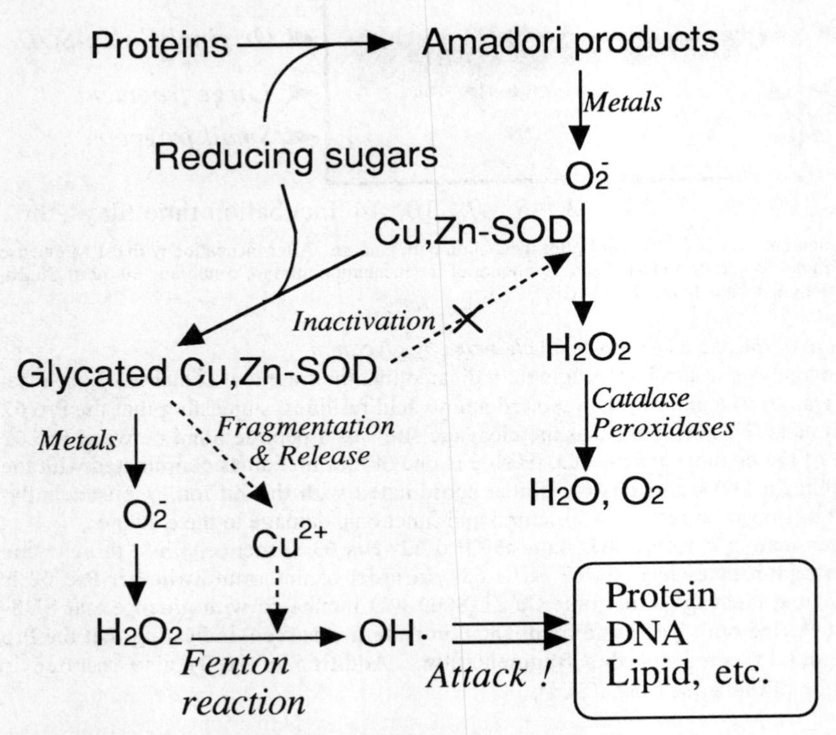

Figure 2. The role of Cu,Zn-SOD in glycation and a Fenton-type reaction.

Discussion

In the present study, we found that Cu,Zn-SOD undergoes a site-specific cleavage between Pro 62 - His 63 following the glycation reaction. Moreover, the lengthy incubation of the enzyme with glucose resulted in random fragmentation as judged by SDS-PAGE. The generation of superoxide anion in the glycation reaction and the subsequent generation of H_2O_2 or hydroxyl radical may play an important role in the site-specific fragmentation reaction.

In the present study we found the cleavage site to be located at Pro 62 - His 63. The mechanism for fragmentation of Cu,Zn-SOD remains unknown. Reactive oxygen species such as superoxide anion, H_2O_2 and hydroxyl radical are likely involved in both site-specific and random fragmentation. It has been reported that the superoxide anion is generated in the presence of trace metals from glycated products (Sakurai et al., 1989) and the dismutation of superoxide anion results in further accumulation of H_2O_2. One possible mechanism in that the enzyme-bound Cu^{2+} is reduced by H_2O_2 or superoxide anion to Cu^+, followed by a Fenton-type reaction of the Cu^+ with additional H_2O_2. This would generate Cu^{2+}-OH•, or ionized equivalent Cu^{2+}-O•$^-$, which could then attack the adjacent Pro - His sequence.

From these results, we suggest that the glycation reaction first produces superoxide anion. The superoxide anion is then converted to H_2O_2 and hydroxyl radical by a combination of Fenton-type reactions, the reverse reaction of superoxide dismutase and the catalytic reaction of Cu,Zn-SOD. The fragmentation of Cu,Zn-SOD under hyperglycemic conditions *in vivo* is now under study in our laboratory.

Conclusions

Glycated proteins produce superoxide anion on one hand, and the glycated Cu,Zn-SOD undergoes inactivation and fragmentation on the other. It is well known that Cu,Zn-SOD plays an important role in preventing Fenton chemistry in vivo. The inactivation and fragmentation of SOD may facilitate the Fenton chemistry to produce hydroxyl radicals in vivo (Figure 2). The hydroxyl radicals may bring about protein and DNA damages in diabetic and aging processes.

References

Arai, K., Iizuka, S., Oikawa, K., and Taniguchi, N. (1987a) Increase in the glucosylated form of erythrocyte Cu,Zn-superoxide dismutase in diabetes and close association of the non enzymatic glucosylation with the enzyme activity. *Biochim.. Biophys. Acta* 924: 292-296.

Arai, K., Maguchi, S., Fujii, S., Ishibashi, H., Oikawa, K., and Taniguchi, N. (1987b) Glycation and inactivation of human Cu,Zn-superoxide dismutase: Identification of the in vitro glycated sites. *J. Biol. Chem.* 262: 16969-16927.

Kawakishi, S., Tsunehiro, J., and Uchida, K. (1991) Autoxidative degradation of Amadori compounds in the presence of copper ion. *Carbohydr. Res.* 211: 167-171.

Kawamura, N., Ookawara, N., Suzuki, K., Konishi, K., Mino, M., and Taniguchi, N. (1992) Increased glycated Cu,Zn-superoxide dismutase levels in erythrocytes of patients with insulin-dependent diabetes mellitus. *J. Clin. Endocrinol. Metab.* 74: 1352-1354.

Ookawara, T., Kawamura, N., Kitagawa, Y., and Taniguchi, N. (1992) Site-specific and random fragmentation of Cu,Zn-superoxide dismutase by glycation reaction: Implication of reactive oxygen species. *J. Biol. Chem.* 267: 18505-18510.

Sakurai, T., and Tsuchiya, S. (1988) Superoxide production from nonenzymatically glycated protein. *FEBS Lett.* 236: 406-410.

Taniguchi, N., Kinoshita, N., Arai, K., Iizuka, S., Usui, M., and Naito, T. (1989) Inactivation of erythrocyte Cu,Zn-superoxide dismutase through nonenzymatic glycosylation, in *The Maillard Reaction in Aging, Diabetes, and Nutrition* (Baynes, J. W., and Monnier, V. M. eds.) pp. 227-290, Alan R. Liss, Inc., New York.

Non-enzymatic Glycation of Type I Collagen: Effects of Aging

K. M. Reiser

DEPARTMENT OF MEDICINE, SCHOOL OF MEDICINE, UNIVERSITY OF CALIFORNIA AT DAVIS, DAVIS, CA 95616, USA

Summary
Recent work concerning the effects of aging on nonenzymatic glycation of collagen have focused on regulation of this process in vivo. In the present review, recent studies from our laboratory investigating regulation and modulation of nonenzymatic glycation of type I collagen in rodents are discussed. Caloric restriction has been found to decrease the accumulation of advanced glycation products in several tissues in both rats and mice; however, such effects are not apparent until late in the life span. Collagen content of the specific Maillard product pentosidine was found to be lower in calorie-restricted rats aged 157 weeks, but not in mice, which have a shorter life span. Additional studies of dietary manipulation have shown that carbohydrate source also affects life span independently of caloric restriction. Rats that were fed sucrose and restricted to 60% of ad libitum intake were found to have longer mean and median life spans than restricted starch-fed rats. Preliminary data suggest that values for collagen-associated fluorescence and pentosidine were lower in restricted sucrose-fed rats after only 5 months of this dietary regimen, while no changes were observed in restricted starch-fed rats. Dietary manipulation does not, however, affect sites of adduct formation, and it appears that site-specificity of glycation remains controlled predominantly by primary structure. These studies provide further evidence that a complex interrelationship exists between aging, life span, and the accumulation of Maillard products on matrix macromolecules.

Introduction

Aging is characterized by structural changes in the extracellular matrix throughout the body in virtually every tissue and organ system. Despite a large literature describing the morphological, physiological, mechanical and biochemical changes that occur in aging connective tissue, the underlying mechanisms responsible for these changes are still poorly understood (Reiser, 1991; Reiser et al., 1992). In some tissues and species, crosslinks mediated by the enzyme lysyl oxidase increase with age and may in part account for physicochemical changes in collagen associated with aging (Yamauchi et al., 1988; Reiser et al., 1987). There is also considerable evidence that some age-associated changes in the properties of collagen result from crosslinks derived from nonenzymatically glycated residues. Several studies have shown that the accumulation of Maillard products in collagen is associated with alterations in mechanical properties, solubility, ligand binding and conformation (Reiser, 1991).

One of the difficulties in studying the relationship between aging and nonenzymatic glycation of collagen is that only a few structurally characterized Maillard products have been identified as collagen adducts in vivo. Such Maillard products include glucitolyllysine and glucitolylhydroxylysine (the initial glucose adducts and their Amadori rearrangement products); carboxymethyllysine and carboxymethylhydroxylysine (the so-called "non-browning intermediate Maillard products") and pentosidine, a fluorescent trifunctional advanced Maillard product. Even though pentosidine comprises only a small fraction of the fluorophores associated with collagen, it has served as a useful marker for studying the accumulation of advanced glycation products in both aging and diabetes (Sell et al., 1992; Reiser, 1991). Alternative methods for measuring accumulation of advanced glycation products in tissues include measuring fluorescence in solubilized collagen extracts. Although this method is nonspecific and subject to limitations in interpretation, measurements of collagen-associated fluorescence have been shown to correlate with clinical and pathological changes in aging and diabetes (Reiser, 1991).

In the present review, we describe recent studies in which we investigated quantitative and qualitative changes in nonenzymatic glycation of type I collagen during aging in vivo. We have been particularly interested in the relationship between life span modulation by dietary manipulation and the effects of such modulation on cumulative changes in nonenzymatic glycation of collagen. We report on the effects of lifetime caloric restriction as well as the effects of dietary carbohydrate source on this relationship in rodents.

Materials and Methods

Animals and tissue sources

In studies of the effects of caloric restriction on collagen crosslinking in F344 X BN rats and C57BL6/N-NIH mice, the animals were maintained by the National Institute on Aging project on Caloric Restriction at the National Center for Toxicological Research, as previously described (Reiser, 1993). The animals were fed NIH-31 rat chow, a grain-based diet, either ad libitum or with intake restricted to 60% of ad libitum intake throughout their life span. In studies of the effects of caloric restriction and dietary carbohydrate source on longevity, male inbred F344 rats, aged 21 days, were obtained from Harlan Sprague Dawley Laboratory (Indianapolis, IN) and maintained in our vivarium for the duration of the study. After 16 weeks of ad libitum access to a semi-purified diet containing corn starch (66% by weight) as the carbohydrate source, animals were assigned to one of four dietary groups; sucrose or starch as the sole carbohydrate source; ad libitum access or restriction to 60% of ad libitum intake. Sucrose and starch diets were isocaloric; that is, all ad libitum fed rats had the same caloric intake, regardless of carbohydrate source, as did all calorie restricted rats. In studies of the effects of aging on sites of glucose adduct formation, type I collagen from tail tendon was obtained from male Sprague-Dawley rats aged 6, 18, and 36 months. This tissue had been obtained during a previous study of aging (Reiser et al., 1987), and had been stored at -70° C until analysis.

Biochemical studies

Analyses of glucitolyllysine content, collagen-associated fluorescence, and pentosidine content were performed using techniques previously described in detail (Buckingham and Reiser, 1990; Reiser, 1993). Techniques for the analysis of glucose adduct location in the CNBr peptides α1(I)CB3 and α2CB3-5 have been described in detail previously (Reiser et al., 1992). In the present review, studies on glucose adduct location in the CNBr peptides α1(I)CB7 and α1(I)CB8 are also discussed. These peptides were prepared and analyzed using techniques previously described (Reiser et al., 1992). The identities of the isolated tryptic peptides were determined by sequence analysis and compositional analysis as previously described (Reiser et al., 1992).

Results

In general, aging had little effect on glucitolyllysine content in most tissues analyzed. However, several differences between rats and mice were observed in this regard. For example, while aging was not correlated with changes in glucitolyllysine content in mouse skin (Reiser, 1993), there was a decrease in glucitolyllysine content with age in rat skin collagen (values at 157 weeks were approximately 25% of values at 26 weeks), although not in tail tendon or aorta. In most tissues, there was considerable variability in mean values between different ages; these differences appear to be due to biological variability, rather than to variability in the assay itself (Reiser, 1993). The effects of caloric restriction on glucitolyllysine content of collagen also varied between rats and mice, but only in some tissues. For example, caloric restriction did not affect collagen content of glucitolyllysine in skin in either rats or mice at any age. In contrast, caloric restriction was associated with significant decreases in mean values of glucitolyllysine in both tail tendon and aorta collagen at several time points in mice, while no such effect was observed in rats.

In contrast to glucitolyllysine, advanced glycation products did increase with age in all of the tissues that were analyzed in both rats and mice. Advanced glycation products were measured as collagen-associated fluorescence in tail tendon and skin. Pentosidine, a specific advanced glycation product, was assayed in aorta collagen. Caloric restriction was associated with significantly decreased tissue levels of collagen-associated fluorescence, but only in animals 2 years of age or older. This delay in detectable effects of caloric restriction on advanced glycation products until late in the life span was observed in both rats and mice. However, the effects of caloric restriction on pentosidine content of aorta collagen differed somewhat between rats and mice. In mice, caloric restriction had no significant effects on pentosidine content at any age (Reiser, 1993), while calorie-restricted rats had decreased values for pentosidine content in aorta collagen at 157 weeks of age, but not at any earlier timepoints (Figure 1).

In a recent study, we observed that the source of dietary carbohydrate appeared to have a significant effect on life span independent of caloric intake (Murtagh and McDonald, 1992; Murtagh et al, 1993). We observed significant differences in survival curves, mean life span, median life span and 90th percentile survival between animals fed starch diets as compared with animals fed sucrose diets. For example, rats fed sucrose diets that were restricted to 60% of ad libitum intake had greater mean and median survival times than did calorie-restricted rats fed starch diets, but a lower maximal life span. In contrast, ad libitum fed animals had lower mean and median survival times when fed sucrose diets in comparison with animals fed starch diets. These differences in life span could not be attributed to differences in timing or severity of the histopathological lesions normally observed with aging in this species. We also determined that sucrose diets did not affect plasma glucose, glucose tolerance, or insulin receptor function.

In order to explore further the relationship between effects of dietary manipulation on life span and effects on nonenzymatic glycation, we compared the effects of sucrose and starch diets on glucitolyllysine, collagen-associated fluorescence, and pentosidine in insoluble skin collagen. Preliminary data suggest that collagen-associated fluorescence is significantly lower in skin collagen from rats fed restricted sucrose diets for 5 months in comparison with rats fed ad libitum sucrose diets; there were no such differences between restricted and ad libitum fed rats on the starch diets. Similarly, pentosidine content of insoluble skin collagen was lower in the rats on the restricted sucrose diet relative to rats on the ad libitum sucrose diet; such differences were not present between restricted and ad libitum fed rats on the starch diet (Figure 2). Interestingly, pentosidine was significantly higher in insoluble skin collagen from rats on the ad libitum sucrose diet as compared with rats on the ad libitum starch diet. There were no differences in glucitolyllysine among the four dietary groups.

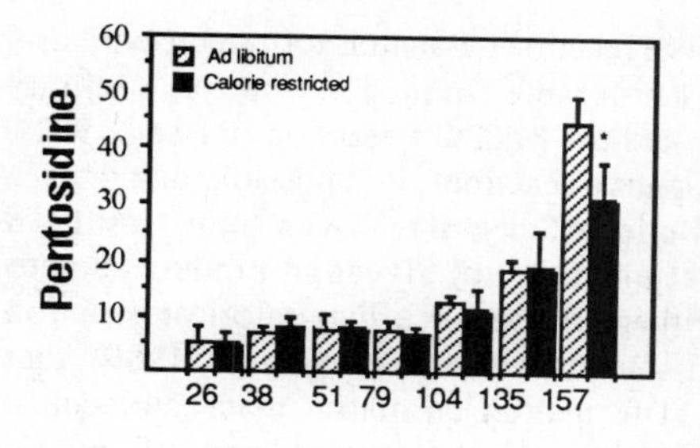

Figure 1. Effects of age and diet on pentosidine content of aorta collagen in F344 X BN rats. Pentosidine content is expressed as mmol/mol of collagen. Legend: Striped bars=ad libitum, solid bars=calorie restricted; t-bars= 1 SD; n=4.

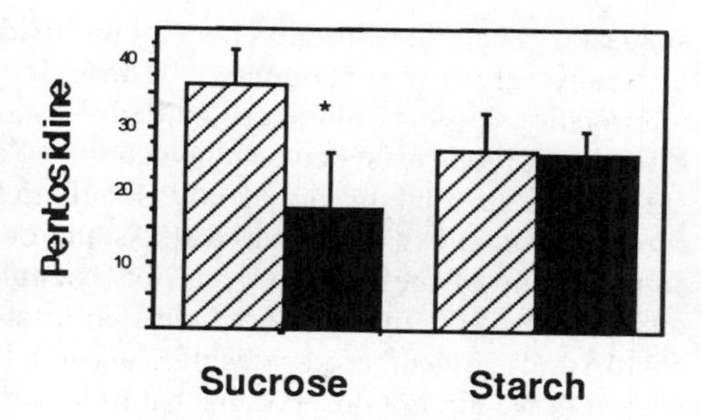

Figure 2. Pentosidine content in insoluble skin collagen from F344 rats fed sucrose or starch diets for 5 months. Legend: same as in Fig. 1; n=4 determinations per group.

We investigated the effects of caloric restriction on sites of adduct formation in the CB3 peptide from 104 week old F344 X BN rats. There were no differences between ad libitum and calorically restricted rats in terms of the relative distribution of glucose adducts among the lysine residues. As part of this study, we extended our investigation of glycation sites to two additional major peptides in the α1(I) collagen chain: CB7 and CB8. As reported previously for CB3 and α2CB3-5, we found marked site specificity of glucose adduct formation. Preferential sites of glycation were characterized by the presence of charged residues, particularly aspartic acid, immediately adjacent to the glycated lysine, or one residue removed. Conversely, nonglycated lysines were never adjacent to aspartic acid residues.

Discussion

Advanced glycation products appear to increase with age in virtually all tissues studied; the accumulation of such compounds is associated with alterations in the properties of collagen, some of which are deleterious to the organism as a whole (Reiser, 1991). In the present review, we describe recent data concerning the effects of aging on nonenzymatic glycation of type I collagen in rats and mice in skin, lung, tendon and aorta. Consistent with other investigators, we observed that glucitolyllysine content of collagen, with a few exceptions, did not change significantly with age, while collagen-associated fluorescence increased significantly with age (Reiser, 1991). We also observed that the accumulation of advanced glycation products may be modulated by dietary manipulation. We report that lifetime caloric restriction results in decreased accumulation of advanced glycation products in skin, tendon and aorta collagen in both rats and mice. In addition, preliminary studies suggest that dietary carbohydrate source may also affect accumulation of advanced glycation products, and that this effect occurs independently of caloric intake.

It is striking that those dietary manipulations that affect accumulation of advanced glycation products also have significant effects on life span. These observations are particularly intriguing in light of suggestions by Kristal and Yu (1992) that aging, and to some extent life span, may result from a synergistic interaction between nonenzymatic glycation and free radical damage. Hence, modulating one aspect of this interaction could affect both glycation and life span. For example, recent studies by Lee and Yu (1990) suggest that caloric restriction may inhibit oxidative reactions essential for the biosynthesis of many glycation products. There is considerable evidence that a complex interrelationship exists between oxidative events and the Maillard reaction (Reiser, 1991). If feedback loops exist between glycation and oxidative reactions, it is possible that dietary manipulations that produce only small metabolic effects may, over time, result in significant changes in oxidative stress and/or accumulation of glycation products due to amplification of the initial effect. For example, there is evidence that caloric restriction results in lower serum glucose levels, at least in some species (Masoro et al., 1989), that could result in decreased rates of glucose adduct formation on matrix macromolecules. Although we did not observe that caloric restriction consistently lowered levels of glucose adducts on collagen, it is possible that small decreases, even though not statistically significant at most time points due to biological variability, may be sufficient to affect the accumulation of advanced glycation products.

Our studies investigating the interrelationship between carbohydrate source, caloric restriction and life span are also intriguing, particularly our preliminary data suggesting that calorically restricted sucrose-fed animals have significant decreases in collagen-associated fluorescence and pentosidine content of insoluble skin collagen after only 5 months on the diet. These results are surprising, given that rats fed sucrose ad libitum had shorter life spans than their starch-fed counterparts. Furthermore, there is evidence in the literature that dietary sucrose provided ad libitum, even in euglycemic animals, is associated with a number of adverse effects (Kanarek et al., 1982). At present, we have insufficient data to speculate further on the interrelationship between carbohydrate source, life span, calorie restriction and nonenzymatic glycation. We are currently pursuing the possibility that the dietary carbohydrate source may affect polyol pathway metabolism such that content and/or availability of key precursors in the biosynthesis of advanced Maillard products are altered.

We have also investigated the effects of aging and diet on site-specificity of nonenzymatic glycation. In a recent study (Reiser et al., 1992), we speculated that primary structure, in particular the location of charged residues, was most likely the major determinant of sites of glucose adduct formation. Our more recent studies, reported herein, on CB7 and CB8, are consistent with our earlier data. Our observation that glycation sites remain highly conserved with aging strongly suggests that even progressive accumulation of advanced glycation products on collagen throughout the life span has less influence on sites of adduct formation than does primary structure. That is, accumulation of glycation products do not appear to accelerate the loss of structural complexity, as these products appear to have little or no influence on sites of adduct formation even in very old animals. Given these findings, it is not surprising that there were no differences in adduct sites between ad libitum fed and calorically restricted animals.

Conclusions

In conclusion, our data provide further evidence for the existence of a complex interrelationship between aging, life span, and the accumulation of advanced glycation products in collagen. In recent studies, we have shown that this relationship may be significantly modulated by caloric restriction and by carbohydrate source. Future studies will include investigation of long-term effects of other carbohydrate sources, such as fructose, in conjunction with caloric restriction, and more detailed investigations of the effects of different dietary carbohydrates on polyol pathway metabolism.

Acknowledgments
This work was supported by grants from the National Institutes of Health (AGO7711, AGO5324, and RR00169), the American Diabetes Association, and by a gift from The Sugar Association, Inc.

References

Buckingham, B. and Reiser, K. M. (1990) Relationship between the content of lysyl oxidase-dependent crosslinks in skin collagen, nonenzymatic glycosylation and long-term complications in type I diabetes mellitus. J. Clin. Invest. 86: 1046-1054.

Kanarek, R.B. and Orthen-Gambill, N.(1982) Differential effects of sucrose, fructose and glucose on carbohydrate-induced obesity in rats. J. Nutr. 112:1546-54.

Kristal, B. S.and Yu, B. P. (1992) An emerging hypothesis: synergistic induction of aging by free radicals and Maillard reactions. J. Gerontol. Biol. Sci.. 47: B107-B114.

Lee, D. W.and Yu, B. P. (1990) Modulation of free radicals and superoxide dismutase by age and dietary restriction. Aging. 2: 357-362.

Masoro, E. J., Katz, M. S., and McMahan, C. A. (1989) Evidence for the glycation hypothesis of aging from the food-restricted rodent model. J. Gerontol. Biol. Sci. 44: B20-B22.

Murtagh,C.M.and McDonald, R.B.(1992) Carbohydrate source and caloric restriction affect lifespan of F344 rats. The Gerontologist 32:31.

Murtagh, C. M., Reiser, K.M., and McDonald, R.B.(1993) Dietary carbohydrate affects lifespan of Fischer 344 rats. J. Gerontology (submitted)

Reiser, K. M. (1993) Influence of age and long-term dietary restriction on enzymatically mediated crosslinks and nonenzymatic glycation of collagen in mice. J. Gerontology (in press).

Reiser, K. M. (1991) Nonenzymatic glycation of collagen in aging and diabetes. Proc. Soc. Exp. Biol. Med. 196: 17-29.

Reiser, K.M., Hennessy, S.M., Last, J.A.(1987) Analysis of age-associated changes i collagen crosslinking in the skin and lung of monkeys and rats. Biochim. Biophys. Acta 926:339-348.

Reiser, K. M., McCormick, R., and Rucker, R. B. (1992) Crosslinking in collagen and elastin. FASEB J. 6: 2439-2449.

Reiser, K., Amigable, M.,and Last, J. (1992) Nonenzymatic glycation of type I collagen: the effects of aging on preferential glycation sites. J. Biol. Chem. 267:24207-24216.

Sell, D. R., Lapolla, A., Odetti, P., Fogarty, J., and Monnier, V.M. (1992) Pentosidine formation in skin correlates with severity of complications in individuals with long-standing IDDM. Diabetes. 41: 1286-92.

Yamauchi, M., Woodley, D. T., Mechanic, G. L. Aging and crosslinking of skin collagen. Biochem. Biophys. Res. Commun. 152: 898-903; 1988.

Glycosylation-induced Modifications of Intact Basement Membrane

Shane S. Anderson, Effie C. Tsilibary, and Aristidis S. Charonis

DEPARTMENT OF LABORATORY MEDICINE AND PATHOLOGY, UNIVERSITY OF MINNESOTA MEDICAL SCHOOL, MINNEAPOLIS, MINNESOTA, USA

Summary
We examined structural changes in bovine kidney tubular basement membrane (TBM) following *in vitro* nonenzymatic glycosylation (NEG). Isolated TBM was incubated for 2 wks at 37° C in the absence of sugar or in the presence of either glucose or ribitol under conditions which minimized degradation and oxidative damage. NEG and crosslink formation in glycated TBM were confirmed by increased specific fluorescence compared to controls. Morphological analysis using high resolution, low voltage scanning electron microscopy (LV-SEM) revealed a complex three-dimensional meshwork of interconnecting strands with intervening openings. Glycated TBM underwent distinct morphological changes including a 58% increase in the amount of image surface area occupied by openings. This was due to an apparent increase in the number of large openings (diameters >12.5nm), whereas the number of small openings (diameters <12.5nm) remained unchanged. These findings corroborate earlier physiological studies which established that the loss of glomerular permselectivity seen in patients with diabetic nephropathy is due to the formation of large pores in the kidney filtration barrier of which the BM is a major component. We conclude that NEG and crosslink formation among BM components lead to modifications of BM ultrastructure which could play a role in loss of barrier function in diabetic microangiopathy and nephropathy.

Introduction

Decreased renal permselectivity is a typical sequela of long standing diabetes mellitus; the cause is unclear. Factors like impaired charge density of the filtration barrier may contribute to the leakage of negatively charged proteins. However, physiological studies have provided evidence that loss of size selectivity is the most crucial element: a slight increase in the proportion of larger pores in the filtration barrier is predicted to account for the altered permselectivity (Myers et al., 1990). The nature of these large pores and the mechanism by which they form are not well established. Structural abnormalities of the sieve-like basement membrane may be important (Østerby et al., 1990).

Nonenzymatic glycosylation (NEG) of long lived proteins, such as the basement membrane components tIV and LM, is one of several mechanisms which may be involved in the generation of diabetic complications. It is a well-described process that becomes highly accelerated in diabetes (Monnier et al., 1989). Isolated tIV and LM, undergo structural and functional changes following glucose incorporation and/or ensuing protein crosslinking (Charonis et al., 1992). The effect of NEG on the structural integrity of whole basement membrane (BM) has not been addressed so far.

Using tubular basement membrane (TBM) as a model system, we examined the structural changes that intact isolated BM undergo following NEG and extensive crosslinking. We have used a novel morphological approach employing ultra high resolution low voltage scanning electron microscopy (LV-SEM) and computer-aided image analysis to identify these changes.

Materials and Methods

Isolation and Glycation of TBM

TBM were isolated using the technique by Carlson et al.(1978) and incubated for 2 weeks at 37° C in solutions containing PBS, 1 mM EDTA, 1 mM PMSF, 1 mM NEM, .02% sodium azide, and one of the following: 1 M glucose, 1 M ribitol, or no sugar. Solutions were degassed under vacuum for 2 h to remove dissolved O_2 and then bubbled with argon to saturate with inert gas. These conditions were designed to limit degradation and oxidative damage. Due to the limited buffering capacity of PBS, the pH of all samples had declined equivalently from 7.4 to 6.6 by the end of incubation.

Fluorometric Analysis

Samples were digested by sequential treatment with collagenase and pepsin. Fluorescence was measured at excitation/emission wavelengths of 335/385 nm and 360/460 nm which detect glycation-induced adducts, including pentosidine (Sell et al., 1989). Values were corrected for protein concentration and expressed as a percent of the negative control.

LV-SEM

TBM were lightly fixed with 4% EM grade paraformaldehyde in PBS for 2 h, dehydrated, dried at critical point, sputter-coated with ~10 Å platinum, and then examined using a Hitachi S-900 high resolution LV-SEM operating at an accelerating voltage of 2 kV.

Image Analysis

Two methods of analysis were used: (1) Images were transferred to an image processor and converted to gray scale histograms in which black and white were assigned values of 0 and 256, respectively. The percentages of image surface area occupied by regions darker than specific levels of grayness were determined. (2) Images were enlarged to a magnification of 104,000X. "Openings" in the TBM structure, defined as dark regions containing no apparent detail or structure underneath, were manually outlined and darkened to facilitate selective detection and measurement in digitized images. The number of openings and amount of image surface areas occupied by openings of specific sizes was determined. Results from each technique, expressed as the mean±SD, were compared using a two-tailed student-*t*-test.

Results

It is well established that incubation with reducing sugars leads to the generation of glycated and crosslinked proteins *in vitro* (Dyer et al., 1991). In order to document such changes in our experiments, TBM were solubilized and examined by fluorometric analysis. Glycated TBM demonstrated a 72% and 67% increase in fluorescence at 335/385 nm and 360/460 nm compared to control, respectively.

LV-SEM examination of TBM revealed a complex meshwork of interconnecting strands with intervening openings of different sizes (Figure 1A-B). The structure was comparable to that recently described by Hironaka et al. (1993). The complexity of the three-dimensional network made it difficult to identify a distinctive structural pattern. However, in some instances semi-regular structures could be visualized in which strands appeared to radiate from a central hub into various planes of the lattice, occasionally having the appearance of a polygonal wheel with spokes.

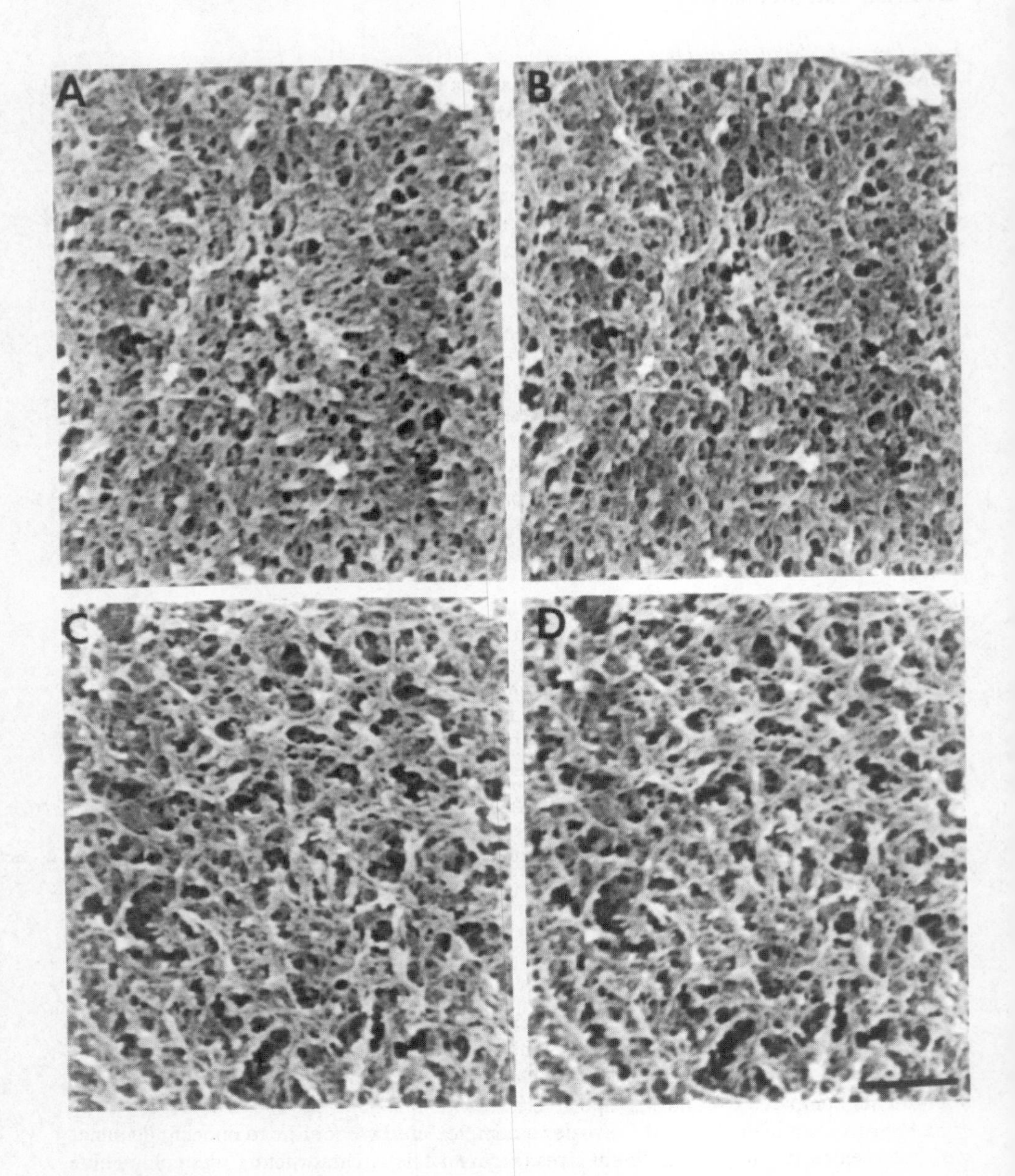

Figure 1. Stereo-Paired LV-SEM images of control and glycated TBM. Representative images of ribitol-treated (A and B) and glucose-treated (C and D) TBM revealed strands of material of varying thickness intertwined to form a quasi-regular network with openings of various sizes. Openings appear to occupy a greater proportion of image surface area and some of these openings appear larger in the glycated matrix compared to the control. Bar equals 250 nm.

Distinctive changes in TBM ultrastructure were apparent following glycation (Figure 1C-D). Visual comparison of LV-SEM images suggested that more surface area was occupied by openings and that some openings were enlarged in glucose-treated TBM compared to the control. Also, there appeared to be more frequent clustering of TBM material in glucose-treated TBM, particularly in regions adjacent to enlarged openings. The differences were more apparent when the TBM were viewed in three dimensions using stereo-paired images (Figures 1A-D).

In order to statistically evaluate the area occupied by openings in the SEM images, computer-aided analysis was performed. First, the gray-scale distribution for images from glucose-treated TBM was skewed towards darker shades of gray when compared control TBM (Figure 2). These differences were statistically significant at gray scale values between 0 and 96, which represented the darkest spots in the SEM images and corresponded to areas visualized as openings. These data indicate that more surface area was occupied by openings in images from glucose-treated TBM compared to control.

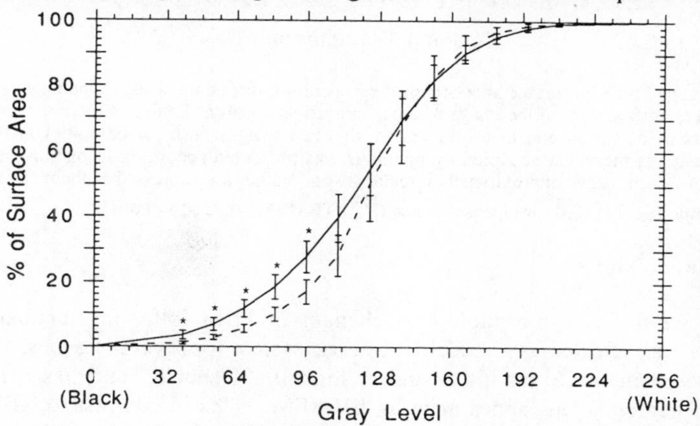

Figure 2. Comparison of gray-scale distribution from images of control and glycated TBM. The percentage of image surface area composed of regions darker than a series of specific shades of gray was determined using computer-aided image analysis of 15 images each of either ribitol-treated *(broken line)* or glucose-treated *(solid line)* TBM. Images of glycated TBM had significantly more surface area composed of dark shades of gray and black (gray scale values between 0 and 96) which correspond to openings in the matrix. Results are expressed as mean±SD (*p<0.001).

A second method of analysis determined that 58% more surface area was occupied by openings in images of glucose-treated TBM compared to control (13.34±0.64% vs. 8.44±1.18%, p<.001). These data were comparable to those generated with the previous image analysis technique in which dark shades of gray (gray level of ~82) were significantly increased in images of glycated TBM.

Using the second analysis technique, the amount of image surface area occupied by openings of specific sizes was also determined and presented according to opening diameter which was calculated assuming each had an approximately circular shape (Figure 3). There were no significant differences in the surface area occupied by "small" openings, with diameters <12.5nm, in images of control and glycated TBM. However, "large" openings with diameters >12.5nm occupied significantly more surface area in the images of glycated TBM. This was due to a statistically significant increase in the number of these large diameter openings (data not shown).

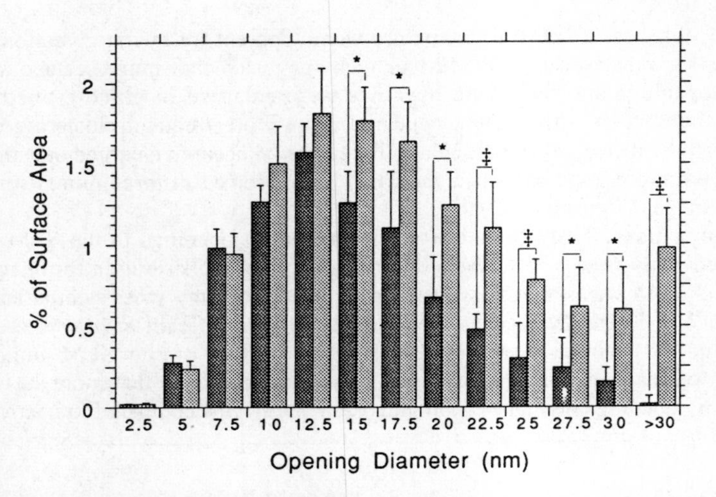

Figure 3. Percentage of image surface area occupied by openings of specific sizes. "Openings" in the TBM structure, defined as dark regions of the image which contained no apparent detail or structure, were darkened to facilitate selective detection and measurement in digitized images used for computer analysis. The percentage of image surface area occupied by openings was presented according to opening diameter as calculated assuming each had an approximately circular shape. Values are expressed as the mean±SD from 6 images each of ribitol-treated (▩) or glucose-treated (▤) TBM (*p<0.02, ‡p<0.001).

Discussion/Conclusions

In this report we describe morphological changes in TBM following incubation with glucose. TBM was selected as a model BM in our system for several reasons. TBM can be easily isolated in large quantities and at high (98%) purity, and has a relatively consistent and homogeneous appearance by LV-SEM. This in contrast to GBM which unavoidably contains mesangial matrix and has a more heterogeneous appearance. In addition, the structure of TBM is representative of other BM (Hironaka et al., 1993). Also, TBM along with GBM has been described to become diffusely thickened in diabetic patients (Steffes et al., 1985). Finally, in physiological studies, TBM and GBM have similar dextran sieving properties (Moran et al., 1985).

We have used a system of "accelerated glycation." *In situ*, most BM components turn over slowly and apparently take several years to become substantially modified in the presence of hyperglycemia (6-15 mM glucose). Our system was designed to produce similar changes in a much shorter time interval (14 days): therefore, it was imperative to use high glucose concentrations. The observed increase in crosslink-associated fluorescence for glycated TBM under our experimental conditions (~70%) was comparable to the relative increase in fluorescence seen for samples of skin collagen from diabetic patients compared to age matched non-diabetic counterparts (Monnier et al., 1986).

Glycation resulted in distinct morphological changes in the TBM. Compared to control, more surface area of images from glycated TBM was occupied by dark regions which correspond to what we defined as openings in the sieve-like BM structure. In order to objectively compare the dimensions of openings in the complex meshwork of control and glycated TBM, we used two different methods of computer aided image analysis.

These demonstrated that the surface area occupied by "openings" was increased by 58% in glucose-treated TBM compared to control. In addition, different populations of openings were revealed which could be clustered into two categories: those with a diameter <12.5nm and those >12.5nm. The "small" openings did not change in number or total surface area in images of control versus glycated TBM. On the contrary, "large" openings showed a small increase in number and a pronounced increase in total surface area in images of glycated TBM. The openings examined in this report represent only a portion of the very tortuous and complicated "pores" through which a molecule must travel in order to traverse the entire thickness of the sieve-like BM. Since the size of a "pore" is apparently limited by the diameter of the smallest opening which occurs along its path, a substantial increase in the amount of large openings could correspond to only a modest increase in the overall porosity of the entire BM.

To our knowledge, we are the first to provide morphological evidence for the existence of more and larger openings in BM chemically modified by glucose (Anderson et al., 1991). If similar changes were to occur in BM structures *in vivo*, these could play a role in the development of changes in glomerular permselectivity such as those seen in diabetic nephropathy (Myers et al., 1990). Several physiological studies in humans have provided evidence that loss of size selectivity through the formation of a small number of larger pores in the filtering sieve of the kidney could account for the loss of permselectivity in diabetes (Myers et al., 1990). Our morphological observations and measurements strongly support the above-mentioned physiological findings, in that we confirmed with high resolution LV-SEM the existence of enlarged openings in glycated BM. Furthermore, our observations indicate that glucose-induced crosslink adducts could be causally related with the appearance of enlarged openings.

Acknowledgments
This work was supported by NIH grants DK-43564 (AC) and DK-43574 (EC) and by a grant from the National Kidney Foundation of the Upper Midwest (SA).

References

Anderson SS, Charonis AS, Tsilibary EC. Diabetic modification of kidney basement membrane ultrastructure and altered interaction with mesangial cells. J. Cell Biol. 1991;115(3,Pt2):438A.

Carlson EC, Brendel K, Hjelle JT, Meezan E. Ultrastructural and biochemical analyses of isolated basement membranes from kidney glomeruli and tubules and brain and retinal microvessels. J. Ultrastruct. Res. 1978;62:26-53.

Charonis AS, Tsilibary EC. Structural and functional changes of laminin and type IV collagen after nonenzymatic glucosylation. Diabetes 1992;41:49-51.

Dyer DG, Blackledge JA, Thorpe SR, Baynes JW. Formation of pentosidine during nonenzymatic browning of proteins by glucose: identification of glucose and other carbohydrates as possible precursors of pentosidine *in vivo*. J. Biol. Chem. 1991;266(18):11654-11660.

Hironaka K, Makino H, Yamasaki Y, Ota Z. Renal basement membranes by ultrahigh resolution scanning electron microscopy. Kidney Int. 1993;43:334-354.

Monnier VM. Toward a Maillard reaction theory of aging. Prog. Clin. Biol. Res. 1989;304:1-22.

Monnier VM, Vishwanath V, Frank KE, Elmets CA, Dauchot P, Kohn RR. Relation between complications of type I diabetes mellitus and collagen-linked fluorescence. N. Engl. J. Med. 1986;314:403-408.

Moran MS, Myers BD. Pathophysiology of protracted acute renal failure in man. J. Clin. Invest. 1985;76:1440-1448.

Myers BD. Pathophysiology of proteinuria in diabetic glomerular disease. J. Hypertens. 1990;8(Suppl. 1):S41-S46.

Østerby R, Parving H-H, Hommel E, Jørgensen HE, Løkkegaard H. Glomerular structure and function in diabetic nephropathy: early to advanced changes. Diabetes 1990;39:1057-1063.

Sell DR, Monnier VM. Isolation, purification, and partial characterization of novel fluorophores from aging human insoluble collagen-rich tissue. Connect. Tissue Res. 1989;19:77-92.

Steffes MW, Sutherland DER, Goetz FC, Rich SS, Mauer SM. Studies of kidney and muscle biopsy specimens from identical twins discordant for type I diabetes mellitus. N. Engl. J. Med. 1985;312:1282-1287.

The Advanced Maillard Reaction in Aging and Age-related Diseases Probed with Pentosidine

V. M. Monnier and D. R. Sell

INSTITUTE OF PATHOLOGY, CASE WESTERN RESERVE UNIVERSITY, CLEVELAND, OHIO 44106, USA

Summary
Pentosidine is an advanced product of the Maillard reaction which is thought to form *in vivo* through glycoxidation of Amadori products of glucose. In vitro, pentoses are its most active precursors and ascorbate is thought to be a major precursor in the lens, and perhaps other cellular systems. Quantitation of pentosidine in extra cellular matrix revealed tissue specific patterns of formation in aging which are likely to reflect protein turnover. Pentosidine is increased in diabetic tissues and skin levels were found to correlate with severity of diabetic complications. Highest levels are found in uremic plasma, and these are largely correctable by renal transplantation. The new association between advanced Maillard reaction and uremia together with its accelerated atherosclerosis, cataract formation and impaired immune function suggests that uremia is a syndrome of accelerated aging. The exact role of oxygen and the nature of precursor sugars in pentosidine biosynthesis needs to be established.

Introduction

Aging is characterized by progressive, irreversible and usually deleterious changes in tissue function. Some of the most pronounced changes occur in tissues with long-lived proteins such as in the extra cellular matrix, the lens and the brain. The extracellular matrix becomes progressively less elastic, less soluble, less digestible by proteolytic enzymes like collagenase, and basement membranes in capillaries, lens and kidney thicken with age. There is also evidence for degradation of the extra cellular matrix in cartilage as well as indication of tissue remodelling in kidney, aorta and arterioles in which progressive sclerosis is a common feature of aging. In the brain, neurofibrillary tangles that are composed of highly insoluble proteins accumulate and are thought to contribute to neuronal loss. Brain, heart muscle, aorta, adrenal gland, retina and other tissues accumulate intracellular pigments of complex composition.

Many of these changes are dramatically worsened by diabetes (Monnier, 1990; Kohn, 1988). Elasticity of the joints, arteries, lung and heart are further decreased, and capillary and renal basement membrane thickness are increased. Bone loss, loss of lens accommodation, cataracts and cardiovascular complications develop at an earlier age. By and large, most of the changes observed in aging occur twice faster in diabetes. Interestingly, uremia is also characterized by an increased rate of cardiovascular complications, impaired T-cell function, impaired ability to fight infections, decreased nerve conduction velocity and a host of changes reflecting accelerated aging (Eknoyan and Knochel, 1984). Thus, both diabetes and uremia are syndromes of accelerated aging.

The changes described above suggest that important aspects of aging may involve 1) postsynthetic nonenzymatic processes which inflict damage to long-lived molecules in form of cross links, blockage of proteolytic digestion sites and possibly protein fragmentation; 2) impaired cellular processing of molecules; 3) a key role for metabolites that can engage in carbonyl amine reactions; and 4) a decreased defense against the formation and the biological effects of post synthetically modified proteins.

Obviously, the mechanisms leading to molecular and cellular lesions in aging and age-related diseases are expected to be highly heterogenous. However, increasing evidence suggests that the Maillard reaction, in combination with oxidation reactions, may explain a number of age-related changes and their acceleration.

The Maillard reaction is initiated by the carbonyl-amine condensation reaction of a sugar with a free amine, whereby the Schiff base formed is stabilized through an Amadori rearrangement. It has become clear that the Amadori product of glucose is in a steady state *in vivo*, and therefore, not useful for assessing chronic, cumulative changes in tissues. As of this day, four products of the advanced Maillard reaction have been used to probe its role in biological tissues: pyrraline, pentosidine, carboxymethyl-lysine (CML) and antibodies to "AGE-protein" (Figure 1). Pyrraline is 5-hydroxymethyl-carbaldehyde norleucine which is formed from glucose. It was detected immunologically in elevated amounts in diabetic human and rat plasma, as well as in the extra cellular matrix of human kidneys, arteries, skin, trachea and lung of young, old and diabetic individuals (Miyata and Monnier, 1992). These results are currently controversial since one group

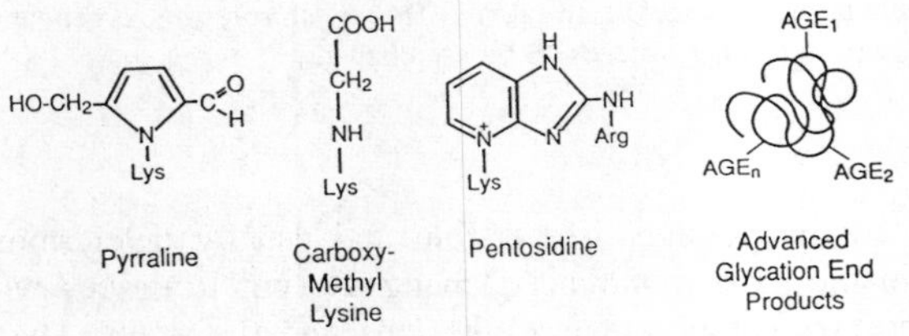

Pyrraline

Carboxy-Methyl Lysine

Pentosidine

Advanced Glycation End Products

Figure 1. Biochemical probes for the advanced Maillard reaction *in vivo*.

has been unable to find pyrraline immunoreactivity in proteins exposed to glucose in vitro or *in vivo* (Smith et al., 1993). Carboxymethyl-lysine (CML), together with pentosidine, have been baptized "glycoxidation" products (Baynes, 1991) because of the requirement of O_2 for their formation from glycated lysine (Dyer et al., 1991). The focus of this chapter is to review recent studies on the significance of pentosidine as a marker of tissue damage by the Maillard reaction in diabetes, aging and uremia. Studies with CML and AGE-proteins are reviewed elsewhere in this volume.

Material and Methods

Pentosidine is quantitated by high performance liquid chromatography in hydrolysates of proteins prepared in 6N HCl, at 110°C for 24 hours. The hydrolysate containing 1-5 mg protein/mL is dried and reconstituted with solvent as previously described (Sell and Monnier, 1990) and injected into a reverse phase C-18 column. Elution is carried out with acetonitrile/H_2O using a gradient system. The eluate is analyzed with a fluorescence detector set at 335/285 nm. For tissues that contain more than 20 pmol/mg protein (human collagen-rich tissues), a single column method is generally accurate, provided strict anaerobic conditions have been used during hydrolysis. Quantitation of low levels of pentosidine (plasma proteins, lens, blood and cultured cells) can be inaccurate due to fluorescent artifacts which are generated during acid hydrolysis (Odetti et al., 1992). In such cases, we have used a dual chromatographic system consisting of chromatography on a C-18 column, automated peak collection, drying by evaporation, reconstitution and injection into the same HPLC instrument equipped with a cation exchange column (Odetti et al., 1992). With such instrumentation, the working range for pentosidine measurements varies from 0.5-10 pmole. This corresponds to protein equivalents of 0.1 to 1.0 mg being directly injected into the system.

Results

Pentosidine formation during aging

The pattern and the contents of pentosidine forming as a function of age varies from tissue to tissue. Five typical patterns are shown in Figure 2. The reasons for the differences in pentosidine accumulation rates in these tissues are unclear. One possibility is that they reflect tissue turnover rates, the latter being presumably slowest in cartilage and dura mater, in a steady state in adult renal glomerular basement membranes, and progressively decreased in aging skin. Another possibility would be that the tissue turnover rate is overall similar in all tissues but that tissue specific factors catalyze pentosidine formation. Such factors could include pentosidine precursor concentration, pH, catalysis by O_2, or protein composition. The very low O_2 tension in the lens is likely to contribute to its very low levels. A third possibility applicable to the extra cellular matrix is the presence of two pools of matrix proteins, one with a very slow, and one with a higher turnover rate. Indeed, extraction experiments by Schnider and Kohn (1982) revealed significant extractability of collagen by salt and acid in skin, but less so in dura mater and cartilage. When pentosidine was measured in highly insoluble fractions of extra cellular matrix from various tissues within a single individual, levels (in pmol/mg collagen) decreased in the following order: tracheal cartilage (152), dura mater (147), kidney medulla (75), aorta (72), kidney cortex (65), skin (41).

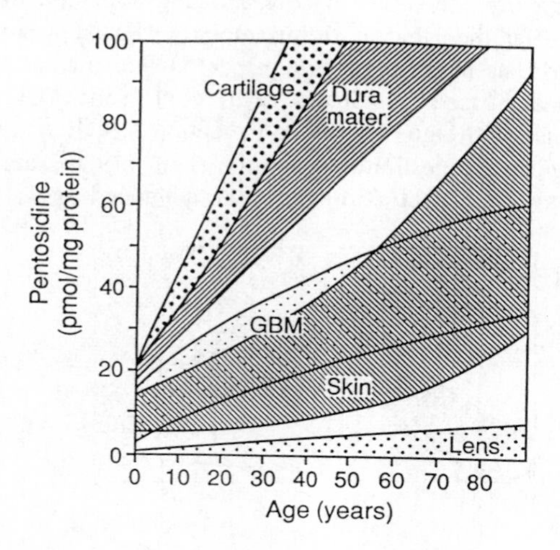

Figure 2. Confidence intervals for pentosidine formation in insoluble matrix from dura mater, glomerular basement membrane, skin and lens (from Sell and Monnier, 1989, 1990, et al. 1993; Nagaraj et al., 1991). The cartilage data are adapted from Uchiyama et al. (1991).

Pentosidine was determined in tunica albuginea and corpus cavernosum of the penis in collaboration with Jiaan et al. (1993). The rationale for this study is that advanced glycosylation products have been found to quench nitric oxide (Bucala et al., 1991), i.e. the mediator responsible for vasodilation and penile erection (Burnett et al., 1992). An exponential pattern of pentosidine formation was found in both tissues which is reminiscent of a similar pattern in skin. Thus, it may be that accelerated formation of advanced Maillard products in the aging penis may contribute to the age-related erectile dysfunction by scavenging nitric oxide.

Pentosidine formation was determined in human intervertebral disc by Uchiyama and colleagues (1991). Its accumulation rate was higher than that of human dura mater (Figure 2). A similar study in intervertebral disc from sand rat showed an 6-fold increase over a period of 18 months (Pokharna et al., 1993). In both studies, levels of pyridinoline were unchanged, reemphasizing thereby that lysyl oxidase cross links are not involved in the age-related insolubilization of ECM proteins. Pokharna et al. (1993) have proposed that the increased rate of formation of advanced Maillard products may stimulate chondrocytes to release the cytokine IL-1 which can trigger release of collagenase from fibroblasts, thereby contributing to degradation of the matrix of intervertebral disc in a process similar to that described by Vlassara et al. (1988). Such process may be important in the pathogenesis of the degenerative disc disease and osteoarthritis.

One of the key questions in gerontology is whether a fundamental aging process exists such that its rate of formation is related to maximum life span among species. Previous experiments by Hamlin and colleagues (1980) revealed that loss of collagen digestibility in aging is inversely related to maximum life span of human, monkey and dog, suggesting thereby that the process underlying collagen aging may represent a biological clock. Experiments were carried out in collaboration with K. Reiser, E. Masoro and others to evaluate the relationship between pentosidine formation and aging in skin of human, rhesus monkey, dog, cow and rat. The results (Figure 3) indicate that pentosidine formation rate was overall similar in human and non-human primates (human and monkey), but dramatically increased in cow and dog skin, the life expectancy of which is considerably shorter than that of rhesus monkey, i.e. approximately 15 years, compared to 40 years in rhesus monkey. Although pentosidine increases with age in tail tendon from rats and is decreased by dietary restriction (Sell et al., 1991), rat skin pentosidine did not increase with age. This is an exception which was also observed in the experiment on collagenase digestion by Hamlin et al. (1980) and which suggests presence of significant turnover of rat skin even at an advanced age.

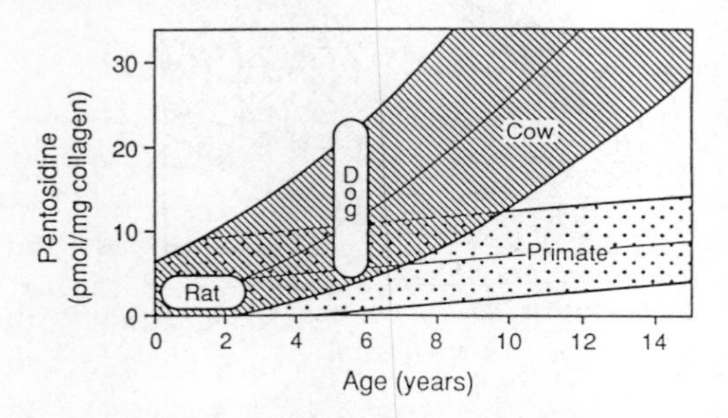

Figure 3. Pentosidine formation in skin from aging human, rhesus monkey, cow, dog and rat. The data in dogs correspond to mean ± S.D. at age 5.5 yrs. The data in rats indicate no change over lifetime.

Pentosidine formation in diabetes
Most studies so far on the relationship between diabetes and pentosidine formation have been carried out with skin from humans with insulin-dependent, so-called type I diabetes. A typical pattern is shown in Figure 4. Levels were elevated in most individuals with long-standing diabetes (Sell et al., 1992). When these individuals were categorized according to severity of complications, pentosidine levels correlated with age ($p < 0.001$), severity of retinopathy ($p < 0.014$), joint stiffness ($p < 0.04$) and the cumulative severity of diabetic complications ($p < 0.005$). Similar experiments by McCance et al. (1993) revealed an association with retinopathy and nephropathy but not with limited joint mobility. Beisswenger and colleagues (1993) found an association between skin pentosidine levels and early nephropathy and retinopathy. These observations do not allow us to conclude that pentosidine or the Maillard reaction itself is implicated in the pathogenesis of these complications. However, they suggest that insult by cumulative

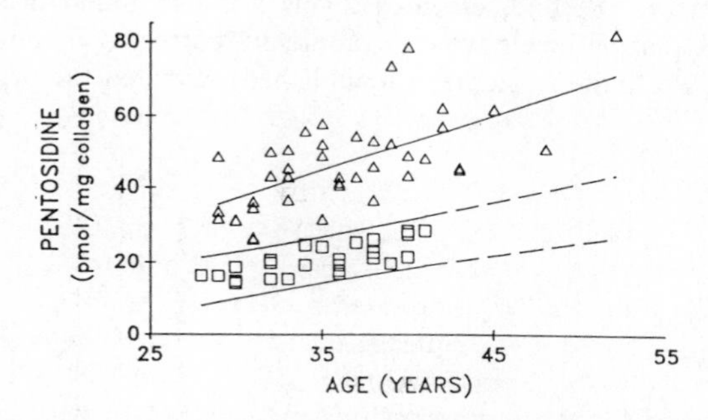

Figure 4. Age-related formation of pentosidine in normal (□) and diabetic (△) subjects (from Sell et al., 1992).

glycemia is higher in subjects who develop more severe complications. Another possible explanation is that these subjects have a propensity to form advanced Maillard or glycoxidation products faster than those without complications because they are exposed to a higher oxidative stress. However, there is no evidence *in vivo* that increased oxidation is implicated in increased glycoxidation (Dyer, 1993).

One unexpected finding was that of normal or near-normal pentosidine levels in skin from non-insulin dependent (Type II) diabetic individuals (Sell et al., 1993). In contrast, elevated levels were found in purified glomerular basement membranes from subjects with type II diabetes. It is thus possible that enhanced turnover rate of skin collagen associated with the high insulin levels occurring in NIDDM may have "normalized" pentosidine levels in skin but not in renal GBM of these individuals.

The question of whether diabetic individuals have been exposed to increased oxidative stress has been addressed by Dyer et al., (1993). Utilizing the carboxymethyllysine:furosine ratio as an index of oxidation they found that the predicted CML levels after several years of diabetes were lower than expected based on glycation alone, suggesting thereby that oxidative stress is decreased rather than increased in diabetes. This observation, however, may not be applicable to cellular systems stressed by hyperglycemia.

Pentosidine formation in uremia

Random determinations of pentosidine in skin obtained at autopsy revealed dramatic elevations in subjects who died of end stage renal disease (Sell and Monnier, 1990). Subsequent studies by Makita et al. (1991) showed that "AGE" immunoreactivity was highly elevated in plasma of uremic subjects and was associated with a low molecular protein fraction of plasma. Thus, one emerging question is whether the advanced Maillard reaction is accelerated in uremia because circulating Maillard reaction precursors are increased, or whether because modified proteins are not excreted. Three observations, we believe, support the notion that Maillard reaction precursors are increased. First, pentosidine is increased not only in plasma, but also in red blood cells from uremic individuals (Figure 5). In contrast, pentosidine was not elevated in diabetic red blood cells although it was elevated in plasma proteins suggesting that a protective mechanism may have been lost in uremia, or that it may have been overridden due to large amounts of pentosidine precursors. Second, elevated levels are also found in skin from uremic individuals and third, plasma levels were incompletely corrected by renal transplantation (Hricik et al., 1993). Preliminary studies (unpublished) revealed the presence of several

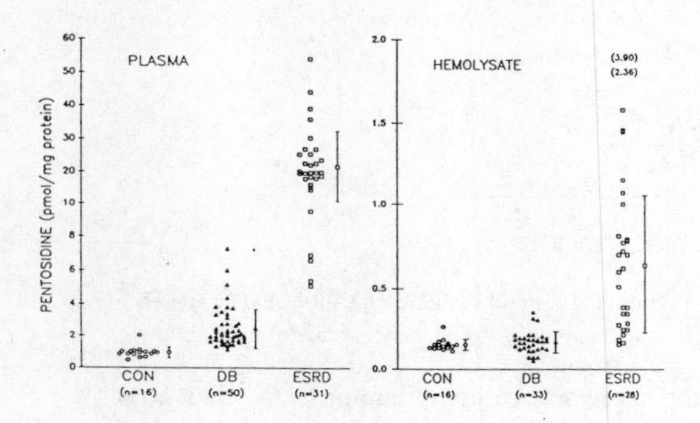

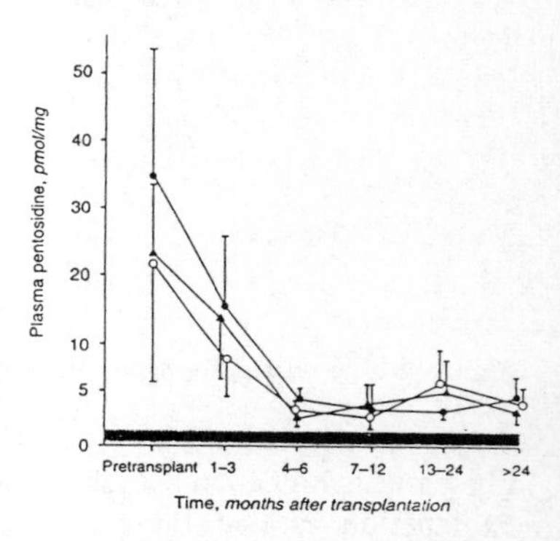

Figure 5. Left: Levels of pentosidine in plasma proteins and erythrocytes from uremic subjects (Odetti et al., 1992), Right: Effect of renal (●) and renal-pancreas (o) transplantation on pentosidine levels (Hricik et al., 1993).

short chain sugars in elevated levels in gas chromatogram of uremic plasma. Such sugars could result in pentosidine formation through multiple pathways (Dyer et al., 1991).

The discovery that the Maillard reaction is accelerated in uremic subjects offers new possibilities for investigation of the biological role of the Maillard reaction, since many features of uremia are encountered in the aging process. In that sense, uremia may represent a new syndrome of accelerated aging.

Conclusions

By all accounts, the absolute levels of pentosidine are very small compared to glycation itself. However, there is little doubt that it represents the tip of the iceberg in terms of overall modification of proteins by the Maillard reaction. Work from Baynes and coworkers (Dyer et al., 1993) so far indicates a high correlation between CML and pentosidine levels, and our own data showed significant correlation between pentosidine levels and skin collagen-linked fluorescence. Short-term control of glycemia revealed that skin pentosidine and CML levels are not reversible (Lyons et al., 1991), reinforcing the notion that a significant quantity of advanced Maillard products is associated with a pool of extra cellular matrix with a very slow turnover. Oxygen tension might play an important role in pentosidine formation since levels in lens which has an anaerobic metabolism are very low and increased in cataracts (Nagaraj et al., 1994). However, it remains to be demonstrated whether glycoxidation, i.e. the formation of pentosidine and CML in extracellular matrix of diabetic and uremic individuals is influenced by other factors than the concentration of pentosidine precursor sugars.

In summary, considerable new information on the relationship between the advanced Maillard reaction and aging and age-related diseases has emerged in recent years. Challenges ahead include the elucidation of the structure of AGE epitopes, AGE ligand(s) to AGE-receptor and the mechanisms of collagen cross linking in aging and age-related diseases. This includes the precise clarification of biosynthetic precursors of pentosidine and the role of "oxidative stress" in their formation.

Acknowledgments
This work was supported by grants from the National Institute on Aging to VMM (AG05601) and DRS (AG11080) and from the National Eye Institute (EY07099). Other support included grants from the American Diabetes Association and the Juvenile Diabetes Foundation International.

References

Baynes, J.W. Role of oxidative stress in development of complications of diabetes. (1991) Diabetes 40: 405-412.

Beisswenger, P.J., Moore, L.L., Brinck-Johnsen, T., and Curphey T.J. (1993) Increased collagen-linked pentosidine levels and advanced glycosylation end products in early diabetic nephropathy. J. Clin. Invest. 92: 212-217.

Burnett, A.L., Lowenstein, C.J., Bredt, D.S., Chang, T.S.K., and Snyder, S.H. (1992) Nitric oxide: a physiological mediator of penile erection. Science 257: 401-403.

Dyer, D.G., Blackledge, J.A., Thorpe, S.R., and Baynes, J.W. (1991) Formation of pentosidine during non enzymatic browning by glucose. J. Biol. Chem. 11654-11660.

Dyer, D.G., Dunn, J.A., Thorpe, S.R., Baillie, K.E., Lyons, T.J., McCance, D.R., and Baynes, J.W. (1993) Accumulation of Maillard reaction products in skin collagen in diabetes and aging. J. Clin. Invest. 91: 2463-2469.

Eknoyan, G. and Knochel, J.P. (Eta.) (1984) The systematic consequences of renal failure. Grune and Stratton, Inc., Orlando.

Harnlin, C.R., Luschin J.H., and Kohn R.R. (1980) Aging of collagen: Comparative rates in four mammmalian species. Exp. Gerontol. 15: 393-398.

Hricik, D.E., Schulak, J.A., Sell, D.R., Fogarty, J.F., and Monnier, V.M. (1993) Effects of kidney and kidney-pancreas transplantation on plasma pentosidine. Kid. Internatl. 43: 398-403.

Jiaan, D.B., Cruz, W., Pomerantz, J., Fogarty, J., Nchemia, H., Monnier, V.M., and Seftel, A. (1993) Advanced glycation end products in penile corporal and tunical tissue correlation with age. (Abstract). American Urological Association, Inc., October.

Kohn, R.R., and Schnider, S.L.. (1981) Glucosylation of human collagen. Diabetes 31 (Suppl.): 47-51.

Lyons T.J., Bailie, K.E., Dyer, D.G., Duon, J.A., and Baynes J.W. (1991) Decrease in skin collagen glycation with improved glycemia control in patients with insulin-dependent diabetes mellitus. J. Clin. Invest. 87: 1910-1915.

Makita, Z., Radoff, S., Rayfield, E.J., Yang, Z., Skolnik, E., Delany, V., Friedman, E.A., Cerami, A., and Vlassara, H. (1991) Advanced glycosylation end products in patients with diabetic nephropathy. N. Engl. J. Med. 325: 836-842.

McCance, D., Dyer D.G., Dunn, J.A., Bailie, K.E., Thorpe, S.R., Baynes, J.W., and Lyons, T.J. (1993) Maillard reaction products and their relation to complications in insulin-dependent diabetes mellitus. l. Clin. Invest. 91: 2470-2478.

Miyata, S., and Monnier, V.M. (1990) Immunocytochemical detection of advanced glycosylation end products in diabetic tissues using monoclonal antibody to pyrraline. J. Clin. Invest. 89: 1102-1112.

Monnier, V.M. (1990) Nonenzymatic glycosylation, the Maillard reaction and the aging process. J. Geront. 45: B105111.

Nagaraj, R.H., Kern, T.S., Engerman, R.L., and Monnier, V.M. (1994) Advanced Maillard reaction in sugar-induced cataract. Studies with molecular markers. This volume.

Nagaraj, R.H., Sell, D.R., Prabhakaram, M., Ortwerth, B.J., and Monnier, V.M. (1991). High correlation between pentosidine protein cross links and pigmentation implicates ascorbate oxidation in human lens senescence and cataractogenesis. Proc. Natl. Acad. Sci. (USA) 88: 10257-10261.

Odetti, P., Fogarty, P., Sell, D.R., and Monnier, V.M. (1992) Chromatographic quantitation of plasma and erythrocyte pentosidine in diabetic and uraemic subjects. Diabetes 41: 153-159.

Pokharna, H.K., Monnier, V.M., Boja, B., and Moskowitz, R. (1993) Role of collagen cross links in sand rat intervertebral disc degeneration. Transactions of the 39th Annual meeting of Orthopedic Research Society 18: 647.

Pokharna H.K., Monnier, V.M., Boja, B., and Moskowitz, R. (1992) Advanced glycosylation end products as a stimulation for cytokine-induced cartilage degeneration. Arthr. Rheumat. 35: A31.

Richard, S., Sell, D.R., Katz, M.S., Masoro, E.J., and Monnier, V.M. (1990) Food restriction in F344 rats delays tail tendon cross linking and formation of advanced glycosylation end products. Gerontol. 321A.

Schnider, S.L., and Kohn, R.R. (1982) Effects of age and diabetes mellitus on the solubility of collagen from human skin, tracheal cartilage and dura mater. Exp. Gerontol. 17: 185-194.

Sell, D.R., Carlson, E.C., and Monnier, V.M. (1993) Differential effect of Type 2 diabetes on pentosidine formation in skin and glomerular basement membrane. Diabetologia 36: 936-941.

Sell, D.R., Lapolla, A., Odetti, P., Fogarty, J., and Monnier, V.M. (1992) Pentosidine formation in skin correlates with severity of complications in individuals with long-standing IDDM. Diabetes 41: 1286-92.

Sell, D.R., and Monnier, V.M. (1989). Structure elucidation of a senescence cross link from human extracellular matrix. J. Biol. Chem. 264: 21547-21602.

Sell, D.R., and Monnier, V.M. (1990) End-stage renal disease and diabetes catalyze the formation of pentose-derived cross link from aging human collagen. J. Clin. Invest. 85: 380-384.

Smith, P.R., Somani, H.H., Thornally, P.J., Benn, J., and Sonksen P.H. (1993) Evidence against the formation of 2-amino-6-(2-formyl-5-hydroxymethyl-pyrrol-1-yl)-hexanoic acid ("Pyrraline") as an early stage product and advanced glycation end product in non-enzymic protein glycation. Clin Sci. 84: 87-93.

Uchiyama, A., Ohishi, T., Takahashi, M., Kushida, K., Inoue, T., and Fujie, M. J. (1991) Fluorophores from aging human cartilage. Biochem. 110: 714-718.

Vlassara, H., Brownlee, M., Manogue, K.R., Dinarello, C.A., and Pasagian A. (1988) Cachectin/TNH and IL-1 induced by glucose-modified proteins: Role in normal tissue remodeling. Science 240: 1546-1548.

Immunological Approach to Advanced Glycation End Products of the Maillard Reaction

Seikoh Horiuchi and Norie Araki

DEPARTMENT OF BIOCHEMISTRY, KUMAMOTO UNIVERSITY SCHOOL OF MEDICINE HONJO, 2-2-1, KUMAMOTO 860, JAPAN

Summary
Long-term incubation of proteins with glucose leads to advanced glycation end products(AGE). Demonstration of AGE *in vivo* as well as determination of its main structure are crucial to elucidate its potential link to aging processes and diabetic complications. We prepared polyclonal and monoclonal anti-AGE antibodies and used to identify and quantitate AGE in human lens proteins. Results suggest that there occurs a common AGE-structure and provide the direct evidence for the presence of AGE in human lens crystallins and its age-related increase.

Introduction

Incubating proteins with reducing sugar, known as the Maillard reaction, proceeds through early stage products such as a Schiff base and Amadori rearrangement products to advanced glycation end products (AGE) with fluorescence, brown color and cross-linking. Although a potential link of AGE to aging and diabetic complications is suggested (Brownlee et al., 1988), evidence for the presence of AGE in vivo is limited. In the present study we characterize polyclonal and monoclonal antibodies to AGE and use them as probes to identify and quantitate AGE in human lens crystallins.

Materials and Methods

AGE-preparations for proteins, lysine derivatives and monoaminocarboxylic acids
AGE-bovine serum albumin (BSA), AGE-human serum albumin (HSA) and AGE-human hemoglobin (Hb) (Takata et al., 1988) as well as AGE-samples from lysine derivatives and monoaminocarboxylic acids were prepared (Horiuchi et al., 1991).

Preparation of human lens proteins
Fresh normal nondiabetic noncataractous human lenses (9 lenses from infants below 4 years of age and 18 lenses from adults from 25 to 89) were purchased through the National Disease Research Interchange and also provided by T. Hirose. Each lens was homogenized and separated into supernatant and precipitate by centrifugation. The supernatant was used as "the water-soluble fraction" and the precipitate was extracted with 0.1 M carbonate buffer (pH 9.5) and used as "alkaline-extracted fraction."

Preparations of Polyclonal and Monoclonal Anti-AGE Antibodies
The polyclonal antibody was raised in rabbits by immunizing with AGE-BSA, and purified by two affinity columns conjugated with BSA and AGE-BSA. The monoclonal antibody was raised in mice by immunizing AGE-BSA. Two antibody-producing cells

positive for AGE-HSA but negative for BSA were screened from fused cells (Horiuchi et al., 1992).

Enzyme-linked Immunosorbent Assay (ELISA) and Immunoblotting
Non-competitive and competitive ELISA systems were used. In the former, the AGE-sample or lens sample to be tested was absorbed to a well, blocked with gelatin and reacted with the anti-AGE antibody. The antibodies bound to wells were reacted with biotinylated anti-IgG antibody and then with avidin-biotin HRP complex (Horiuchi et al., 1991). In the latter, a mixture of a sample and the anti-AGE antibody-HRP conjugate was added to each AGE-BSA-coated well. Each water-soluble fraction was run on SDS-PAGE slab gels, transblotted to nitrocellulose membranes, followed by reaction with these antibodies (Araki, et al. 1992).

Results and Discussion

Immunochemical reactivities of polyclonal and monoclonal anti -AGE antibody
The polyclonal antibody reacted with AGE-BSA, but not with BSA, or BSA obtained by parallel incubation without glucose. AGE-HSA, AGE-Hb and AGE-collagen were also recognized by the antibody, while their parent (unmodified) proteins were not. The monoclonal antibody showed the same reactivity. In addition, both polyclonal and monoclonal antibodies showed a positive reaction to GE-products obtained from lysyl peptides such as Lys-Lys-Lys, Lys-Lys-Lys-Lys and Lys-Ser-Tyr, and lysine derivatives such as α-tosyl-L-lysine and its methylester, and monoaminocarboxylic acids such as ϵ-aminocaproic acid, γ-amino-n-butyric acid and β-alanine. The corresponding unmodified compounds, however, were not recognized by these antibodies. Thus, it is evident that these AGE-samples share a major structure in common.

These anti-AGE antibodies do not react with early products of the Maillard reaction because: (1) the reactivity of these antibodies to GE-BSA was not altered even after its fructosamine level was reduced by treatment with $NaBH_4$; and (ii) these antibodies failed to react with fructosyl-ε-caproic acid, a synthetic Amadori compound. Chemical structures of AGE postulated so far includes FFI (Pongor et al., 1984), pyrrole aldehyde (Njoroge et al., 1987), pentosidine (Sell et al., 1989, Dyer et al., 1991) and crosslines (Nakamura et al., 1992). Neither of the antibodies reacted with FFI or pyrrole aldehyde. Pentosidine and crosslines are not tested, but the former is unlikely to react with the antibody (Araki et al., 1992).

Immunochemical evidence for the presence of AGE in human lens proteins
Neither the water-soluble nor the alkaline-extracted fraction obtained from infant lenses reacted with anti-AGE antibodies. However, both the water-soluble and the alkaline-extracted fraction obtained from adult lenses exhibited significant reactivity with both polyclonal and monoclonal antibody (Figure 1). The reactivities increased steadily with lens age. Statistical analysis of these data yielded a high correlation between immunoreactivity and lens age (see Figure 2). Thus it is evident that AGE-structures are present in the human tissue and that amounts of AGE increase as a direct function of age, suggesting the possibility that the advanced stages of the Maillard reaction could occur ubiquitously *in vivo* in any human tissue.

Upon SDS-PAGE of each water-soluble fraction and subsequent immunoblotting with anti-AGE antibodies, main bands (α, β and γ-crystallin) showed a positive reaction and their immunoreactivities varied with the age of the lenses: none of the bands from infant lenses showed reactivity, whereas the corresponding bands from adult lenses reacted with the antibodies, and the extent of the reactivity increased with age. Another protein of M_r=43K also increased in immunoreactivity in an age-related manner. Available data suggest that the protein could be a polymerized form of crystallin.

Several studies using anti-AGE antibodies have demonstrated the presence of AGE *in vivo*. In diabetic rats, AGE levels in lens crystallins, renal cortexes and aortas were > 10-fold higher than in control rats (Nakayama et al., 1993 and Mitsuhashi et al., 1993). These results taken together makes it conclusive that these tissues contain AGE. The presence of AGE in serum, however, remains controversial. Makita et al. (1992a) reported that human serum contained AGE and that levels of AGE increased significantly in hemodialysis patients with diabetic complications. The same group also demonstrated AGE-human hemoglobin in circulating erythrocytes (Makita et al., 1992). Using similar ELISA systems, however, we have been unsuccessful in detecting AGE in serum. The precise reason for this is not known. Miyata et al. (1993) in collaboration with our group recently shed light on this issue. β_2-Microglobulin is a major constituent of amyloid fibrils in hemodialysis patients. Amyloid fibril deposits isolated from connective tissues forming carpal tunnels in hemodialysis patients with carpal tunnel syndrome were found to contain AGE-modified β_2-microglobulin as a major fraction of β_2-microglobulin. This AGE-β_2-microglobulin was also found at a low concentration in serum and urine from long-term hemodialysis patients. Since it is well known that the serum β_2-microglobulin level is markedly higher (usually >30-fold) in these patients, compared with healthy individuals, these data strongly suggest that proteins accumulated in serum under pathological conditions, like those of hemodialysis patients, could be subjected to AGE-modification. This view may provide a basis for the previous claim for the presence of AGE in serum (Makita et al., 1992a). More information is needed, of course, regarding the structure(s) of AGE found *in vivo* and the relationship between AGE and aging or disease processes. In this connection, we isolated a major fluorescent compound (X1) from AGE-α-tosyl-L-lysine methyl ester. Our preliminary experiment showed that antibody raised against X1-keyhole limpet hemocyanin (KLH) was able to react with

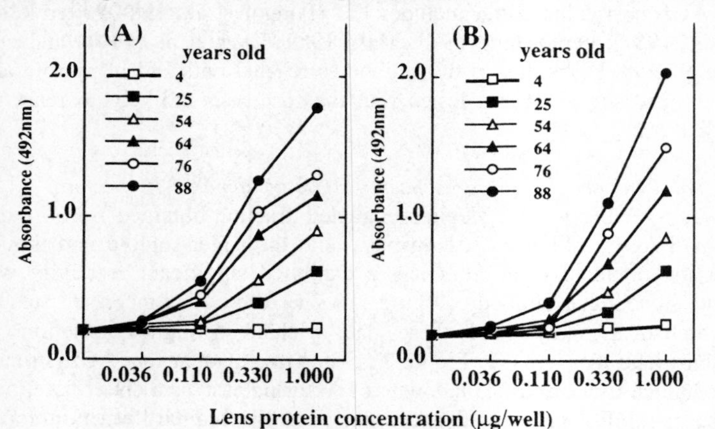

Figure 1. Immunochemical reactivity of human lens proteins to the polyclonal and monoclonal anti-AGE antibody. Reaction of the water-soluble fractions with polyclonal antibody (A) and monoclonal anti-AGE antibody (B). Lens age (years old): <4; 25; 54; 64; 76; 88.

human lens proteins, as well as with AGE-proteins, suggesting the possibility that might represent one of the common AGE structures. Determination of the chemical structure of X1 is under way.

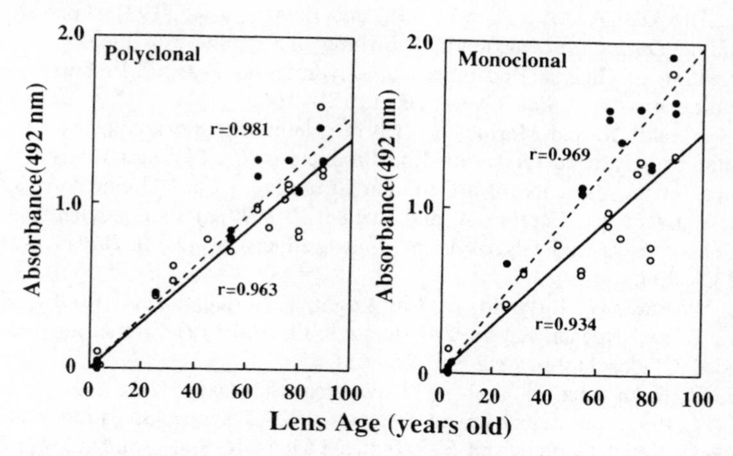

Figure 2. Immunochemical reactivity of human lens proteins and correlation with age. Concentrations of each lens protein of the water-soluble (o) and alkaline-extracted fractions (•) in Figure 1 are fixed at 1.0 mg/system and the extent of their immunoreactivities to the polyclonal (left) and the monoclonal antibody (right) are plotted against lens age. The water-soluble; R=0.963, p<0.0001, alkaline-extracted; R=0.981, p<0.0001 for the polyclonal antibody, and water-soluble; R=0.934, p<0.0001, alkaline-extract; R=0.969, P<0.0001 for the monoclonal antibody.

Conclusions

The present immunological study with polyclonal and monoclonal anti-AGE antibodies revealed the presence of a common structure in AGE-preparations. Experiments with 27 fresh normal nondiabetic noncataractous human lenses clearly demonstrated that human lens crystallins express common AGE structure(s) which increase with tissue age, providing solid evidence for the presence of AGE *in vivo* and its functional link to the aging of human tissue.

Acknowledgments

This work was supported in part by a grant for scientific research from the Ministry of Education, Science and Culture of Japan, and grants from the Sandoz Foundation for Gerontological Research, ONO Medical Research Foundation and Okukubo Memorial Fund for Medical Research in Kumamoto University School of Medicine. We thank Tamami Sakamoto (Panapharm Laboratories, Kumamoto) and Drs. Bireswar Chakrabarti and Norio Ueno at Schepens Eye Research Institute and Harvard Medical School for their collaborative endeavor.

References

Araki, N., Ueno, N., Chakrabarti, B., Morino, Y. and Horiuchi, S. (1992) Immunochemical Evidence for the Presence of Advanced Glycation End Products in Human Lens Proteins and Its Positive Correlation with Aging.*J. Biol. Chem.* 267:10211-10214.

Brownlee, M., Cerami, A. and Vlassara, H. (1988) Advanced Glycosylation End Products in Tissue and the Biochemical Basis of Diabetic Complications. *New Eng. J. Med.* 318:1315-1321.

Dyer, D.G., Blackledge, J.A., Thorpe, S.R., and Baynes, J.W. (1991) Formation of Pentosidine During Nonenzymatic Browning of Proteins by Glucose: Identification of Glucose and other Carbohydrates as Possible Precursors of Pentosidine *in vivo*. . *Biol. Chem.* 266:11654-11660.

Horiuchi, S., Araki, N. and Morino, Y. (1991) Immunochemical Approach to Characterize Advanced Glycation End Products of the Maillard Reaction: Evidence for the Presence of a Common Structure. *J. Biol Chem.* 266:7329-7332.

Makita, Z., Vlassara, H., Cerami, A. and Bucala, R. (1992a) Immunochemical Detection of Advanced Glycosylation End Products *in vivo. J. Biol. Chem.* 267:5133-5138.

Makita, Z., Vlassara, H., Rayfield, E., Cartwright, K., Friedman, E., Rodby, R., Cerami, A. and Bucala, R. (1992b) Hemoglobin-AGE: A Circulating Marker of Advanced Glycosylation. *Science.* 258:651-653.

Mitsuhashi, T., Nakayama, H., Itoh, T., Kuwajima, S., Aoki, S., Atsumi, T., and Koike, T. (1993) Immunochemical Detection of Advanced Glycation End Products in Renal Cortex from STZ-Induced Diabetic Rat. *Diabetes* 42: 826-832.

Miyata, T., Oda, O., Inagi, R., Iida, Y., Araki, N., Yamada, N., Horiuchi, S., Taniguchi, N., Maeda, K., and Kinoshita, T. (1993) β2-Microglobulin Modified with Advanced Glycation End Products is a Major Component of Hemodialysis-Associated Amyloidosis. *J. Clin. Invest.*. 92:1243-1252.

Nakamura, K., Hasegawa, T., Fukunaga, Y., and Ienage, K. (1992) Crosslines A and B as Candidates for the Fluorophores in Age- and Diabetes-Related Cross-Links Proteins, and then Diacetates Produced by Maillard Reaction of α-N-acetyl-L-lysine with D-glucose.*J. Chem. Soc.Chem. Commun.* 14:992-994.

Nakayama, H., Mitsuhashi, T., Kuwajima, S., Aoki, S., Kuroda, Y., Itoh, T., and Nakagawa, S. (1993) Immunochemical Detection of Advanced Glycation End Products in Lens Crystallins from Streptozocin-Induced Diabetic Rat. *Diabetes* 42:345-350.

Njoroge, F.G., Sayre, L.M. and Monnier, V.M. (1987) Detection of D-Glucose Derived Pyrrole Compounds During Maillard Reaction Under Physiological Conditions. *Carbohydr. Res.* 167:211-220.

Pongor, S. Ulrich, P.C., Bencsath, F.A., and Cerami, A. (1984) Aging of Proteins: Isolation and Identification of a Fluorescent Chromophore from the Reaction of Polypeptides with Glucose. *Proc. Natl. Acad. Sci.* USA 81:2684-2688.

Sell, D.R., and Monnier, V.M. (1989) Structure Elucidation of a Senescence Crosslink from Human Extracellular Matrix: Implication of Pentoses in the Aging Process. *J. Biol. Chem.* 264:21597-21602.

Takata, K., Horiuchi, S., Araki, N., Shiga, M., Saitoh, M., and Morino, Y. (1988) Endocytic Uptake of Nonenzymatically Glycosylated Proteins is Mediated by a Scavenger Receptor for Aldehyde-Modified Proteins. *J. Biol. Chem.* 263:14819-14825.

Glucose-mediated DNA Damage and Mutations: *In Vitro* and *In Vivo*

Annette T. Lee, Anthony Cerami, and Richard Bucala

THE PICOWER INSTITUTE FOR MEDICAL RESEARCH, MANHASSET, NEW YORK, 11030, USA

Summary
The nonenzymatic glycosylation of nucleic acids *in vitro* and *in vivo* has been shown to have adverse effects on both their functional activity and biological integrity. *In vitro* and *in vivo* prokaryotic and eukaryotic model systems have demonstrated the effect of glucose-modified DNA on the induction of a number of significant genetic alterations. We have expanded these investigations to a mammalian system; transgenic mice containing the bacterial mutagenesis marker *lacI*. Using *lacI* transgenic mice, we have performed a comparative analysis of age-related DNA mutations in splenic tissue and studied the influence of a hyperglycemic maternal environment on the *lacI* mutation frequency of developing transgenic embryos.

Introduction

Originally proposed to describe the nonenzymatic reaction between sugars and amino acids, the Maillard reaction (Maillard, 1912) has become the central theme in a number of hypotheses to explain the relationship between reducing sugars, such as glucose, and several age-related and diabetic pathologic conditions (Cerami, 1985). The early products of the Maillard reaction, Schiff base and Amadori product, are formed by the direct reaction of the aldehyde group of a reducing sugar with the amino group of a macromolecule. Once formed, the Amadori product itself can then undergo a series of as yet undefined rearrangements and dehydrations to form stable endproducts, collectively referred to as advanced glycosylation endproducts or AGEs (Brownlee et al., 1984). Unlike the early products of the Maillard reaction, AGEs are irreversibly formed endproducts which are yellow-brown in color and have a characteristic fluorescent spectrum with a excitation maxima near 370 nm. In addition, AGEs are able to covalently crosslink proteins both inter- and intra-molecularly.

The adverse biological consequences of nonenzymatic glycosylation are prominent in a number of long-lived proteins. Since the formation of AGEs occurs slowly, over weeks to months, proteins with especially long half-lives are most susceptible to modification by this reaction. AGEs have been found on various long-lived proteins *in vivo* such as lens crystallins, myelin, basement membrane and collagen, which may contribute to cataracts, neuropathy, nephropathy and atherosclerosis, respectively. Due to the ubiquitous nature of reducing sugars, the Maillard reaction occurs gradually and at a constant rate of endproduct formation; however, under a chronic hyperglycemic condition such as in diabetes mellitus, the amount of AGEs present on long-lived proteins is greatly enhanced and may contribute to a number of observed diabetic complications (for review see Bucala and Cerami, 1992).

In vitro model systems
Almost 10 years ago, it was hypothesized that reducing sugars, such as glucose and glucose-6-phosphate could react with the primary amino groups of DNA bases in a

manner analogous to the reaction observed with proteins. The modification of DNA through the direct interaction with glucose or indirectly through a glucose-derived reactive intermediate may contribute to some of the reported age-related genetic dysfunctions such as nucleotide modification (Randerath et al., 1993), DNA crosslinks (Bojanovic et al., 1970) and strandbreaks (Karran and Ormerod, 1973), and chromosomal aberrations (Martin et al., 1985).

The *in vitro* incubation of nucleotides, or single or double stranded DNA with reducing sugars leads to the formation of pigmented products which have absorbance and fluorescent spectra similar to those observed for glucose modified proteins. This similarity in spectra suggests that the primary amino groups of DNA bases can participate in the Maillard reaction analogous to the amino groups of proteins (Bucala, Model and Cerami, 1984). The effect of nonenzymatic glycosylation on biological function was measured using single stranded bacteriophage and double stranded plasmid DNA (Bucala et al., 1985). In both instances, the modification of phage or plasmid DNA by glucose-6-phosphate resulted in a loss in transfection and transformation efficiencies, respectively.

A bacterial model system was developed to measure the effect of elevated intracellular levels of glucose-6-phosphate on the mutation frequency of a host carried target plasmid (Lee and Cerami, 1987). Glycolytic mutants of *E.coli* (DF40 and DF2000) were used to determine the extent that elevated intracellular levels of glucose-6-phosphate affected the mutation rate of plasmid DNA. Compared to the control strain, K10, when *E.coli* strains, DF40 and DF2000 were grown under specific conditions, a 20 and 30 fold increase was measured in glucose-6-phosphate levels, respectively. The significant increase in intracellular glucose-6-phosphate levels corresponded to increases in the mutation frequency of the host carried target plasmid. The DF40 *E.coli* strain which accumulated a 20 fold excess of glucose-6-phosphate demonstrated a 7 fold increase in plasmid mutation rate while the DF2000 strain which accumulated a 30 fold increase in glucose-6-phosphate showed an increase in plasmid mutation rate 14 fold higher than the control K10 strain. The most striking observation from these studies was the high proportion of plasmid mutations which were the result of movement and integration of a host transposon, $\gamma\delta$ into the target plasmid (Lee and Cerami, 1991).

To determine if glucose-induced DNA damage resulting in DNA abnormalities was specific to bacteria alone, or if it was a general pathway for DNA damage and mutation, a eukaryotic system was developed. In this system, mouse lymphoid cells were transfected with shuttle vector DNA which had been modified to different degrees with glucose *in vitro* (Bucala et al., 1993). The transfected cells were selected for G418 resistance, the shuttle vector was then recovered and mutations were detected in a bacterial system. These studies showed the dramatic effect of AGE-modification on the mutation rate of the target shuttle vector in a eukaryotic system. The mutation rate increased from 0.1% in the control sample to 28% in sample with the highest level of AGE-modified vector DNA (Table I). Analysis of recovered mutant vector DNA showed that the majority of the mutations were the result of large (>100bp) insertions or deletions of DNA. Upon sequence analysis of the insertional mutations, it was discovered that a high proportion of this class of mutations were due to the mobilization of an *Alu*-containing sequence (INS-1) from the host chromosome which had then been inserted into the vector.

In vivo model systems

The similarity of these results to those found in the bacterial model system strongly

Table I. Mutation analysis of AGE-modified plasmid pPy35 recovered from murine lymphoid cells (Bucala et al., 1993).

AGE[3], mM	Trans-fections[4]	Plasmids analyzed[5]	Mutation rate[1]%		Mutant plasmids with a size change. [2] % fraction		
			Mean	Range	Insertion>100 bp	Deletion<100 bp	No size change
0	6	11,472	0.1	0.03-0.3	33 (4/12)	50 (6/12)	17 (2/12)
5	5	18,456	5.2	0.2-14	38 (24/64)	50 (32/64)	13 (8/64)
6.5	6	5,574	14.4	1.4-26	75 (44/59)	22 (13/59)	3 (2/59)
7.5	6	2,893	28.0	4.0-68	61 (36/59)	37 (22/59)	2 (1/59)

[1] Calculated from the ratio of the number of mutant plasmids to the number of total plasmids recovered for each DNA modification condition.
[2] Determined in a subset of mutant clones chosen randomly from each transfection , with the exception of the control (0 mM AGE) condition, in which all the recovered mutants were studied by restriction analysis.
[3] Expressed as lysine equivalents
[4] Number of independent transfections analyzed at each corresponding concentration of AGE.
[5] Total number of plasmids pooled for study from each condition of DNA modification.

indicates that *in vitro* or *in vivo* modification of DNA by glucose or glucose-derived intermediates induces DNA mutations in a concentration-dependent manner and has the potential to activate host transposons or transposon-like sequences. Although the absolute level of glucose does not fluctuate within most cells, the continual and long-term exposure of DNA to ambient levels of reducing sugars could lead to the gradual formation and accumulation of AGEs or AGE-derived modifications that result in DNA damage and mutations. This type of constant insult to DNA mediated by glucose and other reducing sugars which lead to significant DNA size changes, suggesting a possible mechanism to account for observed age-related increases in chromosomal aberrations and karyotype instability.

To explore this possibility in further detail in a truly *in vivo* model, we studied age-related DNA mutations of a specific gene in transgenic mice. These mice contain integrated copies of the well-defined bacterial mutagenesis marker gene, *lacI*, in a readily excisable bacteriophage shuttle vector (Kohler et al., 1991). Mutations of the *lacI* gene are easily detected colorimetrically by blue/white screening on X-gal bacterial indicator plates. This system allows for the efficient detection, quantitation and molecular analysis of DNA mutations from *in vivo* samples without prior clonal expansion or selection. Using these *lacI* containing transgenic mice, we performed a comparative analysis of naturally occurring mutations of the *lacI* gene in spleen tissue from animals of different ages (Lee, Cerami and Bucala, 1993). We found that there is approximately a 4-fold increase in mutations of this target gene between birth and 24 months. In addition to an age-related increase in DNA mutations of the *lacI* gene there is also a distinct pattern of types of mutations in older animals. DNA amplification analysis of *lacI* gene mutations showed that in animals less than 3 months, the majority of mutations were due to base substitutions (98%), however, analysis of *lacI* mutations from older animals showed that a higher percentage of them were the result of DNA size changes greater than 25bp (14-22%).

Diabetes induced teratogenesis

Congenital malformations are estimated to be the leading cause of fetal mortality in insulin-dependent diabetic mothers (Kalter, 1990). Previous studies have indicated that the degree of glycemic control during preconception and early pregnancy is essential to normal development of the fetus (Miller et al., 1981). Although there is a positive correlation between the degree of maternal hyperglycemia and serious birth defects, no mechanism of teratogenic action is known.

To study the possible mutagenic effects of a maternal hyperglycemic environment on the developing embryo due to glucose or glucose-derived intermediates, we have transplanted *lacI* containing transgenic embryos into the uteri of streptozotocin-induced diabetic surrogate dams. Two weeks following transplantation the embryos are harvested and the *lacI* mutation frequency of the embryo genomic DNA was measured as a function of maternal hyperglycemia. Preliminary results indicate an almost two-fold increase in *lacI* mutation frequency in embryos which developed in a diabetic host (150-250mg/dL glucose) as compared to those which developed under normoglycemic conditions (<150mg/dL glucose) (Lee et al., 1993). These results suggest the involvement of glucose mediated DNA damage in offspring of diabetic mothers.

Results from our studies using bacteria, cultured eukaryotic cells and transgenic mice model systems suggest that glucose-mediated DNA damage exerts a detrimental effect on DNA function. Glucose-mediated DNA damage, either directly or through a glucose-derived intermediate, has been shown to induce DNA mutations in 3 independent systems. The most striking finding from these studies was the apparent activation of host transposons in both the bacterial and cell culture systems. No other form of DNA damage has been reported to induce transposons in this fashion. Whether similar events occur *in vivo* as a consequence of natural aging or during embryonic development in a hyperglycemic environment is actively being pursued. If transposons or transposon-like sequences are identified as mediators in age-related DNA mutations or embryopathies, then the role of glucose as simply an innocuous carbon source may need to be re-analyzed in terms of age-related genetic dysfunctions, oncogenesis and congenital birth defects.

References

Bojanovic, J.J., Jevtovic, A.D., Pantic, V.S., Dugandzic, S.M. and Jovanovic D.S. (1970). Thymus nucleoproteins. Thymus histones in young and adult rats. *Gerontologia* 16:304-312.

Brownlee, M., Vlassara, H. and Cerami, A. (1984). Nonenzymatic glycosylation and the pathogenesis of diabetic complications. *Ann. Intern. Med.* 101: 527-537.

Bucala, R., Model, P. and Cerami, A. (1984). Modification of DNA by reducing sugars:A possible mechanism for nucleic acid aging and age-related dysfunction in gene expression. *Proc. Natl. Acad. Sci. USA* 81:105-109.

Bucala, R., Model, P., Russel, M. and Cerami, A. (1985). Modification of DNA by glucose-6-phosphate induces DNA rearrangements in an Escherichia coli plasmid. *Proc. Natl. Acad. Sci. USA* 82:8439-9442.

Bucala, R. and Cerami, A. (1992) Advanced Glycosylation: Chemistry, Biology, and Implications for Diabetes and Aging. *Adv. Pharmacol.* 23: 1-34.

Bucala, R., Lee, A.T., Rourke, L. and Cerami, A. (1993) Transposition of an *Alu*-containing element induced by DNA-advanced glycosylation endproducts. *Proc. Natl. Acad. Sci. USA* 90: 2666-2670.

Cerami, A. (1985) Glucose as a mediator of aging. *J. Amer. Ger. Soc.* 33: 626-634.

Kalter, H. (1990) Perinatal mortality and congenital malformations in infants born to women with insulin-dependent diabetes mellitus- United States, Canada and Europe, 1940-1988. *Morbidity and Mortality Weekly Report* 39: 363-365.

Karran, P. and Ormerod, M.G. (1973). Is the ability to repair damage to DNA related to the proliferative capacity of a cell? The rejoining of x-ray produced strand breaks. Biochim. Biophys. Acta. 299:54-64.

Kohler, S.W., Provost, G.S., Fieck, A., Kretz, P.L., Bullock, W.O., Sorge, J.A., Putman, D.L. and Short, J.M. (1991) Spectra of spontaneous and mutagen-induced mutations in the lacI gene in transgenic mice. *Proc. Natl. Acad. Sci. USA* 88: 7958-7962.

Lee, A.T. and Cerami, A. (1987). Elevated glucose-6-phosphate levels are associated with plasmid mutations in vivo. *Proc. Natl. Acad. Sci. USA* 84:8311-8314.

Lee, A.T. and Cerami, A. (1991) Induction of γδ transposition in response to elevated glucose-6-phosphate levels. *Mut. Res.* 249: 125-133.

Lee, A.T., Bucala, R. and Cerami (1993) Comparative analysis of age-related DNA Mutations in lacI transgenic mice. (manuscript in preparation).

Lee, A.T., Plump. A., Bucala, R. and Cerami, A. (1993) Diabetes induced teratogenesis: The role of DNA damage. (manuscript in preparation)

Maillard, L.C. (1912). Action des acides amines sur les sucres;formation des melanoidines par voie methodique. *C.R. Seances Acad. Sci.* 154:66-68.

Martin, G.M., Smith, A.C., Keither, D.J., Ogburn, C.E. and Disteche, C.M. (1985) Increased chromosomal aberrations in the first metaphases of cells isolated from the kidneys of aged mice. *Isr. J. Med. Sci.* 21: 296-301.

Miller, E., Hare, J.W., Cloherty, J.P., Dunn, P.J. Gleason, R.E., Soeldner, J.S. and Kitzmiller, J.L. (1981) *New Eng. J. Med.* 304: 1331-1334.

Randerath, K., Putman, K.L., Osterburg, H.H., Johnson, S.A., Morgan, D.G. and Finch, C. (1993) Age-dependent increases of DNA adducts (I-compounds) in human and rat brain DNA. *Mut. Res.* 295: 11-18.

AGE-receptors and *In Vivo* Biological Effects of AGEs

Helen Vlassara

THE PICOWER INSTITUTE FOR MEDICAL RESEARCH, MANHASSET, NEW YORK, 11030, USA

Summary

Surface receptors for Advanced Glycation Endproducts (AGEs) are found on macrophages, T-lymphocytes, endothelial cells (EC), mesangial cells (MS), fibroblasts, and smooth muscle cells. Binding of AGEs to these receptors leads to a range of cellular responses, including monocyte chemotaxis, activation, growth factor release, increased matrix production, increased EC permeability, and procoagulant activity. A number of these responses can be inhibited by anti-AGE-receptor antibodies, supporting the role of AGE-ligand/receptor interactions in these events. Evidence for similar AGE-mediated biological effects *in vivo* was obtained recently: Short-term (4-8 weeks) exogenous AGE administration to normal rats and rabbits led to multiple vascular defects, including vascular permeability, mononuclear activation, and vasodilatory impairment. Longer treatment with AGEs (3 months) led to arterial basement membrane thickening, mesangial expansion, and glomerulosclerotic changes. These alterations were largely prevented by the simultaneous treatment with Aminoguanidine. These studies suggest that the interaction of de novo implanted reactive AGEs with cellular AGE-receptors of otherwise healthy tissues can generate renal, and vascular pathology similar to that seen in diabetes, in the absence of either the genetic or the metabolic abnormalities linked to diabetes. Progressive loss of kidney function correlates with increasing circulating AGE levels, presumably reflecting tissue AGE-degradation products which are not cleared by the failing kidneys. The pronounced (~8-fold) increase in serum AGEs observed in diabetic anephric patients, a group particularly susceptible to accelerated atherosclerosis, indicates that uncleared "reactive" AGEs may be available for enhanced interaction with cellular AGE-receptors, accelerating existing pathology.

Introduction

Advanced glycosylation is strongly implicated in the initiation and acceleration of multiple organ damage in diabetes, including vasculature, kidneys, and nerves (Bucala et al., 1991). Recent information on the biology of AGEs points to a dynamic process characterized by continuous AGE-formation, deposition in tissues and cells, intracellular degradation, and ultimately by the clearance of AGE-protein breakdown products through the renal system in the form of AGE-peptides (Makita *et al.*, 1991; Makita, *et al.*, 1993). Given the abundance of AGEs in extracellular matrix and in the circulation, investigation into possible cellular receptor-mediated interactions has begun to provide new insights into the role of this ubiquitous process in both health and disease (Table I).

AGE-Receptor Family-Functional Studies

Macrophages have long been implicated in tissue homeostasis and remodeling, but also, in chronic processes such as atherosclerosis. The identification of a system operating for the removal of AGEs and regulated by macrophages has introduced a new interpretation of the role of tissue glycosylation in tissue pathophysiology. Initially, it was observed that both *in vitro* synthesized or *in vivo* formed AGE-modified proteins are recognized by a distinct macrophage AGE-receptor system (Vlassara *et al.*, 1985; Vlassara *et al.*, 1986). The tissue distribution of AGE receptor(s) was examined in the rat, revealing the liver as the organ with the highest AGE-binding activity *in vivo* (Yang *et al.*, 1991). AGE-affinity labeling of cell surface proteins from the mouse monocyte cell line RAW 264.7 revealed a single 90 kD AGE-binding subunit (Radoff *et al.*, 1988; Radoff *et al.*, 1990). More recently, two rat liver membrane AGE-binding proteins were isolated, 60kD and 90kD proteins (Yang *et al.*, 1991) of apparently unique amino acid sequence which, by flow cytometric analysis, were shown to be present on human and rat monocyte/macrophages (Yang *et al.*, 1991), and T-lymphocytes (Imani *et al.*, 1993).

Table I. Potentially pathogenic properties of advanced glycosylation endproducts.

1. Induction of transendothelial migration of monocyte/macrophages
2. Stimulation of cytokine and growth factor release from macrophages
3. Entrapment of non-AGE molecules, leading to: immune complex formation, complement activation, lipoprotein accumulation, etc.
4. Induction of Interferon-γ by T-lymphocytes
5. Increase of endothelial (EC) permeability and procoagulant activity via:
   (a) an increase in Tissue Factor activity, and
   (b) suppression of Thrombomodulin expression on EC surface
6. Enhancement of synthesis of Extracellular Matrix components
7. Enhancement of cellular proliferation (e.g. smooth muscle cell, fibroblast)
8. Enhancement of DNA mutation rate
9. Inhibition of EDRF activity
10. Impaired clearance of plasma lipoproteins modified by AGE

Additional studies have established that *in vivo* or *in vitro* AGE-modified proteins, such as AGE-albumin, AGE-myelin, and AGE-low density lipoprotein are selectively chemotactic for monocytes (Kirstein *et al.*, 1990). These studies indicated that *in vivo* AGE-formation may mediate monocyte recruitment from the circulation, inducing them to migrate through normal endothelium and reach sites of AGE accumulation. This possibility was supported by finding that human monocytes exhibit significant transendothelial migratory activity toward AGE-modified matrix (Kirstein *et al.*, 1990).

AGE-uptake, and intracellular breakdown is associated with cytokine secretion, including TNFα and Interleukin-1β (Vlassara *et al.*, 1988). These responses suggest a complex regulatory system by which macrophages, while selectively removing AGEs, may also communicate to adjacent cells the need for replacement of the AGE-modified proteins. The pathway by which this complex process operates is only partially elucidated, as it involves the synthesis and release of at least two additional mediators, such as Platelet-Derived Growth Factor (PDGF) and Insulin-like Growth Factor-IA (IGF-IA) (Kirstein *et al.*, 1990; Kirstein *et al.*, 1992), and is AGE-receptor dependent, based on anti-receptor antibody experiments (Kirstein *et al.*, 1992). These initial findings suggest that AGEs, abundant in most tissues and fluids, through the coordinate induction of cytokines and growth factors, are likely to participate in tissue remodeling, a process which can also be easily disturbed either by the excessive deposition of AGEs or by altered AGE-receptor function.

For example, studies on the regulation of this receptor system have revealed that insulin may downregulate the number and the affinity of AGE-receptors on monocytes and macrophages (Vlassara *et al*., 1988). Down-regulation of AGE-receptors in certain NIDDM patients with elevated peripheral insulin levels may reduce the rate of tissue AGE removal and thus contribute to the vascular dysfunction commonly associated with hyperinsulinemia.

In contrast to insulin, the potent AGE-inducible cytokine TNF was found to induce a several fold increase in the binding, endocytosis and degradation of AGE-modified albumin by both murine and human macrophage/monocytes *in vitro*, while it enhanced the rate of disappearance of AGE-erythrocytes *in vivo* (Vlassara *et al*.,1989), suggesting that the AGE-receptor system is subject to autocrine regulation.

The presence of AGE-receptors on another member of the hemopoietic system, the T-lymphocyte was an intriguing finding (Imani *et al*., 1993) (Table II). Based on radioligand binding studies and flow cytometry, resting human and rat CD4+ and CD8+ T-cells are found to bind AGE-proteins with a Kd of $7.8X10^{-7}$M, which increased to a Kd of $5.8X10^{-8}$M upon stimulation with PHA (Imani *et al*., 1993). Exposure of pre-stimulated T-cells to AGEs was shown to result in the synthesis and release of IFNγ (Imani *et al*., 1993). An expanded AGE-receptor hypothesis emerges from these preliminary findings, suggesting that AGE-activated IFNγ-producing T-cells may contribute to tissue damage in conjunction with primed macrophages. This is consistent with accumulating evidence suggesting that the activated lymphocytes found in atherosclerotic lesions, for instance, are important participants in an immune-like response underlying atherogenesis (Libby *et al*., 1991; Stemme *et al*., 1991).

Table II. The AGE-Receptor family: cellular expression and function.

Cell type	Species	Function/properties
MONOCYTE/ MACROPHAGE	Mouse, Rat, Human	AGE-ligand binding, endocytosis, degradation Cytokine production (TNFα, IL-1β) Growth factor induction (PDGF,IGF-IA) Chemotaxis Up-regulation by TNFα Down-regulation by insulin
T-LYMPHOCYTE	Mouse, Rat, Human CD4+, CD8+T-cells	Ligand binding Up-regulation by PHA, LPS, TNFα Interferon-γ production
ENDOTHELIAL CELL	Rat, Human, Bovine	Ligand binding, Trancytosis, Degradation ↑ Permeability ↑ Cell surface-associated tissue factor activity ↓ Thrombomodulin
MESANGIAL CELLS	Mouse, Rat, Human	Binding, Endocytosis, Degradation ↑ Fibronectin, Collagen IV, Laminin, HSPG Growth factor induction (PDGF)
FIBROBLAST	Rat, Human	Ligand binding ↑ EGF and EGF-receptor mRNA ↑ Proliferation
SMOOTH MUSCLE CELL	Rat, Human	Ligand binding ↑ Proliferation

The interaction of subendothelial or circulating AGEs with vascular endothelium has been reported to influence endothelial cell (EC) function. A specific binding system for AGE determinants on the EC surface (Kd 100nM) was found to mediate AGE-protein internalization, and transcytosis of mostly unaltered AGE ligand (Esposito *et al.*, 1989) (Table II). Isolation and sequencing of two AGE-binding proteins isolated from bovine lung revealed a 35 kD protein, a member of the immunoglobulin superfamily, and an 80 kD protein, with complete sequence homology to lactoferrin (Schmidt *et al.*, 1992). Subsequent cloning and expression of the 35 kD protein in 293 cells, a tumor fibroblast-like cell line, resulted in an AGE-binding protein of about 50 kD (Neeper *et al.*, 1992). Based on the different protein sequence and ligand specificities found between these molecules and the macrophage AGE receptors, there may be a number of AGE-binding proteins, which, although distinct from one another, may share diverse functions, e.g. cell adhesion, clotting factor activation, and growth factor binding. In this context, evidence has indicated that AGE-endothelial interaction results in enhancement of procoagulant activity and EC permeability (Esposito *et al.*, 1989). The direct role of the AGE-binding proteins in these phenomena has not been established as yet.

Studies on murine and human mesangial cells revealed AGE-binding (Kd= $2.0\pm0.4X10^{-6}$M) and endocytosis (Skolnick *et al.*, 1991) (Table II). Although the intracellular processing and degradation of AGE ligands was observed to be slower than that of macrophages, a role in matrix regulation was suggested for this new group of mesangial cell receptors, based on a significant increase in the production of fibronectin and several other protein-members of the normal mesangial cell matrix (Skolnick *et al.*, 1991; Doi *et al.*, 1992). Exposure of normal mouse mesangial cells to AGE-BSA resulted in an upregulation of α1 type IV collagen, laminin A, laminin B1 and B2, and heparin sulfate proteoglycan mRNAs, while collagen IV secretion increases via the mesangial AGE-receptors, based on antibody-inhibitory studies (Doi *et al.*, 1991). Interestingly, the collagen α1IV mRNA increase appeared to be mediated at least in part by PDGF, based on the inhibitory effect of an antibody to PDGF. Since it has been shown that PDGF is produced in an autocrine fashion in the glomerulus, these experiments suggest that PDGF may represent a key intermediate factor in the development of diabetic glomerulosclerosis contributed to by AGEs.

Human FS4 fibroblasts have been found to express similar AGE-binding receptors (Kd=$3.15X10^{-7}$M) (Kirstein *et al*, 1990) (Table II). The AGE-protein-receptor interaction was shown to induce the expression of Epidermal Growth Factor (EGF) and of EGF-receptor mRNA on human fibroblasts (Kirstein *et al*, 1990). These pilot studies introduced the possibility that AGE-stimulated EGF overexpression, followed by EGF-receptor autocrine upregulation, may contribute to the excessive proliferative changes common in diabetic tissues. Unpublished observations from this laboratory have suggested that specific binding of AGE-modified proteins to cultured rat smooth muscle cells is associated with increased cellular proliferation *in vitro*. While the precise mechanism and *in vivo* of this response is unclear, the growth-promoting effects mediated by this receptor could also be cytokine- or growth factor-mediated, as shown on other cell systems, e.g. macrophages, fibroblasts, and mesangial cells.

In vivo Animal & Clinical Studies

In vivo evidence on the direct pathogenic influence of AGEs, in a manner which distinguishes them from diabetes has been obtained recently. Studies in normal rats and rabbits showed that the intravenous administration of exogenous AGE-albumin over 2-4 weeks results in their deposition within tissues causing multiple vascular defects, including vascular leakage, mononuclear cell extravasation, as well as, unresponsiveness to vasodilatory agents (Vlassara *et al.*, 1992). Longer AGE-albumin administration (2-3 months) to healthy animals induced extensive glomerular and arteriolar basement membrane thickening and mesangial expansion (Fuh *et al.*, 1992). Each of these alterations was largely prevented by the simultaneous treatment with Aminoguanidine, an AGE-crosslink inhibitor shown to prevent numerous diabetic complications in animal models (Vlassara, *et al.*, 1993). These studies demonstrated that the implantation of exogenous AGEs within otherwise healthy tissues is capable of generating pathophysiological perturbations similar to those seen at various stages of diabetic complications, in the absence of genetic or metabolic considerations associated with diabetes.

In normal and diabetic individuals, degradation products of AGE-modified proteins (AGE-peptides) are cleared from the circulation with an efficiency that correlates with renal function (Makita, *et al.*, 1991; Makita, *et al.*, 1993). Thus, with progressive loss of renal function, AGE-peptide levels can increase in plasma up to eight-fold over normal. Furthermore, conventional hemodialysis clears serum AGE-peptides poorly when compared to creatinine and other readily dialyzable substances (Makita, *et al.*, 1991; Makita, *et al.*, 1993). In contrast, renal transplantation can readily restore normal AGE-peptide levels. These AGE-peptides, as found in the human circulation, are chemically "reactive" and can readily crosslink covalently with proteins such as with collagen *in vitro* (Makita, et al., 1993). By virtue of this activity, human serum AGE-peptides could contribute to vascular pathology, as demonstrated in animal studies (Vlassara, *et al.,* 1992; Fuh, *et al.,* 1992).

Conclusions

The evidence obtained thus far points to a broad-based mechanism by which human tissues interact with the spontaneous forming of AGE-proteins during normal development and aging. The data provide a biologically sound basis for the long-speculated role of macrophages in tissue dysfunction in diabetes and aging. Furthermore, the identification of a broad family of AGE-receptors on different tissues, including immune and mesenchymal cells (Table II) point to a system built for the dual purpose of disposing of senescent macromolecules and of tissue remodeling. It is also becoming apparent that under conditions of cumulative AGE deposition, disturbances of AGE-receptor properties can lead to widespread tissue dysfunction, such as one typically sees in clinically complicated diabetes and in aging.

In view of the recent clinical observations, and the animal data on the *in vivo* effects of circulating AGEs, it appears that the presence of reactive AGE-peptides in the circulation (apart from plasma glucose), may act as an additional factor for the rapid evolution of diabetic complications, principally micro- and macroangiopathy. They also introduce a potential mechanism for explaining the rapid progression of diabetic nephropathy once this complication is initiated, as well as, the excessively high morbidity and mortality of diabetic patients with ESRD (United States Rental Data System, 1992; Friedman, 1992).

References

Brownlee, M., Cerami, A., Vlassara, H. (1988) Advanced glycosylation endproducts in tissue and the biochemical basis of diabetic complications. *N Engl J Med* 318: 1315-1321.

Bucala, R., Vlassara, H., Cerami, A. (1991) Advanced glycosylation endproducts, in *Post-translational modification of proteins*, edited by Harding JJ, Crabbe MJC, Boca Raton, CRC Press, p.53.

Doi, T., Vlassara, H., Kirstein, M., Yamada, Y., Striker, G. E., and Striker, L. J. (1992) Receptor-specific increase in extracellular matrix production in mouse mesangial cells by advanced glycosylation endproducts is mediated via platelet-derived growth factor. *Proc Natl Acad Sci USA* 89: 2873-2877.

Esposito, C., Gerlach, H., Brett, J., Stern, D., and Vlassara, H. (1989) Endothelial receptor-mediated binding of glucose-modified albumin is associated with increased monolayer permeability and modulation of cell surface coagulant properties. *J Exp Med* 170: 1387-1407.

Friedman, E.A. (1992) Treatment options for diabetic nephropathy. *Diabetes Spectrum* 5:6-16.

Fuh, H., Yang, D., Striker, L., Striker, G., and Vlassara, H. (1992) *In vivo* AGE-Peptide Injection Induces Kidney Enlargement and Glomerular Hypertrophy in rabbits; Prevention by Aminoguanidine. *Diabetes* 41(1): 9A.

Hayase, F., Nagaraj, R.H., Miyata, S., Njoroge, F.G., and Monnier, V.M. (1989) Aging of proteins: immunological detection of a glucose derived pyrrole formed during Maillard reaction *in vivo*. *J Biol Chem* 264:3758-3764.

Imani, F., Horii, Y., Suthanthiran, M., Skolnik, E., Makita, Z., Sharma, V., Sehajpal, P., and Vlassara, H. (1993) Advanced Glycosylation Endproduct-Specific Receptors on Human and Rat T-Lymphocytes Mediate Synthesis of Interferon-γ: Role in Tissue Remodeling *J Exp Med* 178: 2165-2172.

Kirstein, M., Aston, C., Hintz, R., and Vlassara, H. (1992) Receptor-specific induction of insulin-like growth factor I (IGF-I) in human monocytes by advanced glycosylation endproduct-modified proteins. *J Clin Invest* 90: 439-446.

Kirstein, M., Brett, J., Radoff, S., Ogawa, S., Stern, D., and Vlassara, H. (1990) Advanced protein glycosylation induces transendothelial human monocyte chemotaxis and secretion of PDGF: role in vascular disease of diabetes and aging. *Proc Natl Acad Sci USA* 87(22): 9010-9014.

Kirstein, M., van Deventer, S., and Vlassara, H. (1990) Advanced glycosylation endproducts (AGE) binding to its specific receptor stimulates increase in EGF and EGF receptor mRNAs; role in tissue remodeling. *J Cell Biol* Suppl.14E 89: 2973-77.

Libby, P., and Hansson, G.K. (1991) Involvement of the immune system in human atherogenesis: Current knowledge and unanswered questions. *Lab Invest* 64: 5-15.

Makita, Z., Vlassara, H., Cerami, A., and Bucala, R. (1992) Immunochemical detection of advanced glycosylation endproducts *in vivo*. *J Biol Chem* 267: 5133-5138.

Makita, Z., Radoff, S., Rayfield, E.J., Yang, Z., Skolnik, E., Friedman, E.A., Cerami, A., and Vlassara H. (1991) Clinical assessment and significance of advanced glycation in patients with diabetic nephropathy. *New Engl J Med* 325: 836-842.

Makita, Z., Bucala, R., Rayfield, E., Friedman, E., Kaufman, A., Korbet, S.M., et al. (1993) Circulating "Toxic" Advanced Glycosylation End Products are Resistant to Dialysis Therapy: Role in Mortality of Uremia. (submitted).

Miyata, S., Monnier, V.M. (1992) Immunohistochemical detection of Advanced Glycosylation Endproducts in diabetic tissues using monoclonal antibody to pyrraline. *J Clin Invest* 89(4): 1102-12.

Neeper, M., Schmidt, A. M., Brett, J., Yan, S.D., Wang, F., Pan, Y.C.E, et al. (1992) Cloning and Expression of a Cell Surface Receptor for Advanced Glycosylation End Products of Proteins. *J Biol Chem* 267(21): 14998-15004.

Njoroge, F. G., Sayre, L. M., and Monnier, V. M. (1987) Detection of D-glucose derived pyrrole compounds during Maillard reaction under physiological conditions. *Carbohydrate Res* 167: 211.

Radoff, S., Cerami, A., and Vlassara, H. (1990) Isolation of surface binding protein specific for advanced glycosylation endproducts from mouse macrophage-derived cell line RAW 264.7. *Diabetes* 39: 1510-1518.

Radoff, S., Vlassara, H., and Cerami, A. (1988) Characterization of a solubilized cell surface binding protein on macrophages specific for protein modified nonenzymatically by advanced glycosylation endproducts. *Arch Biochem Biophys* 263: 418-423.

Schmidt, A. M., Vianna, M., Gerlach, M., Brett, J., Ryan, J., and Kao, J. (1992) Isolation and Characterization of Two Binding Proteins for Advanced Glycosylation End Products from Bovine Lung Which Are Present on the Endothelial Cell Surface. *J Biol Chem* 256(21): 14987-14997.

Skolnik, E. Y., Yang, Z., Makita, Z., Radoff, S., Kirstein, M., and Vlassara, H. (1991) Human and rat mesangial cell receptors for glucose-modified proteins: Potential role in kidney tissue remodeling and diabetic nephropathy. *J Exp Med* 174: 931-939.

Stemme, S., Rymo, L., Hansson, G.K. (1991) Polyclonal Origin of T Lymphocytes in Human Atherosclerotic Plaques. *Lab Invest* 65(6): 654-660.

United States Renal Data System: USRDS 1992 Annual Data Report. (August 1992) The National Institutes of Health, National Institute of Diabetes and Digestive and Kidney Disease. Bethesda, MD.

Vlassara, H., Fuh, H., Makita, Z., Krungkrai, S., Cerami, A., and Bucala, R. (1992) Exogenous Advanced Glycosylation Endproducts Induce Complex Vascular Dysfunction In Normal Animals;A Model For Diabetic and Aging Complications. *Proc Natl Acad Sci* 89: 12043-12047.

Vlassara, H., Moldawer, L., and Chan B. (1989) Macrophage/monocyte receptor for non-enzymatically glycosylated proteins is up-regulated by cachectin/tumor necrosis factor. *J Clin Invest* 84: 1813-1820.

Vlassara, H., Brownlee, M., Manogue, K.R., Dinarello, C., and Pasagian, A. (1988) Cachectin/TNF and IL-1 induced by glucose modified proteins: role in normal tissue remodeling. *Science* 240: 1546-1548.

Vlassara, H., Bucala, R., Striker, L. (1993) Pathogenic Effects of Advanced Glycosylation; Biochemical, Biological, and Clinical Implications for Diabetes and Aging. *J Lab Invest* (in press).

Vlassara, H., Brownlee, M., and Cerami A. (1988) Specific macrophage receptor activity for advanced glycosylation endproducts inversely correlates with insulin levels *in vivo*. *Diabetes* 37(4): 456-461.

Vlassara, H., Brownlee, M., and Cerami, A. (1986) Novel macrophage receptor for glucose-modified protein is distinct from previously described scavenger receptors. *J Exp Med* 164: 1301-1309.

Vlassara, H., Brownlee, M., and Cerami, A. (1985) High-affinity receptor-mediated uptake and degradation of glucose-modified proteins: A potential mechanism for the removal of senescent macromolecules. *Proc Natl Acad Sci USA* 82: 5588-5592.

Yang, Z., Makita, Z., Horii Y., Brunelle. S., Cerami, A., Sehajpal, P., et al. (1991) Two novel rat liver membrane proteins that bind Advanced Glycosylation Endproducts: Relationship tomacrophage receptor for glucose-modified proteins. *J Exp Med* 174: 515-524.

Cellular Receptors for Advanced Glycation Endproducts

A. M. Schmidt and D. M. Stern

DEPARTMENTS OF MEDICINE AND PHYSIOLOGY, COLUMBIA UNIVERSITY, COLLEGE OF PHYSICIANS AND SURGEONS, 630 WEST 168TH STREET, NEW YORK, NY 10032, USA

Summary
Advanced glycation endproducts of proteins/lipids (AGEs) which form as the result of nonenzymatic glycation/oxidation, are present in the plasma and tissues in aging, and accumulate more rapidly in diabetes. The interaction of AGEs with cellular elements, such as endothelial cells and mononuclear phagocytes leads to cellular dysfunction which could underlie the development of complications, such as accelerated atherosclerosis in diabetes. To gain insights into the cellular effects of AGEs, we have isolated and characterized two cell surface-associated polypeptides which appear to mediate the interaction with AGEs: Receptor for AGE (RAGE) is a new member of the immunoglobulin superfamily of cell surface molecules, and the Lactoferrin-like Polypeptide (LF-L) is closely related/identical to milk-derived lactoferrin. RAGE and LF-L associate to form a noncovalent complex on endothelial cells and mononuclear phagocytes which mediates the interaction of these cells with AGEs. For example, soluble AGE-bearing proteins induce monocyte migration, which can be blocked by antibody to either RAGE or LF-L, or by addition of soluble RAGE. Future studies using molecular probes for RAGE/LF-L should allow the definition of their contribution(s) to the development of organ dysfunction/complications in diabetes.

Introduction

Exposure of proteins and lipid to aldose sugars leads to nonenzymatic glycation and oxidation (Baynes, 1991; Brownlee et al., 1992; Bucala et al., 1993; Ruderman et al., 1992; Sell and Monnier, 1989). The result of these interactions is termed advanced glycation endproducts (AGEs), and heterogeneous class of glycated structures which interact with cellular binding proteins (Brownlee et al., 1992). The presence and accumulation of AGEs in plasma, on the surface of diabetic red cells, and in the tissues during normal aging is accelerated in patients with diabetes (Baynes, 1991; Brownlee et al., 1992; Makita et al., 1991; Sell and Monnier, 1989; Wautier et al., 1992). AGEs have been speculated to alter elasticity of tissues, and their presence in the basement membrane promotes trapping of plasma constituents, such as lipoproteins (Baynes, 1991; Brownlee et al., Sell and Monnier, 1989). An important means through which AGEs interact and modify cellular properties is through specific cell surface binding sites. Our laboratory has characterized these binding sites on endothelial cells and mononuclear phagocytes (Neeper et al., 1992; Schmidt et al., 1992; Schmidt et al., 1993a), though the results are likely to be applicable to a spectrum of other cell types.

Results

Effect of AGEs on endothelial cell functions
AGEs are present in the vessel wall, indicating the potential importance of understanding their effects on endothelial cell functions regulating vascular homeostasis. Incubation of

AGE albumin with cultured endothelial cells led to a decrease in the expression of anticoagulant cofactor thrombomodulin (Esposito et al., 1989). AGE albumin induced a time-dependent decline in thrombomodulin activity, which was decreased by almost 80% after 8 hrs, whereas native albumin did not alter thrombomodulin activity. Although mechanism through which AGEs suppress thrombomodulin have not been clarified, diminishing the efficacy of vessel wall antithrombotic pathways provides a basis for a potentially prothrombotic state. In addition to suppression of an anticoagulant cofactor, AGE albumin induced expression of the procoagulant cofactor tissue factor (Esposito et al., 1989). AGE-mediated induction of endothelial cell tissue factor was dependent on the incubation time, with increasing tissue factor activity evident up to 60 hrs. The effect of AGE albumin on tissue factor appeared to be half-maximal by ≈100 nM, and reached an apparent maximum at higher concentrations of ligand. These data suggested that AGEs might be interacting with specific binding sites, the latter mediating induction of endothelial procoagulant activity and further promoting a prothrombotic diathesis. AGEs also modulate a spectrum of other endothelial properties, including increased permeability of cell monolayers, cell growth, as well as modulation of mononuclear phagocyte interactions (Esposito et al., 1989; Kirstein et al., 1990).

Purification and characterization of cellular binding proteins for AGEs

Radioligand binding studies with AGE albumin demonstrated that AGE-endothelial interaction was mediated by specific binding sites (Esposito et al., 1989; Schmidt et al., 1992). Endothelial cell binding of ^{125}I-AGE albumin was not competed by either native albumin, glucose, or ribose, alone, and was selective for the advanced products, as shorter incubations of glucose with albumin produced material which did not inhibit the interaction of AGE albumin with the cell surface. Brief exposure of cultured endothelial cells to trypsin blocked the subsequent binding of AGEs, as did extraction of cultures with the detergent octyl-β-glucoside. These data suggested that AGE binding sites could be comprised of trypsin sensitive, detergent-extractable polypeptide(s). Cellular AGE binding proteins were purified to homogeneity, and consisted of ≈35 kDa, ≈80 kDa and ≈20 kDa polypeptides (Schmidt et al., 1992). Each of these polypeptides bound AGEs in a saturable manner, blocked by antibody to the respective AGE binding protein. Sequence analysis indicated that the ≈35 kDa polypeptide was novel, whereas the ≈80 kDa polypeptide was identical to lactoferrin and the ≈20 kDa polypeptide was identical to bovine high mobility group 1 protein (Schmidt et al., 1992).

Molecular cloning studies, leading to the isolation of a full-length cDNA, indicated that the ≈35 kDa AGE binding protein was part of a 404 amino acid mature polypeptide (Neeper et al., 1992). Analysis of the sequence placed the ≈35 kDa polypeptide in the immunoglobulin superfamily based on an extracellular domain comprised of three "immunoglobulin-like" regions: an N-terminal "V-type" followed by two "C-type" domains. There is also a single putative transmembrane spanning domain followed by a short, highly-charged cytosolic tail. Transfection of RAGE into 293 cells enabled these cells to bind ^{125}I-AGE albumin in a dose-dependent manner comparable to endothelial cells and monocytes (Neeper et al., 1992). These findings have led us to assign the name Receptor for AGE (RAGE) to the ≈35 kDa AGE binding protein.

It was important to determine if RAGE was present on endothelial cells and mononuclear phagocytes. Immunofluorescence studies demonstrated RAGE on endothelium derived from large and small vessels, as well as mononuclear phagocytes (Schmidt et al., 1992; Schmidt et al., 1993a, Brett et al., 1993). The central role of

RAGE for binding AGE ligands was shown by inhibition of ^{125}I-AGE albumin interaction with endothelial cells and monocytes in the presence of anti-RAGE IgG (Schmidt et al., 1992; Schmidt et al., 1993a). Thus, although we had initially predicted that AGE ligands would interact with multiple cellular determinants, it appeared, based on studies with AGE albumin and other AGE ligands, that RAGE was a central component of the cellular acceptor site.

The ≈80 kDa AGE binding protein was identical to lactoferrin based on amino terminal sequence analysis, immunoreactivity, and the observation that lactoferrin purified from milk bound AGEs similarly to the material purified from bovine lung (Schmidt et al., 1992). Thus, the ≈80 kDa AGE binding protein was termed the lactoferrin-like polypeptide (LF-L). Immunoblotting of endothelial and monocyte extracts with anti-LF-L IgG showed a band with Mr ≈30 kDa, rather than the expected 80 kDa (Schmidt et al., 1992; Schmidt et al., 1993a). Furthermore, anti-LF-L IgG blocked the binding of ^{125}I-AGE albumin to cultured endothelial cells and monocytes, and microscopic studies co-localized the two molecules on the cell surface. In vitro studies have confirmed that RAGE and LF-L (as well as LF) bind with high affinity, and are associated on the cell surface (Schmidt et al., 1993b).

Effects of AGEs on mononuclear phagocytes: the central role of RAGE/LF-L

Early studies of the cellular interaction of AGEs examined their potential processing by mononuclear phagocytes based on the hypothesis that monocytes would scavenge AGEs deposited in the tissues (Brownlee et al., 1988). Using our reagents to detect AGE binding proteins, we wanted to determine if RAGE/LF-L were important in mediating interaction of these glycated proteins with mononuclear phagocytes. In previous studies, we had shown that soluble AGEs, such as AGE albumin, were chemotactic for mononuclear phagocytes (Kirstein et al., 1990). AGE-induced cell migration was due to true chemotaxis. Addition to antibodies to either RAGE or LF-L blocked AGE-induced monocyte migration (Schmidt et al., 1993a), as did the presence of a soluble truncated form of the extracellular portion of RAGE. These data suggested that both RAGE and LF-L were involved in mediation of the monocyte response to AGEs. In contrast to these results with soluble AGEs, when AGEs were immobilized on a support they slowed monocyte migration (Schmidt et al., 1993a). This was due to immobilized AGE ligands engaging RAGE and LF-L on the monocyte surface, as blocking antibodies to either receptor polypeptide or to AGEs resulted in accelerated migration of monocytes (to levels observed in the absence of immobilized AGEs) (Schmidt et al., 1993a). Taken together, these data lead to the hypothesis that soluble AGEs promote mononuclear phagocyte chemotaxis, whereas immobilized AGEs slow migration. Thus, monocytes would migrate down a concentration gradient to a site of immobilized AGEs in the tissues, coming to rest on a locus of AGEs. The interaction of AGEs with RAGE and LF-L on the monocyte surface appears to mediate the effect of AGEs on cell movement. The consequences of monocyte interaction with AGEs are multiple and include induction of cellular oxidant stress, activation of the transcription factor NF-kB, and induction of growth factors and cytokines (Kirstein et al., 1990; Yan et al., 1993; unpublished observations). Whether these effects of AGEs on monocytes actually contribute to vasculopathy will depend on the concentration and locus of AGEs which interact with RAGE/LF-L in different pathologic states.

Discussion

A central means by which AGEs contribute to normal aging and disease processes includes their interaction with cellular elements. Glycation can mask determinants on molecules which mediate their cellular interactions. In addition, nonenzymatic glycation and oxidation can result in the formation of new epitopes which recognize novel cell surface structures, such as the AGE binding proteins RAGE and LF-L. Although we initially hypothesized that AGE ligands would have multiple cell surface recognition sites, based on studies with AGE albumin, AGE immunoglobulin, AGE prothrombin, AGE collagen, or AGEs immunoisolated from diabetic patient plasma, two cell surface polypeptides have an integral role. RAGE, a novel integral membrane protein, is a new member of the immunoglobulin superfamily of cell surface molecules.

Studies on mononuclear phagocytes and endothelial cells suggest that the AGE binding site consists of a complex of RAGE with LF-L. Such a complex could serve to enhance the affinity of AGEs for the cell surface or to allow the receptor to recognize a greater diversity of AGE-like structures. The existence of RAGE-LF-L complex raises questions as to how it forms, and by what mechanism cell surface LF-L becomes processed to the ≈30 kDa formed observed in Western blotting of detergent extracts from cultured cells. In this context, Yang et al (1991) have reported the identification from liver of two polypeptides, p60 and p90, which bind AGEs. The amino terminal sequences of these proteins and RAGE/LF-L are distinct, although the apparent migration on gels is similar. p60 and p90 appear to be present on monocytes, though their relation to RAGE/LF-L is unclear.

Conclusions

The studies described herein have identified and characterized two cell surface polypeptides which have a central role in the interaction of AGEs with endothelial cells and mononuclear phagocytes. RAGE is a new member of the immunoglobulin superfamily bearing a single transmembrane spanning domain. LF-L, which is noncovalently associated with RAGE, is closely related to lactoferrin. The binding of AGEs to endothelial cells and mononuclear phagocytes is mediated by RAGE/LF-L, as is AGE-mediated induction of monocyte chemotaxis. Studies with probes to detect and to block RAGE/LF-L should define their biology and should delineate consequences of AGE engagement of these binding proteins for mediating the myriad of cellular effects of AGE-bearing proteins/lipids. Taken together, these studies suggest that AGEs, via their interaction with cellular receptors, can modulate properties of endothelium and mononuclear phagocytes setting the stage for the accelerated development of atherosclerosis and other vascular complications.

Acknowledgments
This work was supported by grants from the USPHS (AG00602, HL42833, HL42507, HL21006), the Council for Tobacco Research, Juvenile Diabetes Foundation, American Heart Association (NY).
Inquiries should be addressed to Dr. Ann-Marie Schmidt, Columbia University, Department of Physiology, 630 West 168th Street, New York, NY 10032.

References

Baynes, J. (1991) Role of Glycation in Modification of Lens Crystallins in Diabetic and Nondiabetic Cataracts. *Diabetes* 40:405-412.

Brett, J., Schmidt, A-M., Yan, S-D., Weidman, E., Pinsky, D., Neeper, M., Przysiecki, C., Shaw, A., Migheli, A., and Stern, D. (1993) Tissue Distribution of RAGE. *Am. J. Pathol.* In press.

Brownlee, M., Cerami, A., and Vlassara, H. (1988) Advanced Glycosylation Endproducts in Tissue and the Biochemical Basis of Diabetic Complications. *N. Engl. J. Med.* 318:1315-1320.

Bucala, R., Makita, Z., Koschinsky, T., Cerami, A., Vlassara, H. (1993) Lipid Advanced Glycosylation: Pathway for Lipid Oxidation In Vivo. *PNAS* (USA) 90:6434-6438.

Esposito, C., Gerlach, H., Brett, J., Stern, D., and Vlassara, H. (1989) Endothelial Receptor-Mediated Binding of AGEs is Associated with Increased Monolayer Permeability and Modulation of Cell Surface Coagulant Properties. *J. Exp. Med.* 170:1387-1407.

Kirstein, M., Brett, J., Radoff, S., Stern, D., and Vlassara, H. (1990) AGEs Induce Monocyte Migration Across Endothelium, and Elaboration of Growth Factors. *PNAS* (USA) 87:9010-9014.

Makita, Z., Radoff, S. Rayfield, E., Yang, Z., Skolnik, E., Delaney, V., Friedman, E., Cerami, A., and Vlassara, H. (1991) AGEs in Patients with Diabetic Nephropathy. *N. Engl. J. Med. 325:836*-842.

Neeper, M., Schmidt, A-M., Brett, J., Yan, S-D., Wang, F., Pan, Y-C., Elliston, K., Stern, D., and Shaw, A. (1992) Cloning and Expression of RAGE. *J. Biol. Chem.* 267:14998-15004.

Ruderman, N., Williamson, J., and Brownlee, M. (1992) Glucose and Diabetic Vascular Disease. *FASEB J.* 6:2905-2914.

Schmidt, A-M., Vianna, M., Gerlach, M., Brett, J., Ryan, J., Kao, J., Esposito, C., Hegarty, H., Hurley, W., Clauss, M., Wang, F., Pan, Y-C., Tsang, T.C., and Stern, D. (1992) Isolation and Characterization of Binding Proteins for AGEs. *J. Biol. Chem.* 267:14987-14997.

Schmidt, A-M., Yan, S-D., Brett, J., Mora, R., Nowygrod, R., and Stern, D. (1993a) Regulation of Monocyte Migration by Binding Proteins for AGEs. *J. Clin. Invest.* 92:2155-2168.

Schmidt, A-M., Mora, R., Yan, S-D., Brett, J. Tsang, T-C., Simionescu, M., and Stern, D. (1993b) The EC Binding Site for AGEs Consists of a Complex of RAGE and LF-L. In revision, *JBC.*

Sell, D., and Monnier, V. (1989) Structure Elucidation of a Senescence Cross-Link from Human Extracellular Matrix. *J. Biol. Chem.* 264:21597-21602.

Yan, S-D., Schmidt, A-M., Anderson, G., Zhang, J., Brett, J., Zou, Y-S., and Stern, D. Enhanced Cellular Oxidant Stress by the Interaction of Ages with RAGE/FL-L. In revision, *JBC.*

Yang, Z., Makita, Z., Hori, Y., Brunelle, S., Cerami, A., Suthanthiran, M., and Vlassara, H. (1991) Two Novel Rat Liver Membrane Proteins that Bind AGEs: Relationship to Macrophage Receptor for Glucose-Modified Proteins. *J. Exp. Med.* 174:515-524.

Glycation, Oxidation, and Glycoxidation of Short- and Long-lived Proteins and the Pathogenesis of Diabetic Complications

Timothy J. Lyons[1] and Ralph H. Johnson[2]

[1]DIVISION OF ENDOCRINOLOGY, DIABETES AND METABOLISM, MEDICAL UNIVERSITY OF SOUTH CAROLINA, CHARLESTON, SC 29425, USA;
[2]DEPARTMENT OF VETERANS AFFAIRS MEDICAL CENTER, CHARLESTON, SC 29401, USA

Summary
Evidence is summarized to support the hypothesis that combinations of glycation and oxidation (glycoxidation) result in damage to proteins which contribute to the development of diabetic complications. This implies that oxidative stress, which varies among individuals independent of the presence of diabetes, may modulate the consequences of hyperglycemia, explaining, in part, individual variation in susceptibility to the complications of diabetes.

Introduction

The recently presented results of the Diabetes Control and Complications Trial (DCCT Research Group, 1993) prove conclusively what has long been suspected (Pirart, 1978): that good glycemic control in Type 1 diabetic patients can delay the onset and slow the progression of the microvascular (and perhaps also the macrovascular) complications of diabetes. An attractive explanation for this link between glycemic control and the development of complications is the "glycation hypothesis", which states that enhanced modification of proteins by glucose leads to abnormalities, and that over time these are sufficient to cause disease (Kennedy and Baynes, 1984). Clearly however, neither increased glycation, nor indeed the degree of long-term hyperglycemia, can be the only factor determining the development of diabetic complications. Some patients, despite reasonable glycemic control, develop severe complications at an unusually early stage; others seem resistant to the development of complications despite poor long-term glycemic control. Recent studies by our group have led us to propose an extension of the glycation hypothesis, the "glycoxidant hypothesis states that significant functional abnormalities of proteins are determined not only by glycation, but also, and perhaps to a greater extent, by subsequent oxidative damage. We have shown that free radical oxidation reactions can "fix" glycative damage, leading to the irreversible formation of "glycoxidation products" (Fu et al., 1992; Baynes, 1991). Indeed the formation of advanced glycation end-products appears to be dependent on the presence of an oxidative environment. Studies which we have performed on proteins as different as low density lipoprotein (LDL) (a short-lived circulating species) (Lyons et al., 1992a) and insoluble skin collagen (a long-lived structural protein) (Dyer et al., 1993; McCance et al., 1993) suggest that glycoxidation products, or perhaps the rate of their formation, may be important determinants for the development of diabetic complications. The glycoxidation hypothesis implies that the consequences of hyperglycemia may be modulated by the capacity to resist oxidative stress, which varies among individuals whether or not they are diabetic.

Early and Late Glycation

The chemistry of glycation, which maybe considered as a two stage process, has recently been reviewed (Lyons et al., 1992b) , and is summarized in Figure 1. In this initial stage, glucose (in its open-chain form) reacts with susceptible amino-groups on proteins to form an unstable Schiff base. The Schiff base may either dissociate or undergo an Amadori rearrangement to yield fructose-lysine (FL). Although FL is relatively stable, it has recently been shown that it too may decay, releasing the

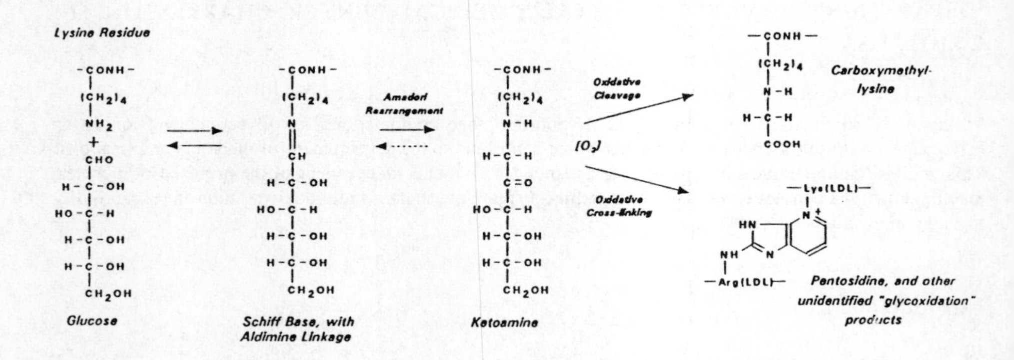

Figure 1. Early glycation results in the formation of the ketoamine, fructose-lysine. Late glycation generates many products of which only carboxymethyllysine and pentosidine have been characterized.

carbohydrate moiety either as glucose or as more reactive hexoses such as glucosone (Kawakishi et al., 1991). Alternatively, it may partake directly in the second stage of glycation, as discussed below. In diabetes, metabolic alterations directly attributable to increase FL (as opposed to the later products) are perhaps most likely to predominate in short-lived proteins, since these do not exist long enough to accumulate high levels of glycation end-products.

The second stage of glycation is vastly more complex: a series of reactions involving FL or its dissociation products leads to the formation of a profusion of stable end-products ("browning products"), "Maillard Reaction products", or "Advanced Glycosylation End-products" (AGE). Several recent review articles have addressed the browning process, which is still only partially understood (Ledl and Schleicher, 1990; Monnier et al., 1992). As mentioned, we have recently shown that the formation or most, if not all, of these products involves free radical-mediated oxidation, and thus the term "glycoxidation products" has been introduced (Fu et al., 1992; Baynes, 1991). This observation is central to the development of the glycoxidation hypothesis. The numerous final products of late glycation include species which are colored, fluorescent, and which constitute cross-links. However, only two of these final products have been identified conclusively: N^{ε}-carboxymethyllysine (CML) (Ahmed et al., 1986) (and the closely related species N^{ε}-carboxymethylhydroxylysine and 3-(N^{ε}-lysino)-lactic acid), and pentosidine (Sell and Monnier, 1989). CML is a product of free radical-mediated oxidative cleavage of FL, while pentosidine is a lysine-arginine cross-link which may form between or within protein molecules.

Glycation, Oxidation and Glycoxidation of a Short-Lived Species: Low Density Lipoprotein

Lipoproteins are of particular interest in the investigation of the glycoxidation hypothesis because their lipid cores are vulnerable to oxidative damage, and because of the close physical association of their apoprotein and lipid constituents. They can therefore serve as models to investigate interactions between protein glycation and lipid peroxidation, and conversely between lipid peroxidation and the generation of glycoxidation products in proteins. There is already a considerable body of evidence implicating oxidized lipoproteins in atherogenesis (Steinberg et al., 1989). In diabetes, studies by ourselves and others suggest that modification of lipoproteins may contribute to accelerated atherosclerosis, and in recent studies we suggest that they may also be implicated in micro-vascular disease (Lyons et al., 1992a). Most investigators have concentrated on LDL: having only one apolipoprotein it is the simplest lipoprotein to study, and is also the most clearly associated with vascular disease.

The consequences of glycation and oxidation of lipoproteins occurring in diabetes have recently been reviewed in some detail (Lyons, 1992c; Lyons, 1991a). Here, a recent study (Lyons et al., 1992a) in which we investigated the possibility that modified lipoproteins may play a role in the development of diabetic retinopathy will be described. We utilized normal human LDL modified *in vitro* to yield mildly glycated, oxidized and glycoxidized LDL. We aimed to simulate the extent of modification occurring *in vivo* in diabetic patients. Mild modification was confirmed by measures of the FL and CML contents of apolipoprotein B, determination of levels of thiobarbituric acid-reacting substances (TBARS), and by the non-recognition of the particles by the human macrophage scavenger receptor. We assessed the effect of the LDL on the viability of cultured bovine retinal capillary cells (endothelial cells and pericytes). Retinal capillary endothelial cells are unusual in that they form tight junctions with one another, constituting the inner blood-retinal barrier which controls the entry of plasma constituents to the retina. In diabetes, the inner blood-retinal barrier becomes leaky (implying endothelial cell injury), resulting in uncontrolled extravasation of plasma. Retinal pericytes are also unusual in that they are much more numerous than those in other capillary beds. They are contractile, responding to endothelin released by endothelial cells, and they regulate retinal capillary flow. Disordered retinal blood flow and pericyte loss are both characteristic features of diabetic retinopathy.

We show that the mildly modified LDL preparations, at physiologically relevant concentrations, were toxic to both types of retinal capillary cell, significantly decreasing total and live cell numbers and total cell protein after either two or three days' exposure. Mild glycation, oxidation, and glycoxidation of LDL, in that order, conferred progressively increased cytotoxicity compared with control (unmodified) LDL. To our knowledge, this is the first study to demonstrate with other cell types and was observed with the retinal cells, we found that mild glycoxidation rendered the particles significantly more toxic than oxidation alone. We believe this to be the first study to show a potential link between qualitative abnormalities of LDL and the development of a microvascular complication of diabetes.

We have demonstrated other effects of glycoxidized LDL which may be relevant to the development of diabetic vascular disease, including retinopathy. We found that LDL isolated from diabetic patients, and also normal LDL oxidized *in vitro*, failed to stimulate the expected release of the fibrinolytic protein, tissue Plasminogen Activator, by cultured human umbilical vein endothelial cells (Jokl et al., 1992a). Furthermore, glycoxidized LDL increased the release of Plasminogen Activator Inhibitor 1 by retinal capillary endothelial cells (Jokl et al., 1992b). Enhanced intraretinal thrombosis is a feature of diabetic retinopathy, and these data suggest that modified LDL may contribute to this phenomenon by shifting the prothrombotic/fibrinolytic balance in favor of thrombosis.

The results of these studies are consistent with the glycoxidation hypothesis. Diabetic patients with combinations of high glycative and high oxidative stress are likely to have increased modification of plasma lipoproteins. Over time, gradual damage to the retinal capillary endothelium may lead to breakdown of the inner blood retinal barrier and capillary leakage, and also enhanced intra-capillary thrombosis. Extravasated LDL, whose sequestration in vessel walls or in the retinal itself may be enhanced by the formation of glucose-mediated cross-links, may undergo more extensive modification leading to pericyte loss, further endothelial damage, and perhaps direct damage to retinal nerve cells.

Glycation and Glycoxidation of a Long-Lived Species: Insoluble Collagen

With advancing age, insoluble collagen undergoes marked physico-chemical changes including decreased solubility, elasticity, and sensitivity to protease digestion, and increased thermal stability (reviewed by Lyons, 1993). These changes are accelerated in diabetic patients, in whom collagen behaves as though prematurely aged. It is of interest that two of the theories to explain the normal aging process are the "glucose theory" (Monnier, 1981) and the "free radical oxidation" theory (Harman, 1981), and clearly the glycoxidation hypothesis of diabetic complications is a closely related concept. Apart from providing a potential long-term indicator of glycoxidation stress, collagen is of special interest because of its ubiquity. Any alterations in its physico-chemical properties may have widespread consequences. Also, the nature of glycoxidative damage means that collagen obtained from convenient sites, e.g. skin, may reflect changes occurring in other sites which are more difficult to access, e.g. retina and kidney.

The identification of the specific glycoxidation products has allowed us to investigate the role of glycoxidation of collagen in aging and in diabetes with greater precision. In *in vitro* studies we have observed that all the measures of collagen browning, including the accumulation of the specific glycoxidation products CML and pentosidine, depend not only on the presence of glucose but also on oxidant conditions (Fu et al., 1992). We have also investigated modifications of collagen occurring *in vivo* in diabetes and aging. We showed that FL in skin collagen increases only slightly with age in non-diabetic individuals, and although elevated in diabetes, it is unrelated to age or duration of diabetes (Dyer, 1993). These findings are consistent with the conclusion that collagen FL levels provide an index of relatively recent glycemic control, even though collagen is a very long-lived protein. In contrast, with advancing age in non-diabetic individuals, we found that collagen CML, pentosidine, and fluorescence all increased in a linear fashion between the ages of 0 and 80 years. The rate of accumulation varied among individuals, some accumulating the products more rapidly than others. Since glycemia in non-diabetic subjects is effectively constant, we attributed the variations in the rate of accumulation of glycoxidation products to variations in long-term oxidative stress. In diabetic patients, levels of collagen glycoxidation products were generally increased compared with age-matched non-diabetic subjects. We made an attempt to determine whether these increased levels should be attributed to a general increase in oxidative stress in diabetes, or whether the increase in glycative stress was a sufficient cause. Accounting for age, duration of diabetes, and long-term glycemic control, we concluded that the increased accumulation of glycoxidation products in diabetic skin collagen can be explained, on average, solely by the increase in glycation: we found no evidence of a general increase in oxidative stress. However, we found that age-corrected collagen FL, CML, pentosidine and fluorescence all increased progressively with increasing severity of diabetic retinopathy (McCance et al., 1993) and nephropathy. Controlling for age, sex and duration of diabetes by logistic regression, FL, CML and fluorescence were each found to be independently associated with retinopathy, and FL CML and pentosidine were independently associated with early nephropathy. When we examined data from patients with long-duration diabetes, there was a general tendency for those under higher oxidative stress, as defined by higher than expected levels of CML (corrected for age, duration of diabetes and glycemic control), to have more severe retinopathy than those with low oxidative stress. This observation is consistent with the glycoxidation hypothesis, but has yet to be proven valid. If proven, the fact that the accumulation of glycoxidation products in collagen occurs only very gradually over periods of years may lend strength to the conclusion that increased oxidative stress may have contributed to the development of retinopathy, rather than being a result of it. It is possible that glycoxidation of collagen itself may mediate the development of complications. Alternatively, or in part, glycoxidation of collagen, although associated with the presence and severity of diabetic complications, may simply reflect long-term glycoxidative stress, while the actual damage may have been mediated by glycoxidation of other, short lived substrates, e.g. lipoproteins. At present, attribution of cause and effect from these observations is impossible.

Finally, we investigated the effect of a period of "tight" glycemic control in Type 1 diabetic patients on early and late products of glycation. We found that while levels of collagen FL could be reduced significantly by improving glycemic control, there was no effect on levels of the late products of glycation (total fluorescence, CML and pentosidine) (Lyons, 1991b). This implies that once glycative damage is "fixed" by oxidation, permanent, irreversible damage to collagen may result. Such damage could manifest as disease at a later date despite a subsequent improvement in glycemic control.

Conclusions

Evidence has been summarized that glycoxidation of both short- and long-lived proteins may contribute to both macro- and micro-vascular disease in diabetes. Levels of glycoxidation products in collagen may provide a very-long-term index of cumulative glycoxidative stress: the pathogenic effects of this stress may be mediated through damage to a variety of substrates, including collagen itself, and also short-lived species. Among these short-lived substrates, lipoproteins may be particularly significant because of their susceptibility to oxidation. In the glycoxidation hypothesis, we suggest that variations in oxidative stress/antioxidant defenses, which occur among normal individuals, may determine their susceptibility to complications should they develop diabetes.

References

Ahmed, M.U., Thorpe, S.R., and Baynes, J.W. (1986) Identification of Carboxymethyllysine as a Degradation Product of Fructose-Lysine in Glycosylated Protein. *J. Biol. Chem.* 261:4889-4994.

Baynes, J.W. (1991) Role of Oxidative Stress in Development of Complications in Diabetes. *Diabetes* 40:405-412.DCCT Research Group (1993) The Effect of Intensive Treatment of Diabetes on the Development and Progression of Long-Term Complications in Insulin-Dependent Diabetes Mellitus. *N. Engl.J. Med.* 329:977-986

Dyer, D.G., Dunn, J.A., Thorpe, S.R., Bailie, K.E., Lyons, T.J., McCance, D.R., and Baynes, J.W. (1993) Accumulation of Maillard Reaction Products in Skin Collagen in Diabetes and Aging. *J. Clin. Invest.* 91:2463-2469.

Fu, M-X., Knecht, K.J., Thorpe, S.R., and Baynes, J.W. (1992) Role of Oxygen in the Cross-Linking and Chemical Modification of Collagen by Glucose. *Diabetes* 41, Suppl. 2:42-48.Harman, D. (1981) The Aging Process. *Proc. Natl Acad. Sci.* 78:7124-7128.

Jokl, R., Klein, R.L., Colwell, J.A., and Lopes-Virella, M.F. (1992a) Low Density Lipoproteins Isolated from Diabetic Patients Alter the Release of Tissue Plasminogen Activator (tPA) by Cultured Human Endothelial Cells.*Diabetes* 41, Suppl. 1:113A.

Jokl, R., Li, W., Colwell, J.A., Klein, R.L., Lopes-Virella, M.F. and Lyons, T.J. (1992b) Glycoxidized LDL Modifies PAI-1 Release by Retinal Endothelial Cells. *Diabetologia* 35, Suppl. 1:A46.

Kawakishi, S., Tsunehiro, J., and Uchida, K. (1991) Autoxidative Degradation of Amadori Compounds in the Presence of Copper Ion. *Carbohyd. Res.* 211:167-171.

Kennedy, L., and Baynes, J.W. (1984) Non-Enzymatic Glycosylation and the Chronic Complications of Diabetes: An Overview. *Diabetologia* 26:93-98.

Ledl, F., and Schleicher, E. (1990) New Aspects of the Maillard Reaction in Foods and in the Human Body. *Angew Chem (Int. Ed. Engl.)* 29:565-594.

Lyons, T.J. (1991a) Oxidized Low Density Lipoproteins - A Role in the Pathogenesisof Atherosclerosis in Diabetes? *Diabetic Medicine* 8:411-419.

Lyons, T.J., Bailie, K., Dunn, J.A., Dyer, D.G., Thorpe, S.R., and Baynes, J.W. (1991b) Decrease in Skin Collagen Glycosylation with Improved Glycemic Control in Patients with Insulin-Dependent Diabetes Mellitus. *J. Clin. Invest.* 87:1910-1915.

Lyons, T.J., Li, W., and Jokl, R. (1992a) Toxicity of Low Density Lipoproteins to Retinal Capillary Pericytes: An Effect Enhanced by Lipoproteins Glycation and Oxidation. *Diabetes* 41, Suppl. 1:111A.

Lyons, T.J., Thorpe, S.R., and Baynes, J.W. (1992b) Glycation and Autoxidation of Proteins In Aging and Diabetes. In: *Hyperglycemia, Diabetes, and Vascular Disease.* Ruderman, N., Williamson, J., Brownlee, M. (eds.) Oxford University Press, New York, pp. 197-217.

Lyons, T.J. (1992c) Glycation of Low Density Lipoprotein and its Metabolic Consequences. *Diabetes* 41, Suppl. 67-73.

Lyons, T.J. (1993) Glycation and Oxidation: A Role in the Pathogenesis of Atherosclerosis. *Am. J. Cardiol.* 71:26B-31B.

McCance, D.R., Dyer, D.G., Dunn, J.A., Bailie, K.E., Thorpe, S.R., Baynes, J.W., and Lyons, T.J. (1993) Maillard Reaction Products and their Relation to Complications in Insulin Dependent Diabetes Mellitus. *J. Clin. Invest.* 91:2470-2478.

Monnier, V.M., and Cerami, A. (1981) Non-Enzymatic Browning *In Vivo*. Possible Process for Aging of Long-Lived Proteins. *Science* 211:491-493.

Monnier, V.M., Sell, D.R., Nagarai, R.H., Miyata, S., Grandhee, S., Odetti, P., Ibrahim, S.A. (1992) Maillard Reaction -Mediated Molecular Damage to Extracellular Matrix and other Tissue Proteins in Diabetes, Aging, and Uremia. *Diabetes* 41, Suppl. 2:36-41.

Pirart, J. (1978) Diabetes Mellitus and its Degenerative Complications: A Prospective Study of 4400 Patients Observed Between 1947 and 1973. *Diabetes Care* 1:168-188.

Sell, D.R., and Monnier, V.M. (1989) Structure Elucidation of a Senescence Cross-Link from Human Extracellular Matrix. Implication of Pentoses in the Aging Process. *J. Biol. Chem.* 264:21597- 21602.

Steinberg, D., Parthasarathy, S., Carew, T.E., Khoo, J.C., and Witztum, J.L. (1989) Beyond Cholesterol. *N. Engl. J. Med.* 320:915-924.

Pathways of the Maillard Reaction *In Vitro* and *In Vivo*

D. V. Zyzak,[1] K. J. Wells-Knecht,[1] J. A. Blackledge,[1] J. E. Litchfield,[1] M. C. Wells-Knecht,[1] M-X. Fu,[1] S. R. Thorpe,[1] M. S. Feather,[2] and J. W. Baynes[1,3]

[1]DEPARTMENT OF CHEMISTRY AND [3]SCHOOL OF MEDICINE, UNIVERSITY OF SOUTH CAROLINA, COLUMBIA, SC 29208, USA; [2]DEPARTMENT OF BIOCHEMISTRY, UNIVERSITY OF MISSOURI, COLUMBIA, MO 65211, USA

Summary
Oxygen is a catalyst of the Maillard reaction *in vitro* and a fixative of damage resulting from glycation of proteins *in vivo*. Oxidative pathways of the Maillard reaction are initiated both by autoxidative glycosylation and by oxidative fragmentation of Schiff base or Amadori adducts to protein, followed by continued reaction of sugar-derived oxidation products with protein. Work in our laboratory suggests that autoxidative glycosylation is the initial step in the Maillard reaction *in vitro* at physiological pH and temperature, and that the major dicarbonyl product formed in this reaction is glyoxal, rather than glucosone. At later stages in the reaction, rearrangement, hydrolysis and both oxidative and non-oxidative fragmentation reactions of Amadori adducts yield deoxyglucosones, pentoses, tetroses and smaller sugars which propagate the Maillard reaction. The kinetics of formation of glycoxidation and advanced glycosylation end-products in tissue proteins in diabetic animals suggests that autoxidation of glucose or other carbohydrates also contributes to the Maillard reaction *in vivo*.

Introduction

In previous original articles (Fu et al., 1992, 1993) and reviews (Dyer et al., 1991; Baynes, 1992) we have emphasized the role of oxidation in the Maillard reaction, particularly in the formation of glycoxidation products [N^{ε}-(carboxymethyl)lysine (CML) and pentosidine] and in the crosslinking of collagen by glucose. The oxidation chemistry may begin with autoxidation of glucose (autoxidative glycosylation: Wolff and Dean, 1987; Wolff et al., 1991), by autoxidative degradation of Schiff base adducts (Hayashi and Namiki, 1986), or at some point following the formation of Amadori adducts (Hodge, 1953). All of these processes may occur simultaneously, or different mechanisms may dominate the course of the reaction at various times and in different environments. In current work, we have characterized products formed during autoxidation of glucose, decomposition of Amadori compounds, and browning of collagen by glucose *in vitro* at physiological pH and temperature (Fu et al., 1992, 1993). These studies, summarized below, provide insight into mechanisms of the Maillard reaction *in vitro* and *in vivo*.

Autoxidation of Glucose

Dicarbonyl sugars, such as deoxyglucosones, are generally recognized as reactive intermediates in the Maillard reaction (Hodge, 1953). Wolff and colleagues (Wolff and Dean, 1987; Wolff et al., 1991) have proposed that the autoxidation of glucose to glucosone, followed by formation of a keto-imine adduct to protein, sets the stage for oxidative damage to proteins during the Maillard reaction (Figure 1). We have observed, however, that glucosone is not formed on oxidation of glucose in aqueous buffers at physiological pH and temperature, nor is there evidence of formation of ribulose, the characteristic pentulose product formed on degradation of glucosone (Zyzak et al., in preparation) (Figure 1). In addition, using aminoguanidine (Hirsch et al., 1991) and Girard-T reagent (Mitchel and Birnboim, 1977) as dicarbonyl traps, we have identified glyoxal as the major dicarbonyl product formed during metal-catalyzed autoxidation of glucose (Wells-Knecht, K.J. et al., in preparation).

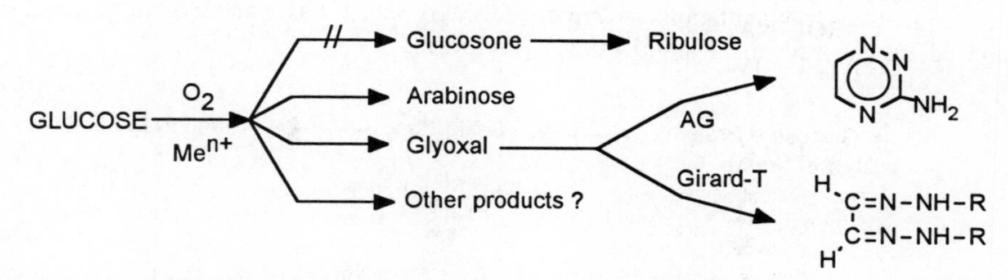

Figure 1. Pathways of autoxidation of glucose in phosphate buffer, pH 7.4, at 37°C. Glucose is oxidized to arabinose, glyoxal and unidentified split-products. Glyoxal can be trapped as either the aminoguanidine (AG) or Girard-T adduct. Glucosone is readily detected by these reagents, but neither glucosone nor ribulose, the degradation product of glucosone, is formed on autoxidation of glucose in phosphate buffer at physiological pH.

Preincubation of glucose under oxidative conditions in phosphate buffer also yields products which accelerate the glycoxidation and crosslinking of collagen, i.e., $k_2 >> k_1$ in Schemes I and II, below.

$$\text{(I) glucose} \xrightarrow[k_1]{[\text{protein} + O_2]} \text{glycoxidized/crosslinked protein}$$

$$\text{(II) glucose} \xrightarrow{[O_2]} \text{oxidation products} \xrightarrow[k_2]{[+ \text{protein}]} \text{glycoxidized/crosslinked protein}$$

During inhibition of the Maillard reaction by aminoguanidine, a glyoxal-aminoguanidine adduct, 3-amino-1,2,4-triazine, is a major reaction product. A triose-bis-aminoguanidine adduct, amino-guanidine triazine, is also formed during autoxidation of glucose in the presence of aminoguanidine (Blackledge et al., 1993), but its origin is uncertain. If the spontaneous autoxidation of glucose to glyoxal also occurs *in vivo*, this would support the proposed role for the glyoxalase system in limiting tissue damage from the Maillard reaction in diabetes and aging (Thornalley, 1990). Arabinose has also been detected as the five-carbon sugar formed on autoxidation of glucose, suggesting a route for formation of pentoses and pentosidine (Sell and Monnier, 1989) during reactions between glucose and protein under oxidative conditions.

Autoxidative Glycosylation

In studies on the chemical modification of rat tail collagen by glucose, we found that chelators, sulfhydryl compounds and antioxidants have negligible effect on glycation of the protein, but significantly inhibit the development of fluorescence, formation of glycoxidation products and crosslinking of collagen by glucose (Fu et al., 1992, 1993). The kinetics of these reactions indicate that the formation of glycoxidation products and crosslinks in this *in vitro* system proceeds either directly from autoxidation of glucose or by oxidative fragmentation of rapidly-formed Schiff base adducts to protein, as proposed by Hayashi and Namiki (1986) (Figure 2).

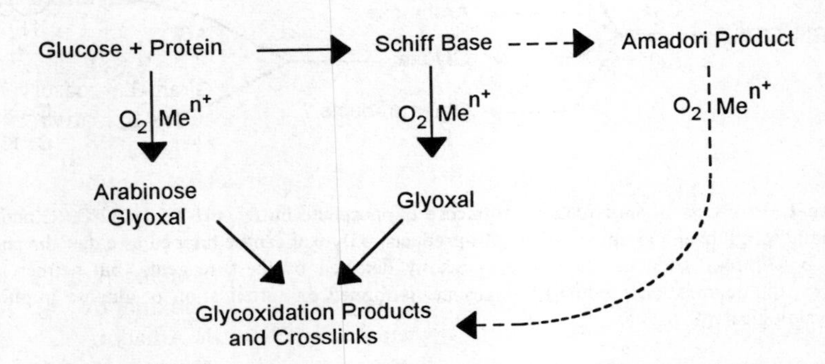

Figure 2. Alternate pathways for formation of glycoxidation products and crosslinks in protein during glycation of collagen under oxidative conditions *in vitro*. Arabinose and glyoxal are formed from glucose, or glyoxal from the Schiff Base adduct (Hayashi and Namiki, 1986). The Amadori product is formed more slowly and is probably a less important source of glycoxidation products and crosslinks in this reaction system (dotted line).

Amino groups on protein may also catalyze the autoxidation of glucose, but the Amadori product is formed too slowly to serve as an intermediate in glycoxidation and crosslinking reactions (Fu et al., 1993). Glyoxal may be a common intermediate formed either by autoxidation of glucose or of Schiff base adducts formed in the presence of protein (Hayashi and Namiki, 1986) (Figure 1). It reacts with both lysine and arginine residues in protein and is a potent protein crosslinking agent (Bowes and Cater, 1968). On reaction with lysine, it also forms the glycoxidation product, CML, probably by an intramolecular Cannizzaro reaction (Scheme III). Glyoxal may also serve as a precursor of other advanced glycosylation end-products (AGEs), glycoxidation products and crosslinks in protein.

$$\text{(III) Lysine-NH}_2 + \begin{matrix}\text{CHO}\\ |\\ \text{CHO}\end{matrix} \xrightarrow{-H_2O} \overset{+}{\text{Lys-NH=CH-CHO}} \xrightarrow{+H_2O} \overset{+}{\text{Lys-NH}_2}\text{-CH}_2\text{-COOH}$$

Role of Amadori Adducts in Formation of Glycoxidation Products and AGEs

Although the Amadori product may be a by-stander or side-product during glycoxidation and crosslinking of protein by glucose *in vitro*, it is more easily oxidized than glucose and readily participates in advanced Maillard reactions. Thus, collagen may be glycated under anti-oxidative conditions, such that Amadori adducts are formed, but do not proceed to formation of glycoxidation products, generation of fluorescence or crosslinking reactions. When this collagen is washed to remove free glucose and Schiff base adducts, then incubated under oxidative conditions in the absence of glucose, both CML and pentosidine are formed, accompanied by an increase in protein-bound fluorescence (IV). However the extent of these advanced Maillard

$$\text{(IV) protein} \xrightarrow{[+\text{ glucose, } N_2]} \text{glycated protein} \xrightarrow{[-\text{ glucose, } +O_2]} \text{glycoxidized protein}$$

reactions from pre-glycated collagen is only about 10% of that observed when the collagen is incubated at the same glucose concentration under oxidative conditions for the same time period. These experiments, in addition to kinetic data discussed above, suggest that autoxidative glycosylation, rather than oxidative degradation of Amadori adducts, is the primary route of the Maillard reaction *in vitro*. However, both processes are likely to occur simultaneously, and their relative rates may depend on the respective concentrations of glucose and Amadori compounds or on the concentration of metal ions or reducible substrates. Regardless of whether the glycoxidation and crosslinking reactions proceed from free glucose, or from Schiff base or Amadori adducts to protein, these reactions are accompanied by collateral oxidative damage to the protein as a result of production of reactive oxygen species (Wolff et al., 1991). We have, in fact, detected ortho-tyrosine in glycoxidized collagen, formed by reaction of hydroxyl radicals with phenylalanine residues in the protein. o-Tyrosine in glycoxidized collagen is readily quantified by gas chromatography - mass spectrometry (Wells-Knecht, M.C. et al., in preparation) and may serve as a useful indicator of the extent of damage to the protein backbone during glycoxidation reactions *in vivo*. Protein carbonyls are also formed by oxidation of other amino acids (Stadtman, 1993), but may be difficult to distinguish from products derived from reactions of carbohydrate with protein.

Decomposition of Amadori compounds is a complex process, yielding a number of products which may propagate the browning stages of the Maillard reaction (Figure 3). Ledl and Schleicher (1991) reviewed pathways leading to 1,3 and 4-deoxyglucosones by dehydration and hydrolysis reactions, and Kawakishi et al. (1991) have reported the formation of glucosone by oxidation of Amadori adducts by cupric ion. Both of these pathways may proceed at physiological pH and temperature. 3-Deoxyglucosone has, in fact, been detected in human blood and urine (Knecht et al. 1992), although it may be derived from fructose-3-phosphate (Szwergold et al., 1992), as well as from Amadori compounds. In recent work (Zyzak et al., submitted) we have shown that the model Amadori compound, N^{α}-formyl-N^{ε}-fructose (fFL), decomposes under both oxidative and anti-oxidative conditions to yield 3-deoxyglucosone and a range of pentose and tetrose sugars, including aldoses, alduloses and aldosuloses (Figure 3). Similar products were formed on decomposition of Amadori adducts on collagen under anti-oxidative conditions, suggesting that Amadori adducts may contribute to Maillard reactions *in vivo* where oxidative reactions are suppressed by powerful metal chelation and antioxidant systems.

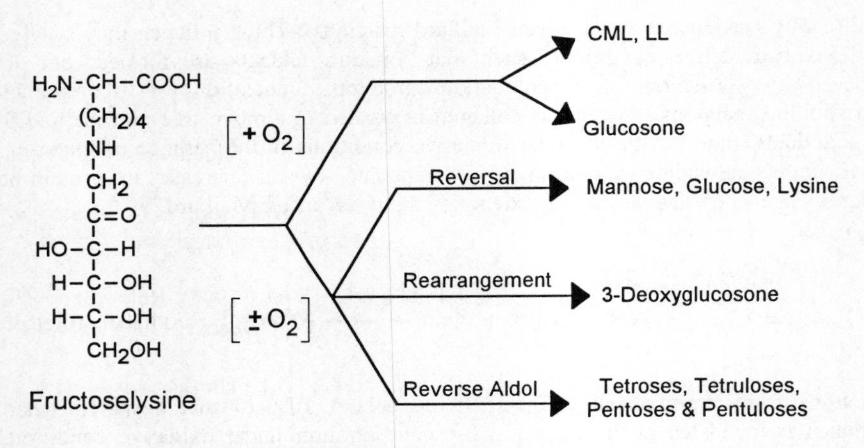

Figure 3. Pathways for decomposition of fructoselysine at physiological pH and temperature. CML is the major product under oxidative conditions (Ahmed et al., 1986), while in the presence of reducible substrates, such as cupric ion, glucosone is formed by oxidation and hydrolysis reactions (Kawakishi et al., 1991). Numerous reversal, rearrangement and fragmentation products are formed by enolization, hydrolysis and reverse aldol reactions under both oxidative and anti-oxidative conditions.

Summary and Conclusions

Recent experiments suggest that autoxidation of glucose is the initial step in formation of glycoxidation products and protein crosslinks during glycation of proteins *in vitro*. Glyoxal, rather than glucosone, has been identified as the dicarbonyl intermediate in these reactions. Decomposition of Schiff base and Amadori adducts may also be an important source of reactive intermediates for the Maillard reaction, depending on reaction conditions and the relative concentration of sugars and adducts. The rapid appearance of advanced glycosylation end-products following induction of diabetes in experimental animals (Mitsuhashi et al., 1993; Leonard et al., 1993) suggests that AGEs formed *in vivo* during the early stages of diabetes may also be derived primarily from autoxidation of sugars, rather than from the accumulation and subsequent degradation of Amadori adducts (Fu et al., 1993).

Abbreviations
AGE, Advanced Glycosylation End-product; CML, N^{ε}-(carboxymethyl)lysine; FL, N^{ε}-fructoselysine.

Acknowledgments
This work was supported in part by research grants from the National Institute for Diabetes and Digestive and Kidney Diseases (DK-19971) and the Juvenile Diabetes Foundation International (192198).

References

Ahmed, M.U., Thorpe, S.R., and Baynes, J.W. (1986) Identification of N^{ε}-carboxymethyllysine as a degradation product of fructoselysine in glycated protein. J. Biol. Chem. 261: 2889-4894.

Baynes, J.W. (1991) Role of oxidative stress in development of complications in diabetes. Diabetes 40: 405-412.

Blackledge, J.A., Fu, M-X., Thorpe, S.R., and Baynes, J.W. (1993) Characterization of a triazine (2-[3-amino-1,2,4-triazin-5-yl) methylene]-hydrazinecarboximideamide) formed from glucose and aminoguanidine under oxidative conditions. J. Org. Chem. 58: 2001-2002.

Bowes, J.H., and Cater, C.W. (1968) The interaction of aldehydes with collagen. Biochim. Biophys. Acta 168:341-352.

Dyer, D.G., Blackledge, J.A., Katz, B.M., Hull, C.J., Adkisson, H.D., Thorpe, S.R., Lyons, T.J., and Baynes, J.W. (1991) The Maillard reaction *in vivo*. Z. Ernahrungwiss. 30: 29-45.

Fu, M-X., Knecht, K.J., Thorpe, S.R., and Baynes, J.W. (1992) Role of oxygen in cross-linking and chemical modification of collagen by glucose. Diabetes 41 (Suppl. 2): 42-48.

Fu, M-X., Wells-Knecht, K.J., Blackledge, J.A., Lyons, T.J., Thorpe, S.R., and Baynes, J.W. (1993) Glycation, glycoxidation and crosslinking of collagen by glucose: kinetics, mechanisms and inhibition of late stages of the Maillard reaction. Diabetes (in press).

Hayashi, T., and Namiki, M. (1986) Role of sugar fragmentation in the Maillard reaction. In: Amino-Carbonyl Reactions in Food and Biological Systems (Fujimaki, M., Namiki, M., and Kato, H., editors), Elsevier, Amsterdam, pp. 29-38.

Hirsch, J., Baynes, J.W., Blackledge, J.A., and Feather, M.S. (1991) The reaction of 3-deoxy-D-glycero-pentos-2-ulose ("3-deoxyxylosone") with aminoguanidine. Carbohyd. Res. 220: c-5-c7.

Hodge, J. E. (1953) Dehydrated foods: chemistry of browning reactions in model systems. Agric. Food Chem. 1: 928-943.

Kawakishi, S., Tsunehiro, J., and Uchida, K. (1991) Autoxidative degradation of Amadori compounds in the presence of copper ion. Carbohyd. Res. 211: 167-171.

Knecht, K.J., Feather, M.S., and Baynes, J.W. (1992) Detection of 3-deoxyfructose and 3-deoxyglucosone in human urine and plasma: evidence for advanced stages of the Maillard reaction *in vivo*. Arch. Biochem. Biophys. 294: 130-137.

Ledl, F., and Schleicher, E. (1990) New aspects of the Maillard reaction in foods and in the human body Angew. Chem. 29:565-594.

Leonard, C, Yamin, M.A., and Mallon, V.M. (1993) Measurement of advanced glycosylation end-products (AGEs) in urine of STZ-diabetic rats. Diabetes 42 (Suppl. 1): 245A.

Mitchel, R.E., and Birnboim, H.C. (1977) The use of Girard-T reagent in a rapid and sensitive method for measuring glyoxal and certain other α-dicarbonyl compounds. Anal. Biochem. 81: 47-56.

Mitsuhashi, T., Nakayama, H., Itoh, T., Kuwajima, S., Aoki, S., Atsumi, T., and Kioke, T. (1993) Immunochemical detection of advanced glycation end products in renal cortex from STZ-induced diabetic rat. Diabetes 42:826-832.

Sell, D.R., and Monnier, V.M. (1989) Structure elucidation of a senescence cross-link from human extracellular matrix: implication of pentoses in the aging process. J. Biol. Chem. 264: 21597-21602.

Stadtman, E. R. (1993) Oxidation of free amino acids and amino acid residues in proteins by radiolysis and by metal-catalyzed reactions. Ann. Rev. Biochem. 62:797-821.

Thornalley, P. J. (1990) The glyoxalase system: new developments toward functional characterization of a metabolic pathway fundamental to biological life. Biochem. J. 269: 1-11.

Wolff, S. P., and Dean, R. T. (1987) Glucose autoxidation and protein modification: the potential role of 'autoxidative glycosylation' in diabetes. Biochem. J. 245: 243-250.
Wolff, S. P., Jiang, Z. Y., and Hunt, J. V. (1991) Protein glycation and oxidative stress in diabetes mellitus and ageing. Free Rad. Biol. Med. 10: 339-352.

Biomimic Oxidation of Glycated Protein and Amadori Product

S. Kawakishi, R. Z. Cheng, S. Sato, and K. Uchida

DEPARTMENT OF APPLIED BIOLOGICAL SCIENCES, FACULTY OF AGRICULTURE, NAGOYA UNIVERSITY, FURO-CHO, CHIKUSA-KU, NAGOYA 464-01, JAPAN

Summary
The oxidative degradation of N^{α}-t-Boc-N^{ε}-fructoselysine (FL) and glycated protein in phosphate buffer solution have been investigated under physiological condition during long time incubation. FL was degraded to mainly lysine, carboxymethyllysine (CML) and α-dicarbonyl sugars which contained glucosone, 3-deoxyglucosone (3-DG) and 1-deoxy-2,3-hexodiulose (1-DG). Glucosone was a major sugar product which corresponded to lysine formation, and 3-DG and 1-DG were minors. Glycated bovine serum albumin (G-BSA) was also degraded in phosphate buffer solution like FL, and glucosone was also a main α-carbonyl compound.

Introduction

It is well known that protein glycation in blood plasma of diabetes and hyperglycemia arises rapidly and may contribute to diabetic complications. Glycation means the formation of Amadori compounds by the reaction between terminal amino group in protein and glucose. Amadori compounds gradually decompose to give the advanced glycation end products (AGE) in senile or diabetic tissues. N^{α}-Formyl-N^{ε}-fructoselysine as a model compound of glycated protein gradually decomposes to carboxy- methyllysine (CML) and erythronic acid (Ahmed et al., 1986), lysinolactic acid and glyceric acid (Ahmed et al., 1988) in biomimic condition (in phosphate buffer solution, pH 7.4, at 37°C). Since these oxidative degradation of FL were inhibited by the addition of radical scavenger, catalase and metal ion chelator, and since the CML formation was significantly increased with phosphate concentration, these reaction could arise with some oxygen radical formed by the action of trace metal ion in phosphate buffer solution. We also observed the degradation of Amadori compounds under physiological condition in the presence of cupric ion to amino acid and glucosone (D-*arabino*hexosulose), not 3-DG, in about 70-80% yields (Kawakishi et al., 1991). Moreover, glycated protein in phosphate buffer solution containing micromolar level of cupric ion were quickly degraded to form α-carbonyl sugars and fragmented polypeptides. The major α-carbonyl found was D-glucosone (Cheng et al., 1991). In the present work, we have investigated in detail the oxidative degradation of fructoselysine and glycated proteins under physiological condition without addition of transition metal ion. Especially, sugar moieties formed from fructoselysine and glycated proteins were quantitatively determined.

Materials and Methods

Reaction Mixtures and Conditions

N^{α}-t-Boc-N^{ε}-fructoselysine was prepared from N^{α}-t-Boc-lysine followed by the method of Njoroge et al. (1988) and glycated bovine serum albumin (G-BSA) was also prepared from BSA and glucose in phosphate buffer (Cheng et al., 1991). FL (10 or 20 mM) and o-phenylenediamine (o-PD, 10 mM) in phosphate buffer (200 mM, pH 7.4) was incubated at 37°C for 30 days, and G-BSA (0.4%) and o-PD (10 mM) in phosphate buffer (67 mM, pH 7.2) was incubated at 37°C for 20 days.

Determination of the Degradation Products from FL and G-BSA
FL and its degradation products (lysine derivatives) were determined by HPLC using a Develosil ODS-5 column eluted with 50 mM CH_3COONH_4/CH_3OH (13/1, v/v). α-Dicarbonyl sugars, glucosone, 1-deoxy-2,3-hexodiulose (1-DG) and 3-deoxyglucosone (3-DG) were determined as quinoxaline derivatives, which were formed by the condensation reaction with α-dicarbonyls and o-PD, by HPLC using a Develosil ODS-5 or ODS HG-5 columns eluted with 0.1% $TFA/CH_3OH/CH_3CN$ (8/2/1, v/v). Since α-dicarbonyls, especially glucosone, were so labile in phosphate buffer and quickly degraded to C_2 and C_3 compounds, o-PD was added to the reaction mixture at starting time of incubation to trap α-carbonyls at the same time with its formation. Quinoxaline derivative of glucosone was very stable in phosphate buffer solution for long time.

Results

The Formation of Lysine and Carboxymethyllysine from FL
The degradation of FL was observed in long time incubation of FL in phosphate buffer solution to four peaks mainly by HPLC. The major product was lysine and minor one was CML, and other two peaks have been unidentified. The time course changes of FL degradation are shown in Figure 1. About 80% of FL has been decomposed during 26 days, and the formation amounts of lysine and CML corresponded to about 50% and 12% of FL degradation, respectively.

The Formation of α-carbonyl Sugars during FL Degradation
Upon degradation of FL, the formation of lysine suggests C-N bond cleavage and simultaneous formation of sugar moieties corresponding to lysine regeneration. When micromolar level of copper ion was added to FL or glycated protein solution, glucosone was the major α-dicarbonyl sugar formed.

Using the same reaction conditions as in Figure 1, α-dicarbonyls (glucosone and 3-DG) were determined by the quinoxaline method in a time course experiment. However, their formation amounts were a little compared to lysine formation. This was because α-dicarbonyls, especially glucosone, gradually decomposed through the side effects of trace amounts of transition metal ion contaminating the phosphate buffer solution. Then, o-PD was added to FL solution at the initial stage of incubation and α-dicarbonyls were trapped to quinoxaline derivatives simultaneously with their formation. FL (10 mM) was incubated in phosphate buffer (200 mM, pH 7.4) containing o-PD (10 mM) at 37°C during 30 days and α-dicarbonyls were determined in a time course experiment by HPLC. The formation of each α-dicarbonyls has been confirmed by GC-MS analysis of TMS ether of quinoxaline derivative obtained from FL degradation (Figure 2). As shown in Figure 2, the main product was glucosone, and 1-DG and 3-DG were minors. The time course formation of the α-dicarbonyls is shown in Figure 3. The amounts of glucosone formed corresponded to the amount of lysine liberated in FL degradation. Moreover, it was also examined whether the formation of α-dicarbonyl sugars may change through the reduction of pH. At lower pH (pH 5.3), the formation of glucosone markedly decreased and 3-DG increased a little, however, glucosone was a major dicarbonyl compound.

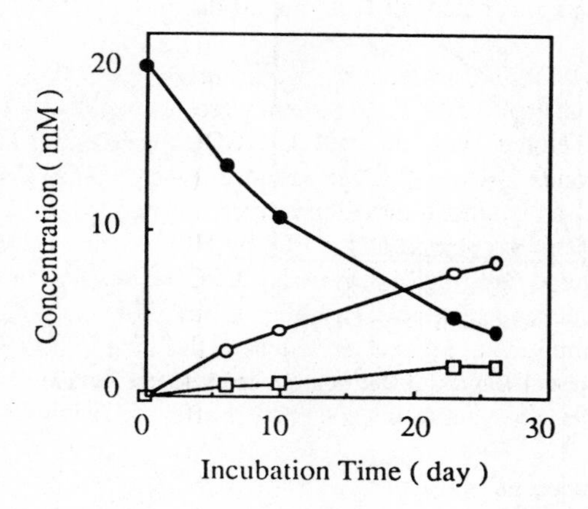

Figure 1. Degradation of N^{α} -t-Boc-N^{ε} -fructoselysine (FL, 20 mM) in phosphate buffer solution (200 mM, pH 7.4) at 37°C and formation of lysine and carboxymethyllysine. ● - FL, o - Lysine, □ - CML.

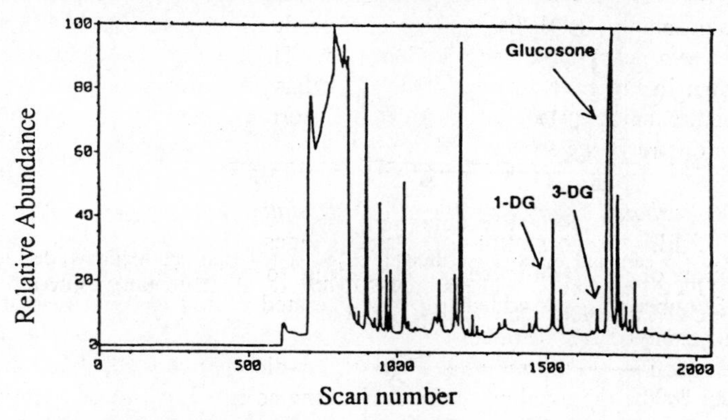

Figure 2. GC-MS chromatogram of TMS ethers of quinoxaline derivatives prepared from the degradation products of FL. FL (10 mM) and o-PD (10 mM) were incubated in phosphate buffer solution (200 mM, pH 7.4) for 30 days. Quinoxaline derivatives were transformed to TMS ether and submitted to GC-MS (JMS-DX-750L mass spectrometer, column: Fused silica DB-1).

The Degradation of G-BSA in Phosphate Buffer Solution

FL gradually decomposed to lysine and α-dicarbonyls, mainly glucosone, in phosphate buffer solution during long time incubation. We have then examined degradation of G-BSA under similar conditions.

G-BSA (0.4%) was incubated in 67 mM phosphate buffer (pH 7.2) containing o-PD (10 mM) for 20 days. α-Dicarbonyls as quinoxaline derivative were determined by HPLC. As shown in Figure 4, there was a same tendency for the formation of each α-dicarbonyls with the case of FL. Their time course changes are shown in Figure 5 and it suggests clearly that glucosone was also a major dicarbonyls on G-BSA similarly to the case of FL.

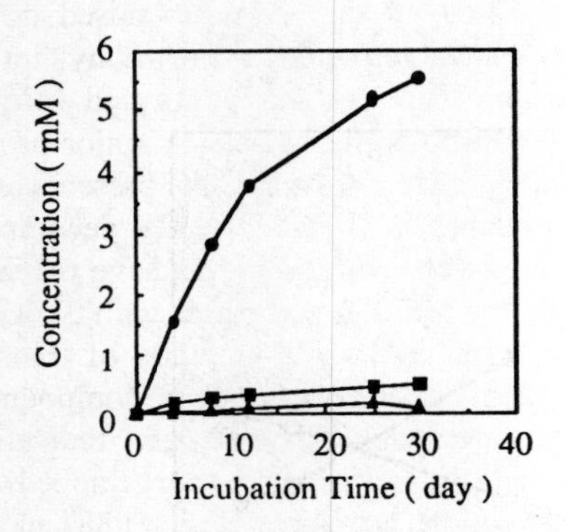

Figure 3. Formation of glucosone, 1-DG and 3-DG from FL in phosphate buffer. ● - Glucosone, ■ - 1-DG, ▲ - 3- DG.

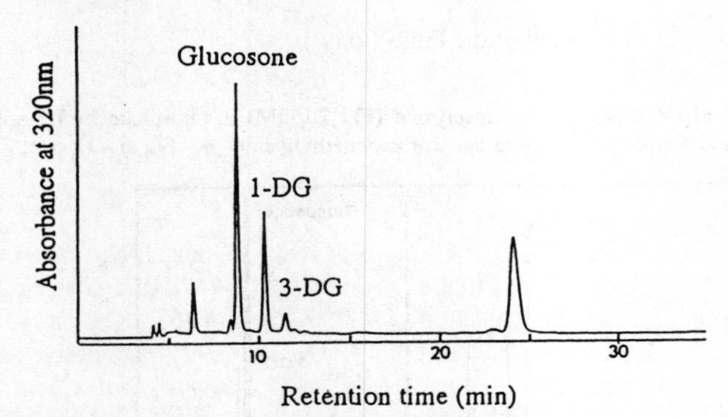

Figure 4. HPLC profile of α-carbonyl quinoxaline derivatives generated from the oxidative degradation of glycated BSA. HPLC conditions: column, ODS-HG-5 (8 x 250 mm); solvent, 0.1% TFA/CH3OH/CH3CN = 8/2/1; detection, 320 nm; flow rate, 2 mL/min.

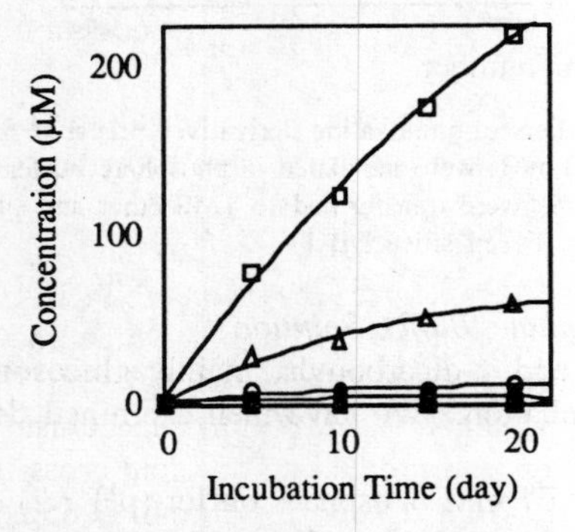

Figure 5. α-Dicarbonyl compounds generated from the oxidative degradation of glycated BSA in the presence of EDTA. Reaction conditions: glycated BSA (0.4%) was incubated with 10 mM o-PD in phosphate buffer (pH 7.2, 67 mM) at 37 °C. □ - Glucosone (+ o-PD), △ - 1-DG (+o-PD), o - 3-DG (+ o-PD), ■ - Glucosone (- o-PD), ▲ - 1-DG (- o-PD), ● - 3-DG (- o-PD).

Discussion

When Amadori compounds and glycated protein are decomposed to form AGE *in vivo* or *in vitro*, 3-DG was proposed as a most important α-dicarbonyl intermediate (Kato et al., 1962, 1970). Moreover, presence of 3-DG was confirmed in diabetic tissues and urine. We have presented here the data that glucosone is a major product, not 3-DG, in the model experiments using FL and G-BSA. However, the presence of glucosone in any tissues is not yet confirmed. Glucosone seems to be easily decomposed in biological medium, especially on the presence of trace metal ion. We have revealed that glucosone is quickly degraded in phosphate buffer solution compared with 3-DG (data not shown), and this degradation is thought to arise with trace amounts of transition metal ion in phosphate buffer solution. The degradation of Amadori compounds and glucosone conjugated with transition metal ion generates some oxygen radicals which cause the oxidative damage to biological bodies unless they are trapped by some radical scavengers. The addition of glucose-copper ion (Hunt et al., 1988) and glucosone-copper ion (Cheng et al., 1992) systems to BSA solution leads to marked oxidative changes of BSA which include the fragmentation to lower molecular weight peptides and the oxidative modification of special amino acid residues.

References

Ahmed, M.U., Thorpe, S.R. and Baynes, J.W. (1986) Identification of N^{α}-carboxymethyllysine as a degradation product of fructoselysine in glycated protein. *J. Biol .Chem.* 261:4889-4894.

Ahmed, M.U., Dunn, J.A., Walla, M.D., Thorpe, S.R. and Baynes, J.W. (1988) Oxidative degradation of glucose adducts to protein: formation of 3-(N^{ε}-lysino)-lactic acid from model compounds and glycated proteins. *J. Biol. Chem.* 263:8816-8821.

Cheng, R.Z., Tsunehiro, J., Uchida, K. and Kawakishi, S. (1991) Oxidative damage of glycated protein in the presence of transition metal ion. *Agric. Biol. Chem.* 55:1993-1998.

Cheng, R.Z., Uchida, K. and Kawakishi, S. (1992) Selective oxidation of histidine residues in proteins or peptides through the copper(II)-catalyzed autoxidation of glucosone. *Biochem. J.* 285:667-671.

Hunt, J.V., Dean, R.T. and Wolff, P. (1988) Hydroxyradical production and autoxidative glycosylation. *Biochem. J.* 256:205-212.

Kato, H. (1962) Chemical studies on aminocarbonyl reaction. I. isolation of 3-deoxypentosone and 3-deoxyhexosone formed by browning degradation N-glycosides. *Agric. Biol. Chem.* 26:187-192.

Kato, H., Tsusaka, N., and Fujimaki, M. (1970) Isolation and identification of α-ketoaldehydes in calf and rabbit livers. *Agric. Biol. Chem.* 34:1541-1548.

Kawakishi, S., Tsunehiro, J., and Uchida, K. (1991) Autoxidative degradation of Amadori compounds in the presence of copper ion. *Carbohyd. Res.* 211:167-171.

Njoroge, F.G., Fernandes, A.A. and Monnier, V.M. (1988) Mechanism of formation of the putative advanced glycosylation end product and protein cross-link 2-(2-furoyl)-4(5)-2(2-furanyl)-1H-imidazole. *J. Biol. Chem.* 263:10646-10652.

Advanced Glycosylation of LDL: Role in Oxidative Modification and Diabetic Complications

Richard Bucala, Zenji Makita, Theodor Koschinsky,[1] Anthony Cerami, and Helen Vlassara

THE PICOWER INSTITUTE FOR MEDICAL RESEARCH, NY, USA; [1]DIABETES RESEARCH INSTITUTE, DUSSELDORF, GERMANY

Summary

We investigated the possibility that amine-containing phospholipids react directly with glucose to form advanced glycosylation endproducts (AGEs). Phospholipid-linked AGEs formed readily in vitro, mimicking the absorbance, fluorescence, and immunochemical properties of AGEs that result from the advanced glycosylation of proteins. Lipid-AGE formation was accompanied by fatty acid oxidation, which occurred in the absence of added, transition metal ions. In additional experiments, incubation of low-density lipoprotein (LDL) with glucose was found to produce AGE moieties that were attached to both the lipid and the apoprotein (ApoB) components. Oxidized-LDL formed concomitantly with AGE-modified LDL. AGE-specific ELISA analysis of LDL specimens isolated from diabetic individuals revealed increased levels of both apoprotein- and lipid-linked AGEs when compared to specimens obtained from normal, non-diabetic controls. Circulating levels of oxidized-LDL were elevated in diabetic patients, and correlated significantly with apoB-AGE and lipid-AGE levels. Lipid-advanced glycosylation may result from intra-molecular redox reactions that occur during advanced glycosylation and presents a facile mechanism to explain lipid oxidation in vivo, where free metals are present in too low a concentration to induce oxidative modification.

Introduction

The oxidative modification of lipids is believed to play a central role in atherogenesis and in the vascular sequelae of diabetes and aging (Witztum and Steinberg, 1991). Oxidation of the lipid component of low-density lipoprotein (LDL) for example, leads to the loss of LDL recognition by cellular LDL receptors and in the preferential uptake of oxidized-LDL by macrophage "scavenger" receptors. Vascular wall macrophages become transformed into lipid-laden "foam" cells. This leads to the development of fatty streaks and the complex, proliferative lesions that characterize atherosclerotic vascular disease (Witztum and Steinberg, 1991; Goldstein *et al.*, 1979; Fogelman *et al.*, 1980; Sparrow *et al.*, 1989; Ross, 1986).

Despite intensive interest in the biological effects of lipid oxidation, there has been little insight into the biochemical processes that initiate oxidative modification in vivo. In vitro studies have demonstrated that metal-catalyzed peroxidation reactions occur readily at the unsaturated bonds within fatty acids. Mono-unsaturated and especially polyunsaturated fatty acids are sensitive to oxidation because bis-allylic hydrogens are easily abstracted by free radical processes. Once lipid oxidation is initiated, fatty acids decompose to a variety of reactive aldehydes that rapidly propagate oxidative processes (Dix & Aikens, 1993; Kanner *et al.*, 1987). Nevertheless, low concentrations of trace metals, the high availability of ligands that form tight coordination complexes with metals, and the abundant anti-oxidant capacity of plasma suggest that metal-catalyzed oxidative processes and reactive oxygen species play little, if any role in mediating lipid oxidation in vivo (Klaassen, 1985; Frei *et al.*, 1988a, 1988b).

The presence of reactive, primary amino groups on phospholipids led us to consider the possibility that glucose may react directly with lipids to initiate advanced glycosylation. We hypothesized that intermolecular oxidation-reduction reactions might then occur to oxidize fatty acid residues - in the absence of added transition metals or exogenous, free-radical generating systems (Bucala *et al.*, 1993).

Materials and Methods

We incubated anaerobic, buffered suspensions of phosphatidylethanolamine (PE) or phosphatidylcholine (PC) (14 mM) with glucose (500 mM) and EDTA (1 mM) at 37°. PE contains a free amino group that can react with glucose-derived carbonyls to form Schiff base and Amadori products. PC was utilized as a control and in contrast to PE, contains a blocked, tertiary amine that cannot react with glucose to form the initial Schiff base. Lipid-derived products were isolated by extraction into chloroform/methanol and then analyzed for advanced glycosylation. PE but not PC was observed to react with glucose to form products with the absorbance and the fluorescence properties of AGEs. Phospholipid-derived fluorophores showed an excitation maximum of 360 nm and an emission maximum of 440 nm. Immunoreactive AGEs, as detected by AGE-specific ELISA (Makita *et al.*, 1992), formed in incubations which contained PE and glucose, but not in incubations which contained PE alone, PC and glucose, or PC alone (Figure 1A-C).

Results

To assess the contribution of advanced glycosylation reactions to lipid oxidation, the PE and PC utilized in these experiments contained esterified oleic acid, a mono-unsaturated fatty acid. After incubation with glucose, lipid-soluble material then was assayed for the presence of reactive, fatty acid-derived aldehydes by reaction with thiobarbituric acid (TBA) (Kikugawa *et al.*, 1992). Lipid oxidation products were found to form at a rate that was slightly delayed but parallel to the rate of AGE formation (Figure 1D). It is noteworthy that these products formed in the presence of metal chelators and in the absence of exogenously-added transition metals, which frequently are used to catalyze lipid oxidation in model systems. In further studies, the inclusion of aminoguanidine, an advanced glycosylation inhibitor, prevented the formation both of lipid-AGEs and lipid oxidation products.

We then sought to investigate the contribution of advanced glycosylation to the oxidative modification of LDL. LDL was incubated with glucose in vitro and analyzed for advanced glycosylation and oxidative modification. At intervals, the LDL was fractionated into the lipid and apoprotein (ApoB) components, and AGEs measured by AGE-specific ELISA. Incubation of LDL with 200 mM glucose for 3 days resulted in the formation of readily measurable levels of AGEs on both lipid and apoprotein (Figure 2A, B). These measurements indicate that lipid-linked AGEs are present at a specific activity that is 100-fold greater than apoprotein-linked AGEs. Although differences in the immunoreactivity between lipid-AGEs and apoprotein-AGEs may not make it possible to compare quantitatively the degree of AGE modification on lipid versus protein, it can be concluded that AGEs form at a more rapid rate on lipid than on apoprotein. Measurements of oxidative modification revealed that LDL was oxidized concomitantly with the formation of AGEs. Lipid oxidation products also were observed to form more rapidly than was observed during the incubation of PE and glucose (Figure 1D) and this is due most likely to the presence in LDL of polyunsaturated fatty acids that rapidly propagate oxidative reactions.

To begin to define the relationship between advanced glycosylation and LDL oxidation in vivo, LDL was isolated from both non-diabetic and diabetic individuals and analyzed for the presence of lipid-AGEs, apoprotein-AGEs, and oxidative modification (Figure 3). Both the lipid- and the apoprotein-linked AGEs in the diabetic LDL specimens were found to be markedly elevated when compared to LDL obtained from non-diabetic individuals. Lipid-AGE levels were elevated almost 4-fold in diabetic patients [NL: (n=8) 0.11±0.03 Units of AGE/μg lipid; DM: (n=16) 0.41±0.25 Units of AGE/μg lipid, P<0.005)]. ApoB-AGE levels were increased more than 2-fold in the diabetic samples [NL: (n=8) 0.0028±0.0006 Units of AGE/μg ApoB; DM: (n=16) 0.0068 ±0.004 Units of AGE/μg ApoB, P<0.0001)]. These measurements revealed a similar quantitative ratio between LDL oxidation and the level of lipid-AGEs and ApoB-AGEs that was observed during LDL advanced glycosylation in vitro (Figure 2). There also appeared to be a marked increase in the level of lipid-AGEs relative to the level of ApoB-associated AGEs. In agreement with prior studies (Nishigaki *et al.*, 1981), LDL from diabetic individuals showed significantly greater oxidative modification than the LDL from non-diabetic individuals [NL: (n=8) 3.7±1.25 pm MDA equivalents/μg LDL; DM: (n=16) 6.8±1.2 pm MDA equivalents/μg LDL (Mean ± SD), P< 0.0001]. Linear regression analysis of these data revealed a significant correlation between the level of AGE modification and LDL oxidation. For the measurement of ApoB-AGE versus LDL oxidation, this analysis showed a correlation coefficient of r=0.52 and P<0.01. For lipid-AGEs versus LDL oxidation, the corresponding values were r=0.63 and P<0.005.

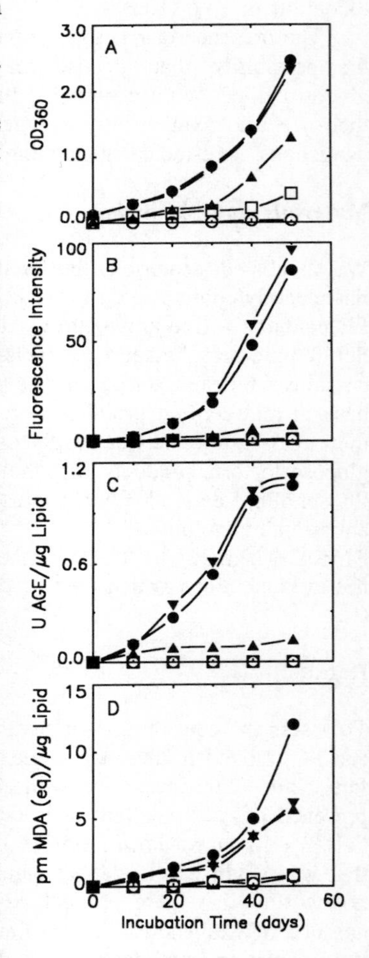

Figure 1. Time course for the production of lipid-bound chromophores (**A**), fluorophores (**B**), AGE immunoreactive products (**C**), and lipid oxidation products (measured as malondialdehyde equivalents) (**D**). PE (14 mM) incubated with glucose (500 mM) (●), PE incubated alone (o), PE incubated with glucose and aminoguanidine (100 mM) (▲), PE incubated with glucose and BHT (0.2 mM) (▼), PC incubated with glucose (□), PC incubated alone (same as o). All incubations contained EDTA (1 mM). From Bucala *et al.*, 1993.

Discussion

The advanced glycosylation reaction between phospholipid and glucose provides a facile mechanism

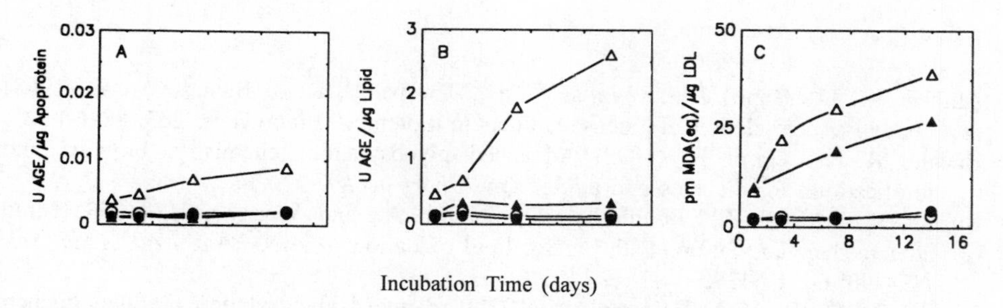

Figure 2. Time-dependent reaction of human LDL (2.5 mg/ml) with glucose (200 mM). Samples were dialyzed and portions separated into lipid and apoprotein components for AGE (**A**, **B**) or oxidized-LDL determination (**C**). LDL incubated with 200 mM glucose (▵). LDL incubated with 200 mM glucose and 300 mM aminoguanidine (▴). LDL incubated alone (o). LDL incubated with aminoguanidine (•). From Bucala *et al.*, 1993.

to explain the progressive, oxidative modification of lipids in vivo. The oxidative interaction between AGEs and unsaturated fatty acid residues, which we term "AGE-oxidation", appears to involve inter- or intramolecular oxidation-reductions and possibly free radical formation, processes which have been noted to occur during advanced glycosylation (Njoroge and Monnier, 1989; Ahmed *et al.*, 1988; Namiki and Hayashi, 1981). In addition to initiating oxidative modification, lipid-AGEs may contribute to several of the pathological effects that generally have been attributed to protein-linked AGEs. An increase in the subendothelial density of lipid-AGEs, as might occur in atherosclerotic plaque for example, would enhance protein deposition and crosslinking, stimulate macrophage-monocyte chemotaxis, and inactivate the vasodilatory effects of nitric oxide (Esposito *et al.*, 1989, Bucala *et al.*, 1991). Measurement of lipid-linked AGEs may serve as a convenient, surrogate marker for lipid oxidation and prove useful in assessing the contribution of advanced glycosylation and lipid oxidation to both diabetic and non-diabetic vascular sequelae.

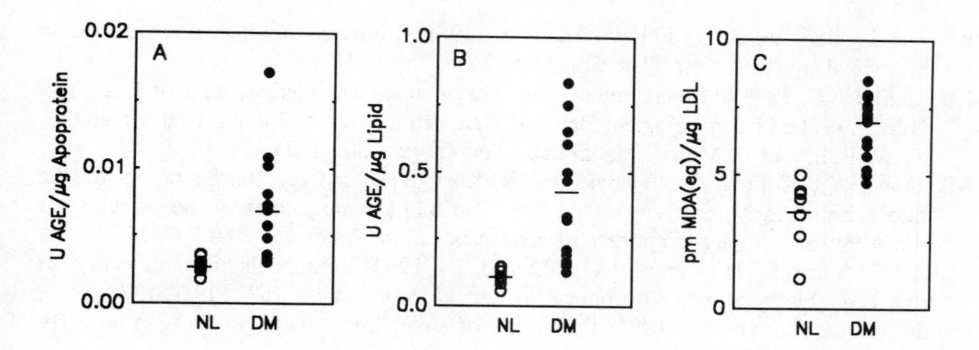

Figure 3. Measurement of AGE and oxidative modification of LDL from non-diabetic individuals (NL) and patients with diabetes mellitus (DM). (**A**): AGE modification of LDL apoprotein, (**B**): AGE modification of LDL lipid, and (**C**): oxidative modification of LDL. (From Bucala *et al.*, 1993)

References

Ahmed, M. U., Dunn, J. A., Walla, M. D., Thorpe, S. R. & Baynes, J. W. (1988) Oxidative degradation of glucose adducts to protein. *J. Biol. Chem.* 263: 8816-8821.

Bucala, R. & Cerami A. (1992) Advanced glycosylation: chemistry, biology, and implications for diabetes and aging. *Adv. Pharmacol.* 23: 1-34.

Bucala, R., Makita, Z., Koschinsky, T., Cerami, A., and Vlassara, H. (1993) Lipid advanced glycosylation: Pathway for lipid oxidation *in vivo*. *Proc. Natl. Acad. Sci. USA* 90: 6434-6438.

Bucala, R., Tracey, K. J. & Cerami, A. (1991) Advanced glycosylation products quench nitric oxide and mediate defective endothelium-dependent vasodilatation in experimental diabetes. *J. Clin. Invest.* 87: 432-438.

Dix, T. A. & Aikens, J. (1993) Mechanisms and biological relevance of lipid peroxidation initiation. *Chem. Res. Toxicol.* 6: 2-18.

Esposito, C., Gerlach, H., Brett, J., Stern, D. & Vlassara, H. (1989) Endothelial receptor-mediated binding of glucose-modified albumin is associated with increased monolayer permeability and modulation of cell surface coagulant properties. *J. Exp. Med.* 170: 1387-1407.

Fogelman, A. M., Schecter, J. S., Hokom, M., Child, J. S. and Edwards, P. A. (1980) Malondialdehyde alteration of low density lipoprotein leads to cholesterol accumulation in human monocyte-macrophages. *Proc. Natl. Acad Sci. USA* 77: 2214-2218.

Frei, B., Yamamoto, Y., Niclas, D. & Ames, B. N. (1988a) Evaluation of an isoluminol chemiluminescence assay for the detection of hydroperoxides in human blood plasma. *Anal. Biochem.* 175: 120-130.

Frei, B., Stocker, R. & Ames, B. N. (1988b) Antioxidant defenses and lipid peroxidation in human blood plasma. *Proc. Natl. Acad. Sci. USA* 85: 9748-9752.

Goldstein, J. L., Ho, Y. K., Basu, S. K. and Brown, M. S. (1979) Binding site on macrophages that mediates uptake and degradation of acetylated low density lipoprotein, producing massive cholesterol deposition. *Proc. Natl. Acad. Sci. USA* 76: 333-337.

Kanner, J., German, J. B. & Kinsella, J. E. (1987) Initiation of lipid peroxidation in biological systems. *Crit. Rev. Food Sci. Nutr.* 25(4): 317-364.

Klaassen, C. D. (1985) Heavy metals and heavy metal antagonists, *In* Goodman and Gilman's The Pharmacological Basis of Therapeutics. A. G. Gilman, L. S. Goodman. T. W. Rall, and F. Murad. Macmillan, New York. 1605-1627.

Kikugawa, K., Kojima, T., Yamaki, S. & Kosugi, H. (1992) Interpretation of the thiobarbituric acid reactivity of rat liver and brain homogenates in the presence of ferric ion and ethylenediaminetetraacetic acid. *Anal. Biochem.* 202: 249-255.

Makita, Z., Vlassara, H., Cerami, A. & Bucala, R. (1992) Immunochemical detection of advanced glycosylation end products *in vivo*. *J. Biol. Chem.* 267: 5133-5138.

Namiki, M. & Hayashi, T. (1981) Formation of novel free radical products in an early stage of Maillard reaction. *Prog. Fd. Nutr. Sci.* 5: 81-91.

Nishigaki, I., Hagihara, M., Tsunekawa, H., Maseki, M. & Yagi, K. (1981) Lipid peroxide levels of serum lipoprotein fractions of diabetic patients. *Biochem. Med.* 25: 373-378.

Njoroge, F. G. & Monnier, V. M. (1989) The chemistry of the Maillard reaction under physiological conditions: A review. *Prog. Clin. Biol. Res.* 304: 85-107.

Ross, R. (1986) The pathogenesis of atherosclerosis. An update. *New Eng. J. Med.* 314: 488-500.

Sparrow, C. P., Parthasarathy, S. and Steinberg, D. (1989) A macrophage receptor that recognizes oxidized LDL but not acetylated LDL. *J. Biol. Chem.* 264: 2599-2604.

Witztum, J. L., and Steinberg, D. (1991) Role of oxidized low density lipoprotein in atherogenesis. *J. Clin. Invest.* 88: 1785-1792.

The Aldose Reductase Activity in Lens Extracts Protects, but Does Not Prevent, the Glycation of Lens Proteins by L-Threose

Beryl J. Ortwerth, Jane Anne Speaker, and Malladi Prabhakaram

MASON INSTITUTE OF OPHTHALMOLOGY AND DEPARTMENT OF BIOCHEMISTRY, UNIVERSITY OF MISSOURI, COLUMBIA, MISSOURI 65212, USA

Summary
L-Threose, a breakdown product of ascorbic acid, rapidly crosslinks lens proteins in vitro. This rapid glycation reaction, however, may possibly be prevented in lens tissue by the reduction of sugars to polyols by aldose reductase (AR). L-Threose was an excellent substrate for purified aldose reductase measured both spectrophotometrically and by the formation of L-threitol by TLC. Human lens contains AR activity, even in the lens core. When L-threose was incubated with increasing lens extracts, both protein incorporation and threitol formation were increased. The addition of high levels of NADPH increased threitol formation, but did not completely prevent protein incorporation. Sorbinil addition inhibited AR, blocked threitol formation and stimulated protein incorporation. AR at the levels present in human lens, therefore, was able to reduce but not eliminate the glycation of lens proteins by L-threose. This reaction could be important in causing protein crosslinks as well as the introduction of fluorophores into the lens crystallins.

Introduction

The lens crystallins, due to their possibly life-long stability, accumulate damaging modifications which may lead to protein aggregation, protein crosslinking and eventually to age-onset cataract. The glycation of lens proteins has been suggested to be one of the major protein modifications present in older lenses (Stevens et al, 1978; Monnier and Cerami 1981, 1983), and indeed diabetic patients are at greater risk for cataract formation.

Glucose readily glycates protein, but the demonstration of significant crosslinking likely requires the presence of both oxygen and heavy metal ions (Fu et al., 1992). A much more rapid crosslinking of lens proteins can be seen with either ascorbic acid in the presence of oxygen (Ortwerth and Olesen, 1988b) or by dehydroascorbic acid in the absence of oxygen and in the presence of a heavy metal chelator (Prabhakaram and Ortwerth, 1991). Therefore, while ascorbic acid is not able to undergo a glycation reaction, the oxidation products of ascorbic acid are able to glycate and crosslink proteins without any further need for oxygen or oxygen free radicals.

The extent to which ascorbic acid is degraded in lens tissue is not known, but the products of ascorbate decomposition could play an important role in forming the protein crosslinks seen in age-onset cataract. The current work was undertaken to evaluate the role of aldose reductase in inhibiting or completely preventing the glycation of lens proteins by L-threose, the most active of the ascorbate degradation products.

Materials and Methods

All procedures including the synthesis of [1-^{14}C]tetrose (a mixture of [1-^{14}C]threose and [1-^{14}C]erythrose have been described previously (Ortwerth et al, 1994).

Incorporation of [1-^{14}C]L-tetrose into lens protein

Calf lens extract (30,000 x g supernatant of a total homogenate) were incubated with 4mM [1-^{14}C]L-tetrose (2.5mCi/mMol) and various levels of NADPH in 90 mM phosphate buffer, pH 7.0, or with and without Sorbinil. A 0.2 mL reaction mixture, containing 50 mg/mL of lens protein, was incubated at 37ºC, and 10 μL aliquots were removed in triplicate at various times, spotted on 25mm filter paper discs and subjected to the wash procedure of Mans and Novelli (1961). In a similar experiment 0 to 50 mg/mL of lens homogenate was incubated with 4 mM [1-^{14}C]L-tetrose and 20 μM NAPH and aliquots taken in triplicate after 24, 48 and 72 hours to determine the incorporation into protein. Similar aliquots were spotted on TLC plates, developed with isopropanol:ethyl acetate:water (83:11:6) as irrigant and each lane scanned with an AMBIS radioactivity scanner. Each spot was quantified to determine the loss of threose and the formation of threitol. Erythrose and erythritol were not sufficiently separated to quantify these compounds individually.

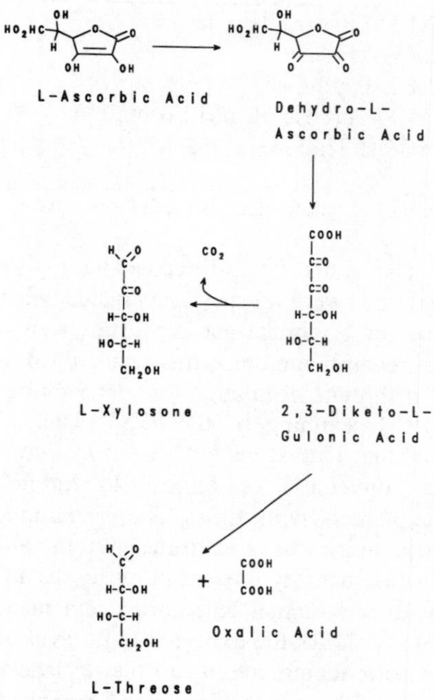

L-Ascorbic Acid
Dehydro-L-Ascorbic Acid
CO_2
L-Xylosone
2,3-Diketo-L-Gulonic Acid
L-Threose
Oxalic Acid

Figure 1. Degradation reactions of ascorbic acid.

Results

Crosslinking of lens protein

When ascorbic acid (ASA) is incubated with lens proteins in the presence of air, the crosslinking of lens crystallins readily occurs via a glycation mechanism (Ortwerth and Olesen, 1988a). The oxidation of ASA, leads to the formation of the intermediates dehydroascorbic acid (DHA) and diketogulonic acid (DKG) and finally to L-xylosone or L-threose (Figure 1). Of these compounds threose was the most active as a crosslinking agent with lens proteins when assayed under a nitrogen atmosphere with a chelating agent (Ortwerth, et al., 1994).

Aldose reductase activity in human lens

The ability of L-threose to serve as a substrate for aldose reductase was measured in assays with a pure preparation of recombinant human aldose reductase (Table I). The L-tetroses were excellent substrates, with L-threose having an activity equal to that of D-glyceraldehyde. In every case the activity measured was completely inhibited by the addition of Sorbinil. To rule out the fact that the activity was not an artifactual oxidation of NADPH (Wolff and Crabbe, 1985) the enzyme was incubated with [1-^{14}C]tetrose and the products separated by thin layer chromatography.

Table I. Aldose reductase activity with various substrates.

Substrate	Activity (units)[1]
5mM D-Glyceraldehyde	4.4
5mM L-Threose	5.0
5mM L-Erythrose	6.1
5mM L-Threose +1mM Sorbinil	0
0.2mM L-Threose	1.2

[1]One unit of aldose reductase activity is equal to 1.0 nmole of NADPH oxidized/min.

The spots for L-threose and L-threitol standards were clearly separated, whereas those for erythrose and erythritol were not. The recombinant enzyme resulted in [1-^{14}C]L-threitol formation as determined by AMBIS scanning of the TLC plate. The individual lane scans are shown in Figure 2. The conversion of threose to threitol is clearly seen with time. The spectrophotometric assay was used to measure the aldose reductase activity present in rat lenses and in bovine and human lens cortex and nucleus. Table II shows the expected high level of L-threose-reducing activity in the rat lens with considerably less in bovine and human lens. While the bovine lens had no activity in the nucleus, definite activity was seen in the human lens nucleus. Assays of four different frozen human lenses of 60-78 yrs of age (data not shown) showed only slightly less aldose reductase than fresh lens. There is, therefore, sufficient enzyme activity to account for the 3.4 ± 0.8μg of threitol seen in the TCA supernatants from human lens extracts (Ortwerth et al., 1994), even if the formation of L-Threose occurred only in the lens nucleus.

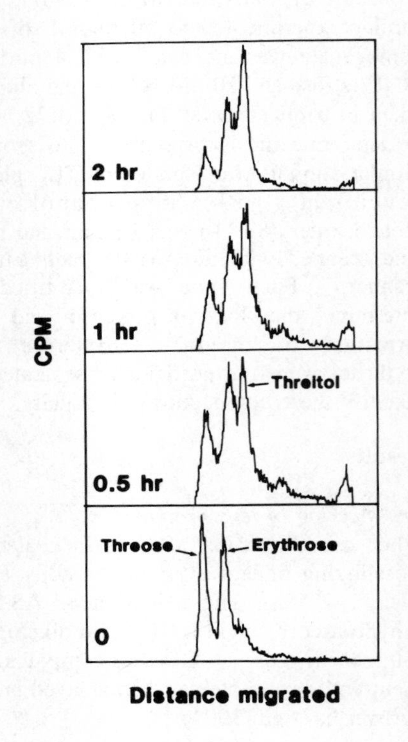

Figure 2. Radioactivity scans of TLC lanes of an AR assay.

The fate of [1-^{14}C]tetrose was measured in reaction mixtures containing increasing levels of bovine lens homogenate from 0 to 50 mg/mL protein. Aliquots separated by TLC showed increasing loss of threose with subsequent threitol formation as the homogenate protein was increased. At 50 mg/mL homogenate, there was a loss of 1.2 mmoles of threose with a net production of only 0.24 mmoles of threitol (data not shown).

The incorporation of radioactivity into protein increased with increasing protein in almost a linear fashion (Figure 3). After 24 hours with a 50 mg/mL total homogenate,

20% of the original [1-^{14}C]tetrose remained unchanged, 14% was incorporated into protein, 15% was reduced to threitol and the remainder was converted to an unknown compound.

Table II. Aldose reductase activity in rat, bovine and human lens.

Lens	Section	Total AR activity (Units[1]/lens)	Specific Activity (Units/mg protein)
Rat	whole lens	40	5.0
Bovine (calf)	cortex	41	0.27
	nucleus	8.5	0.045
Bovine (mature)	cortex	117	0.28
	nucleus	0	0
Human	cortex	7.4	0.70
	nucleus	1.7	0.15

[1]One unit of aldose Reductase (AR) activity is equal to 1.0 nmole of NADPH oxidized/min. All assays were carried out with 5.0 mM L-threose as substrate.

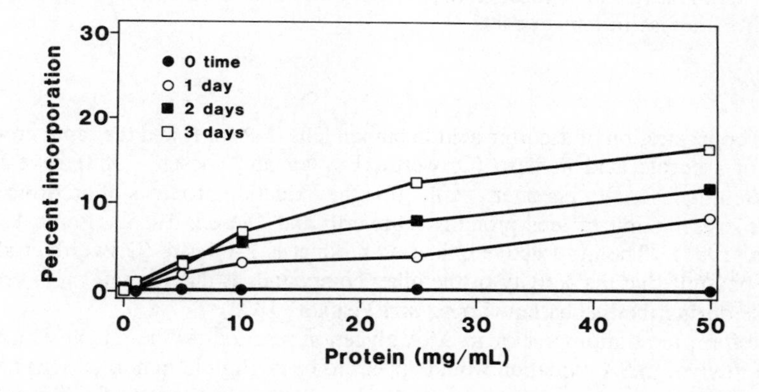

Figure 3. The incorporation of [L-^{14}C]tetrose into protein by increasing amounts of a bovine lens cortical homogenate.

Since the levels of AR activity in the extracts of bovine lens cortex exceeded those in the human lens nucleus, the protein glycation in the human lens nucleus would be at least those present in the reaction mixtures in Figure 3. It was possible, however, that the AR activity in these assays was markedly limited by the lack of NADPH, which may be regenerated very slowly in this in vitro system. Therefore, the effect of increasing NADPH was measured in assays containing 50mg/mL bovine lens cortical extracts. The incorporation into protein and the levels of L-threitol and L-threose after 24 hrs are presented in Figure 4. Increasing NADPH caused greater conversion of threose to threitol, and correspondingly less incorporation into protein. Significant, however, was the fact that a plateau was reached at which no more threose could be diverted to threitol. Therefore, optimum AR activity could not prevent the glycation of lens proteins by threose.

The enzyme controlling threitol formation was AR as shown by experiments with increasing Sorbinil addition to an incorporation assay. Sorbinil caused a decrease in threitol almost to the levels present in the original [1-^{14}C]tetrose preparation (Figure 5).

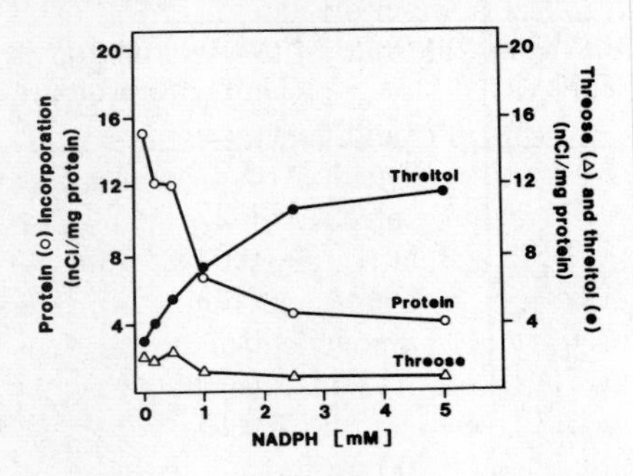

Figure 4. The effect of increasing NADPH on protein incorporation and L-threitol formation.

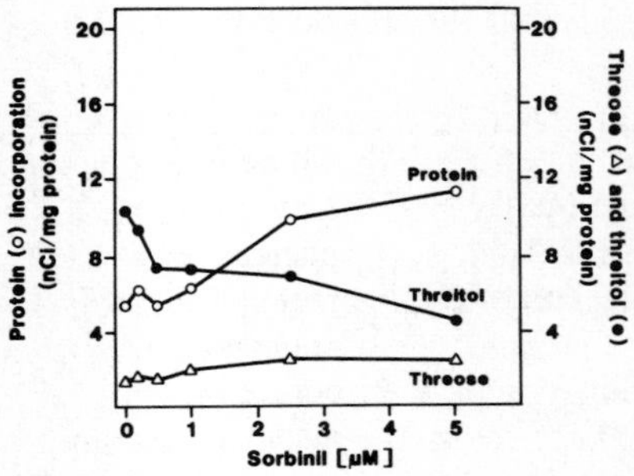

Figure 5. The effect of increasing Sorbinil on protein incorporation and L-threitol formation.

As expected, there was a corresponding increase in protein incorporation. In both experiments the levels of unreacted threose were the same after 24 hrs, suggesting a short half life for threose in lens extracts.

Discussion

The high concentration of ascorbic acid in human lens (1-2 mM) and the rapid crosslinking activity of ascorbic acid in vitro (Ortwerth, Feather and Olesen, 1988) have led us to investigate this reaction in greater detail. It is the oxidation products of ascorbic acid that cause the crosslinking of lens proteins (Ortwerth and Olesen, 1988b; Prabhakaram and Ortwerth, 1991). The most active oxidation product is L-threose (Ortwerth et al. 1994), and it is possible that the activity of the other compounds is due to their rapid breakdown to threose during the incubation (Lopez and Feather, 1992).

Therefore, the limiting factor in ASA glycation reactions is the rate at which ASA is oxidized in vivo. ASA oxidation would appear to be negligible in normal lens because of 1) the low levels of oxygen in the lens (20 μM), 2) the high levels of GSH, which will readily reduce DHA to ASA in solution (Winkler, 1987; Ortwerth and Olesen, 1988b) and 3) by the possible presence of an enzyme which catalyzes the GSH-dependent reduction of DHA to ASA (Rose and Bode, 1992). In spite of these apparent limitations, measurements have shown the presence of 0.1 mM DHA in human lenses and cataracts (Lohmann et al. 1986), and the presence of 0.09 mM DHA and 0.07 mM DKG in bovine lenses (Kern and Zolot, 1987). Even higher levels of DHA were measured by Taylor et al. in human lens (1991). This is significant because DHA and DKG represent intermediates in ASA breakdown. Human lens was shown to contain 3.4 μg of threitol per lens, or greater than 0.1 mmoles/g lens tissue, present almost exclusively as threitol within the lens (Ortwerth et al., 1994). This is consistent with the proposed protective role of aldose reductase to nullify potentially Maillard-reactive carbonyl compounds (Richard, et al., 1991).

The fact that the threose produced by ASA oxidation is in the L-configuration suggested that it, like L-galactose, may not be a substrate for aldose reductase (Lin et al. 1991). L-Threose was an excellent substrate for AR as demonstrated by the sugar dependent reduction of NADPH and by the formation of L-threitol as detected by TLC. Aldose reductase activity was high in rat lens as observed by others (Jedzeniak et al., 1981), but considerably lower in bovine and human lens. Whether the activity is sufficient to reduce all the threose formed cannot be predicted until the rate of ASA oxidation is known.

Protein crosslinking by L-threose in the lens, therefore, would depend upon the rate of reduction by aldose reductase, compared to the rate of glycation of protein amino groups by threose. Since L-threose has a high percentage of molecules in the open chain form it should be very active in the Maillard reaction (Bunn and Higgins, 1981). This is confirmed by the rapid incorporation of [^{14}C]tetrose into lens proteins (Figure 3). The actual amount of incorporation into protein using a whole homogenate of bovine lens increased with increasing protein and constant [^{14}C]tetrose. The linear increase in incorporation in Figure 3, if extrapolated, argues that most of the threose would go towards glycation at the high levels of protein in the lens ($>$300 mg/mL).

The assay should represent conditions within the lens as a whole homogenate supernatant from freshly obtained lenses was used. The AR levels in the fresh bovine cortex was slightly greater than those present in the human lens nucleus, and the NADP levels employed were much higher than those in human lens (Giblin and Reddy, 1980). At high NADPH concentrations, the AR activity was highest as measured by threitol formation, yet protein labeling was still observed. This argues strongly that a given fraction of threose will escape reduction by AR and glycate proteins within the lens.

The inhibition of threitol production by Sorbinil confirms AR as the reducing enzyme in the lens extract, and results in a marked increase in glycation by threose. This appears to be inconsistent with the observation that feeding Sorbinil to rats causes a delay in sugar cataract formation (Hu, Datiles and Kinoshita, 1983). The inhibition of AR, however, would initially act to delay cataract by preventing the precipitous drop in NADPH and glutathione (Sippel, 1966; Reddy et al., 1976). This is supported by the observation that the feeding of reducing agents are also able to delay sugar cataract (Creighton and Trevithick, 1979; Srivastava, 1989). While increased glycation may occur with Sorbinil feeding, this may not represent a significant increase in the accumulation of glycated protein over the short time frame employed in these cataract models. Also, the retention of the reducing environment in the lens may reduce the oxidation of ascorbate thereby producing less threose to glycate protein.

Therefore, the glycation of lens protein by L-threose will not likely be prevented by the levels of AR in human lens nuclei. The rapid rate of protein crosslinking observed in vitro with only 5 mM L-threose argues that low levels of L-threose could make a significant contribution to the non-disulfide crosslinks seen in aged and cataractous human lens protein. The limiting reaction would be the rate of oxidation in the lens. While this may be slow due to the reducing atmosphere and limiting oxygen levels, the presence of DHA, threitol and xylitol in human lens argues that ascorbate breakdown does occur.

Acknowledgments
The authors would like to express their appreciation to Dr. J. Mark Petrash for his kind gift of recombinant aldose reductase and Sorbinil and to Ms. Sherry DiMaggio for typing this manuscript. This work was supported by NIH grant EY 07070 and by Research to Prevent Blindness, Inc.

References

Bunn, H.F. and Higgins, P.J. (1981) Reaction of monosaccharides with proteins: Possible evolutionary significance. Science 213, 222-224.

Creighton, M.O. and Trevithick, J.R. (1979) Cortical cataract formation prevented by vitamin E and glutathione. Exp. Eye Res. 29, 689-93.

Fu, M.X., Knecht, K.J., Thorpe, S.R. and Baynes, J.W. (1992). Role of oxygen in cross-linking and chemical modification of collagen by glucose. Diabetes 41, 42-8.

Giblin, F.J. and Reddy, V. (1980) Pyridine nucleotides in ocular tissues as determined by the cycling assay. Exp. Eye Res. 31, 601-9.

Hu, T.S., Datiles, M. and Kinoshita, J.H. (1983) Reversal of galactose cataract with Sorbinil in rats. Invest. Ophthalmol. Vis. Sci. 22; 174-9.

Jedziniak, J.A., Chylack Jr., L.T., Cheng, H-M., Gillis, M.K., Kalustian, A.A. and Tung, W.H. (1981) The sorbitol pathway in the human lens: aldose reductase and polyol dehydrogenase. Exp. Eye Res. 20, 314-326.

Kern, H.L. and Zolot, S.L. (1987) Transport of vitamin C in the lens. Curr. Eye Res. 6, 885-96.

Lin, L.R., Reddy, V.N., Giblin, F.J. Kador, P.F. and Kinoshita, J.H. (1991) Polyol accumulation in cultured lens epithelial cells. Exp. Eye Res. 52, 93-100.

Lohmann, W., Schmehl, W. and Strobel, J. (1986) Nuclear cataract: oxidative damage to the lens. Exp. Eye Res. 43, 859-62.

Lopez, M.G. and Feather, M..S. (1992) The production of threose as a degradation product from L-ascorbic acid. J. Carbohydr. Chem. 11, 799-806.

Mans, R.J. and Novelli, G.D. (1961) Measurement of the incorporation of radioactive amino acids into protein by a filter-paper disk method. Arch. Biochem. Biophys. 94, 48-53.

Monnier, V.M., and Cerami, A. (1981) Nonenzymatic browning in vivo: Possible process for aging of long-lived proteins. Science 211, 491-493.

Monnier, V.M. and Cerami, A. (1983) Detection of nonenzymatic browning products in the human lens. Biochim. et Biophys. Acta, 760, 97-103.

Ortwerth, B.J., Feather, M.S. and Olesen, P.R. (1988). The precipitation and cross-linking of lens crystallins by ascorbic acid. Exp. Eye Res. 47, 155-68.

Ortwerth, B.J. and Olesen, P.R. (1988a) Ascorbic acid-induced crosslinking of lens proteins: evidence supporting a Maillard reaction. Biochim. Biophys. Acta 956, 10-22.

Ortwerth, B.J. and Olesen, P.R. (1988b) Glutathione inhibits the glycation and crosslinking of lens proteins by ascorbic acid. Exp. Eye Res. 47, 737-50.

Ortwerth, B.J., Speaker, J.A., Prabhakaram, M., Lopez, M.G., Li, E.Y. and Feather, M.S. (1994) Ascorbic acid glycation: the reactions of L-threose in lens tissue. Exp. Eye Res. submitted.

Prabhakaram, M. and Ortwerth, B.J. (1991). The glycation-associated crosslinking of lens proteins by ascorbic acid is not mediated by oxygen free radicals. Exp. Eye Res. 53, 261-8.

Reddy, V.N., Schwass, D., Chakrapani, B. and Lim, C.P. (1976) Biochemical changes associated with the development and reversal of galactose cataracts. Exp. Eye Res. 23, 483-93.

Richard, S., Tamas, C., Sell, D.R. and Monnier, V.M. (1991). Tissue-specific effects of aldose reductase inhibition on fluorescence and cross-linking of extracellular matrix in chronic galactosemia. Diabetes 40m 1049-1056.

Rose, R.C. and Bode, A.M. (1992). Tissue-mediated regeneration of ascorbic acid: is the process enzymatic? Enzyme 46, 196-203.

Sippel, T.O. (1966) Changes in the water, protein and glutathione contents of the lens in the course of galactose cataract development in rats. Exp. Eye Res. 5, 568-75.

Srivastava, S.K., Ansari, N.H., Bhatnagar, A., Hair, G. Liu, S. and Das, B. (1989) Activation of aldose reductase by nonenzymatic glycosylation in The Maillard Reaction in Aging, Diabetes and Nutrition (Eds. Baynes, J.W. and Monier, V.M.) Pp. 171-184. Alan R. Liss, Inc. New York, NY.

Stevens, V.J., Rouzer, C.A., Monnier, V.M. and Cerami, A. (1978) Diabetic cataract formation: potential role of glycosylation of lens crystallins. Proc. Nat. Acad. Sci. U.S.A. 75, 2918-22.

Taylor, A., Jacques, P.F., Nadler, D., Morrow, F., Sulsky, S.I. and Shepard, D. (1991) Relationship in humans between ascorbic acid consumption and levels of total and reduced ascorbic acid in lens, aqueous humor and plasma. Curr. Eye Res. 10, 751-9.

Winkler, B.S. (1987). In vitro oxidation of ascorbic acid and its prevention by GSH. Biochim. Biophys. Acta 925, 258-64.

Wolff, S.P. and Crabbe, M.J.C. (1985) Low apparent aldose reductase activity produced by monosaccharide autooxidation. Biochem. J. 226, 625-30.

Advanced Maillard Reaction in Sugar-induced Cataract: Studies with Molecular Markers

R. H. Nagaraj and V. M. Monnier

INSTITUTE OF PATHOLOGY, CASE WESTERN RESERVE UNIVERSITY, CLEVELAND, OH 44106, USA

Summary
The advanced Maillard reaction was investigated in lens proteins of galactosemic rats and dogs. The relationship between the degree of metabolic control and advanced Maillard reactions was investigated in long-term diabetes in dogs. The effect of polyol pathway activation on advanced Maillard reactions was studied by treating the animals with an aldose reductase inhibitor, sorbinil. Advanced Maillard reactions were studied using three different parameters: (1) protein-bound fluorescence, (2) pentosidine, and (3) fluorophore LM-1. In galactosemic rats, both protein-bound fluorescence and pentosidine accumulated significantly in comparison to controls, and sorbinil treatment almost completely prevented these changes. A close association between the degree of metabolic control and advanced Maillard reactions tested by all three parameters was noted in diabetic dogs. In addition, in diabetic dogs, pentosidine formation revealed the requirement of a glycemic threshold for enhancement of advanced Maillard reactions.

Introduction

The eye lens is one of the best models to study the advanced stages of the Maillard reaction (advanced glycation) *in vivo*. Proteins in the lens, known as crystallins, hardly turn over and thus accumulate glycation end products in aging and diabetes. Recently, several advanced Maillard reaction products (Advanced Glycation End Products, AGEs) have been isolated and characterized and shown to be present in human lenses. These include, carboxymethyllysine (Lyons et al., 1991), pentosidine (Nagaraj et al., 1991, Lyons et al., 1991) and fluorophore LM-1 (Nagaraj and Monnier, 1992). An association between lens crystallin pigmentation and pentosidine formation was recently demonstrated in human lenses (Nagaraj et al., 1991). Studies with antibodies have also established a role for advanced Maillard reaction products in lens aging and cataractogenesis (Araki et al., 1992). These observations strongly suggest that the Maillard reaction plays an important role in pigmentation and perhaps cataractogenesis. On the other hand, studies using aldose reductase inhibitors have established polyol pathway activation in sugar-induced cataractogenesis. We have therefore investigated the relationship between these two pathways in experimentally-induced cataracts in rats and dogs.

Several studies have shown that strict glycemic control from the very onset of diabetes can delay the onset of some of the complications of diabetes (Engerman and Kern, 1989). In order to understand the effect of long-term glycemic control on advanced Maillard reactions, we have quantified fluorophore LM-1 and pentosidine in chronic diabetic dogs maintained on three different metabolic control levels (Nagaraj et al., 1993).

Materials and Methods

Treatment of rats

Sprague-Dawley rats were made galactosemic by feeding 33% galactose in the diet and maintained for one year. A group of rats on galactose diet received in addition, sorbinil in the diet (25 mg/Kg diet). Control and control+sorbinil groups of animals were maintained simultaneously.

Treatment of dogs

Dogs were treated and maintained at the facility of Drs. T. Kern and R. Engerman at the University of Wisconsin, Madison, WI. Dogs (mixed breed) were made diabetic by alloxan injection and maintained for 5 years at three metabolic control levels, i.e., good control (GC, average plasma glucose 6 mM; control= 4 mM), moderate control (MC, 12 mM) and poor control (PC, 18 mM). Galactosemia was induced by feeding 30% galactose in the diet. Sorbinil at a dosage of 20 and 60 mg/Kg/Day was given to a group of diabetic and galactosemic dogs, respectively.

Incubation experiments

Bovine lens crystallins were incubated with ascorbate and its oxidation products, and sugars at 37^{0}C. Aliquots were withdrawn at fixed time intervals and assayed for pentosidine.

Assay of advanced Maillard reaction products

Pentosidine and fluorophore LM-1 were estimated by the HPLC methods as described elsewhere (Nagaraj et al., 1991; Nagaraj and Monnier, 1992).

Results

A significant increase in crystallin-bound fluorescence was observed in galactosemic rat lens compared to control ($p<0.0001$, Figure 1). Sorbinil treatment prevented fluorescence accumulation up to 67 and 52% in water soluble and insoluble fractions, respectively. Control and control+sorbinil had almost similar levels.

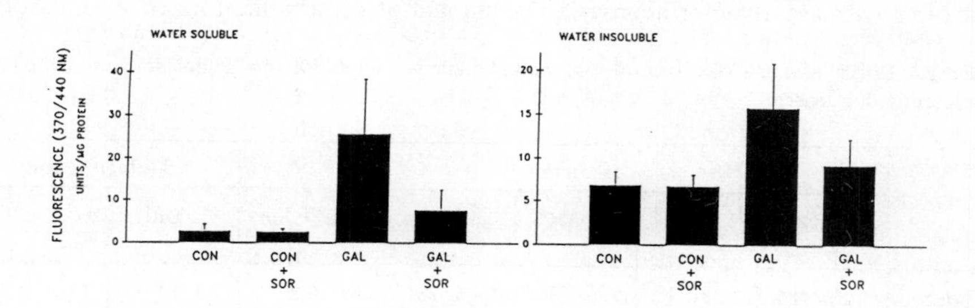

Figure 1. Effect of chronic galactosemia and inhibition of polyol pathway on protein-bound fluorescence (excit/emiss-370/440 nm) in rat lens crystallins.

Pentosidine levels were 4-6 times higher in galactosemic rats compared to controls (p<0.0001, Figure 2). Inhibition of aldose reductase completely prevented excessive synthesis of pentosidine.

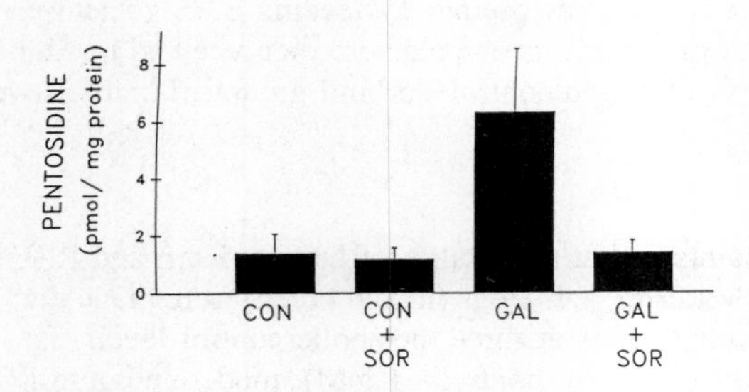

Figure 2. Effect of chronic galactosemia and polyol pathway inhibition on pentosidine in water insoluble crystallins of rats.

No significant changes in pentosidine levels were observed in good and moderate glycemic control animals, compared to controls. However, lenses from diabetic dogs in poor control, which were cataractous and highly pigmented, had a dramatic elevation of pentosidine, 4 and 6 times higher than controls in both water soluble and insoluble crystallins, respectively (p<0.0001). (Table I). Sorbinil treatment resulted in about 30% and 60% inhibition of pentosidine synthesis in water soluble and insoluble crystallins, respectively. These results suggest the presence of glycemic threshold, possibly linked to polyol pathway activation, for enhancement in the synthesis of pentosidine. Galactosemia induced only a marginal increase in pentosidine levels. Sorbinil treatment in galactosemia suppressed pentosidine synthesis only in water soluble crystallins (30%) but had no effect in water soluble crystallins. A more pronounced effect of glycemic control was observed for fluorophore LM-1. The levels were 3.9 and 5.4 times higher in water soluble and 3.8 and 6.2 times higher in water insoluble fractions of animals with moderate and poor control when compared with animals in good control, respectively (Table II).

Levels in water insoluble fractions in general were 2-4 times higher than water soluble crystallins. On the other hand, galactosemia had no effect on fluorophore LM-1 level. The nature of pentosidine precursors in galactosemia was investigated by incubating lens crystallins with ascorbate or hexoses. The amount of pentosidine formed from ascorbate

Table I. Effect of metabolic control and polyol pathway inhibition on pentosidine in diabetic and galactosemic dog lenses.

	Diabetic*					**Galactosemic***	
	Con	**GC**	**MC**	**PC**	**MC + Sor**	**Gal**	**Gal + Sor**
Water Soluble	1.2±0.48[a]	1.3±0.3[a]	1.4±.73[a]	5.1±3.2[b]	0.96±.29[a]	1.48±1.0[a]	1.03±.63[a]
Water Insol.	0.9±0.58[a]	0.8±0[a]	2.7±2.4[a]	13.8±13.9[b]	1.07±.26[a]	1.18±.41[a]	1.64±1.23[a]

Values are expressed as pmol/mg protein
*(mean ± SD). Values which do not share common superscript are statistically significant (P < 0.05)

Table II. Effect of metabolic control and polyol pathway inhibition on fluorophore LM-1 in diabetic and galactosemic dog lenses.

	Diabetic*					Galactosemic*	
	Con	**GC**	**MC**	**PC**	**MC + Sor**	**Gal**	**Gal + Sor**
Water Soluble	0.7±.4[a]	0.6±.11[a]	2.6±1.0[b]	3.6±1.8[b]	2.6±.61[b]	1.41±.52[c]	1.27±.52[c]
Water Insol.	0.8±.4[a]	1.7±.7[a]	6.7±2.7[b]	10.9±5.2[c]	6.36±1.8[b]	2.13±.69[c]	2.15±.69[c]

*Relative peak area/mg protein (mean ± SD)
Values which do not share common superscript are statistically significant ($P < 0.05$)

was 40 times higher than that from galactose after 3 weeks of incubation. Galactitol, on the other hand, was unable to form pentosidine. Glutathione at 8mM concentration, both in the oxidized and the reduced state, inhibited pentosidine formation from ascorbate up to 60%. This inhibition may be due to the reaction of the free amino group of glutathione with ascorbate oxidation products, since ϵ-aminocaproic acid at equimolar concentration showed inhibition similar to glutathione. Ascorbate oxidation products formed pentosidine when incubated with lens crystallins, the highest rate being with threose followed by xylosone, 2,3-diketogulonate and dehydroascorbate.

Discussion

The purpose of these studies was to investigate the relationship between advanced Maillard reactions, the degree of glycemic control and cataractogenesis utilizing polyol pathway activation as a cataractogenic stress. Significant accumulation of protein-linked fluorescence and pentosidine in galactosemic rat lenses and the total inhibition of pentosidine synthesis by the aldose reductase inhibitor suggest an association between polyol pathway activity and advanced Maillard reactions in sugar-induced cataracts. Ascorbate appears to be the most likely precursor of pentosidine in sugar-induced cataract based on the following reasons: (1) Ascorbate concentration in the lens is very high (almost 2 mM), and it can easily undergo oxidation and fragmentation under oxidative conditions to form highly potent pentosidine precursors; (2) decreased synthesis and leakage of glutathione from fiber cells, and accumulation of metal ions may accelerate ascorbate oxidation in hyperglycemia: and, (3) galactose may not be the primary precursor of pentosidine, since in galactosemic animals treated with sorbinil pentosidine levels were dramatically reduced, even though galactose levels were comparable to untreated animals. However, glycoxidation of "galactated" lysine residues (Dunn et al., 1989) cannot be completely excluded at this time. The observations above imply that in hyperglycemia, ascorbate undergoes enhanced oxidation to dehydroascorbate, which undergoes spontaneous hydrolysis to form 2,3-Diketogulonate. 2,3-Diketogulonate undergoes further oxidation to form threose or decarboxylation to form xylosone. These products react with lysyl and arginyl residues in crystallins to form pentosidine. Inhibition of pentosidine synthesis by sorbinil may be due to its ability to restore cellular glutathione levels or its ability to bind to metal ions which promote oxidation of ascorbate. It should be noted, however, that in spite of its attractiveness, the hypothesis involving ascorbate oxidation products is not yet supported by direct evidence. In summary, studies in hyperglycemic dogs revealed a relationship between the degree of glycemic control and advanced Maillard reactions in lens proteins.The most significant finding in this study is the requirement of a glycemic threshold for the excessive synthesis of pentosidine in diabetes. Interestingly, fluorophore LM-1 did not appear to require such a threshold, suggesting that the two fluorophores are synthesized by two different mechanisms. One possibility is that in poorly controlled diabetes, significant oxidative stress in the lens favors oxidative Maillard reaction leading thus to excessive synthesis of pentosidine, unlike fluorophore LM-1 which does not require oxidative steps to form from ascorbate.

Conclusions

These studies revealed that in hyperglycemia, polyol pathway activation can lead to acceleration of advanced Maillard reactions in lens crystallins. In diabetic dogs, advanced Maillard reactions closely follow the degree of glycemic control except for pentosidine formation in diabetic lens for which a glycemic threshold was observed. This suggests that, during formation of pentosidine in diabetic lens, a dramatic increase in oxidative stress occurs in poorly controlled diabetic dogs.

Acknowledgement
These studies were supported from NIH grants EY 07099 and AG 05601.

References

Araki, N., Ueno, N., Chakrabarti, B., Morino, Y., and Horiuchi, S. (1992). Immunochemical evidence for the presence of advanced glycation end products in human lens proteins and its positive correlation with aging. J. Biol. Chem. 267: 10211-10214.

Dunn, J.A., Patrick, J.S., Thorpe, S.R. and Baynes, J.W. (1989). Oxidation of glycated proteins: Age-dependent accumulation of N^{ϵ}-(carboxymethyl)lysine in lens proteins. Biochemistry 28: 9464.

Engerman., R.L. and Kern., T.S. (1989). Hyperglycemia and development of glomerular pathology: Diabetes compared with galactosemia. Kidney Int. 36: 41-45.

Lyons, T.J., Silverstri, G., Dunn, J.A., Dyer, D.G., and Baynes J.W. (1991). Role of glycation in modification of lens crystallins in diabetic and non-diabetic senile cataracts. Diabetes 40: 1010-1015.

Nagaraj, R.H., Sell, D.R. Prabhakaram, M., Ortwerth, B.J., and Monnier, V.M. (1991). High correlation between pentosidine protein crosslinks and pigmentation implicates ascorbate oxidation in human lens senescence and cataractogenesis. Proc.Natl. Acad. Sci.88: 10257-10261.

Nagaraj, R.H. and Monnier V.M. (1992). Isolation and characterization of a blue fluorophore from human lens crystallins: In vitro formation from the Maillard reaction with ascorbate and ribose. Biochim. Biophys. Acta. 1116: 34-42.

Nagaraj, R.H., Kern, T.S., Engerman, R.L. and Monnier, V.M., (1993). Advanced Maillard reaction in the lens of hyperglycemic dogs follows the degree of plasma glucose level. Invest. Ophthalmol. Vis. Sci. 34: 1372.

A Receptor Recognizing ε-Fructosyllysine in Glycated Albumin

S. Krantz, R. Salazar, and R. Brandt

INSTITUTE OF BIOCHEMISTRY, ERNST-MORITZ-ARNDT-UNIVERSITY OF GREIFSWALD, KLINIKUM D-17487 GREIFSWALD, GERMANY

Summary

On rat macrophages, human monocytes and human monocyte-like cell lines fructosyllysine specific binding sites could be determined, which are not identical with AGE, mannose/fucose and general scavenger receptors, as well as with binding sites for formaldehyde-modified albumins.

Introduction

It is well established, that glycated proteins may play a significant role in aging and in the development of diabetic late complications. Studies by Vlassara and coworkers (for review see Vlassara, 1992) have demonstrated, that AGE-modified proteins can be bound by several cell types via specific binding proteins. In the following, experiments are described, which have been performed to characterize a fructose-lysine specific receptor on rat macrophages and human monocytes.

Materials and Methods

Preparation of glycated albumins

Rat and human albumins were incubated at pH 7.4 and 37°C with 100 mM glucose for 3 days. Glycated proteins were purified by boronic acid affinity chromatography. These proteins containing 0.94 mol ketoamine/mol albumin were bound to bovine aortic endothelial cells only nonspecifically and were not able to displace AGE-modified albumin from its binding sites on these cells. Thus, it was assumed, short-term glycated albumin not to be significantly contaminated by AGEs. AGE-albumin was formed by incubation of the protein with 100 mM glucose at pH 7.4 during 5 months.

Cells

For binding studies rat peritoneal cells (40% macrophages), human monocytes, prepared according to Gilcrease and Hoover, 1990, purity 60 to 80%, and the monocyte-like cell lines U937, MonoMac 6, THP-1 and HL-60 were used. U937, HL-60 and THP-1 cells were grown in RPMI 1640, containing 10% heat-inactivated fetal calf serum and 2% kanamycin. MonoMac 6 cells were grown in Optimem 1 containing 10% myoclone plus fetal calf serum, nonessential amino acids, 200 U/mL penicillin and 200 μg/mL streptomycin.

Binding assays

Binding assays were carried out in collagen coated polyethylene tubes as described (Krantz et al., 1993).

Cross-linking experiments

Cross-linking of ^{125}I-labeled short-term glycated albumin (specific activity 8000 cpm/ng)

to rat macrophages or the monocyte-like cells was performed according to Radoff et al., 1988. Solubilized membrane proteins were separated by SDS-PAGE on 4-15% gradient gels. Labeled protein fractions were detected by autoradiography.

Ligand blotting
Crude membrane protein preparations from 2 x 10^8 monocyte-like cells were separated by SDS-PAGE using 4-15% gradient gels under nonreducing conditions. Proteins were electrophoretically transferred onto nitrocellulose paper. Thereafter, nonspecific binding sites were blocked with 5% milk powder. Nitrocellulose strips were next incubated with ^{125}I-labeled glycated albumin for 16 hours at 4°C. After intensive washings autoradiography was performed.

Production of cytokines by monocyte-like cells in response to glycated albumin
Incubation of MonoMac 6, U937 and HL-60 cells (10^6 cells/mL) was performed as described by Vlassara et al., 1988. IL-1 and TNF production was determined in cell lysates and culture fluids by ELISA (Biochrom).

Results and Discussion

Rat macrophages
As recently reported (Krantz et al., 1993), it was found, that after incubation of rat peritoneal cells with increasing amounts of glycated albumin binding increased in a saturable fashion. From analysis of the binding data an average association constant of 3.15 10^7 M^{-1} and 11,000 binding sites per cell could be calculated.

Glycated rat and human albumin were able to compete for binding with ^{125}I-labeled glycated rat albumin in a concentration dependent manner. Native albumin did not displace glycated albumin. Low concentrations of glycated polylysine (m.w. 3500), fructosyllysine and fructosyl-ß-alanine increased the binding of glycated albumin, higher concentrations were competitive. Fructose-lysine was more effective than fructose-ß-alanine. Hexitollysine was not competitive. For the determination of the specificity of cell binding for Amadori adducts experiments comparing the competition of short-term glycated albumin by AGE-modified albumin and vice versa were performed. AGE-albumin due to its content of 1.6 mol fructosyllysine/mol could compete for binding of radiolabeled glycated albumin in the same fashion as short-term glycated albumin. But in contrast, short-term glycated albumin failed to compete for binding of AGE-albumin to the same extent as AGE-albumin, indicating that AGE-albumin may be bound to at least two distinct sites by macrophages. Maleylated albumin, chondroitin sulfates, glucose, fructose and L-lysine were not competitive.

At 4°C, degradation of bound glycated albumin by macrophages could not be demonstrated. During incubation at 37°C increasing amounts of non-acid precipitable radioactivity appeared in the culture fluid, indicating internalization and degradation of the bound protein.

Chemical cross-linking of ^{125}I-labeled glycated albumin to its receptor revealed a main binding protein of 290 kDa and an additional protein fraction with a molecular mass of 230 kDa. Native albumin did not prevent the cross-link, whereas an excess of glycated albumin inhibited cross-linking.

Human monocytes and monocyte-like cell lines

The binding reaction of glycated albumin with sites on freshly isolated monocytes was similar to the interactions described for macrophages. Data from Scatchard analysis were: $K_a = 9.3\ 10^6\ M^{-1}$ and N = 5000 to 10,000 sites per cell. Monocytes, which did not show binding sites for fructosyllysine - a fact, that will be discussed later, bound radiolabeled AGE-albumin. Short-term glycated albumin was not competitive for AGE. This result is consistent with the binding studies on macrophages, confirming that sites for AGE and short-term glycated albumin are not identical.

Test systems, which require monocytes, freshly prepared from blood, have limited applicability, because their preparation is time-consuming and labor-intensive. Therefore, the monocyte-like cell lines MonoMac 6, U937, THP-1 and HL-60, which retain some monocyte characteristics, have been investigated. THP-1 and HL-60 cells did not show specific binding for glycated albumin. The binding behaviour of U937 was very similar to blood monocytes. The binding parameters were $K_a = 9.6\ 10^6\ M^{-1}$ and N = 8000 to 10,000 sites per cell. MonoMac 6 cells possessed similar affinities, but the number of binding sites, 2000 per cell, was low.

Fructosyllysine was competitive for binding of glycated albumin. But in comparison with monocytes, the initial increased binding could not be observed. Hexitollysine and mannose-albumin (neoglycoprotein) were not competitive.

Ligand-receptor cross-linking, using MonoMac 6 and U937 cells, showed two binding proteins of 190 and 90 kDa. In THP-1 and HL-60 cells these proteins were not demonstrable. After direct ligand blotting also two protein fractions of 190 and 90 kDa could be detected in MonoMac 6 and U937 cell membranes, but not in HL-60 and THP-1 cells. Ligand blot detection of fructosyllysine specific binding proteins was only possible, when proteins were not reduced by SH-reagents prior to electrophoresis.

Vlassara et al., 1988, have reported on a stimulated TNF and IL-1 production by human monocytes in response to AGE-binding. We have tested such a response with short-term glycated albumin by the monocytic cell lines MonoMac 6, U937 and HL-60. The stimuli, which release cytokines from these cells, are less well characterized than those for macrophages or monocytes. U937 and HL-60 cells did not show any reactions in response to LPS, AGE and glycated albumin on the production of IL-1 and TNF. MonoMac 6 cells did not secrete IL-1, but intracellularly, dependent on the stimulus used, different amounts of IL-1 accumulated (Table I). Cell associated TNF was only moderately increased after exposure to AGE and glycated albumin. On the basis of these observations it may be suggested that not only AGE, but also Amadori adducts, stimulate monocytes to produce cytokines.

Table I. IL-1 production by MonoMac 6 cells after exposure to glycated albumins.

Ligand	IL-1ß (pq/mL)
LPS (20 ng/mL)	274 ± 237
AGE-albumin (250 μg/mL)	1489 ± 401 $p<0.02$
native albumin (250 μg/mL)	104 ± 91
glycated albumin (250 μg/mL)	265 ± 324
glycated albumin (500 μg/mL)	1631 ± 421 $p<0.02$
no ligands	61 ± 22

Data are means ± S.D. from 2 experiments. p was calculated from a t-test.

In contrast to AGE-receptors, which seem to be constitutively expressed, we found in only 40% of the investigated rats and in 80% of investigated humans (as a preliminary result) a specific binding of glycated albumin to macrophages or monocytes. A hypothesis, that different rates of development of complications in diabetic patients with similar degrees of chronic hyperglycemia can be the result of genetic variability in the expression of fructosyllysine specific binding sites, is intriguing, but demands further investigations.

Acknowledgements
The studies with macrophages and monocytes were financially supported by the Bundesministerium für Forschung und Technologie (promotion No. 07NBL02). The investigations with monocyte-like cells were supported by the Deutsche Forschungsgemeinschaft, Bonn-Bad Godesberg.

References

Gilcrease, M.Z., and Hoover, R.L. (1990) Activated human monocytes exhibit receptor-mediated adhesion to a non-enzymatically glycosylated protein substrate. *Diabetologia* 33, 329-333.

Krantz, S., Brandt, R., and Gromoll, B. (1993) Binding sites for short-term glycated albumin on peritoneal cells of the rat. *Biochim. Biophys. Acta* 1177, 15-24.

Radoff, S., Vlassara, H., and Cerami, A. (1988) Characterization of a solubilized cell surface binding protein on macrophages specific for proteins modified nonenzymatically by advanced glycosylation end products. *Arch. Biochem. Biophys.* 263, 418-423.

Vlassara, H. (1992) Receptor-mediated interactions of advanced glycosylation end products with cellular components within diabetic tissues. *Diabetes* 41 (Suppl. 2), 52-56.

Vlassara, H., Brownlee, M., Manogue, K.R., Dinarello, C.A., Pasagian, A. (1988) Cachectin/TNF and IL-1 induced by glucose-modified proteins: Role in normal tissue remodeling. Science 240, 1546-1548.

2-Oxoaldehyde Metabolizing Enzymes in Animal Tissues

H. Kato, H. S. Shin, Z. Q. Liang, T. Nishimura, and F. Hayase[1]

DEPARTMENT OF AGRICURAL CHEMISTRY, THE UNIVERSITY OF TOKYO, BUNKYO-KU, TOKYO 113, JAPAN; [1]DEPARTMENT OF AGRICURAL CHEMISTRY, MEIJI UNIVERSITY, KAWASAKI 214, JAPAN

Summary

In the advanced stage of protein glycation, Amadori compounds are degraded to 2-oxoaldehydes such as 3-deoxyglucosone (3DG) and glucosone, which react as cross-linkers. In the present paper, NADPH-dependent aldo-keto reductases in porcine tissues were investigated using methylglyoxal (MG), 3DG, and acetaldehyde as substrates. Three reductases, enzyme (I), (II), and (III), were purified from porcine liver. Enzyme (I) named 2-oxoaldehyde reductase was major enzyme metabolizing 3DG, but it was inactive toward alcohols, indicating that enzyme (I) is different from alcohol dehydrogenase (aldehyde reductase). Enzyme (II), which was identified as aldose reductase, was also active toward 3DG. Although enzyme (III) was inactive toward 3DG, it was much active toward acetaldehyde and MG, and showed typical property of aldehyde reductase, EC 1.1.1.2. 2-Oxoaldehyde reductase was purified from chicken liver, and this enzyme inhibited effectively polymerization of glycated lysozyme in vitro.

Introduction

Amadori-modified proteins generate 2-oxoaldehydes such as 3-deoxyglucose (3DG) under physiological condition (Kato et al., 1987). 3DG is toxic and cross-links protein molecules (Igaki et al.,1990). 3DG is also an intermediate in the formation of pyrraline, which is one of the advanced stage products of glycation (Hayase et al., 1989).

Simplest 2—oxoaldehyde, methylglyoxal (MG), has been known to be metabolized by glyoxalase system, and also by dehydrogenase and reductase. However, glyoxalase is inactive toward 3DG. Oimomi et al. (1989) reported purification of a NAD-dependent dehydrogenase from human liver, which was able to act toward 3DG. When 3DG was administered to rats orally or intravenously, 3-deoxyfructose was excreted in the urine as a major reduced product (Kato et al., 1990).

Accordingly, it was considered that 3DG was mainly metabolized by a reducing enzyme. In the present paper, we investigate 2-oxoaldehyde reducing enzymes in animal tissues.

Results and Discussion

Distribution and purification of 2-oxoaldehyde reducing enzymes in porcine tissues

We surveyed NADH- and NADPH-dependent reductase activity in porcine organs using MG, 3DG, and acetaldehyde as substrates (Hayase et al., 1991). As seen in Table I, activity of NADPH-dependent enzymes was higher than that of NADH-dependent ones. Activity toward acetaldehyde localized in liver, whereas activities toward MG and 3DG were widely distributed in each organ. NADH-dependent activity toward acetaldehyde in liver would be responsible for aldehyde reductase, EC 1.1.1.1. Kidney and liver contained much enzyme activity toward MG and 3DG.

Then, we fractionated 3DG-reducing activity in liver. After ammonium sulfate fractionation of acetone powder of porcine liver, the active fraction was applied to a column of DEAE-cellulose.

Table I. Distribution of methylglyoxal (MG), 3-deoxyglucosone (3DG), and acetaldehyde reducing enzyme activity in internal organs of porcine*.

Organ	NADH-dependent			NADPH-dependent		
	MG	3DG	acetal dehyde	MG	3DG	acetal dehyde
liver	192.1	31.3	41.4	395.3	191.2	148.9
kidney	18.6	20.1	1.4	321.0	358.7	3.4
lung	12.2	4.3	1.1	41.7	42.9	1.9
heart	4.3	3.5	1.1	19.9	16.2	0.8
pancreas	4.2	3.5	1.8	51.2	71.3	1.4
spleen	4.3	4.7	1.3	41.4	59.5	0.9
S.intestine	405	4.7	0.5	53.0	68.6	0.8
L.intestine	3.4	1.3	0.5	11.1	12.9	0.6
stomach	2.6	2.6	0	18.8	24.6	0.6
muscle	4.5	3.4	0.5	9.2	6.7	0.8
brain	2.6	2.7	0.5	13.3	12.2	0.5
eyeball	2.7	0.9	1.1	4.7	3.7	0.5

*reducing activity, μM/min/100g tissue

Major and minor peaks were separated. The major fraction was further purified with a 2nd DEAE-cellulose column and then with a hydroxyapatite column. There were two activity peaks. The major fraction was electrophoretically homogeneous. The enzyme was purified about 156-fold from acetone powder with 16.6% activity yield (Liang et al., 1991). The purified enzyme was designated as enzyme (I). The same enzyme was obtained from porcine kidney.

The minor activity fractions eluted from hydroxyapatite column mentioned above, were collected and further purified through chromatographies with Sephadex G-100, Blue Sepharose CL-6B, and 2nd Sephadex G-100 columns. Purification was about 125-fold from acetone powder with 0.8% activity yield (Liang et al., 1992). This enzyme was designated as enzyme (II).

As shown in Table I, much higher activity of NADPH-dependent reductase toward acetaldehyde was observed than that of NADH-dependent one. The major fractions of corresponding activity from the column of DEAE-cellulose were collected and purified through hydroxyapatite, Blue Sepharose CL-6B, and NADP-agarose columns. There were two activity peaks in NADP-agarose column chromatography, and the major fraction was electrophoretically homogeneous (Liang et al., 1991). This enzyme was designated as enzyme (III).

Substrate specificity and classification of three NADPH-dependent reductases
Table II shows substrate specificity rand some Km values of each purified enzyme. Enzyme (I) was highly active toward 2-oxoaldehydes such as 3DG, MG, and phenylglyoxal and also active toward hydrophilic monoaldehydes such as glucuronate and glyceraldehyde, whereas it was only weakly active toward alkylaldehydes such as acetaldehyde and propanal. This enzyme is different from aldose reductase, because it is inactive toward glucose. We named enzyme (I) 2-oxoaldehyde reductase. Enzyme (II) was also highly active toward 2-oxoaldehyde including 3DG. However, this enzyme was highly active toward dihydroxyacetone, and further, weakly active toward glucose. Therefore, the enzyme was identified as aldose reductase, EC 1.1.1.21. 2-Oxoaldehyde reductase reduced 3DG to 3-deoxyfructose and MG to acetol, indicating that the aldehyde

Table II. Substrate specificity and Km values of three NADPH-dependent reductases purified from porcine liver.

Substrate	Relative activity,% (Km, mM)		
	(I) 2-oxoaldehyde reductase	(II) aldose reductase	(III) aldehyde reductase
3-Deoxyglucose	100.0(2.1)	100.0(2.5)	0
Glucosone	31.3	36.3	-
Methylglyoxal	76.2(3.3)	132.9(1.9)	91.4(6.0)
Phenylglyoxal	115.0	180.9	20.0
Glyoxal	20.7	-	701
Glyceraldehyde	20.7	69.7(4.9)	-
Benzaldehyde	17.1	1704	-
Propanal	1.5	1.2	100.0
Acetaldehyde	1.3	1.2	85.7(1.5)
Glucuronate	58.7	-	-
Diacetyl	4.6	-	-
Dihydroxyacetone	2.6	141.9	0
Acetol	3.3	12.1	-
Glucose	0	1.1	-
Fructose	0	1.1	-

Each purified enzyme was incubated with 10mM of different substrate in 0.1 M sodium phosphate buffer (pH 7.4; in case of aldose reductase, pH 7.0) contianing 0.3 mM NADPH. - : not determined.

group was reduced to alcohol (Liang et al., 1991). In case of aldose reductase, however, reduced products remain unknown.

Enzyme (III) was inactive toward 3DG, whereas it was highly active toward acetaldehyde, propanal and MG. This enzyme showed reverse reaction toward alcohols in the presence of NADP+ . These properties indicate that enzyme (III) should be classified as alcohol dehydrogenase (aldehyde reductase), EC 1.1.1.2.

Recently, Kanazu et al. (1991) reported identity of 2-oxoaldehyde reductase with already reported aldehyde reductase (Bohren et al., 1989). Furthermore, Takahashi et al. (1993) determined the amino acid sequence of 2-oxoaldehyde reductase purified from rat liver, and indicated its identity with that of aldehyde reductase. However, we have found that this enzyme is inactive toward alcohols in the presence of NADP, indicating that the enzyme is not responsible to EC l.l.l.Z (Liang et al., 1992). Since enzyme (III) is responsible to EC 1.1.1.2 as mentioned above, enzyme (I), 2-oxoaldehyde reductase, should be classified differently from aldehyde reductase.

We also surveyed NADPH-dependent 2-oxoaldehyde reducing enzyme activity in chicken tissues, and the corresponding three reductases were purified from liver (Shin et al., 1991; Shin, 1992). In case of chicken liver, another enzyme, aldehyde reductase II, which was moderately active toward 3DG, was partially purified.

Inhibition of protein cross-linking by 2-oxoaldehyde reductase
By using 2-oxoaldehyde reductase purified from chicken liver, inhibition of cross-linking was examined in vitro. Lysozyme-(l0 mg/ ml) was incubated with glucose (200mM) in 0.lM phosphate buffer (pH 7.4) at 50°C for 7 days, followed by dialysis and lyophilization. Thus prepared glycated lysozyme was further incubated without glucose under physiological condition. The glycated lysozyme could polymerize even if glucose was not present. When 2-oxoaldehyde reductase was added, glycated lysozyme did not show any further polymerization. We analyzed carbonyl compounds in the incubation mixture, and it was found that 3DG was generated from glycated lysozyme and, in the presence of 2-oxoaldehyde reductase, the enzyme eliminated 3DG (Shin, 1992). These experimental results propose that 2-oxoaldehyde metabolizing enzymes would effectively control the advanced stage of protein glycation in vivo.

Conclusions

Animal tissues contain many NADPH-dependent aldo-keto reductases. 2-Oxoaldehyde reductase is a major enzyme for detoxification of 2-oxoaldehydes such as 3DG and glucosone generated from glycated proteins, and in addition, aldose reductase is also active toward 2-oxoaldehydes. The role of 2-oxoaldehyde metabolizing enzymes in the control of advanced glycation and aging needs to be elucidated. Aldehyde reductase, which is identical with EC 1.1.1.2, is highly active toward alkylaldehydes such as propanal and acetaldehyde, although this enzyme is inactive toward 3DG. Since alkylaldehydes are generated from lipid hydroperoxides, the role of alkylaldehyde metabolizing enzymes should not be overlooked.

References

Bohren, K.M., Bullock, B., Wermuth, B., and Gabbay, K.H. (1989). The aldoketo reductase superfamily. J. Biol. Chem. 264: 9547—9551.

Hayase, F., Nagaraj, R.H., Mivata, S., Njoroge, F.G., and Monnier, V.M. (1989). Aging of proteins: immunological detection of a glucose—derived pyrrole formed during Maillard reaction in vivo. J. Biol. Chem. 264: 3758-3764.

Hayase, F., Liang, Z.Q., Suzuki, Y., Chuyen, N.V., Shinoda, T., and Kato, H. (1991). Enzymatic metabolism of 3—deoxyglucosone, a Maillard intermediate. Amino Acids 1: 307—318.

Igaki, N., Sakai, M., Hata, H., Oimomi, M., Baba, S., and Kato, H. (1990). Effect of 3—deoxyglucosone on the Maillard reaction. Clin. Chem. 36: 631-634.

Kanazu, T., Shinoda, M., Nakayama, T., Deyashiki, Y., Hara, A., and Sawada, H. (1991). Aldehyde reductase is a major protein associated with 3—deoxyglucosone reductase activity in rat, pig and human livers. Biochem. J. 279: 903-906.

Kato, H., Cho, R.K., Okitani, A., and Hayase, F. (1987). Responsibility of 3-deoxyglucosone for the glucose—induced polymerization of proteins. Agric. Biol. Chem. 51: 683—689.

Kato, H., Chuyen, N.V., Shinoda, T., Sekiya, F., and Hayase, F. (1990). Metabolism of 3—deoxyglucosone, an intermediate compound in the Maillard reaction, administered orally or intravenously to rats. Biochim. Biophys. Acta 1035: 71-76.

Liang, Z.Q., Hayase, F., and Kato, H. (1991). Purification and characterization of NADPH-dependent 2-oxoaldehyde reductase from porcine liver. Eur. J. Biochem. 197: 373—379.

Liang, Z.Q., Hayase, F., and Kato, H. (1992). Aldose reductase from porcine liver metabolizing 3—deoxyglucosone, a Maillard reaction intermediate. Biosci. Biotech. Biochem. 56:1074-1078.

Oimomi, M., Hata, F., Igaki, N., Nakamichi, T., Baba, S., and Kato, H. (1989). Purification of α-ketoaldehyde dehydrogenase from the human liver and its possible significance in the control of glycation. Experientia 45: 463-466.

Shin, H.S., Nishimura, T., Hayase, F., and Kato, H. (1991). Purification and characterization of NADPH-dependent 2-oxoaldehyde reductase, from chicken liver. Agric. Biol. Chem. 55: 957-966.

Shin, H.S. PhD Thesis (1992). A liver enzyme, 2-oxoaldehyde reductase, effective to control the Maillard reaction in vivo. Faculty of Agriculture, The University of Tokyo.

Takahashi, M., Fujii, J., Teshima, T., Suzuki, K., Shiba, T., and Taniguchi, N. (1993). Identity of a major 3-deoxyglucosone-reducing enzyme with aldehyde reductase in rat liver established by amino acid sequencing and cDNA expression. Gene 127: 249—253.

Ketoaldehyde Detoxification Enzymes and Protection Against the Maillard Reaction

David L. Vander Jagt,[1] Lucy A. Hunsaker,[1] Lorrain M. Deck,[2] Brian B. Chamblee,[2] and Robert E. Royer[1]

[1]DEPARTMENT OF BIOCHEMISTRY, UNIVERSITY OF NEW MEXICO SCHOOL OF MEDICINE, ALBUQUERQUE, NM 87131, USA, [2]DEPARTMENT OF CHEMISTRY, UNIVERSITY OF NEW MEXICO, ALBUQUERQUE, NM 87131, USA

Summary

D-3-Deoxyglucosone (1), D-glucosone (2), L-3-deoxyxylosone (3), and L-xylosone (4) were compared with methylglyoxal as substrates for aldose reductase, aldehyde reductase and glyoxalase I. All of these 2-ketoaldehydes are substrates for aldose reductase and aldehyde reductase but only methylglyoxal is a substrate for glyoxalase I, suggesting that only methylglyoxal exists in solution in a form with a free 2-keto group.

Introduction

Dicarbonyls, especially 2-ketoaldehydes (2-oxoaldehydes), appear to play an important role in the complex chemistry of the Maillard reaction. Considerable attention has been directed to 3-deoxyglucosone (Kato *et al.*, 1989), glucosone (Cheng *et al.*, 1992), 3-deoxyxylosone and xylosone (Hirsch *et al.*, 1992; Nagaraj and Monnier, 1992), all of which are 2-ketoaldehydes (Figure 1) that are capable of crosslinking proteins (Kato *et al.*, 1989). In particular, 3-deoxyglucosone has received special attention. This ketoaldehyde has both intracellular and extracellular origins. 3-Deoxyglucosone is formed during hydrolysis of fructose 3-phosphate which is present in diabetic lens (Szwergold *et al.*1990). 3-Deoxyglucosone is also produced as an intermediate in the Maillard reaction between hexoses and proteins (Kato *et al.*, 1989) and has been detected in vivo (Knecht *et al.*, 1992). Related ketoaldehydes may be produced from sugars by autoxidative and glycoxidative processes (Hunt *et al.*, 1993; Wolff and Dean, 1987).

Less attention has been given to the role that methylglyoxal (Figure 1) may play in protein modification in aging and diabetes. Methylglyoxal has the basic structure of a 3-deoxyosone. Methylglyoxal is produced at increased levels in type I diabetes where both hyperglycemia and ketoacidosis contribute (Thornalley, 1990; Aleksandrovskii, 1992). Increased glycolytic flux produces increased methylglyoxal as a byproduct of the triosephosphate isomerase reaction (Richard, 1991). Methylglyoxal can modify proteins to form fluorescent products that are similar spectrally to those produced during non-enzymatic glycation by hexoses (Vander Jagt *et al.*, 1992).

In the present study, we have addressed the question of the metabolism of osones and 3-deoxyosones by aldo-keto reductases and by glyoxalase I. The ubiquitous glyoxalase system protects cells from methylglyoxal toxicity; however, the enzyme shows broad specificity for 2-ketoaldehydes (Vander Jagt, 1989). Methylglyoxal is also an excellent substrate for aldose reductase, a member of the aldo-keto reductase multigene family that has been implicated in the development of diabetic complications (Vander Jagt *et al.*, 1992). We have compared the substrate properties of methylglyoxal and of the other ketoaldehydes in Figure 1 with human aldose reductase and aldehyde reductase, and with glyoxalase I.

Materials and Methods

Human aldose reductase and aldehyde reductase were purified from psoas muscle and kidney, respectively, as described previously (Vander Jagt *et al.*, 1990a,b). Yeast glyoxalase-I GSH and NADPH were from Sigma. Osones 1 - 4 were synthesized according to published procedures

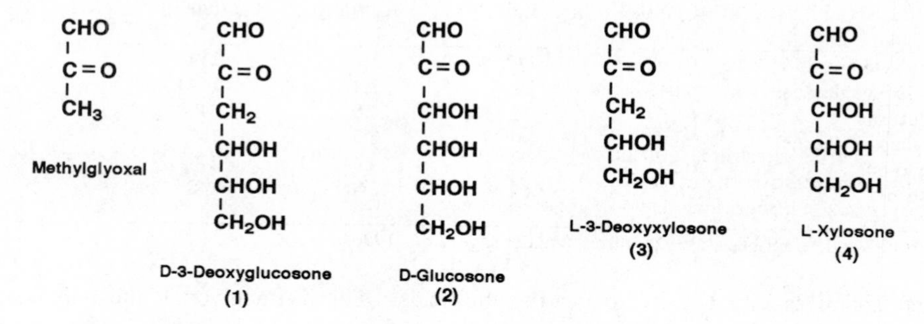

Figure 1. Structures of some physiologically important 2-ketoaldehydes.

(Hirsch *et al.*, 1992). Methylglyoxal was generated from methylglyoxal dimethylacetal (Aldrich) by acid hydrolysis of the acetal followed by distillation. Kinetic studies of aldose and aldehyde reductases were carried out at pH 7, 25°C. K_m and k_{cat} values were determined by nonlinear regression analysis of initial rate data using the Enzfitter program.

Results

Substrate Specificity of Human Aldose Reductase

Osones 1 - 4 were compared with methylglyoxal as substrates for aldose reductase (Table I). All five of these 2-ketoaldehydes are good substrates. The k_{cat} values are very similar, consistent with previous reports (Vander Jagt *et al.*, 1993). K_m values are lowest for methylglyoxal and 3-deoxyglucosone. These two substrates are 10^3 to 10^4 times better than glucose.

Substrate Specificity of Human Aldehyde Reductase

Osones 1 to 4 and methylglyoxal are also substrates for aldehyde reductase (Table I). However, K_m values are consistently much higher than for aldose reductase, consistent with previous observations. Under physiological conditions where methylglyoxal and, presumably, osones 1 to 4 are present at low concentrations, aldose reductase appears more likely than aldehyde reductase to function as a detoxification enzyme, especially in tissues like muscle that express mainly aldose reductase.

Substrate Specificity of Glyoxalase I

Among the compounds in Figure 1, only methylglyoxal is a substrate for glyoxalase I.

Discussion

The broad substrate specificity of glyoxalase I is well established and suggests that almost any 2-ketoaldehyde will be a substrate for this enzyme. Glyoxalase I and glyoxalase II provide a two reaction pathway that utilizes glutathione as coenzyme to convert 2-

Table I. Kinetic constants for human aldose reductase and aldehyde reductase with 2-Kktoaldehyde substrates produced physiologically.

Substrate	Aldose Reductase		Aldehyde Reductase	
	K_m (mM)	k_{cat} (min^{-1})	K_m (mM)	k_{cat} (min^{-1})
methylglyoxal[a]	0.008	142	1.2	214
3-deoxyglucosone	0.047	122	2.4	1443
glucosone	0.167	92	4.0	341
3-deoxyxylosone	0.078	167	2.7	1473
xylosone	0.234	123	2.6	277
glucose[b]	70	64	2500	246

[a]From Vander Jagt *et al.*, 1992; [b] From Vander Jagt *et al.*, 1990.

ketoaldehydes into D-2-hydroxycarboxylic acids. For methylglyoxal, the only known physiological substrate, this pathway results in formation of D-lactate. The fact that compounds 1 to 4 are not substrates for glyoxalase I suggests that they exist in solution entirely cyclized, with essentially no open chain 2-ketoaldehyde. It has been reported previously that 3-deoxyglucosone is not a substrate for glyoxalase I (Jellum, 1968).

It has also been reported that 3-deoxyglucosone is a substrate for an aldo-keto reductase from liver (Liang *et al.*, 1991, 1992). The similarity between reported K_m values and the data in Table I suggest that this reported aldo-keto reductase is aldehyde reductase. However, aldose reductase shows much higher affinity than aldehyde reductase for all of the 2-ketoaldehydes. Some physiological sources of 2-ketoaldehydes are summarized in Figures 2 and 3.

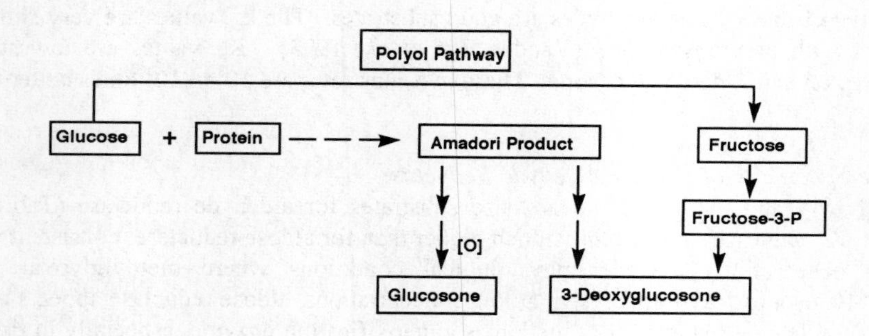

Figure 2. Both intracellular reactions, such as the polyol pathway, and extracellular reactions involving non-enzymatic glycosylation may contribute to the production of 2-ketoaldehydes, such as glucosone and 3-deoxyglucosone. As an intracellular enzyme, aldose reductase may only protect against 2-ketoaldehydes produced intracellularly.

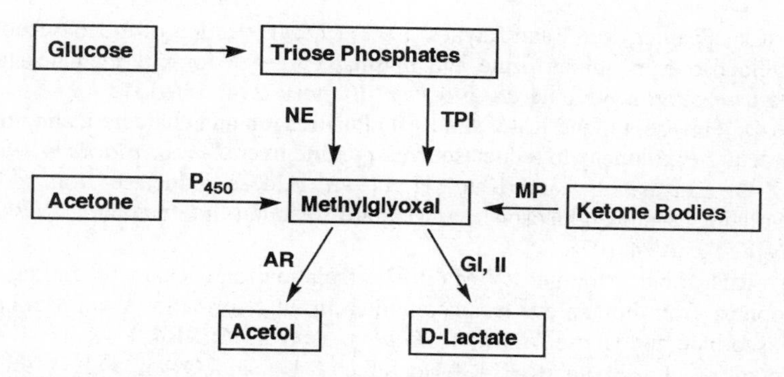

Figure 3. Methylglyoxal is formed non-enzymatically (NE) from triose phosphates and is produced as a byproduct of triose phosphate isomerase (TPI). In addition, acetone and ketone bodies are precursors of methylglyoxal through cytochrome P450 and myeloperoxidase reactions (Casazza *et al.*, 1984; Aleksandrovskii, 1992). Reduction of methylglyoxal to acetol increases in diabetes (*Reichard* et al., 1986). However, acetol may be toxic (Skutches *et al.*, 1990). Therefore, detoxification of methylglyoxal through conversion to D-lactate by the glyoxalase system may be the normal protective mechanism against methylglyoxal, and reduction of methylglyoxal to acetol through aldose reductase (polyol pathway) may be an undesirable pathway.

Conclusions

It is apparent from the data in Table I that most 2-ketoaldehydes are better substrates for aldose reductase than for aldehyde reductase, and that aldose reductase is more likely to be involved in protection against Maillard intermediates than is aldehyde reductase.

Acknowledgements
This work was supported by grant DK43238 from the National Institutes of Health.

References

Aleksandrovskii, Ya. A. (1992) Antithrombin III, CI inhibitor, methylglyoxal, and polymorphonuclear leukocytes in the development of vascular complications in diabetes mellitus. *Thrombosis Res.* 67:179-189.

Casazza, J.P., Felver, M.E., and Veech, R.L. (1984) The metabolism of acetone in rat. *J. Biol. Chem.* 259:231-236.

Cheng, R-Z., Uchida, K., and Kawakishi, S. (1992) Selective oxidation of histidine residues in proteins or peptides through the copper(II)-catalysed autoxidation of glucosone. *Biochem. J.* 285:667-671.

Hirsch, J., Petrakova, E., and Feather, M.S. (1992) The reaction of some dicarbonyl sugars with amino-guanidine. *Carbohydr. Res.* 232:125-130.

Hunt, J.V., Bottoms, M.A., and Mitchinson, M.J. (1993) Oxidative alterations in the experimental glycation model of diabetes mellitus are due to protein-glucose adduct oxidation. *Biochem. J.* 291:529-535.

Jellum, E. (1968) Metabolism of the ketoaldehyde 2-keto-3-deoxyglucose. *Biochim. Biophys. Acta* 165:357-363.

Kato, H., Hayase, F., Shin, D.B., Oimoni, M., and Baba, S. (1989) 3-Deoxyglucosone, an intermediate of the Maillard reaction. *Prog. Clin. Biol. Res.* 304: 69-84.

Knecht, K.J., Feather, M.S., and Baynes, J.W. (1992) Detection of 3-deoxyfructose and 3-deoxyglucosone in human urine and plasma: evidence for intermediate stages of the Maillard reaction in vivo. *Arch. Biochem. Biophys.* 294: 130-137.

Liang, Z-Q., Hayase, F., and Kato, H. (1991) Purification and characterization of NADPH-dependent 2-oxo-aldehyde reductase from porcine liver. *Eur. J. Biochem.* 197: 373-379.

Liang, Z-Q., Hayase, F., and Kato, H. (1992) Aldose reductase from porcine liver metabolizing 3-deoxyglucosone, a Maillard reaction intermediate. *Biosci. Biotech. Biochem.* 56: 1074-1078.

Nagaraj, R.H. and Monnier, V.M.(1992) Isolation and characterization of a blue fluorophore from human eye lens crystallins: In vitro formation from Maillard reaction with ascorbate and ribose. *Biochim. Biophys. Acta* 1116: 34-42.

Reichard, G.A., Jr., Skutches, C.L., Hoeldtke, R.D., and Owen, O.E. (1986) Acetone metabolism in humans during diabetic ketoacidosis. *Diabetes* 35:668-674.

Richard, J.P. (1991) Kinetic parameters for the elimination reaction catalyzed by triosephosphate isomerase and an estimation of the reaction's physiological significance. *Biochem.* 30:4581-4585.

Skutches, C.L., Owen, O.E., and Reichard, G.A., Jr., (1990) Acetone and acetol inhibition of insulin-stimulated glucose oxidation in adipose tissue and isolated adipocytes. *Diabetes* 39:450-455.

Szwergold, J. S., Kappler, F., and Brown, T.R. (1990) Identification of fructose 3-phosphate in the lens of diabetic rats. *Science* 247:451-454.

Thornalley, P.J. (1990) The glyoxalase system: new developments towards functional characterization of a metabolic pathway fundamental to biological life. *Biochem. J.* 269: 1-11.

Vander Jagt, D.L. (1989) The Glyoxalase System, in Glutathione: Chemical, Biochemical, and Medical Aspects- Part A, D. Dolphin, R. Poulson, and O. Avramovic, eds., J. Wiley, New York, pp. 598-641.

Vander Jagt, D.L., and Hunsaker, L.A. (1993) Substrate specificity of reduced and oxidized forms of human aldose reductase. *Adv. Exp. Med. Biol.* 328:279-288.

Vander Jagt, D.L., Robinson, B., Taylor, K.K., and Hunsaker, L.A. (1992) Reduction of trioses by NADPH-dependent aldo-keto reductases: Aldose reductase, methylglyoxal, and diabetic complications. *J. Biol. Chem.* 267:4364-4369.

Vander Jagt, D.L., Robinson, B., Taylor, K.K., and Hunsaker, L.A. (1990) Aldose reductase from human skeletal and heart muscle: Interconvertible forms related by thiol-disulfide exchange. *J. Biol. Chem.* 265:20982-20987.

Vander Jagt, D.L., Hunsaker, L.A., Robinson, B., Stangebye, L.A., and Deck, L.M. (1990) Aldehyde and aldose reductases from human placenta: Heterogeneous expression of multiple enzyme forms. *J. Biol. Chem.* 265:10912-10918.

Wolff, S.P., and Dean, R.T. (1987) Glucose autoxidation and protein modification: The potential role of "autoxidative glycosylation" in diabetes. *Biochem. J.* 245: 243-250.

In Vivo Effects of Aminoguanidine

G. Jerums, T. Soulis-Liparota, S. Panagiotopoulos, and M. E. Cooper

ENDOCRINE UNIT, AUSTIN HOSPITAL AND DEPARTMENT OF MEDICINE, HEIDELBERG REPATRIATION HOSPITAL, HEIDELBERG, 3084, AUSTRALIA

Summary
Over the last 10 years, the process of advanced glycation endproduct (AGE) formation has been implicated in the pathogenesis of many of the longterm complications of diabetes. Since the restoration of blood glucose levels to normal remains difficult, any treatment which can modify the effects of longterm hyperglycemia represents an attractive option for the prevention of diabetic complications. This report describes the effects of aminoguanidine on micro- and macroangiopathy in experimental diabetes. Acute effects of aminoguanidine in the diabetic state have included prevention of the increase in blood flow and permeability in a number of tissues including the eye and kidney. Aminoguanidine has also largely prevented the reduction in motor and sensory nerve conduction velocity and the increased arterial wall protein cross-linking that is observed in diabetes. Several studies have shown that aminoguanidine prevents the accumulation of AGEs in the kidney, with a concomitant reduction in albuminuria and renal structural changes. With respect to retinopathy, it has been shown that aminoguanidine prevents the diabetes-induced increase in number of acellular capillaries and reduces the magnitude of pericyte loss. Studies on experimental diabetic neuropathy have demonstrated that aminoguanidine prevents axonal atrophy and a decrease in nerve conduction velocity. Recently it has also been shown that aminoguanidine may prevent atherosclerosis in a non-diabetic model of atherosclerosis.

Introduction

Aminoguanidine, which is an inhibitor of advanced but not early glycation, is a nucleophilic hydrazine compound. Traditionally, the effect of aminoguanidine on tissue AGEs has been monitored by measuring collagen related fluorescence (Monnier et al., 1986). However, not all AGEs are fluorescent and the fluorescence assay is not entirely specific for AGEs. For example, tissue fluorescence may be increased by products of oxidation (Baynes, 1991) and may be decreased by aldose reductase inhibitors (Suárez et al., 1988; Odetti et al., 1990). Recent studies have evaluated the acute effects of aminoguanidine on vascular permeability. This is a complex process which is probably mediated by a cascade of reactions involving diacylglycerol, protein kinase C, phospholipase A_2, prostaglandins and nitric oxide (Williamson et al., 1993). There has been recent controversy about the role of aminoguanidine as an inhibitor of nitric oxide synthesis. It has been suggested that acute effects of aminoguanidine on vascular permeability may be linked to the nitric oxide pathway since aminoguanidine, as well as the nitric oxide inhibitor, n-monomethylarginine (Corbett et al., 1992) and an aminoguanidine analogue, methylguanidine, (Tilton et al., 1993) can all prevent increases in vascular permeability but only aminoguanidine can reduce AGE production. However, a contrasting effect of aminoguanidine has been suggested by studies in which the quenching of nitric oxide production by AGE products can be prevented by aminoguanidine (Bucala et al., 1991; Vlassara et al., 1992). Furthermore, a recent study in experimental neuropathy suggests that aminoguanidine may not inhibit nitric oxide synthesis under all conditions. For example, although aminoguanidine can prevent the defect in motor and sensory nerve conduction velocity in experimental diabetes (Cameron et al., 1992) nitric oxide inhibition accentuates this abnormality, therefore suggesting opposing effects of nitric oxide inhibitors and aminoguanidine (Cameron et al., 1993).

Aminoguanidine effects on extracellular matrix
The original study of the effects of aminoguanidine on collagen was performed in the alloxan diabetic rat, in which aminoguanidine was injected daily over a 16 week period (Brownlee et al., 1986). Assessment of arterial wall protein cross-linking by 3 separate methods revealed an increase in this parameter in diabetes which was prevented by aminoguanidine therapy. Shortly afterwards it was shown that the aldose reductase inhibitor, sorbinil, could also decrease collagen fluorescence in diabetic rats and this suggested that not only advanced glycation but also fructosylation could contribute to protein-bound fluorescence (Suarez et al., 1988). Subsequent studies revealed that both aminoguanidine and the aldose reductase inhibitor, rutin, could protect against the formation of fluorescent products. It was suggested that these two compounds were acting via different mechanisms since only rutin affected nerve sorbitol levels (Odetti et al., 1990). More recently, aminoguanidine has been shown to reduce the diabetes-related increase in collagen stability in rats treated for 120 days (Oxlund and Andreassen, 1992). Aminoguanidine has also been shown to improve reduced erythrocyte deformability in alloxan diabetic New Zealand white rabbits (Brown et al., 1993). These studies are all consistent with the hypothesis that alterations in collagen and other extracellular matrix proteins in experimental diabetes are caused by cross-links derived from advanced glycation products and that aminoguanidine prevents these changes (Brownlee, 1992).

Effects of aminoguanidine in diabetic nephropathy
In the streptozotocin-diabetic mouse, seven months of aminoguanidine therapy reduced AGEs in renal basement membranes as measured by tissue fluorescence per unit of hydroxyproline (Nicholls and Mandel, 1989). However, no functional or structural correlations were feasible in that model. Studies in our own laboratory in the streptozotocin diabetic rat have shown that 32 weeks of aminoguanidine therapy, from the time of induction of diabetes, retards the development of albuminuria and glomerular fluorescence (Figure 1. Soulis-Liparota et al., 1991). Fractional mesangial volume increased by 40% in diabetic glomeruli at 32 weeks and this was completely prevented by aminoguanidine therapy. However, glomerular basement membrane thickening was not influenced by aminoguanidine. In contrast, another study using aminoguanidine in a different rat strain has documented prevention of glomerular basement membrane thickening (Ellis and Good, 1991). In a model of diabetic hypertensive rats, an effect on albuminuria was also observed with aminoguanidine (Edelstein and Brownlee, 1992). In a 10 week study of proteinuria in diabetic rats, aminoguanidine reduced the excretion of albumin and high molecular weight proteins without influencing total proteinuria (Itakura et al., 1991). These studies suggest that aminoguanidine can prevent both functional and structural abnormalities of experimental diabetic nephropathy. Studies assessing fluorescence in the kidney confirm that aminoguanidine reduces this parameter not only in isolated glomeruli but also in glomerular basement membranes.

Effects of aminoguanidine in diabetic eye disease
As previously mentioned, Williamson's group has shown a reduction in diabetes-related vascular permeability with aminoguanidine therapy. Recent studies in our own laboratory have shown that retinal and anterior and posterior uveal albumin clearances are reduced by

aminoguanidine. By contrast, in vivo assessments of cataract formation by a retroillumination technique did not detect any beneficial effects of aminoguanidine after 12 weeks of streptozotocin diabetes (Panagiotopoulos et al., 1993A). There is only one publication on the effects of aminoguanidine on the development of experimental retinopathy (Hammes et al., 1991). After 75 weeks of diabetes untreated diabetic animals had a dramatic increase in the number of acellular capillaries and in capillary microaneurysms. Aminoguanidine therapy largely prevented the increase in acellular capillaries and completely prevented microaneurysm formation. In addition, aminoguanidine prevented the diabetes-induced proliferation of endothelial cells and reduced the magnitude of pericyte loss.

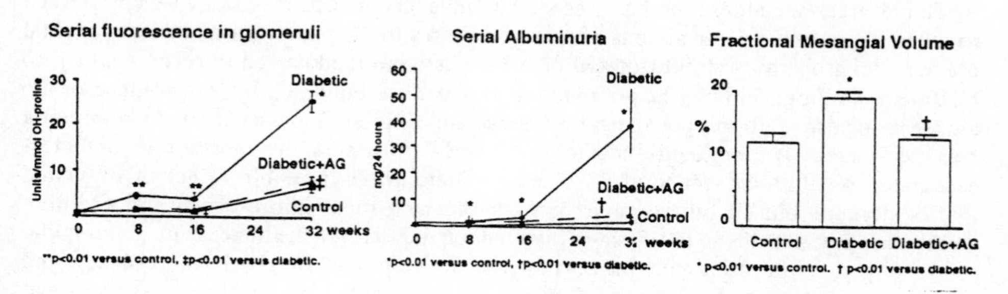

Figure 1.

Effects of aminoguanidine on diabetic peripheral neuropathy

Studies in rat sciatic nerve have shown that aminoguanidine prevents the increase in albumin vascular permeability that is observed in streptozotocin-induced diabetes. Nerve blood flow was normalised within 8 weeks and conduction velocity was also significantly improved in a dose-dependent manner by 16 and 24 weeks in sciatic, tibial and caudal nerves respectively (Kihara et al., 1991). It was suggested that aminoguanidine was acting by reversal of nerve ischaemia and that the improved electrophysiology was via an action on nerve microvessels. A morphometric study recently revealed that longterm aminoguanidine therapy inhibited the accumulation of fluorescent AGE in diabetic nerves and that this was associated with improvements in myelinated fibre size and degree of axonal atrophy (Yagihashi et al., 1992). A third study showed that 8 weeks of aminoguanidine therapy can prevent the diabetes-associated defects in sciatic motor nerve conduction velocity and sensory saphenous conduction velocity. These effects occurred in the absence of any effect of aminoguanidine on the polyol pathway or on myoinositol levels, suggesting that the effects of aminoguanidine are not mediated by aldose reductase inhibition but may involve a vascular mechanism (Cameron et al., 1992).

Studies in Progress

Recent studies in our laboratory have assessed whether aminoguanidine is more effective when administered early or late in the evolution of experimental diabetic nephropathy. Four groups of diabetic rats were studied over 32 weeks, one receiving no aminoguanidine, the second receiving aminoguanidine throughout the study period, the third receiving aminoguanidine for the first 16 weeks only and the fourth receiving aminoguanidine for the last 16 weeks. The greatest increase in albuminuria was observed in untreated diabetic rats and almost complete prevention of increases in urinary albumin excretion was seen in the group that received aminoguanidine throughout the study. Urinary albumin excretion in the early and late intervention groups was intermediate between the results of the fully treated and untreated animals. Glomerular and tubular fluorescence was increased approximately 3-fold in diabetic rats and was completely prevented in all aminoguanidine-treated groups irrespective of the duration or timing of therapy. The urinary albumin data confirm the hypothesis that the effects of aminoguanidine are related to duration of treatment. The lack of difference between early and late intervention was not consistent with the concept of hyperglycemic memory (Soulis-Liparota et al., 1993A).

Since the in vivo mechanism of action of aminoguanidine may include inhibition of oxidative stress and aldose reductase as well as advanced glycation, a study was performed to assess the relative contributions of these processes to the pathogenesis of experimental diabetic nephropathy. Streptozotocin diabetic rats were randomised to receive either no treatment, aminoguanidine, the antioxidants probucol or butylated hydroxytoluene or the aldose reductase inhibitor, ponalrestat. Assessment of tissue fluorescence and albuminuria was performed. Aminoguanidine therapy largely prevented the increase in diabetes-associated renal fluorescence at 32 weeks. There was a possible effect of butylated hydroxytoluene but all other agents had no effect on renal fluorescence. In addition, aminoguanidine was the most effective intervention in retarding the increase in albuminuria, suggesting that the dominant process in the development of renal fluorescence and albuminuria is advanced glycation. The lack of significant effects of the antioxidants and the inability of the aldose reductase inhibitor to prevent increases in renal fluorescence or urinary albumin excretion, despite reducing aortic fluorescence, is consistent with the hypothesis that aminoguanidine is most likely acting via advanced glycation rather than through inhibition of oxidative stress or aldose reductase activity (Soulis-Liparota et al., 1993B).

More recently, we have evaluated the effects of aminoguanidine at three dose levels (26-100mg/kg/day) on atherogenesis in the non-diabetic cholesterol-fed rabbit (Panagiotopoulos et al., 1993B). Aminoguanidine reduced atherosclerotic plaque formation as measured by Sudan IV staining in the arch and thoracic and abdominal aorta. These studies suggest that advanced glycation may be an important biochemical process in the genesis of atherosclerosis in the absence of hyperglycemia. The role of aminoguanidine in diabetes-associated atherosclerosis obviously warrants further study.

Acknowledgements
This work was supported by grants from the Juvenile Diabetes Foundation International (No. 191009), Commonwealth Health & Medical Research Council and Alteon Inc. We thank Mrs. B.Richmond for typing the manuscript.

References

Baynes, J.W. (1991) Role of oxidative stress in development of complications in diabetes. *Diabetes*. 40: 405-412.

Brown, C.D., Zhao, Z.H., DeAlvaro, F., Chan, S., and Friedman, EA. (1993) Correction of erythrocyte deformability defect in ALX-induced diabetic rabbits after treatment with aminoguanidine. *Diabetes*. 16: 590-593.

Brownlee, M. (1992) Nonenzymatic glycosylation of macromolecules. Prospects of Pharmacologic Modulation. *Diabetes*. 41: 57-60.

Brownlee, M., Vlassara, H., Kooney, T., Ulrich, P., Cerami, A.(1986) Aminoguanidine prevents diabetes-induced arterial wall protein cross-linking. *Science*. 232: 1629-1632.

Bucala, R., Tracey, K.J., Cerami, A. (1991) Advanced glycosylation products quench nitric oxide and mediate defective endothelium-dependent vasodilatation in experimental diabetes. *J. Clin. Invest.* 87: 432-438.

Cameron, N.E., Cotter, M.A., Dines, K., and Love, A. (1992) Effects of aminoguanidine on peripheral nerve function and polyol pathway metabolites in streptozotocin-diabetic rats. *Diabetologia* .35: 946-950.

Cameron, N.E., Cotter, M.A., Dines, K., and Maxfield, E.K. (1993) Pharmacological manipulation of vascular endothelium function in non-diabetic and streptozotocin-diabetic rats: effects on nerve conduction, hypoxic resistance and endoneurial capillarization. *Diabetologia* .36: 516-522.

Corbett, J.A., Tilton, R.G., Chang, K., Hasan, K.S., Ido, Y., Wu, J.L., Sweetland, M.A., Lancaster Jr, J.R., Williamson, J.R., McDaniel, M.L. (1992) Aminoguanidine, a novel inhibitor of nitric oxide formation, prevents diabetic vascular dysfunction. *Diabetes* .41: 552-556.

Edelstein, D., and Brownlee, M. (1992) Aminoguanidine ameliorates albuminuria in diabetic hypertensive rats. *Diabetologia* .35: 96-97.

Ellis, E.N., and Good, B.H. (1991) Prevention of glomerular basement membrane thickening by aminoguanidine in experimental diabetes mellitus. *Metabolism* .40: 1016-1019.

Hammes, H-P., Martin, S., Federlin, K., Geisen, K., and Brownlee, M. (1991) Aminoguanidine treatment inhibits the development of experimental diabetic retinopathy. *Proc. Natl. Acad. Sci.*. 88: 11555-11558.

Itakura, M., Yoshikawa, H., Bannai, C., Kato, M., Kunika, K., Kawakami, Y., Yamaoka, T., Yamashita, K. (1991) Aminoguanidine decreases urinary albumin and high-molecular weight proteins in diabetic rats. *Life Sciences* .49: 889-897.

Kihara, M., Schmelzer, J.D., Poduslo, J.F., Curran, G.L., Nickander, K.K., Low, P.A. (1991) Aminoguanidine effects on nerve blood flow, vascular permeability, electrophysiology, and oxygen free radicals. *Proc. Natl. Acad. Sci.*. 88: 6107-6111.

Monnier, VM., Vishwanath, V., Frank, KE., Elmets, C.A., Dauchot, P., Kohn, R.R. (1986). Relation between complications of type I diabetes mellitus and collagen-linked fluorescence. *N. Engl. J. Med.* 314: 403-408.

Nicholls, K., and Mandel, T.E. (1989) Advanced glycosylation end-products in experimental murine diabetic nephropathy: Effect of islet isografting and of aminoguanidine. *Laboratory Investigation* .60: 486-491.

Odetti, P.R., Borgoglio, A., DePascale, A., Rolandi, R., Adezati, L. (1990) Prevention of diabetes-increased aging effect on rat collagen-linked fluorescence by aminoguanidine and rutin. *Diabetes* .39: 796-801.

Oxlund, H., and Andreassen, T.T. (1992) Aminoguanidine treatment reduces the increase in collagen stability of rats with experimental diabetes mellitus. *Diabetologia* .35: 19-25.

Panagiotopoulos, S., Lim Joon, T., Foo, K., Jerums, G., Taylor, H. (1993A) In vivo assessment of an animal model of diabetic cataract: medical intervention studies. (Abstract) *Proceedings EASD,* September.

Panagiotopoulos, S., O'Brien, R.C., Cooper, M.E., Jerums, G., Bucala, R. (1993B) Aminoguanidine decreases atherosclerotic plaque formation in the cholesterol-fed rabbit. (Abstract). *5th Int Symp on Maillard Reaction*, Minnesota, USA.

Soulis-Liparota, T., Cooper, M., Papazoglou, D., Clarke, B., Jerums, G. (1991) Retardation by aminoguanidine of development of albuminuria, mesangial expansion, and tissue fluorescence in streptozocin-induced diabetic rat. *Diabetes* .40: 1328-1334.

Soulis-Liparota, T., Cooper, M., Jerums, G. (1993A) Roles of Advanced Glycation, oxidation and aldose reductase inhibition in diabetic nephropathy. (Abstract). *Proceedings* EASD, September.

Soulis-Liparota, T., Cooper, M.E., and Jerums, G.(1993B) Effects of early and late aminoguanidine therapy on experimental diabetic nephropathy. (Abstract). *5th Int Symp on Maillard Reaction*, Minnesota, USA.

Suárez, G., Rajaram, R, Bhuyan, K.C., Oronsky, A.L., Goidl, J.A. (1988) Administration of an aldose reductase inhibitor induces a decrease of collagen fluorescence in diabetic rats. *J. Clin. Invest.*. 82: 624-627.

Tilton, R.G., Chang, K., Hasan,K.S., Smith, S.R., Petrash, J.M., Misko, T.P., Moore, W.M., Currie, M.G., Corbett, J.A., McDaniel, M.L., Williamson, J.R. (1993) Prevention of diabetic vascular dysfunction by guanidines. Inhibition of nitric ozide synthase versus advanced glycation end-product formation. *Diabetes*. 42: 221-232.

Vlassara, H., Fuh, H., Makita, Z., Krungkrai, S., Cerami, A., Bucala, R. (1992) Exogenous advanced glycosylation end products induce complex vascular dysfunction in normal animals: A model for diabetic and aging complications. *Proc. Nat. Acad. Sci*.. 89: 12043-12047.

Williamson, J.R., Chang, K., Frangos, M., Hasan, K.S., Ido, Y, Kawamura, T, Nyengaard, J.R., Van den Enden, M., Kilo, C., Tilton, R.G. (1993) Hyperglycemic pseudohypoxia and diabetic complications. *Diabetes* .42: 801-813.

Yagihashi, S., Kamijo, M., Baba, M., Yagihashi, N., and Nagai, K. (1992) Effect of aminoguanidine on functional and structural abnormalities in peripheral nerve of STZ-induced diabetic rats. *Diabetes* .41: 47-52.

Aminoguanidine as an Inhibitor of the Maillard Reaction

J. Hirsch[1] and M. S. Feather[2]

[1]INSTITUTE OF CHEMISTRY, SLOVAK ACADEMY OF SCIENCES, BRATISLAVA, SLOVAKIA; [2]DEPARTMENT OF BIOCHEMISTRY, UNIVERSITY OF MISSOURI, COLUMBIA, MO 65211, USA

Summary

Aminoguanidine (guanylhydrazine) reacts rapidly and completely with sugar-derived dicarbonyl intermediates, produced during the Maillard reaction. The final reaction products are 5- and 6-substituted 3-aminotriazine derivatives.

Introduction

During the *in vivo* glycation stage of Maillard reactions, reducing sugars interact with protein amino groups to give 1-amino-1-deoxy-2-ketose derivatives (Amadori compounds). The latter undergo further degradation to give highly reactive, dicarbonyl intermediates, which are thought to play a role in the formation of protein crosslinks, and, possibly, some of the further reaction products observed for this process (advanced glycation end products). For example, Kato and his group (Kato, et al., 1987) have shown that 3-deoxy-**D**-*erythro*-glycos-2-ulose (3-deoxyglucosone), a product produced during Maillard reactions involving **D**-glucose, greatly accelerates the rate of protein crosslinking, and Baynes and co-workers have recently reported its detection and quantitative measurement in human serum and urine (Knecht, et al., 1992). Aminoguanidine (guanylhydrazine) has been shown to inhibit the Maillard reaction at *in vivo* conditions (Brownlee, et al., 1986), but the precise mechanism for the inhibition is not understood. The most plausible explanation of the inhibition is that aminoguanidine reacts with highly reactive, carbonyl-containing Maillard intermediates to give hydrazones, which are more unreactive and, hence, cannot react further in the latter stages of the reaction. Accordingly, we prepared several of the dicarbonyl intermediates that are known to be produced from carbohydrates, either during Maillard reactions or as a result of spontaneous degradation of sugars at physiological conditions. The reactants chosen for study were: 3-deoxy-**D**-*erythro*-hexos-2-ulose (3-deoxyglucosone, 1a), 3-deoxy-**D**-*glycero*-pentose-2-ulose (3-deoxyxylosone, 1b), **D**-*erythro*-hexos-2-ulose (glucosone, 1c), and **D**-*glycero*-pentose-2-ulose (xylosone, 1d). Compounds 1b and 1d are produced from ascorbic acid during its degradation at physiological conditions. Compound 1a, 3-deoxyglucosone is the most common degradation intermediate produced from a hexose-derived Amadori compound, and 1c, glucosone may be produced from hexose sugars via oxidation by molecular oxygen.

Material and Methods

Compound 1a was prepared as described in an earlier report (Madson and Feather, 1981), compound 1b, as described by Shin and Feather (1990), compound 1c as described by Bayne (1963) and 1d by the direct oxidation of **D**-xylose using cuprous oxide as described by Salomon and coworkers (1952). All compounds were obtained as slightly yellow syrups, which were chromatographically homogeneous by thin layer chromatography. Compounds were reacted with aminoguanidine (bicarbonate salt) as described below.

All incubations were carried out under essentially the same conditions. In a typical reaction, 1a (500 mg) in 25 mL of 0.2 M phosphate buffer (pH=7.0) was reacted with a 1.2 molar equivalent of aminoguanidine at 37 °C. Aliquots were removed from time to time and examined by TLC ($CHCl_3$-MeOH-H_2O, 7:3:0.3, v/v) using silica gel plates containing a fluorescent UV indicator. When the reaction was complete, the reaction solution was evaporated to dryness and the residue separated by conventional chromatography using 4 by 25 cm silica gel columns irrigated with the above TLC irrigant.

Results

The compounds produced in the reaction were UV absorbers, as evidenced by the TLC experiments, and all were obtained as crystalline materials. Compounds 1a and 1b both reacted very rapidly (within 40 minutes time) and each gave two chromatographically distinct reaction products, while 1c and 1d reacted slower, requiring several hours for completion of the reaction and gave only one UV absorbing reaction product each. The following data were collected on all compounds: ^1H NMR, decoupled ^{13}C NMR, as well as DEPT editing and HETCOR, electron impact mass spectrometry, elemental analysis, melting point and UV spectral data. These data have been published in an earlier report (Hirsch, et al., 1992) and serve, collectively, to identify the compounds as 5- and 6-substituted 3-amino-triazines, derived from the reaction of one mole of dicarbonyl sugar with one mole of aminoguanidine.

The analytical data collected on the compounds clearly show that all are 3-aminotriazine derivatives, and that the two triazines derived from 1a and 1b both have the same basic triazine ring structure, but have substituent groups at either positions 5 and 6 of the ring. This accounts for the fact that two distinct products were obtained. That the compounds were as assigned was further confirmed by crystal structure X-ray analysis of the reaction products derived from 1a. These structures appear in an earlier paper (Hirsch, et al., 1992). In some parallel experiments, we found that D-glucose reacts with aminoguanidine to give a syrupy hydrazone, which, on acetylation (pyridine-acetic anhydride) give a crystalline heptaacetate.

Discussion

Figure 1 shows a scheme that accounts for the formation of the triazines. We hypothesize that aminoguanidine can react with either of the two carbonyl groups of the osones to give either a hydrazone residue at C-1 or at C-2. Further cyclization and rearrangement of the hydrazone then gives the triazine derivative. Therefore, an initial reaction at C-1 of either 1a or 1b gives rise to a 5-substituted triazine, while an initial reaction at C-2 of the osone gives the 6-substituted isomer. It is not clear why 1c and 1d give only the 6-substituted isomer, but this may be due to the formation of a stable hydrogen bonded C-2 hydrazone (when formed) with the hydroxyl group situated at C-3 for these compounds. Computational work is now underway to examine this possibility. It is surprising that no evidence could be found for the formation of a bis-hydrazone. Wolfrom and co-workers (1964) reported the synthesis of the aminoguanidyl bis-hydrazone derived from 1c some time ago and we repeated this preparation, but, using it as a TLC standard, we could not confirm its presence in the incubation mixtures under the conditions used in this study.

Because hydrazone formation is an equilibrium condition, we incubated the **D**-glucose aminoguanidyl hydrazone with 1a at the same conditions under which triazines are formed and found that, as expected, an exchange reaction occurs, with triazines derived from 1a being formed, as evidenced by TLC.

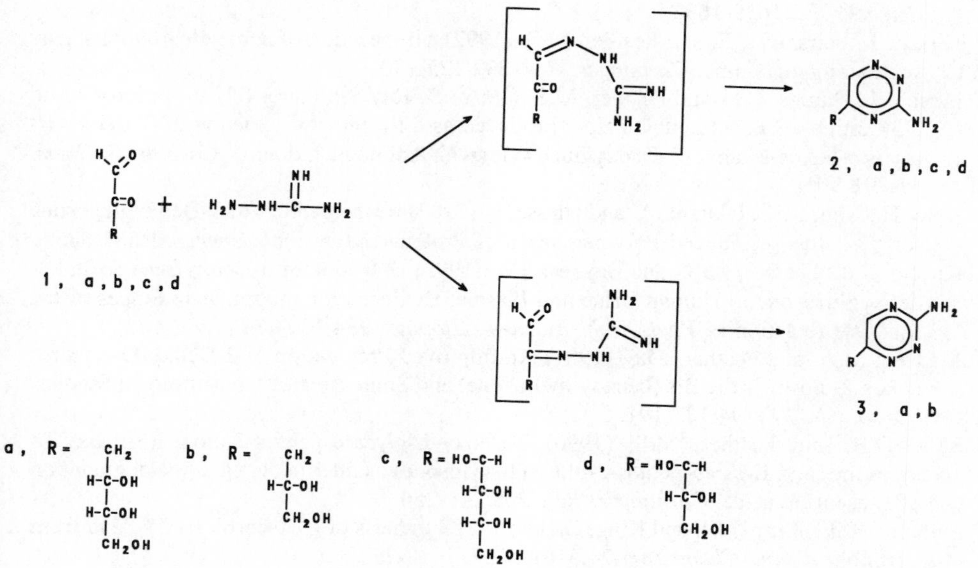

Figure 1. The mechanism for the formation of the triazines from 1a, 1b, 1c and 1d.

Conclusions

The collected data suggests that the mechanism by which aminoguanidine inhibits in vivo Maillard reactions is via reaction with the highly reactive dicarbonyl intermediates to convert them to the comparatively more stable and unreactive triazines. This essentially removes them from the reaction as they are formed, thereby preventing them from participating in further secondary stages of the Maillard reaction.

Acknowledgements
This work was supported by grant NO. 190821 from the Juvenile Diabetes Foundation International.

References

Bayne, S. (1963) Aldosuloses (Osones). *Methods in Carbohydr. Chem.* 2:421-423.

Brownlee, M., Vlassara, H., Kooney, A., Ulrich, P., and Cerami, A. (1986). Aminoguanidine Prevents Diabetes-Induced Arterial Wall Protein Cross-Linking. *Science.* 232:1629-1632.

Hirsch, J., Petrakova, E. and Feather, M.S. (1992) The reaction of some dicarbonyl sugars with aminoguanidine. *Carbohydr. Res.* 232:125-130.

Hirsch, J., Barnes, C.L. and Feather, M.S. (1992) X-Ray Structures of a 3-Amino-5- And a 3-Amino-6 Substituted Triazine, Produced as a Result of a Reaction of 3-Deoxy-**D**-*erythro*-hexos-2-ulose (3-Deoxyglucosone) with Aminoguanidine. *J. Carbohydr. Chem.* 11:891-901.

Kato, H., Cho, R.K., Okitani, A. and Hirase, F. (1987) Responsibility of 3-Deoxyglucosone for the Glucose-Induced Polymerization of Proteins. *Agr. Biol. Chem.* 51:683-685

Knecht, K.J., Feather, M.S., and Baynes, J.W. (1992) Detection of 3-Deoxyfructose and 3-deoxyglucosone in Human Urine and Plasma: Evidence for Intermediate Stages of the Maillard Reaction *in Vivo. Arch. Biochem. Biophys.* 294:130-137.

Madson, M.A. and Feather, M.S. (1981) An Improved Preparation of 3-Deoxy-D-erythro-hexos-2-ulose vis the Bis (benzoylhydrazone) and Some Related Constitutional Studies. *Carbohydr. Res.* 94:183-191.

Shin, D.B. and Feather, M.S. (1990) 3-Deoxy-**L**-*glycero*-pentos-2-ulose (3-deoxy-**L**-xylosone) and **L**-*threo*-pentos-2-ulose (**L**-xylosone) as intermediates in the degradation of L-ascorbic acid. *Carbohydr. Res.* 208:246-250.

Salomon, L.L., Burns, J.J. and King, O.G. (1952) Synthesis of L-Ascorbic-1-C^{14} Acid from Sorbitol. *J. Am. Chem. Soc.* 74:5161-5162.

Wolfrom, M.L., El Khadem, H.S. and Alfes, H. (1964) Amido and Carbamoyl Osazones of Sugars. *J. Org. Chem.* 29:3074-3076.

Aminoguanidine Increases Hydrogen Peroxide Formation during Glycation *In Vitro*

Peimian Ou and Simon P. Wolff

DIVISION OF CLINICAL PHARMACOLOGY AND TOXICOLOGY, DEPARTMENT OF MEDICINE, UNIVERSITY COLLEGE LONDON MEDICAL SCHOOL, THE RAYNE INSTITUTE, 5 UNIVERSITY STREET, LONDON WC1E 6JJ, UK

Summary
Aminoguanidine, proposed as an in vivo inhibitor of diabetes-associated browning reactions, accelerates hydrogen peroxide production during glycation in vitro. Since AG has also recently been shown to be an inhibitor of catalase there is the possibility that aminoguanidine may increase oxidative stress. If AG is confirmed as useful in diabetes then this result could indicate that the complications of diabetes are not related to increased tissue oxidation.

Introduction

Aminoguanidine (AG) has been proposed as an agent of potential benefit in prophylaxis of the diabetic complications of kidney, nerve and eye [Brownlee et al, 1986; 1988; Edelstein and Brownlee, 1992; Hammes et al, 1991; Kumari et al, 1991]. In vivo, AG has been shown to lessen the formation of collagen-linked fluorescence associated with the production of Advanced Glycosylation Endproducts (AGE). AG has been postulated to block pathological tissue alterations via inhibition of deleterious long-term changes to proteins associated with hyperglycemia.

Preliminary results indicated that AG seemed to increase hydrogen peroxide (H_2O_2) production from glucose, which is prone to a slow metal-catalyzed oxidation process of possible relevance to protein modification by glucose in diabetes [Wolff et al, 1991]. This result was surprising since it had been shown previously that agents which lessen browning reactions in vitro also lower H_2O_2 production during the glycation process. In this paper we show that (1) AG increases H_2O_2 production from glucose, including during glucose autoxidation in the presence of protein; (2) that rates of H_2O_2 production depend upon transition metal availability and that (3) high glucose enhances H_2O_2 formation in cells, which might be exacerbated by AG.

Material and Methods

Ammonium ferrous sulphate, xylenol orange, H_2O_2, sorbitol, catalase (Type C-10) and AG bicarbonate were obtained from Sigma (Poole, Dorset). AG bicarbonate salt was dissolved rapidly in 1N HCl and adjusted to pH 7.4 with sodium hydroxide. All solutions were prepared with chelex-treated double distilled water.

Determination of H_2O_2

The concentration of H_2O_2 in purchased batches was determined using its extinction coefficient of $39.4 M^{-1} cm^{-1}$ at 240nm [Nelson and Kiesow, 1972]. Standard solutions of

H_2O_2 were then used to calibrate the FOX (Ferrous Oxidation in Xylenol orange) assay for hydroperoxide determination as described previously [Jiang et al, 1990]. Briefly, 50µl of test sample or H_2O_2 standard were mixed with 950µl of FOX1 reagent (composed of 250µM ammonium ferrous sulphate, 100µM xylenol orange, 100mM sorbitol in 25mM H_2SO_4) and incubated for 30 minutes at room temperature before reading absorbance at 560nm. Where appropriate, flocculated protein was first removed by centrifugation (12000g x 5 minutes). Absorbance data were gathered on a Pye Unicam Series 8700 Spectrophotometer.

Determination of catalase activity in erythrocytes

Human blood was drawn from healthy volunteers and centrifuged to separate the plasma and buffy layer. Erythrocytes were washed three times in ice-cold phosphate-buffered saline. Washed erythrocytes were incubated with various concentrations of glucose and AG at a final cell volume of 5 percent. At varying time intervals, 50µL of the cell suspension was removed and added to 950µL of lysis buffer (potassium phosphate buffer, pH 7.4; 10mM) containing 200µM H_2O_2 and incubated for 10 minutes at room temperature with shaking. Catalase activity was determined by reference to the concentration of H_2O_2 (determined by the FOX assay) remaining at the end of the incubation period. 50µl aliquots of the H_2O_2-exposed homogenate were added to 950µl of the FOX reagent (which also halted H_2O_2 consumption by catalase) and then incubated for 30 minutes at room temperature in 1ml microcentrifuge vials with occasional vortexing. The vials were centrifuged to remove flocculated material (12000g x 3 minutes) and the supernatants were read at 560 nm against H_2O_2 standards. Catalase activity was calculated in terms of H_2O_2 consumed per minute. Under the conditions used, all incubations contained residual H_2O_2. H_2O_2 consumption was approximately linear over the 10 minutes observed.

Results

AG and H_2O_2 Production in Relation to Glycation

We were originally interested in AG in relation to its proposed use as a potential inhibitor of protein modifications associated with enhanced glycation reactions in diabetes. Our preliminary observations led us to postulate that AG could act as a precursor for H_2O_2 since we found that glucose appeared to generate consistently greater amounts of H_2O_2 in the presence of AG (Figure 1 and Figure 2). This result was unexpected. Although AG is believed to block AGE formation we had previously shown that AGE formation could generally be inhibited by strategies which lowered H_2O_2 formation during glycation [Hunt et al, 1988]. We had thus wrongly postulated that AG would also lower H_2O_2 formation from glucose.

H_2O_2 Formation by AG

We subsequently observed (Ou and Wolff, 1992) that AG incubated in chelex-treated potassium phosphate buffer slowly accumulated H_2O_2 in a reaction which was catalyzed by the addition of copper or iron ions and was inhibited by the transition metal chelator diethylenetriaminepenta-acetic acid. Similar reactions occur with the reaction of glucose with AG since copper ion increased H_2O_2 formation from AG and glucose in all cases (Figure 3).

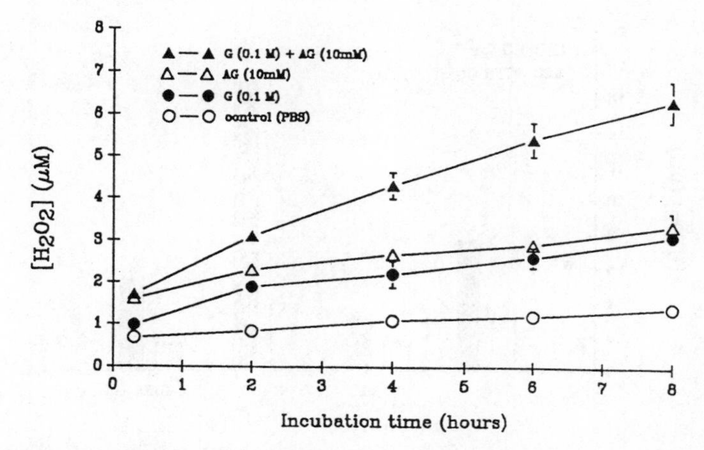

Figure 1. Aminoguanidine (AG) increases H_2O_2 production from glucose (G). Glucose (100mM), AG (10mM) and glucose with AG were incubated in potassium phosphate buffer, pH 7.4 at 37°C. After incubation at various time intervals, 100 μL samples were added to 900μL of FOX reagent. Values are the means +/- SD of 3 determinations.

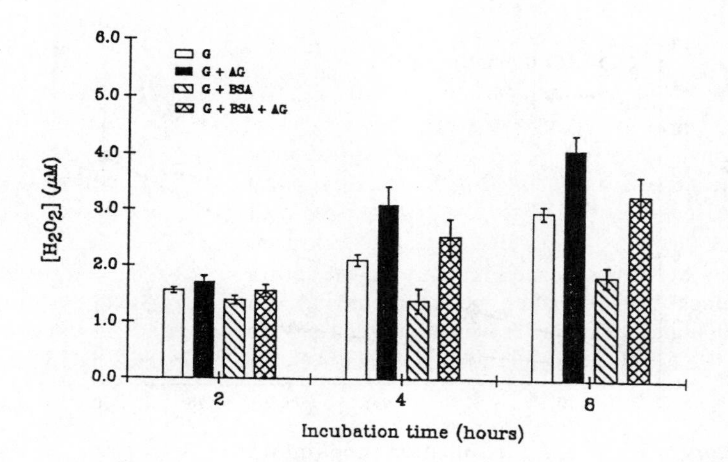

Figure 2. Effects of aminoguanidine (AG) on glucose with or without bovine serum albumin (BSA) induced H_2O_2 production. AG (10 mM) was incubated with glucose(G) (100 mM) with or without 1 mg/ml bovine serum albumin (BSA) in potassium phosphate buffer (pH 7.4, 50 mM) at 37°C for up to 8 hr. Samples were taken at various time intervals and analyzed for $H_2O_2$by using FOX reagent. Values given represent the means ± SD of triplicate measurements.

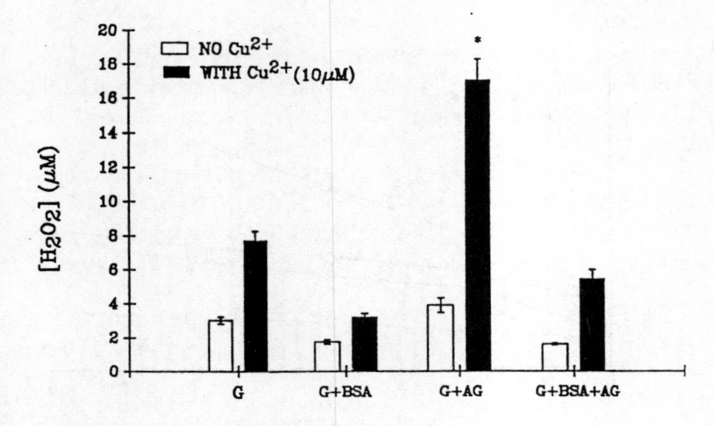

Figure 3. Effects of transition metal (Cu2+) on glucose (G) and aminoguanidine (AG) with or without bovine serum albumin (BSA) induced H_2O_2 production. AG (10 mM) and G (0.1 mM) with and without BSA (1 mg/ml) were incubated with or without Cu2+ (10 μM) in potassium phosphate buffer (pH 7.4, 50 mM) for 8 hrs at 37°C. H_2O_2 concentrations were measured as described in Figure 1. (* represents only one third of real value).

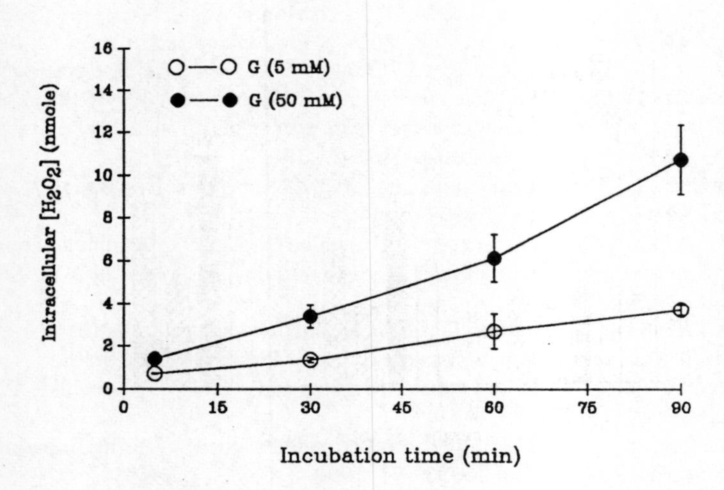

Figure 4. Glucose increases intracellular H_2O_2 production in human erythrocytes, amounts of H_2O_2 from glucose(50 mM) was approximate 15 % of H_2O_2 from ascorbate (0.1 mM) under the same condition (not shown). Glucose (5, 50 mM) was incubated with human RBCs in the presence of AMT (50 mM) with constant shaking at 37°C for 90 min. Endogenous catalase inactivation is correspond to intracellular H_2O_2 production that was calibrated by glucose/glucose oxidase (H_2O_2 generating system).

Inhibition of Catalase by AG and Aminotriazole

We also observed that AG inhibited catalase activity in a time- and dose-dependent manner. The IC50 for catalase inhibition by AG was found to be 15.33mM compared with 4.11mM for aminotriazole (AMT), used as a reference inhibitor. Using AG to inhibit catalase in rat liver slices and human erythrocytes we observed that AG shares the characteristic of AMT of being capable of inactivating catalase irreversibly in a manner which is dependent upon the continued supply of H_2O_2. AG, like AMT, can thus act as a "suicide substrate" for catalase. Since hyperglycemia leads to increased H_2O_2 production in erythrocytes (Figure 4) it is conceivable that AG increases oxidative stress in diabetes.

Discussion

AG generates H_2O_2 at a low rate in vitro directly and also increases H_2O_2 production from glucose. This may be of relevance to model studies of AGE formation in which proteins are incubated with glucose, in the presence of AG, for extended periods in vitro. H_2O_2 could conceivably be involved in chromophore bleaching. AG also inhibits catalase in a manner reminiscent of AMT. This inhibition, like AMT, is probably reversible in the absence of a constant source of H_2O_2 and irreversible in its presence. If AG protects against diabetic complications and increases oxidative stress in vivo then this would suggest that oxidative stress is not a pathophysiological factor in the diabetic complications.

References

Brownlee M, Cerami A, Vlassara H. Advanced glycosylation end-products in tissue and the biochemical basis of diabetic complications. N Engl J Med 318: 1315-1321, 1988

Brownlee M, Vlassara H, Kooney T, Ulrich P, Cerami A. Aminoguanidine prevents diabetes-induced arterial wall protein cross-linking. Science 232: 1629-1632, 1986

Edelstein D, Brownlee M. Mechanistic studies of advanced glycosylation end product inhibition by aminoguanidine. Diabetes 41: 26-29, 1992

Edelstein D, Brownlee M. Aminoguanidine ameliorates albuminuria in diabetic hypertensive rats. Diabetologia 35: 96-97, 1992

Hammes HP, Martin S, Federlin K, Geisen K, Brownlee M. Aminoguanidine treatment inhibits the development of experimental diabetic retinopathy. Proc Natl Acad Sci. 88: 11555-11558, 1991

Hunt JV, Dean RT, Wolff SP. Hydroxyl Radical Production and Autoxidative Glycosylation: Glucose Autoxidation as the Cause of Protein Damage in the Experimental Glycation Model of Diabetes Mellitus and Ageing. Biochem. J. 256: 205-212, 1988

Jiang Z-Y, Woollard ACS, Wolff SP. Hydrogen Peroxide Production During Experimental Protein Glycation. FEBS Letters 268: 69-71, 1990

Kumari K, Umar S, Bansal V, Sahib MK. Inhibition of diabetes-associated complications by nucleophilic compounds. Diabetes 40: 1079-1084, 1991

Nelson DP, Kiesow LA. Enthalpy of decomposition of hydrogen peroxide by catalase at 25°C (with molar extinction coefficients of H_2O_2 solutions in the UV). Anal Biochem 49: 474-478, 1972

Ou P, Wolff SP. Aminoguanidine: A drug proposed for prophylaxis in diabetes inhibits catalase and generates hydrogen peroxide in vitro. Biochemical Pharmacology 46: 1139-1144,1992

Wolff SP, Jiang ZY, Hunt JV. Protein glycation and oxidative stress in diabetes mellitus and ageing. Free Rad Biol Med 10: 339-352, 1991

Specific Maillard Reactions Yield Powerful Mutagens and Carcinogens

J. H. Weisburger

AMERICAN HEALTH FOUNDATION, VALHALLA, NY 10595, USA

Summary
Sugimura and colleagues reported in 1977 that the surface of fried or grilled meats or fish contain powerful mutagens. Isolation and identification of these products show them to be a new class of chemicals, heterocyclic amines (HCAs). They are not only mutagenic, but are active in other rapid bioassays, such as induction of DNA repair in hepatocytes, typical of genotoxins. The mode of formation involve classical Maillard reactions with the important point that creatinine, uniquely present in meat or fish, participates in an essential manner as origin of the aminomethylimidazo part of the molecule. Foods with a higher creatinine content yield more HCAs. On the other hand, decreasing creatinine by short microwave heating of a food yields lower HCA formation. Addition of soy protein, antioxidants such as butylated hydroxyanisole (BHA), or tryptophan and proline prior to cooking also gives reduced HCAs. Black tea and green tea, and the corresponding polyphenols epigallocatechin gallate (EGCG) and theaflavin gallate (TFG), are potent inhibitors of the expression of genotoxicity of HCAs. Tea may reduce the carcinogenicity in experimental systems. Thus, Maillard reactions with foods containing creatinine lead to the production of powerful mutagens and genotoxic carcinogens, associated with a risk of important types of cancer in the breast, colon and pancreas. Laboratory and practical approaches have been developed to reduce the formation and the genotoxic effects of HCA's produced during cooking.

Historical Background and Introduction

In the field of nutrition and cancer much has been learned through studies in humans and in animal models as to causative elements. It was established by 1976, that most dietary fats played a major role in the occurrence of cancer in the breast, colon, pancreas, prostate, ovary and endometrium (see Ip et al., 1986). Studies in which a fat effect was not found involved fat levels of 30% of calories or higher, where an effect is not expected, or used types of fat such as monounsaturated olive oil, where no enhancing action occurs (see Micozzi and Moon, 1992). The underlying mechanism involved promotion of neoplasms at these specific sites. However, and importantly, the causative genotoxic carcinogens were unknown in 1977. A major finding bearing on this lack of knowledge was made by Nagao, Sugimura and staff (Nagao et al., 1977; Sugimura et al., 1977) at the Japanese National Cancer Center Research Institute with their seminal discovery that the surface of cooked protein foods, essentially meats or fish, contain mutagenic activity. The Tokyo group and my laboratory collaborated in the isolation and identification of the first such mutagens from foods, 2-amino-3-methylimidazo[4,5-*f*]quinoline (IQ) (Kasai et al., 1980). We were impressed by this chemical structure, since other o-methylamino compounds were carcinogens and affected often the mammary gland and the intestinal tract. Bioassays in rats indicated that IQ, indeed, induced cancer of the colon, mammary gland and select other tissues (Tanaka et al., 1985; Ohgaki et al., 1991). The Tokyo group, furthermore, identified other chemicals of this type, heterocyclic amines (HCAs), a new class of "naturally occurring" chemicals in cooked foods that are carcinogenic (see Wakabayashi et al., 1992; this volume).

Another group at the Lawrence Livermore Laboratory at the University of California, discovered that the major product of about 17 HCAs found in cooked foods was 2-amino-1-methyl-6-phenylimadazo[4,5-*b*]pyridine (PhIP) (Felton et al., 1992). This chemical produced mainly cancer of the mammary gland and colon in rats (Ito et al., 1991). Depending on the type of cooking and type of foods, appreciable amounts of HCAs are generated (Spingarn and Weisburger, 1979; Gross et al., 1993).

Thus, it would seem that the mode of cooking of foods, consumed by most people in the world, can generate mutagens that also display activity in other rapid bioassays denoting genotoxicity (Yoshimi et al., 1988). In rats, they caused cancer mainly at target organs, such as mammary gland and colon that are major types of cancer in humans in the Western world. The 1992 Symposium of the Princess Takamatsu Cancer Research Fund dealt with the topic HCAs (Adamson et al., 1994).

Mechanisms of Formation of HCAs

Reviewing the structures of HCAs, Jägerstad et al. (see 1991; this volume) initially suggested and then demonstrated that the critical aminomethylimidazo part of all HCAs stemmed from creatinine, a major component of muscle in meats and fish. Jägerstad developed model systems involving heating to 128°-150°C mixtures of glycine, or other appropriate amino acids, glucose and creatinine in diethyleneglycol-water, that yielded HCAs. The combination of amino acids and glucose are typically used in studying Maillard type reactions, as reviewed in this and previous Conferences on this subject. This discovery explained why foods containing creatinine were good sources of HCAs (see also Furihata and Matsushima, 1986). Addition of extra creatinine to the surface of meat prior to frying increased the yield of HCAs.

The other part of the molecules of HCAs, consisting mainly of heterocyclic rings involving quinoline, quinoxaline and pyridine structures, is derived from complex and specific Maillard reactions. Investigators of the Maillard reactions concerned with the production of food flavors during cooking have often identified heterocyclic compounds responsible for the flavor. These were smaller monocyclic chemicals like pyrazine derivatives. The HCAs are produced through similar Maillard type mechanisms, not fully known, that yield more complex polycyclic ring structures. In any case, Maillard reactions are clearly involved, yielding reactive intermediates, probably with aldehyde reactive terminal groups, that can combine, under the influence of heat, with creatinine to form HCAs. We have shown that reaction of a simple aldehyde like acetaldehyde with creatinine can yield one of the simplest mutagens in this series of compounds (Jones and Weisburger, 1988).

Prevention of the Formation of HCAs

In view of the probable adverse effects in humans, as demonstrated also in animal models of HCAs (see Adamson et al., 1994), it seems important to prevent their formation during cooking of foods containing creatinine.

Felton's group (1992) demonstrated that brief microwave heating of meats lowered the presence of creatinine, since it was removed with the hot water phase eliminated from the meat. Subsequent frying or grilling yielded a lower formation of HCAs, because of the decreased amount of available creatinine.

Our laboratory observed that adding of 10% or more of soy protein to meat, prior to frying, completely abolished the formation of HCAs, as measured by mutagenicity testing (Weisburger, 1991). Also, addition of antioxidants like BHA decreased the formation of HCAs through mechanisms possibly related to interference with the formation of Maillard intermediates, or their reaction with creatinine. A more specific means of inhibition of formation involved the use of L-tryptophan and L-proline individually or in mixtures, in relatively small amounts. The underlying mechanism appears to be a competition between creatinine as acceptor of the Maillard reactive products, and tryptophan or proline.

Thus, there are a number of simple, practical procedures to lower or totally prevent the formation of HCAs during cooking.

Prevention of the Action or Activation of HCAs

The activity of HCAs as mutagens and as carcinogens depends on metabolic activation through N-hydroxylation to yield a proximate carcinogen (Kato and Yamazoe, 1987; Aeschbacher and Turesky, 1991; Alexander and Wallin, 1991; Hayatsu, 1991; Yamazoe and Ozawa, 1991; McKinnon et al., 1992; Synderwine et al., 1992). The N-hydroxy compound is further converted to a reactive electrophilic metabolite, through acetylation, prolylation, or conversion to a sulfuric acid ester. These reactions are dependent on the presence in specific organs and tissues of the necessary enzymes, such as acetyltransferases or sulfotransferases. The sulfotransferases are present chiefly in liver. Therefore, those HCAs where the N-hydroxy compound has a structure such as to be a substrate for the sulfotransferase are liver carcinogens. Similar considerations account for the organotropism of specific HCAs in relation to their terminal activation by acetyltransferases.

All the HCAs also undergo detoxification, in great part by hydroxylation on carbon. The resulting phenolic compounds are metabolized further by phase II conjugation reactions, mainly by forming glucuronides or sulfate esters, that are readily excreted in urine. With IQ, a major detoxified product is the 5-hydroxy compound (Luks et al., 1989) and with PhIP, it is the 4'-hydroxy compound (Buonarati et al., 1992).

We have observed that black tea and green tea lowered the mutagenicity in Salmonella typhimurium TA98 of IQ and of PhIP. These teas also inhibited the development of DNA repair in freshly explanted liver cells exposed to IQ and PhIP. These two tests are indicators of genotoxic activity. Their inhibition by tea suggests that these naturally occurring plant extracts can modify the metabolic activation of HCAs. Tea polyphenols, specifically epigallocatechin gallate from green tea and theaflavin gallate from black tea, displayed similar inhibiting action in both tests (Weisburger, et al. 1993).

Role of HCAs in Human Disease Causation

Cancer

The Lawrence Livermore Laboratory has done extensive research on the quantitation and dosimetry of HCAs formed during cooking (Felton et al., 1992). Even in well done meat, HCAs are present in relatively small amounts, of the order of nanograms per gram of cooked meat (Gross et al., 1993). While the HCAs are powerful genotoxic carcinogens, the presence of such small amounts may not necessarily represent a human cancer risk (see Felton et al., 1992). Nonetheless, consideration needs to be given to the fact that humans are exposed daily to these chemicals in their food from childhood onward. Even a few doses given to newborn mice elicited a carcinogenic effect (Dooley et al., 1992), and HCAs are secreted in breast milk and yield DNA adducts in the pups (Goshal and Snyderwine,

1993). More important, however, is the fact that at least in the Western world, meat eaters consume relatively large amounts, 38-45% of calories, of various types of fat from the meat itself and also from other foods (Micozzi and Moon, 1992). There was a higher level of DNA binding, reflecting increased metabolic formation of reactive products, in the liver of mice fed high fat diets and given a dose of labeled HCA (Alldrick et al., 1993). Also, as noted in the introduction, fats have a powerful promoting effect of cancer at specific organs in humans, and in animal models initiated by genotoxic carcinogens such as the HCAs. Therefore, it seems reasonable to suggest that the HCAs, formed through Maillard reactions during cooking, are the initiating carcinogens for major types of cancers seen in humans, consuming at the same time high amounts of diverse fatty foods. In the case of the HCAs, risk analysis necessarily requires consideration of promoting effects due to fats in the usual Western diets.

Coronary heart disease
Adamson and colleagues have observed lesions in the vascular system and the heart of non-human primates given specific HCAs and especially after IQ. Also, radioactivity from labeled IQ was found at that critical site (see Adamson, 1994). This might suggest that HCAs play a role in the etiology of coronary heart disease. In this instance also, dietary fats and specifically saturated fats, enhance or promote the process through their well known action in generating cholesterol and related fatty substances, involved in atherosclerosis.

Conclusions

Specific types of Maillard reactions occurring during the frying, grilling or other cooking methods of meats and fish, involving temperatures above 130°C, generate a new class of chemicals, heterocyclic amines. About 17-19 such chemicals have so far been identified in cooked foods or in model reactions. The formation entails Maillard intermediates, reacting with creatinine, a component of meat or fish muscle. Their formation or biochemically mediated action can be inhibited through specific methods, described in this paper. Current views are that HCAs are likely to play a major role as the initiating genotoxic carcinogens for specific types of important human cancers. They may also be an etiological factor as regards coronary heart disease. Even so and interestingly, regulatory agencies, the media and the public in most countries have great concerns with food additives and contaminants, especially chemicals present in tiny amounts as residues from agricultural products as causes of cancer. This fear of such environmental chemicals is not justified, especially since most of such xenobiotic chemicals are not genotoxic. In contrast, the active, proven genotoxicity and carcinogenicity in many species, including non-human primates, of the HCAs indicate that they constitute cancer risks for humans (see Adamson et al., 1994, for overview). We conclude, therefore, that the control of the formation and action of HCAs may contribute substantially to preventive medicine, health promotion, and reduced medical care expenditures.

Acknowledgments
Research described was supported by grants from USPHS-NIH CA24217, CA29602, and CA42381, and from the American Cancer Society, CN-29. I am indebted to Beth-Alayne McKinney for excellent editorial support.

References

Adamson, R.H., Gustafsson, J.A., Ito, N., Nagao, M., Sugimura, T., Wakabayashi, K., and Yamazoe, Y. (1994). Heterocyclic amines in cooked foods. Proceedings, 23rd International Symposium of the Princess Takamatsu Cancer Research Fund. Princeton Scientific Publishing Co., Inc., Princeton, NJ.

Aeschbacher, H.U. and Turesky R.J. (1991) Mammalian cell mutagenicity and metabolism of heterocyclic aromatic amines. Mutation Res. 259:235-250.

Alexander, J. and Wallin, H. (1991) Metabolic fate of heterocyclic amines from cooked food. In: Mutagens in Food: Detection and Prevention. (H. Hayatsu, ed). CRC Press, Boca Raton, FL, pp. 144-152.

Alldrick, A.J., Ho, T.A., Rowland, I.R., Phillips, D.H. and Ni She, M. (1993) Influence of dietary fat on DNA binding by 2-amino-3,8-dimethylimidazo[4,5-f]quinoxaline (MeIQx) in the mouse liver. Fd. Chem. Toxic. 31:483-489.

Buonarati, M.H., Roper, M., Morris, C.J., Happe, J.A., Knize, M.G. and Felton, J.S. (1992) Metabolism of 2-amino-1-methyl-6-phenylimidazo[4,5-b]pyridine (PhIP) in mice. Carcinogenesis 4:621-627.

Dooley, K.L., Von Tunglen, L.S., Bucci, T., Fu. P.P. and Kadlubar, F.F. (1992) Comparative carcinogenicity of 4-aminobiphenyl and the food pyrolysates, Glu-P-1, IQ, PhIP, and MeIQx in the neonatal $B6CF_1$ male mouse. Cancer Lett. 62:205-209.

Felton, J.S., Knize, M.G., Roper, M., Fultz, E., Shen, N.H., and Turteltaub, K.W. (1992) Chemical analysis, prevention, and low-level dosimetry of heterocyclic amines from cooked food. Cancer Res. 52:2103-2107.

Furihata, C., and Matsushima, T. (1986) Mutagens and carcinogens in foods. Ann. Rev. Nutr. 6:67-94.

Ghoshal, A. and Snyderwine E.G. (1993) Excretion of food-derived heterocyclic amine carcinogens into breast milk of lactating rats and formation of DNA adducts in the newborn. Carcinogenesis 14:2199-2203.

Gross, G.A., Turesky R.J., Fay, L.B., Stillwell, W.G., Skipper, P.L. and Tannenbaum, S.R. (1993) Heterocyclic aromatic amine formation in grilled bacon, beef and fish and in grill scrapings. Carcinogenesis 14:2313-2318.

Hayatsu, H. (1991) Mutagens in Food: Detection and Prevention. CRC Press, Boca Raton, Florida.

Ip, C., Birt, D.F., Rogers, A.E., Mettlin C. (eds.) (1986) Dietary fat and cancer. Progress in clinical and biological research. Liss, New York, Vol. 222.

Ito, N., Hasegawa, R., Sano, M., Tamano, S., Esumi, H., Takayama, S. and Sugimura, T. (1991) A new colon and mammary carcinogen in cooked food, 2-amino-1-methyl-6-phenylimidazo[4,5-b]pyridine (PhIP). Carcinogenesis 12:1503-1506.

Jägerstad, M., Skog, K., Grivas, S. and Olsson, K. (1991) Formation of heterocyclic amines using model systems. Mutation Res. 259:219-233.

Jones, R.C. and Weisburger, J.H. (1988) Inhibition of aminoimidazoquinoxaline-type and aminoimidazo-4-one-type mutagen formation in liquid reflux models by L-tryptophan and other selected indoles. Jpn. J. Cancer Res. 79:222-230.

Kasai, H., Yamaizumi, Z., Wakabayashi, K., Nagao, M., Sugimura, T., Yokoyama, S., Miyazawa, T., Spingarn, N.E., Weisburger, J.H. and Nishimura, S. (1980) Potent novel mutagens produced by broiling fish under normal conditions. Proc. Jpn. Acad. 56B:278-283.

Kato, R., Yamazoe, Y. (1987) Metabolic activation and covalent binding to nucleic acids of carcinogenic heterocyclic amines from cooked foods and amino acid pyrolysates. Jpn. J. Cancer Res. 78:297-311.

Luks, H.J., Spratt, T.E., Vavrek, T., Roland, S.F. and Weisburger, J.H. (1989) Identification of the sulfate and glucuronic acid conjugates of the 5-hydroxy derivative as major metabolites of 2-amino-3-methylimidazo[4,5-f]quinoline (IQ) in rats. Cancer Res. 49:4407-4411.

McKinnon, R.A., Burgess, W.M., de la M. Hall, P., Abdul-Aziz, Z. and McManus, M.E. (1992) Metabolism of food-derived heterocyclic amines in human and rabbit tissues by P4503A proteins in the presence of flavonoids. Cancer Res. 52:2108-2113.

Micozzi, M.S., Moon, T.E. (eds) (1992) Macronutrients: Investigating their Role in Cancer. Marcel Dekker, New York.

Nagao, M., Honda, M., Seino, Y., Yahagi, T. and Sugimura, T. (1977) Mutagenicities of smoke condensates and the charred surface of fish and meat. Cancer Lett. 2:221-226.

Ohgaki, H., Takayama, S. and Sugimura, T. (1991) Carcinogenicities of heterocyclic amine in cooked food. Carcinogenicities of heterocyclic amines in cooked food. Mutation Res. 259:399-410.

Snyderwine, E.G., Schut, H.A.J., Adamson, R.H., Thorgeirsson, U.P., Thorgeirsson, S.S. (1992) Metabolic activation and genotoxicity of heterocyclic arylamines. Cancer Res. 52:2099-2102.

Spingarn, N.E. and Weisburger, J.H. (1979) Formation of mutagens in cooked foods. I. Beef. Cancer Lett. 7:259-264.

Sugimura, T., Kawachi, T., Nagao, M., Yahagi, T., Seino, Y., Okamoto, T., Shudo, K., Kosuge, T., Tsuji, K., Wakabayashi, K., Iitaka, Y., and Itai, A. (1977) Mutagenic principle(s) in tryptophan and phenylalanine pyrolysis products. Proc. Jpn. Acad. 53:58-61.

Tanaka, T., Barnes, W.S., Williams, G.M., Weisburger, J.H. (1985) Multipotential carcinogenicity of the fried food mutagen 2-amino-3-methylimidazo[4,5-f]quinoline (IQ) in rats. Jpn J Cancer Res (Gann) 76:570-576.

Wakabayashi, K., Nagao, M., Esumi, H. and Sugimura, T. (1992) Food-derived mutagens and carcinogens. Cancer Res. 52:2092-2098.

Weisburger, J.H. (1991) Prevention of heterocyclic amine formation in relation to carcinogenesis. In: Mutagens in Food: Detection and Prevention. CRC Press, Boca Raton, Florida pp. 193-203.

Weisburger, J.H., Arrigale, R., and Reinhart, L. (1993) Inhibition by tea and tea polyphenols of mutagenicity and DNA repair by heterocyclic arylamines. Proc. Am. Assoc. Cancer Res. 34:124 (abstract).

Yamazoe, Y. and Ozawa, S. (1991) Activation of food mutagens. In: Mutagens in Food: Detection and Prevention. CRC Press, Boca Raton, Florida pp. 125-141.

Yoshimi, N., Sugie, S., Iwata, H., Mori, H., Williams, G.M. (1988) Species and sex differences in genotoxicity of heterocyclic amine pyrolysis and cooking products in the hepatocyte primary culture/DNA repair test using rat, mouse, and hamster hepatocytes. Env. Molec. Mutagen 12: 53-64.

Mechanism of Antimutagenic Effect of Maillard Reaction Products Prepared from Xylose and Lysine

Gow-Chin Yen and Ping-Ping Hsieh

DEPARTMENT OF FOOD SCIENCE, NATIONAL CHUNG HSING UNIVERSITY, 250 KUOKUANG ROAD, TAICHUNG, TAIWAN, REPUBLIC OF CHINA

Summary

The possible mechanism of antimutagenicity of Maillard reaction products (MRPs) prepared by heating xylose and lysine (molar ratio 1:2) at pH 9.0 and 100 °C for 1 h were investigated using the Salmonella/microsome assay. The mutagenicity of 2-amino-3-methylimidazo[4,5-f]quinoline (IQ) towards *Salmonella typhimurium* TA 98 and TA100 was markedly reduced by the addition of xylose-lysine MRPs, whereas the mutagenicity of 4-nitroquinoline-N-oxide was not inhibited. The xylose-lysine MRPs exhibited no inhibitory activity to IQ on the bio-antimutagenic assay; this result indicated that the antimutagenic effect of xylose-lysine MRPs is a desmutagenic action, not by modification of DNA repair processes in the bacterium cell. According to a further study to elucidate the mechanisms of antimutagenic activity, xylose-lysine MRPs reduced the mutagenicity of IQ by interaction with proximate metabolites of IQ, not by direct inhibition of hepatic microsome activation. The mutagenicity of IQ was decreased by increasing duration of reaction between xylose-lysine MRPs and metabolites of IQ but not with S9 mixture, or intact IQ, or by masking the mutagen-binding sites of DNA. In conclusion, the antimutagenic effects of xylose-lysine MRPs towards IQ might be due to interaction with proximate metabolites of IQ to form inactive adducts.

Introduction

The Maillard reaction between carbonyl and amino compounds is an important reaction that occurs in foods during processing and storage. Numerous reviews of this reaction from various aspects of food science are available (Namiki, 1988; O'Brien and Morrissey, 1989). Reactions of amino acids and sugars have been used to study the antimutagenicity of Maillard reaction products (MRPs) in model systems (Chan et al., 1979; Kato et al., 1985; Yen et al., 1992). The nature of antimutagenic compounds in MRPs are poorly understood. The antimutagenic activity of MRPs may be due partially to reactive substances in melanoidin that scavenge free radicals, inactivate mutagens or inhibit enzyme activity of the S9 mixture (Kim et al., 1986; Hayase et al., 1989). We demonstrated the antimutagenic effect of MRPs to correlate well with their antioxidative activity and reducing power (Yen et al., 1992).

Antimutagens are classified into desmutagens and bio-antimutagens according to modes of action (Kada et al., 1985). The former inactivates mutagens by chemical or enzymatic modifications, whereas the latter suppresses the process of mutagenesis ('mutation fixation') after DNA is damaged by mutagens. Yen and Tsai (1993) indicated that the inhibitory effect of a glucose-tryptophan MRP is due to a desmutagenic effect but not to a bio-antimutagenic effect. The role of MRPs in the inhibition of mutagenicity induced by mutagens at the microsomal activation step, by direct interaction with proximate metabolites of mutagens, or by modification of DNA metabolism remains unclear.

The objective of this study was to investigate the possible mechanisms of antimutagenicity of Maillard reaction products prepared from xylose and lysine.

Materials and Methods

Materials

D-Xylose, L-lysine monohydrochloride, 7,8-benzoflavone and 4-nitroquinoline N-oxide (NQNO) were obtain from Sigma Chemical Co. (St. Louis, MO). 2-Amino-3-methylimidazo[4,5-f]quinoline (IQ) was purchased from Wako Pure Chemical Co. (Tokyo).

Preparation of Xylose-Lysine Maillard Reaction Products (XL MRPs)

The XL MRPs were prepared by dissolving D-xylose (0.1 mol) and L-lysine monohydrochloride (0.2 mol) in potassium phosphate buffer (0.1 M, pH 9.0), adjusted to 100 mL, and then refluxed in an oil bath at 100 °C for 1 h. After reaction, the XL MRPs were stored at -20 °C until use.

Assay for Antimutagenic and Bio-Antimutagenic Effects

The antimutagenic effect of XL MRPs was examined according to the Ames method (Maron and Ames, 1983), and the detail procedure was described in our previous study (Yen and Lii, 1992). The bio-antimutagenicity test was conducted as described by Sato et al. (1987), and as used in our previous study (Yen and Tsai, 1993).

Effect of XL MRPs on the Hepatic Microsomal Activation of IQ

The two experimental protocols to study effects of XL MRPs on the hepatic microsomal activation of IQ were designed by Alldrick and Rowland (1988) and adapted to our optimal assay conditions. In the first experiment, IQ (0.025 μg/plate), S9 mix (0.5 mL) and XL MRPs (0.5-4 mg/plate) were incubated together at 37 °C for 30 min. Overnight -cultured TA98 (0.1 mL) and 7,8-benzoflavone (10 μg/plate) were added to the reaction mixture and the assay was conducted as described in antimutagenicity test (route 1). In the second experiment, the procedure was slightly modified. IQ and S9 mix were incubated together for 20 min; XL MRPs and 7,8-benzoflavone were then added and the mixture was incubated for another 20 min. Finally, overnight-cultured TA98 (0.1 mL) was added to the reaction mixture and the assay was performed as before (route 2).

Effects of XL MRPs on the Hepatic Microsomal Enzyme, IQ, IQ Metabolites and Salmonella typhimurium Strain TA98

To investigate the effect on S9 mix, XL MRPs (0.1 mL, 2 mg/plate) and S9 mix (5 mL) were pre-incubated together at 37 °C for 0, 10, 20 or 40 min. Overnight-cultured TA98 (0.1 mL) and IQ (0.1 mL, 0.1 μg/plate) were then added and the mixture was incubated at 37 °C for another 20 min, assay was performed as before. To examine the effect of XL MRPs on the intact IQ, XL MRPs (0.1 mL, 2 mg/plate) and IQ (0.1 mL, 0.1 μg/plate) were incubated together at 37 °C for 0, 10, 20 or 40 min. Overnight cultured TA98 (0.1 mL) and S9 mix (0.5 mL) were then added and the mixture was incubated at 37 °C for another 20 min, and the assay was performed as before. To study the effect of XL MRPs on IQ metabolites, S9 mix (5 mL) and IQ (0.1 mL, 0.1 μg/plate) were pre-incubated together at 37°C for 20 min. XL MRPs (0.1 mL, 2 mg/plate) was then added and the mixture was incubated for 0, 10, 20 and 40 min. Finally, overnight-cultured TA98 (0.1 mL) was added and the mixture was incubated at 37 °C for another 20 min; the assay was then conducted as before. To discover the effect of XL MRPs on strain TA98, overnight-cultured TA98

(5 mL) and XL MRPs were incubated together at 37 °C for 1 h. The bacteria were washed three times with nutrient broth with centrifugation (6000 rpm, 20 °C, 10 min). Finally, treated TA98 (0.1 mL), IQ (0.1 mL, 0.1 μg/plate) and S9 mix (0.5 mL) were incubated together at 37 °C for another 20 min, the assay was then conducted.

Results and Discussion

Antimutagenic and Bio-antimutagenic Effects of XL MRPs

Two mutagens, direct-acting mutagen NQNO and indirect-acting mutagen IQ, were used to evaluate the antimutagenic activity of XL MRPs. No antimutagenic activity was found for XL MRPs towards NQNO in *Salmonella typhimurium* TA98 and TA100 in the absence of S9 mix at the dosage of 0.5-5 mg/plate. However, the mutagenicity of NQNO towards TA100 was enhanced by increasing concentration of XL MRPs. The mutagenicity of IQ towards both TA98 and TA100 with S9 mix was decreased with increasing concentration of XL MRPs. The inhibitory effect of XL MRPs on the mutagenicity of IQ toward TA98 was stronger than toward TA100. The bio-anti-mutagenic effect of XL MRPs on the mutagenicity of IQ is also evaluated. The results indicate that the mutagenicity of IQ to TA98 and TA100 was not decreased by XL MRPs in the bio-antimutagenic test. Hence the XL MRPs may directly inactivate mutagens or metabolic enzymes to inhibit mutagenicity before the DNA reacted with mutagen. The DNA was not repaired by treatment with XL MRPs when it had been damaged by IQ. Thus the antimutagenicity of XL MRPs on IQ is the desmutagenic effect, not a bio-antimutagenic effect.

Effect of XL MRPs on the Hepatic Microsomal Activation of IQ

In order to investigate at the level at which XL MRPs inhibit IQ mutagenicity (e.g. at the microsomal activation level, or by direct interaction with proximate metabolites of IQ), we performed experiments based on modifications of the method of Alldrick and Rowland (1988). Because 7,8-benzoflavone can inactivate cytochrome P-450 in S9 mix and inhibit the mutagenicity of IQ (Kato and Yamazoe, 1987), we used it to inactivate the S9 mix. About 95 % of mutagenicity of IQ was inhibited by 7,8-benzoflavone at a concentration 5-20 μg/plate, and therefore, a concentration 10 μg/plate was used in the test. The inhibitory effect of XL MRPs on the mutagenicity of IQ through routes 1 and 2 is shown in Table I. In general, there are similar inhibitory effects for XL MRPs on IQ through these two routes ($P>0.05$). Thus we predict that the antimutagenic activity of XL MRPs on IQ was not due to inactivation of the S9 mix. Alldrick and Rowland (1988) indicated the antimutagenicity of caffeine was due to inactivation of S9 mix, as the inhibitory effect was found only in route 1, not in route 2.

Effect of XL MRPs on the Hepatic Microsomal Enzyme, IQ, IQ Metabolites and Salmonella typhimurium Strain

The inhibitory effect of XL MRPs on the mutagenicity of IQ by pre-incubation with S9 mix for various periods was studied. In the control group, the mutagenicity of IQ was increased with increasing pre-incubation period with S9 mix up to 20 min. When the S9 mix was pre-incubated with XL MRPs for 0-40 min, the inhibitory effect of XL MRPs on the mutagenicity of IQ was about 60 %; the inhibitory effect was not increased with the pre-incubation period.

Table I. Effect of xylose-lysine Maillard reaction products (XL MRPs) on the hepatic metabolic activation of IQ.

XL MRPs (mg/plate)	His+ revertants/plate Route 1	Route 2
Control	1178±204 (100.0)	1085±65 (100.0)
0.5	949± 53 (80.5)	943±92 (86.9)
2.0	772± 49 (65.5)	527±52 (48.5)
4.0	234± 30 (19.8)	252±38 (23.2)

Control plates were with IQ but without MRPs. Spontaneous revertants was determined without IQ and MRPs. Data are means ±SD of three plates. Values in parentheses are percentage relative to control value (100%).

Therefore, the antimutagenic effect of XL MRPs was not due to inactivation of the activity of hepatic microsomal enzymes.

The inhibitory effect of XL MRPs on the mutagenicity of IQ by pre-incubation of XL MRPs with IQ for various periods was studied. The mutagenicity of IQ was decreased by 50 % by added XL MRPs; however, the inhibitory effect of XL MRPs on the mutagenicity of IQ was not increased with increased pre-incubation period of XL MRPs with IQ. This result clearly indicates that intact IQ did not directly interact with XL MRPs to diminish its mutagenicity.

The inhibitory effect of XL MRPs on the mutagenicity of IQ pre-incubated with S9 mix for various periods is shown in Table II. In the control group (without added XL MRPs), the mutagenicity of IQ increased with incubation period. Hence, the content of IQ metabolites was increased with the incubation period. The result showed 22.8-40.9 % of the mutagenicity of IQ metabolites was inhibited following reaction with XL MRPs. The inhibitory effect of XL MRPs on proximate metabolites of IQ was significantly ($P<0.05$) increased with increased period of incubation. Therefore, the antimutagenic effect of XL MRPs on IQ is clearly attributed to interaction between XL MRPs and IQ metabolites to form adducts. When the bacteria (TA98) were exposed to XL MRPs at various concentrations and the XL MRPs were subsequently removed by washing with fresh nutrient broth before interaction with IQ, a decreased mutagenic response was not observed. This result indicated that XL MRPs do not interact with DNA in such a way to protect it from the ultimate mutagen.

Conclusions

Based on the results of the present study, it can be concluded that the antimutagenicity of XL MRPs on IQ is a desmutagenic effect, not a bio-antimutagenic effect. The antimutagenic mechanism of XL MRPs on IQ may be attributed to the interaction of XL MRPs with proximate IQ metabolites to form inactive adducts, not to inhibition of the activity of hepatic microsomal enzymes, or to direct reaction with intact IQ, or to interaction with DNA.

Acknowledgements

This research work was partially supported by the National Science Council, Republic of China, under Grant No. NSC 81-0409-B005-515.

Table II. Effect of xylose-lysine Maillard reaction products (XL MRPs)* on the mutagenicity of IQ which was pre-incubated with S9 mix for various periods.

Incubation time (min)*	His+ revertants/plate**		
	Control	XL MRPs	% of Control
0	2690± 25A***	2384±140A	87.2A
10	2807±107AB	2120±144AB	75.5AB
20	3048± 26AB	2080±200AB	68.2BC
40	3205±244B	1879±205B	59.1C
Spontaneous revertants	40± 2	40± 2	

* The IQ (0.1 μg/plate) was pre-incubated with S9 mix at 37 °C for various periods and then mixed with XL MRPs and *Salmonella typhimurium* TA98 and incubated at 37°C for an additional 20 min.
** Results are presented as means ±SD for three plates. The number of IQ-induced revertants in the absence of XL MRPs (control group) was expressed as 100%.
*** Values in a column with a different upper case letters are significantly different ($P<0.05$).

References

Alldrick, A.J. , and Rowland, I.A. (1988). Caffeine inhibits hepatic-microsomal activation of some dietary genotoxins. Mutagenesis. 3: 423-427.

Chan, R.I.M., Stich, H.F., Rosin, M.P., and Powrie, W.D. (1979). Antimutagenic activity of browning reaction products. Cancer Lett. 15: 27-33.

Hayase, F., Hirashima, S. , Okamoto, G., and Kato, H. (1989). Scavenging of active oxygen by melanoidins. Agric. Biol. Chem. 53: 3383-3385.

Kada, T., Inoue, T., Ohta., and Shirasu, Y. (1985). Antimutagens and their modes of action. In Antimutagenesis and Anticarcinogenesis Mechanisms; Shankle, D.M, Hartman, P.E.,

Kada, I., and Hollaender, A., Eds.; Basic Life Sciences, Vol. 39, Plenum, New York, pp. 181-196.

Kato, R., Yamazoe, Y. (1987). Metabolic activation and covalent binding to nucleic acids of carcinogenic heterocyclic amines from cooked food and amino acid pyrolysates. Jpn. J. Cancer Res. 78: 297-311.

Kato, H., Kim, S.B., Hayase, F., and Chuyen, N.V. (1985). Desmutagenicity of melanoidins against mutagenic pyrolysates. Agric. Biol. Chem. 49: 3093-3095.

Kim, S.B., Hayase, F., and Kato, H. (1986). Desmutagenicity effect of melanoidins against amino acid and protein pyrolysates. Dev. Food Sci. 13: 383-392.

Maron, D.M. , and Ames, B.N. (1983). Revised methods for the Salmonella mutagenicity test. Mutat. Res. 113: 173-215.

Namiki, M. (1989). Chemistry of Maillard reactions: Recent studies on the browning reaction mechanism and the development of antioxidants and mutagens. Adv. Food Res. 32: 115-184.

O'Brien, J., and Morrissey, P.A. (1989). Nutritional and toxicological aspects of the Maillard browning reaction in Foods. Crit. Rev. Food Sci. Nutri. 28: 211-248.

Sato, T., Ose, Y., Nagase, H., and Hayase, K. (1987). Mechanism of the desmutagenic effect of humic acid. Mutat. Res. 199-204.

Yen, G.C., Tsai, L.C., and Lii, J.D. (1992). Antimutagenic effect of Maillard browning products obtained from amino acids and sugars. Food Chem. Toxicol. 30: 127-132.

Yen, G.C., and Lii, J.D. (1992). Influence of the reaction conditions on the antimutagenic effect of Maillard reaction products derived from xylose and lysine. J. Agric. Food Chem. 40: 1034-1037.

Yen, G.C., and Tsai, L.C. (1993). Antimutagenic of a partially fractionated Maillard reaction product. Food Chem. 47: 11-15.

Elucidation of 3-Deoxyglucosone Modified Structure in Glycated Proteins

F. Hayase, Y. Konishi,[1] H. Hinuma,[1] and H. Kato[1]

DEPARTMENT OF AGRICULTURAL CHEMISTRY, MEIJI UNIVERSITY, KANAGAWA, JAPAN; [1]DEPARTMENT OF AGRICULTURAL CHEMISTRY, THE UNIVERSITY OF TOKYO, TOKYO, JAPAN

Summary

In the glucose-lysozyme reaction system, unknown peaks were detected on an amino acid chromatogram of the acid hydrolysates. The unknown peaks were also detected in acid hydrolysates of a N-α-benzoylarginine amide ($BzArgNH_2$) and 3-deoxyglucosone (3DG) reaction system. Several peaks from reversed-phase HPLC were purified and subjected to mass spectrometry. Mass spectral data showed that major compounds were formed from the reaction of one molecule of $BzArgNH_2$ with one or two molecules of 3DG. A major product was identified as R-4-imidazolone (I); R=2-(4-benzoyl-5-pentamide)amino-5-(2,3,4-trihydroxy-butyl). Intermediate products were identified as R-4(5H)-imidazolone (II), R-4(5-hydroxy)-imidazolone (III), and R-4-dihydroxy-2-imidazoline (IV). Formation of compound I via compound II is considered to be the major pathway, followed by compound I via, compound III. In further studies, a fluorescent product was purified from the butylamine-3DG reaction system. The compound was formed by the loss of five molecules of water from the reaction between 2 molecules of butylamine and 2 molecules of 3DG, resulting in the formation of a pyridinium ring.

Introduction

Amadori compounds in Glycated proteins are mostly decomposed to generate dicarbonyls such as 3-deoxyglucosone (3DG). In an advanced stage of the Maillard reaction of proteins, the intermediate dicarbonyls result in damage to proteins in biological systems as well as in foods, such as the impairment of amino acids residues, the reduction of digestibility, loss of solubility, the formation of fluorescent compounds, browning, polymerization and degradation (Ledl and Schleicher, 1990; Wolff *et al.*, 1991). Among such chemical changes, glucose-mediated crosslinking reactions have been hypothesized to be responsible for the aging of tissues and for the complications of diabetes (Cerami, 1985; Monnier *et al.*, 1991). Our group (Kato *et al.*,1987a; Shin *et al.*, 1988) indicated that 3DG liberated through the degradation of ϵ-deoxyfructosyllysine residues, attacks reactive amino acids such as arginine, lysine and tryptophan in secondary reactions to crosslink proteins in the solid state at 50°C and 75% relative humidity and under physiological conditions. However, the mechanisms of reaction of arginine and lysine residues with 3DG have not been elucidated, with the exception of the formation of pyrraline (Hayase *et al.*, 1989).

In the present study, we identified the major products formed by the reaction between 3DG and Nα-benzoyl-L-arginine amide ($BzArg\text{-}NH_2$) as a model of the reaction of arginine residues, and between 3DG and butylamine as a model of the reaction of lysine residues.

Materials and Methods

Lysozyme (10 mg/mL, Seikagaku Co., Japan) or $BzArgNH_2$ (0.1 M, Sigma) and 3DG (0.1 M for lysozyme and 0.2 M for $BzArgNH_2$) were dissolved in 0.1 M sodium phosphate

buffer at pH 7.4 and incubated at 50°C. Butylamine (0.5 M)-glucose(1 M), butylamine (0.5 M)-3DG (0.25 M) and butylamine (0.5 M)-glucose (1 M)
3DG (0.02 M) systems in 0.2 M sodium phosphate buffer at pH 7.4 were incubated at 37°C for 48 h.

The reaction products from the BzArgNH_2-3DG system were purified by HPLC using an ODS column with detection at 235 nm. The major fluorescent compounds formed from the reaction systems of butylamine with glucose or 3DG were purified by HPLC using an ODS column and fluorescence detection (Ex: 370nm and Em: 460nm).

Mass spectra were recorded using a JEOL SX102 mass spectrometer. The ^{13}C-NMR and ^{1}H-NMR spectra were recorded using a Bruker ACP-300 (300 MHz) and a Bruker AM-600 (600 MHz).

Results and Discussion

3DG-Nα -Benzoylarginine Amide Reaction System

Figure 1 shows that four unknown peaks, indicated by arrows on amino acid chromatograms of acid hydrolysates of proteins, were detected in the 3DG-lysozyme reaction system incubated under physiological conditions of 50°C at pH 7.4. These peaks were identical to those detected in the 3DG-BzArgNH_2 reaction system as well as the lysozyme-glucose reaction system (Cho *et al.*, 1986; Kato *et al.*, 1987b). The results indicate that 3DG-modified arginine residues generated four unknown peaks upon HCL - hydrolysis of lysozyme incubated with glucose. Seventeen peaks were found on an HPLC chromatogram obtained from a 3DG-BzArgNH_2 reaction system and named S1 to S17. The peaks S6, S12 and S17 were isolated and purified as major products by HPLC. The peaks S4, S5, S7, S10 and S11 were also purified as minor products. Each purified product was subjected to amino acid analysis after acid hydrolysis. Each peak appeared in the elution position shown by the arrows in Figure 1. S10 and S17 regenerated arginine after acid hydrolysis.

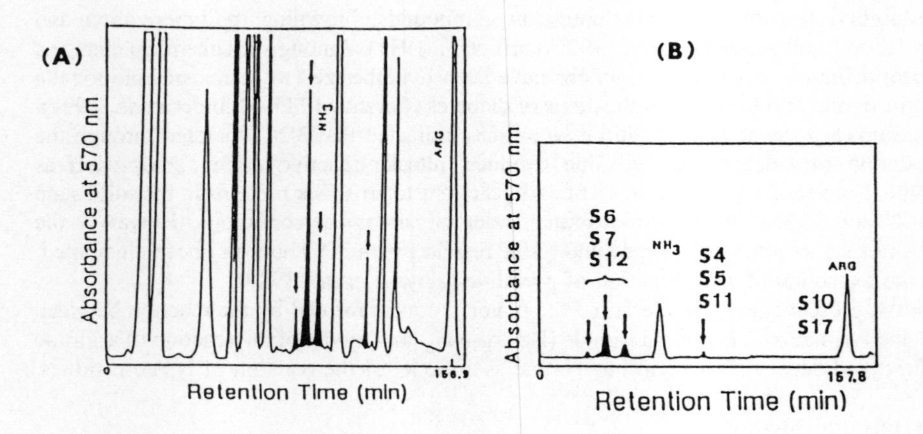

Figure 1. Amino acid chromatograms of hydrolysates of lysozyme (A) and Bz-ArgN_2 (B) incubated with 3DG in 0.1 M sodium phosphate buffer (pH 7.4) at 50°C for 96 and 132 h, respectively.

Isolated compounds were also subjected to fast atom bombardment (FAB)-mass spectrometry. Table I gives information on the chemical formulae of the isolated products. These compounds were formed from one molecule of $BzArgNH_2$ and one or two molecules of 3DG. The purified products S6, S7, S10, S12 and S17 were subjected to high-resolution FAB-MS, ^{1}H-NMR and ^{13}C-NMR spectroscopies including ^{1}H-^{1}H COSY, ^{1}H detected ^{1}H-^{13}C COSY (HMQC) and ^{1}H detected long range ^{1}H-^{13}C COSY (HMBC). S17 was increased with incubation time and identified as R-4-imidazolone, [R=2-(4-benzoylamino-5-pentamide) amino-5-(2,3,4-trihydroxybutyl)]. S10 and S12 were identified as R-4(5-hydroxy)-imidazolone and R4(5H)-imidazolone, respectively. S6 and S7 showed similar mass spectral and NMR data; however, they showed symmetrical optical rotatory dispersion (ORD) spectra. Consequently, S6 and S7 were identified as diastereomers of R-4-dihydroxy-2-imidazoline.

Figure 2 shows the chemical structures of the isolated compounds and the proposed formation scheme. The formation of S17 via S12 is considered to be a major pathway with dehydration followed by oxidation; the formation of S17 via S10 is less important. S6 and S7 are apparently hydration products of S12. S4, S5 and S11 are estimated to be formed by the reaction of an additional molecule of 3DG with S6, S7 or S12. Since S11 increased and then decreased over incubation time, S11 is speculated to participate in further reactions such as the formation of crosslinks.

3DG-Butylamine Reaction System

Sell and Monnier (1989) isolated a fluorescent compound from human extracellular matrix and identified it as pentosidine. We also examined the detection of fluorescent compounds formed by the reaction of lysine residues with glucose. The results of HPLC analysis revealed that three major fluorescent compounds (FL-A, FL-B, FL-C) were produced in a butylamine-glucose reaction system under physiological conditions (37°C at pH 7.4). If 3DG was added to this reaction system, FL-C was increased. In a butylamine-3DG reaction system, FL-C was generated as a major fluorescent compound.

Since FL-A and FL-B were increased under alkaline condition, the fluorescent compounds were isolated from the reaction mixture of butylamine with glucose at pH 10. From FAB mass and NMR data, FL-A and FL-B were identified as epimers of 3,4-dihydroxy-5-(1,2,3,4-tetrahydroxybutyl)-1,7-dibutyl-1,2,3,4-tetrahydro-1,7-naphthyridinium as shown in Figure 3. The fluorescent naphthyridinium compounds, named crosslines were recently identified by Nakamura *et al.* (1992) from a reaction mixture of α-N-acetyllysine and glucose. FL-C was isolated from a butylamine-3DG reaction system

Table I. Chemical formulae of arginine adducts formed by the reaction of Nα-benzoylarginine amide ($BzArgNH_2$) with 3-deoxyglucosone (3DG), estimated from FAB mass spectrometry.

Products	Mw
S4, S5, SlI	582 = 2x(3DG) + $BzArgNH_2$ - H_20
S6, S7	439 = 3DG + $BzArgNH_2$
S10	437 = 3DG + $BzArgNH_2$ - 2H
S12	421 = 3DG + $BzArgNH_2$ - H_20
S17	419 = 3DG + $BzArgNH_2$ - H_20 -2H

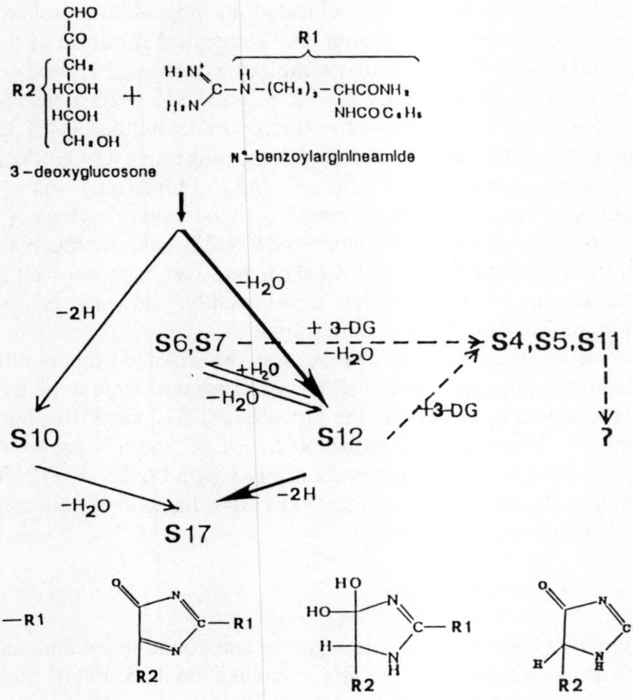

Figure 2. Proposed scheme for the formation of imidazolone compounds through the reaction of Bz-ArgNH$_2$ with 3DG. The major pathway is represented by the heavy arrows.

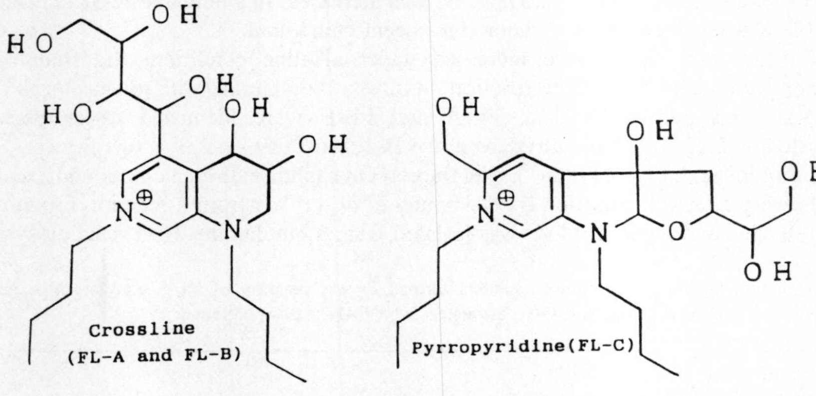

Figure 3. Chemical structure of crossline and pyrropyridine formed by butylamine-glucose reaction system under physiological conditions at pH 7.4.

and identified as 6,8-dibutyl-3,3a,8,8a-tetrahydro-3a-hydroxy-2-(1,2-dihydroxyethyl)-5-hydroxymethyl-2*H*-furo[3',2':4,5]pyrrolo-[2,3-c]- pyridinium as shown in Figure 3. The fluorescent compound named pyrropyridine was formed following the loss of five molecules of water from the reaction between 2 molecules of butylamine and 2 molecules of 3DG.

References

Cerami, A. (1985) Hypothesis, Glucose as a mediator of aging. J. Am. Geriatr. Soc. 33: 626-634.

Cho, R.K., Okitani, A. and Kato, H. (1986) Polymerization of proteins and impairment of their arginine residues due to intermediate compounds in the Maillard reaction. Dev. Food Sci., 13: 439-448.

Hayase, F., Nagaraj, R.H., Miyata, S., Njoroge, F.G. and Monnier, V.M. (1989) Aging of proteins: Immunological detection of a glucose-derived pyrrole formed during Maillard reaction *in vivo*. J. Biol. Chem. 264: 3758-3764.

Kato, H., Cho. R.K., Okitani. A. and Hayase, F. (1987a) Responsibility of 3-deoxyglucosone for the glucose-induced polymerization of Agric. Biol. Chem. 51: 683-689.

Kato, H., Shin, D.B., and Hayase, F. (1987b) 3-Deoxyglucosone proteins under physiological conditions. Agric. Biol. Chem. 51: 2009-2011.

Ledl, F., and Schleicher E. (1990) New aspects of the Maillard reaction foods and in human body. Angew. Chem. Int. Ed. Engl. 29: 565-594.

Monnier, V.M., Sell, D.R., Nagaraj, R.H., and Miyata, S. (1991) Mechanisms of protection against damage mediated by the Maillard reaction in aging. Gerontol. 37: 152-165.

Nakamura, K., Hasegawa, T., Fukunaga, Y., and Ienaga, K. (1992) Crosslines A and B as candidates for the fluorophores in age- and diabetes-related cross-linked proteins, and their diabetes produced by Maillard reaction of α-N-acetyl-L-lysine with D-glucose. J. Chem. Soc.,Chem. Commun. 992-994.

Sell, D.R., and Monnier V.M. (1989) Structure elucidation of a senescence cross-link from human extracellular matrix. J. Biol. Chem. 264: 21597-21602.

Shin, D.B. Hayase, F., and Kato, H. (1988) Polymerization of proteins caused by reaction with sugars and the formation of 3-deoxyglucosone under physiological conditions. Agric. Biol. Chem. 52: 1451-1458.

Wolff, S.P., Jiang, Z.Y., and Hunt, J.V. (1991) Protein glycation and oxidative stress in diabetes mellitus and ageing. Free Radical Biol. Med. 10: 339-352.

Identification of Three New Mutagenic Heterocyclic Amines and Human Exposure to Known Heterocyclic Amines

K. Wakabayashi, R. Kurosaka, I.-S. Kim, H. Nukaya,[1] H. Ushiyama,[2] M. Ochiai, K. Fukutome, H. Nagaoka, T. Sugimura, and M. Nagao

CARCINOGENESIS DIVISION, NATIONAL CANCER CENTER RESEARCH INSTITUTE, TOKYO 104; [1]SCHOOL OF PHARMACEUTICAL SCIENCE, UNIVERSITY OF SHIZUOKA, SHIZUOKA 422; [2]THE TOKYO METROPOLITAN RESEARCH LABORATORY OF PUBLIC HEALTH, TOKYO 169, JAPAN

Summary

We identified three new mutagenic heterocyclic amines (HCAs), 2-amino-1-methyl-6-(4-hydroxyphenyl)imidazo[4,5-*b*]pyridine (4'-OH-PhIP) in broiled beef, and 2-amino-4-hydroxymethyl-3,8-dimethylimidazo[4,5-*f*]quinoxaline (4-CH_2OH-8-MeIQx) and 2-amino-1,7,9-trimethylimidazo[4,5-*g*]quinoxaline (7,9-DiMeI*g*Qx) in beef extract. The level of 4'-OH-PhIP was 21 ng/g in broiled beef, and the levels of 4-CH_2OH-8-MeIQx and 7,9-DiMeI*g*Qx were 6.0 and 53 ng/g, respectively, in beef extract. The mutagenicity of 4'-OH-PhIP to TA98 with S9 mix was 180 revertants/100 μg. 4-CH_2OH-8-MeIQx and 7,9-DiMeI*g*Qx induced 326,000 and 13,800 revertants of YG1024, and 99,000 and 670 revertants of TA98 per mg, respectively, in the presence of S9 mix. Furthermore, 4'-OH-PhIP was demonstrated to be formed by heating a mixture of creatine, tyrosine and glucose. Similarly, creatine, threonine and sugars were found to be involved in the formations of 4-CH_2OH-8-MeIQx and 7,9-DiMeI*g*Qx. Thus, the Maillard reaction is associated with the formation of the newly identified mutagenic HCAs, like other known 2-amino-3-methylimidazo[4,5-*f*]quinoline (IQ)-type HCAs.

Quantitative data on the known HCAs, previously reported to be carcinogenic, in cooked foods and cigarette smoke, and in urine samples from healthy volunteers indicated that the daily levels of exposure to 2-amino-3,8-dimethylimidazo[4,5-*f*]quinoxaline (MeIQx) and 2-amino-1-methyl-6-phenylimidazo[4,5-*b*]pyridine (PhIP) are 0.2-2.6 μg and 0.1-13.8 μg per person, respectively. ^{32}P-Postlabeling analysis demonstrated that MeIQx and PhIP produced adducts primarily at the C-8 position of guanine residues in DNA of rats *in vivo*. It is very important to analyze DNA adducts of HCAs in human tissues for estimation of the actual contributions of HCAs to human cancer.

Introduction

Twenty heterocyclic amines (HCAs) in cooked foods, pyrolysates of amino acids and proteins, and heated mixtures of creatin(in)e, amino acids and sugars have been identified as mutagens in *Salmonella* (Sugimura, 1986; Wakabayashi et al., 1992). Ten of these mutagenic compounds have been demonstrated to be carcinogenic in rodents and one in monkeys. However, the mutagenic potencies of HCAs in *Salmonella* are not quantitatively correlated with their carcinogenic activities in rodents. This suggests that even though their mutagenicities are low, compounds present in high amounts in cooked foods should not be overlooked. Furthermore, cooked foods probably contain many other as yet unidentified mutagenic HCAs. Thus, to better estimate the risk of cooked foods to known cancer, cooked foods should be scrutinized for new mutagenic HCAs, and human exposures to known HCAs should be quantified.

Identification of New Mutagenic Heterocyclic Amines (HCAs)

1) 2-Amino-1-methyl-6-(4-hydroxyphenyl)imidazo[4,5-b]pyridine (4'-OH-PhIP)
Various kinds of mutagenic and carcinogenic HCAs are present in cooked meat and fish (Sugimura, 1986). Among them, 2-amino-1-methyl-6-phenylimidazo[4,5-*b*]pyridine(PhIP) was isolated as the most abundant mutagen by weight in fried beef (Felton et al., 1986). The 4'-hydroxy derivative of PhIP has been detected as one of its metabolites in feces of rats given PhIP (Watkins et al., 1991). 4'-OH-PhIP was reported to be non-mutagenic to *Salmonella typhimurium* TA98 (Turteltaub et al., 1990a), but, as in other HCAs, an exocyclic primary amino group on this compound would be expected to be involved in mutagenicity. The structure of 4'-OH-PhIP is shown in Figure 1. We synthesized 4'-OH-PhIP from 2-chloro-1-methyl-6-phenylimidazo[4,5-*b*]pyridine, and tested carefully its mutagenicity (Kurosaka et al., 1992). The synthesized 4'-OH-PhIP induced 180 revertants of *S. typhimurium* TA98 with S9 mix per 100 μg. This mutagenicity was lower than that of PhIP. It was not mutagenic to TA98 in the absence of S9 mix.

Since the precursors of PhIP are known to be creatin(in)e, phenylalanine and glucose (Shioya et al., 1987), 4'-OH-PhIP should be formed by heating a mixture containing creatin(in)e, tyrosine and glucose. Actually, the 4'-hydroxy derivative of PhIP was produced in a yield of 1 nmol/mmol of original creatine when creatine (10 mmol), L-tyrosine (10 mmol) and glucose (5 mmol) were heated in 30 ml of diethylene glycol containing 14% water at 128°C for 2h (Wakabayashi et al., 1993). We, therefore, next examined whether 4'-OH-PhIP is present in broiled beef. Broiled beef was extracted with 0.1 N HCl, and 4'-OH-PhIP in the extract was purified by blue cotton treatment and then cation exchange fiber column chromatography. 4'-OH-PhIP was finally isolated and characterized by HPLC. The amount of 4'-OH-PhIP was estimated to be 21.0 ng per g of broiled beef (Kurosaka et al., 1992). This level was comparable to that of PhIP in the same broiled beef and was higher than the levels of MeIQx and other HCAs in broiled beef reported previously (Wakabayashi et al., 1992).

2) 2-Amino-4-hydroxymethyl-3,8-dimethylimidazo[4,5-f]quinoxaline (4-CH_2OH-8-MeIQx) and 2-amino-1,7,9-trimethylimidazo[4,5-g]quinoxaline (7,9-DiMeIgQx)
We tried to detect other new mutagens in beef extract using a new *Salmonella* tester strain, YG1024, which has a higher *O*-acetyltransferase level and is much more sensitive to various kinds of mutagens than TA98 (Watanabe et al., 1990). Mutagens in bacteriological-grade beef extract were separated by blue cotton treatment, a cation exchange fiber column and then HPLC on a semi-preparative ODS column with a gradient solvent system. On HPLC, four major mutagenic fractions, accounting for 53% of the total mutagenicity adsorbed to blue cotton, were detected. These mutagenicities were found to be due mainly to four known HCAs, 2-amino-3-methylimidazo[4,5-*f*]quinoline (IQ), 2-amino-3,4-dimethylimidazo[4,5-*f*]]quinoline (MeIQ), 2-amino-3,8-dimethylimidazo[4,5-*f*]]quinoxaline (MeIQx) and 2-amino-3,4,8-trimethylimidazo[4,5-*f*]quinoxaline (4,8-DiMeIQx). In addition, two mutagenic peaks, one (peak I) eluted at a retention time of 4 min before that of MeIQx and the other (peak II) with a retention time of 4 min after 4,8-DiMeIQx, were clearly detected. These elution times did not correspond to those of any known HCAs. The mutagenicities in peaks I and II constituted 2.1 and 2.0 %, respectively, of the total mutagenicity. The mutagens in peaks I and II were further purified by HPLC on cation exchange, SP-2SW and ODS columns and mutagenic compounds I and II, respectively, were obtained. The amounts of compounds I and II in bacteriological-grade beef extract

were estimated to be 6.0 and 53 ng per g of beef extract from their UV absorbances at 260 nm.

Since the levels of compounds I and II in beef extract were too low to allow measurements of various spectra for their structure determinations, we obtained increased amounts of these compounds using creatine-supplemented model systems as reported previously (Jägerstad et al., 1991). A nine-fold increase in the level of compound I over that in the original beef extract was obtained when a mixture of beef extract (50 g), creatine (17 mmol) and L-threonine (17 mmol) was heated at 200°C for 5h, and a 860-fold increase in the level of compound II over that in the original beef extract was achieved by heating creatine (1 mol) with L-threonine (1 mol) and glucose (0.5 mol) in 1000 ml of ethylene glycol at 200°C for 5h.

Compounds I and II were isolated from the above creatine-supplemented heated materials by various column chromatographies, and their structures were determined to be 4-CH_2OH-8-MeIQx and 7,9-DiMeI*g*Qx, respectively (Figure 1) (Wakabayashi et al., 1993). 4-CH_2OH-8-MeIQx induced 326,000 revertants of YG1024 and 99,000 revertants of TA98 per μg with S9 mix and 7,9-DiMeI*g*Qx gave 13,800 revertants of YG1024 and 670 revertants of TA98 per μg with S9 mix. In the absence of S9 mix, these compounds showed no mutagenicity to either strain.

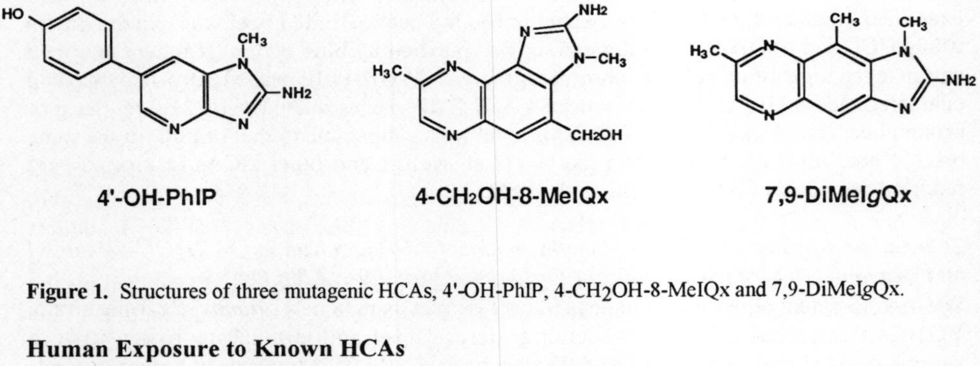

4'-OH-PhIP 4-CH_2OH-8-MeIQx 7,9-DiMeI*g*Qx

Figure 1. Structures of three mutagenic HCAs, 4'-OH-PhIP, 4-CH_2OH-8-MeIQx and 7,9-DiMeIgQx.

Human Exposure to Known HCAs

1)Quantification of HCAs in cooked foods and cigarette smoke.

Quantification of the extent of human exposure to HCAs is essential for estimation of their carcinogenic risks. We measured the amounts of HCAs in various kinds of cooked meat and fish, using blue cotton treatment and HPLC in combination. By this method, nine HCAs were detected at levels of 0.03 - 69.2 ng/g in cooked foods (Wakabayashi et al., 1992). PhIP was the most abundant, its levels being 0.56 - 69.2 ng/g in cooked foods. MeIQx was the second most abundant at levels of 0.64 - 6.44 ng/g in cooked foods.

By a similar quantification method to that used for detection of HCAs in cooked foods, the amounts of five HCAs, 3-amino-1,4-dimethyl-5*H*-pyrido[4,3-*b*]indole (Trp-P-1), 3-amino-1-methyl-5*H*-pyrido[4,3-*b*]indole (Trp-P-2), 2-amino-9*H*-pyrido[2,3-*b*]indole (AαC), 2-amino-3-methyl-9*H*-pyrido[2,3-*b*]indole (MeAαC) and PhIP, in mainstream and sidestream cigarette smoke condensates were analysed (Wakabayashi et al., 1993). Trp-P-1, Trp-P-2, AαC and MeAαC were detected in all the samples of mainstream smoke condensates of five Japanese brand cigarettes (four filter-tipped cigarettes and a non-filter cigarette), at levels of 0.02 - 0.51 ng, 0.63 - 1.54 ng, 2.22 - 13.5 ng and 0.37 - 1.92 ng per cigarette, respectively. These four HCAs were also detected in sidestream smoke condensates of three brands of cigarettes, the levels per cigarette being 0.11 - 0.28 ng for Trp-P-1, 0.22 - 0.48 ng for Trp-P-2, 1.24 - 2.72 ng for AαC and 0.11 - 0.39 ng for MeAαC. These levels were lower than those in mainstream smoke condensates. On the other hand, PhIP was not detected in any sample of mainstream or sidestream smoke condensate, so its level must be less than 0.01 ng per cigarette. These data are not consistent with a report by Manabe et al. (1991) that PhIP was detected in mainstream cigarette smoke condensates of six brand-name cigarettes, including three Japanese brands, at levels of 11 - 23 ng per cigarette.

We reported previously that four HCAs, MeIQx, PhIP, Trp-P-1 and Trp-P-2, were detected in all urine samples from ten healthy volunteers eating a normal diet at levels of 0.03 - 47 ng/24 hr urine, but not in the urine of three in-patients receiving parenteral alimentation (Ushiyama et al., 1991). The daily intakes of HCAs were estimated from the quantitative data on their levels in cooked foods, cigarette smoke condensates and human urine; for example, the intake of MeIQx was estimated as 0.2 - 2.6 μg/person/day and that of PhIP as 0.1 - 13.8 μg/person/day. These levels are comparable with those of other typical carcinogens including *N*-nitrosodimethylamine and benzo[*a*]pyrene (Wakabayashi et al., 1993).

2) ^{32}P-Postlabeling methods for detection of MeIQx-and PhIP-DNA adducts.

The above observations indicate that humans are continuously exposed to low levels of HCAs in daily life. A linear relationship has been found between DNA adduct levels and a wide range of doses of MeIQx in animals (Yamashita et al., 1990; Turteltaub et al., 1990b). Thus, even at low doses such as those to which humans are exposed, HCAs could form DNA adducts in human tissues. Since the HCA-DNA adduct levels in human tissues are expected to be very low, a sensitive method to detect DNA adducts is necessary. We, therefore, tested ^{32}P-postlabeling methods for detecting MeIQx-and PhIP-DNA adducts under several conditions.

DNA adducts formed in the liver of rats fed 400 ppm MeIQx were analyzed by the ^{32}P-postlabeling method under standard, nuclease P1 or butanol extraction conditions. The nuclease P1 method was found to be much more sensitive than the standard or butanol extraction methods and gave one major and four minor spots of MeIQx-DNA adducts. The major adduct was identified as N^2-(deoxyguanosin-8-yl)-MeIQx 3',5'-diphosphate(3',5'-pdGp-C8-MeIQx) (Ochiai et al., 1993). Four spots of PhIP-DNA adducts were detected in DNA from organs of rats fed 400 ppm PhIP under adduct-intensification conditions, but after additional treatment with nuclease P1 and phosphodiesterase I, a single adduct spot was detected. The material in this spot was identified as N^2-(deoxyguanosin-8-yl)-PhIP 5'-monophosphate(5'-pdG-C8-PhIP). With the above methods, the detection limits of 3',5'-pdGp-C8-MeIQx and 5'-pdG-C8-PhIP were 1 - 5 per 10^{10} nucleotides. Therefore, these methods will be very useful for detecting MeIQx- and PhIP-DNA adducts in human tissues.

Acknowledgements

This study was supported by Grants-in-Aid for Cancer Research from the Ministry of Health and Welfare and the Ministry of Education, Science and Culture of Japan, and grants from the Bristol-Myers Squibb Foundation and the Smoking Research Foundation.

References

Felton, J. S., Knize, M. G., Shen, N. H., Lewis P. R., Andresen, B. D., Happe, J. and Hatch, F. T. (1986) The isolation and identification of a new mutagen from fried ground beef: 2-amino-1-methyl-6-phenylimidazo[4,5-*b*]pyridine (PhIP). *Carcinogenesis*, 7: 1081-1086.

Jägerstad, M., Skog, K., Grivas, S. and Olsson, K. (1991) Formation of heterocyclic amines using model systems. *Mutat. Res.*, 259: 219-233.

Kurosaka, R., Wakabayashi, K., Ushiyama, H., Nukaya, H., Arakawa, N., Sugimura, T. and Nagao, M. (1992) Detection of 2-amino-1-methyl-6-(4-hydroxyphenyl)imidazo[4,5-*b*]pyridine in broiled beef. *Jpn. J. Cancer Res.*, 83: 919-922.

Manabe, S., Tohyama, K., Wada, O. and Aramaki, T. (1991) Detection of a carcinogen, 2-amino-1-methyl-6-phenylimidazo[4,5-*b*]pyridine (PhIP), in cigarette smoke condensate. *Carcinogenesis*, 12: 1945-1947.

Ochiai, M., Nagaoka, H., Wakabayashi, K., Tanaka, Y., Kim, S.-B., Tada, A., Nukaya, H., Sugimura, T. and Nagao, M. (1993) Identification of N^2-(deoxyguanosin-8-yl)-2-amino-3,8-dimethylimidazo[4,5-*f*]quinoxaline 3',5'-diphosphate, a major DNA adduct, detected by nuclease P1 modification of the ^{32}P-postlabeling method, in the liver of rats fed MeIQx. *Carcinogenesis* (in press).

Shioya, M., Wakabayashi, K., Sato, S., Nagao, M. and Sugimura, T. (1987) Formation of a mutagen, 2-amino-1-methyl-6-phenylimidazo[4,5-*b*]pyridine (PhIP) in cooked beef, by heating a mixture containing creatinine, phenylalanine and glucose. *Mutat. Res.*, 191: 133-138.

Sugimura, T. (1986) Studies on environmental chemical carcinogenesis in Japan. *Science*, 233: 312-318.

Turteltaub, K. W., Knize, M. G., Buonarati, M. H., McManus, M. E., Veronese, M. E., Mazrimas, J. A. and Felton, J. S. (1990a) Metabolism of 2-amino-1-methyl-6-phenylimidazo[4,5-*b*]pyridine (PhIP) by liver microsomes and isolated rabbit cytochrome P450 isozymes. *Carcinogenesis*, 11: 941-946.

Turteltaub, K. W., Felton, J. S., Gledhill, B. L., Vogel, J. S., Southon, J. R., Caffee, M. W., Finkel, R. C., Nelson, D. E., Proctor, I. D. and Davis, J. C. (1990b) Accelerator mass spectrometry in biomedical dosimetry: relationship between low-level exposure and covalent binding of heterocyclic amine carcinogens to DNA. *Proc. Natl. Acad. Sci. U.S.A.*, 87: 5288-5292.

Ushiyama, H., Wakabayashi, K., Hirose, M., Itoh, H., Sugimura, T. and Nagao, M. (1991) Presence of carcinogenic heterocyclic amines in urine of healthy volunteers eating normal diet, but not of inpatients receiving parenteral alimentation. *Carcinogenesis*, 12: 1417-1422.

Wakabayashi, K., Nagao, M., Esumi, H. and Sugimura, T. (1992) Food-derived mutagens and carcinogens. *Cancer Res. (Suppl.)*, 52: 2092s-2098s.

Wakabayashi, K., Kim, I.-S., Kurosaka, R., Yamaizumi, Z., Ushiyama, H., Takahashi, M., Koyota, S., Tada, A., Nukaya, H., Goto, S., Sugimura, T. and Nagao, M. (1993) Identification of new mutagenic heterocyclic amines and quantification of known heterocyclic amines. In Adamson, R.H., Gustaffson, J.A., Ito, N., Nagao, M., Sugimura, T., Wakabayashi, K. and Yamazoe, Y. (eds.) *Heterocyclic Amines in Cooked Foods:Possible Human Carcinogens, Proceedings of the 23rd International*

Symposium of the Princess Takamatsu Cancer Research Fund, Princeton Scientific Publishing, Princeton (in press).

Watanabe, M., Ishidate Jr., M. and Nohmi, T. (1990) Sensitive method for the detection of mutagenic nitroarenes and aromatic amines: new derivatives of *Salmonella typhimurium* tester strains possessing elevated *O*-acetyltransferase levels. *Mutat. Res.*, 234: 337-348.

Watkins, B. E., Suzuki, M., Wallin, H., Wakabayashi, K., Alexander, J., Vanderlaan, M., Sugimura, T. and Esumi, H. (1991) The effect of dose and enzyme inducers on the metabolism of 2-amino-1-methyl-6-phenylimidazo[4,5-*b*]pyridine (PhIP) in rats. *Carcinogenesis*, 12: 2291-2295.

Yamashita, K., Adachi, M., Kato, S., Nakagama, H., Ochiai, M., Wakabayashi, K., Sato, S., Nagao, M. and Sugimura, T. (1990) DNA adducts formed by 2-amino-3,8-dimethylimidazo[4,5-*f*]quinoxaline in rat liver: dose-response on chronic administration. *Jpn. J. Cancer Res.*, 81: 470-476.

Balance Experiments on Human Volunteers with ϵ-Fructoselysine (FL) and Lysinoalanine (LAL)

K. Lee[1] and H. F. Erbersdobler[2]

[1]DEPARTMENT OF FOOD AND NUTRITION, CHANG-WON NATIONAL UNIVERSITY, #9, SALIM-DONG, CHANG-WON, KYUNGNAM-DO, 641-773, KOREA; [2]DEPARTMENT OF HUMAN NUTRITION AND FOOD SCIENCE, UNIVERSITY OF KIEL, DUESTERNBROOKER WEG 17. D-24105 KIEL, GERMANY

Dedicated to Prof. Dr. Dr. h.c. Kurt Heyns on the occasion of his 85th birthday on 13. 12. 1993

Summary

To evaluate FL and LAL excretion, urine and fecal samples were collected from 19 healthy students for up to 84 hours after consuming a single test meal enhanced with either FL or LAL. In study one 12 volunteers received a test meal of a FL rich protein containing 2860 mg FL. In the second study 7 volunteers were fed a test meal with 300 mg LAL.

The recovery of FL was extremely low (1.4-3.5 % of the intake in urine and 0.1-2.3 % in feces). In contrast, LAL was recovered in amounts of 0.6-1.2 % in urine and 33-83 % in feces. The small amounts of FL absorbed were very rapidly excreted. Almost 80 % of the total amount of FL in the urine was found in the first 12 hours after the test meal. In contrast to the FL excretion the urinal excretion of LAL was delayed and not fully completed after 24 hours. Moreover, the details of the experiments with FL and LAL show that the timing for feces collection up to 72-84 hours was not sufficient in all cases to fully recover particularly the ingested LAL. The results suggest that FL is easily decomposed by the intestinal flora while LAL is more resistant.

Introduction

Main changes during heat processing of food products are due to the Maillard reaction (Erbersdobler and Hupe, 1991). The presence of Maillard products alters the nutritive value of protein and reduces the bioavailability of amino acids. Moreover, the metabolic transit of Maillard products in humans is poorly defined. However, the metabolic break down of several Amadori products especially of ϵ-fructoselysine, the Amadori product of the reaction of lysine with glucose, but also of lysinoalanine (LAL) have been tested in rats as is discussed later.

Previous studies with student volunteers have indicated that the absorption and urinal elimination of FL is very low in humans because only 1-3 % of the ingested amount were excreted in urine. In addition preliminary results of feces sampling from 3 volunteers revealed that only 3-6 % of the ingested FL were excreted in the feces. On the other hand, balance studies in humans with LAL have not been performed and no data exist on urinary and fecal excretion of LAL.

Materials and methods

In order to obtain a test material enriched in FL, acid precipitated casein was heated with glucose and tap water. The concentration of FL in the so prepared casein was quite high (143.000 ppm in the protein). For the LAL containing protein, casein was heated with tap water which was alkalized with NaOH to a pH of 12 for 2 h at 90°C and precipitated afterwards with 1M HCl at pH of 4.0 as was described by Karayannis et al. (1979). The LAL content of the so treated protein was 47.000 ppm.

Prior to the two experiments the student volunteers were fed three days a defined diet extremely low in both FL and LAL. During the third day a 24 h-pooled sample of either urine or feces was collected from each person to determine that all dietary sources of FL or LAL were washed out. In fact FL was not detectable at all and only traces of LAL were found in the feces (for details see Lee. 1992). On the morning of the fourth day the volunteers received at breakfast the above described caseins in amounts providing either 2.860 mg FL (study 1) or 300 mg LAL (study 2). The FL and LAL poor diet was continued until fecal and urine collection was completed. Urine was collected for 24 hours (in two cases samples were collected up to 72 hours) and feces up to 84 hours in single portions. After weighing and recording the collected material, an aliquot of the samples was frozen and stored at -20° C. In experiment 1 a total of 12 volunteers were fed the FL test meal and in experiment 2 a total of 7 subjects received the meal containing LAL. The details of the experimental design are described elsewhere (Lee, 1992). The volunteers gave signed consent after explaining the experimental procedure which was approved by the Ethic Commission of Kiel University.

FL, which itself is very unstable during acid hydrolysis, was assessed as furosine and analyzed using an amino acid analyser after hydrolysis of 100-250 mg of the test diet, 500 mg of the foods, 1 g feces or 100 ml urine with 7.8 M HCl for 20 h under reflux as described previously (Erbersdobler et al., 1987). Furosine was calculated using arginine as external standard and afterwards corrected to the recently available pure furosine standard according to comparative experiments (Hartkopf and Erbersdobler, 1993). For the calculation of FL from furosine the conversion factor of 2.5 was used. LAL was determined using an amino acid analyser after acid hydrolysis of similar amounts as have been used for furosine analysis with 6 M HCl in flasks with screw caps placed in a laboratory oven for 24 h.

Results and Discussion

The results for the excretion of FL and LAL are summarized in table I. Less than 3% of the ingested FL was excreted in the urine. In the feces only 0.6-1.4% FL were excreted. This suggests that the fate of about 96% of the ingested FL remains unexplained. Metabolic studies in rats including experiments with ^{14}C-labeled FL have shown no indication for an utilization (Erbersdobler et al., 1981, 1989, Finot, 1973, 1983). However, microorganisms

Table I. Excretion of fructoselysine (FL) and lysinoalanine (LAL) in % of intake after a test meal of 2860 mg FL (n=12) or 300 mg LAL (n=7), respectively.

	fructoselysine	lysinoalanine
excretion in urine	2.69 ± 0.72	0.90 ± 0.20
output in feces	0.96 ± 0.69	63.6 ± 16.2
total recovery	3.68 ± 1.09	64.5 ± 16.3

in the intestinal tract are able to break down FL as was shown earlier in studies in which FL was incubated with gut contents from rats or pigs respectively (Erbersdobler, 1989, Erbersdobler et al., 1991). In contrast, two third of the ingested amount of LAL was excreted in the feces with single values up to 83% of the intake. However, the urinary excretion of LAL was also extremely low.

In Figure 1 the time course of the excretion of FL and LAL is depicted for several representative cases. Depending on the time of urination and defecation different profiles for the individual subjects were observed. As described previously (Erbersdobler et al., 1989, Erbersdobler et al., 1991) the peak of urinal excretion of FL was reached between 4 and 8 hours after consuming the test meal. For LAL a similar peak can be observed although on a much lower level. Smaller amounts of FL but considerably more of LAL were found in later stages suggesting an accumulation in the urine during the night or indicating a late absorption in the lower part of the small intestine after microbial release from protein as was proposed by Finot (1973) from experiments with rats.

An apparent peak of FL and LAL excretion in the feces could be seen in several subjects between 24 and 36 hours after the test meal even though partially the concentrations especially in LAL remained high up to 84 hours. The time of feces collection was obviously not long enough particularly for LAL. This is further demonstrated by the LAL balance data. In the 4 volunteers where the fecal LAL values had significantly decreased until 84 hours the recovery in the feces was 65-83 % of the ingested LAL. In the three other subjects the LAL recovery was only 33-59%. Figure 1 gives examples for both groups. However, the urinal concentration of LAL decreased rapidly after 25 hours in spite of the high concentrations of LAL in the gut at this time. This was also seen in two cases in which high fecal LAL levels continued up to 84 hours. Figure 1 contains an example of one of these two volunteers.

The FL-rich casein used as test component in the present study was additionally tested in a balance trial with growing rats (Lee, 1992). Although the results should be compared with certain reservation (balance trial vs. a single test meal) they are presented in Figure 2. In fact, the rats excreted 3-times more FL in the urine than the human volunteers whereas the recovery of FL in the feces was quite similar. On the other hand, the difference was smaller than the 30 % of urinary excretion observed in previous trials with rats which received a single test meal containing ^{14}C-labeled FL (see in: Erbersdobler et al., 1991).

Conclusions

There is no evidence in the literature that FL is available as source of lysine in any way. In contrast, there are reports about a limited utilization of LAL in growing chicken (Robbins et al., 1980). On the other hand, Kawamura and Hayashi (1987) found a low LAL degrading activity in crude kidney extracts from humans expressing practically no availability. However, the authors suggest that humans could be more sensitive to LAL than

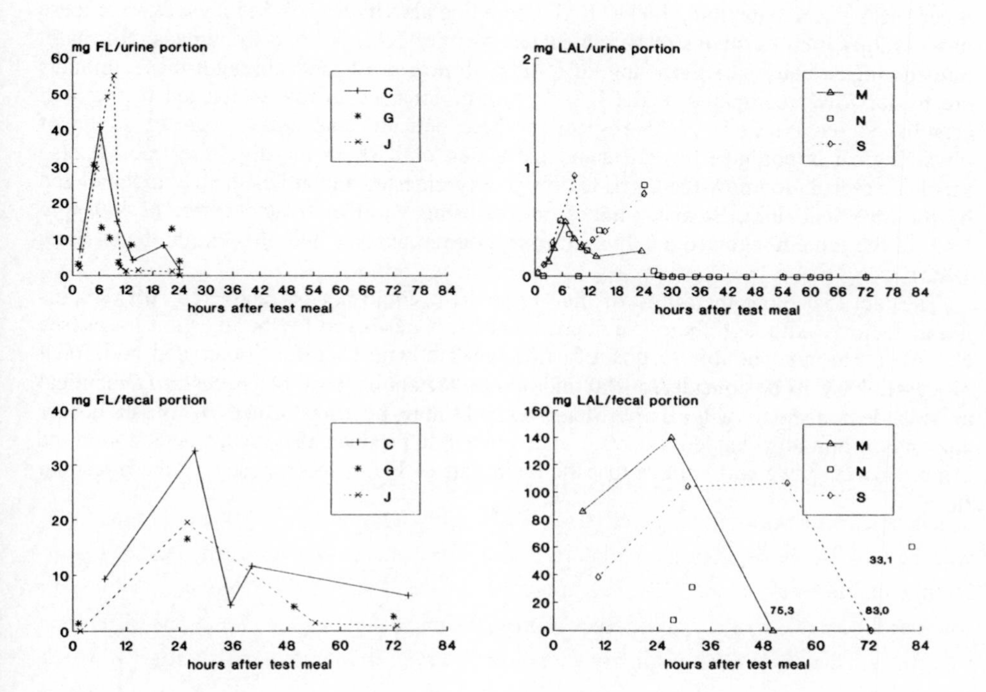

Figure 1. Profiles of the excretion of fructoselysine (FL) and lysinoalanine (LAL) in the urine and feces of selected persons of the experiments.
Capitals in the figures = indication of the different volunteers (n = 19 ; A-S)
The figures in the part: LAL/fecal portion = recovery of LAL in % of the intake

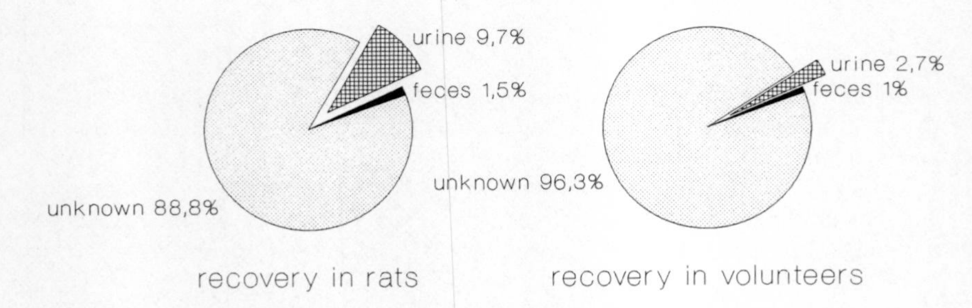

Figure 2. A comparison of the recoveries of fructoselysine (FL) in urine and feces after giving the same test protein in a balance period to growing rats and as single test meal to human volunteers.

other species which exhibit a higher LAL degrading activity in their kidneys. Despite these observations it still remains to be elucidated whether LAL is toxic to humans. Normally humans ingest daily small amounts of LAL with processed food although these amounts are by far lower compared to the LAL concentrations fed to rats which led to the well established renal lesions. With respect to these delicate questions it appears to be of physiological importance to understand the action of LAL in the digestive tract. It also seems reassuring to know that LAL is only poorly digested and at best poorly metabolized by the intestinal flora. Balance studies on rats using a milder treated casein (2.400 ppm LAL in the protein) showed 3.7 % LAL excretion in the urine and 36 % in the feces (Lee, 1992).

The fact that more than 90% of the ingested FL could not be recovered still remains unsatisfactory. Although there is evidence that FL is destroyed by the intestinal flora (see above) it was not possible to quantify this break down. Therefore other and additional processes have to be considered. Rat and human metabolism are not necessarily identical as was demonstrated with LAL. Moreover, FL may be transformed by brush border enzymes to non-absorbable products or metabolized to products that escape detection in the urine. However, we still believe that the main part of FL is decomposed by the intestinal flora.

References

Erbersdobler, H.F. (1989) Changes due to food processing in low-protein diets for renal patients. *In: Low Protein Diets in Renal Patients: Composition and Absorption (Eds. N. Gretz, S. Giovanetti, M. Strauch). Contrib. Nephrol., Karger, Basel, pp. 49-65*

Erbersdobler, H.F., Brandt, A., Scharrer, E., von Wangenheim, B. (1981) Transport and metabolism studies with fructose amino acids. *Prog. Fd. Nutr. Sci. Vol. 5 ("Maillard Reaction in Food and Nutrition Science" Ed. C. Eriksson) Pergamon Press, Oxford,* pp. 257-263.

Erbersdobler, H.F., Dehn, B. Nangpal, A. and Reuter, H. (1987) Determination of furosine in heated milk as a measure of heat intensity during processing. *Journal of Dairy Research* 54:147-151.

Erbersdobler, H.F., Groß, A., Klusmann, U., and Schlecht, K. (1989) Absorption and metabolism of heated protein-carbohydrate mixtures in humans. *In: Absorption and Utilization of Amino Acids. M. Friedmann, Ed., CRC Press, INC, Boca Raton,* pp. 91-102.

Erbersdobler, H.F., Hupe, A. (1991) Determination of lysine damage and calculation of lysine bio-availability in several processed foods. *Zeitschrift für Ernährungswissenschaft* 30:46-49.

Erbersdobler, H.F., Lohmann, M., Buhl, K. (1991) Utilization of early Maillard reaction products by humans. *In: Nutritional and Toxicological Consequences of Food Processing. (Ed. M. Friedman), Advances in Experimental Medicine and Biology, Vol. 289, Plenum Press, New York,* pp. 363-370.

Finot, P.A., (1973) Non-enzymatic browning. *In: Proteins in Human Nutrition. J.W.G. Porter and B.A. Rolls eds., Academic Press, London and New York,* pp. 501-514.

Finot, P.A., (1983a) Chemical modifications of the milk proteins during processing and storage. Nutritional, metabolic and physiological consequences. *Kieler Milchwirtschaftliche Forschungsberichte* 35:357-369.

Finot, P.A., (1983b) Lysinoalanine in food proteins. *Nutrition Abstracts and Reviews* 53:67-80.

Hartkopf, J. and Erbersdobler, H.F. (1993) Stability of furosine during ion exchange chromatography in comparison with reversed-phase high-performance liquid chromatography. *J. Chromatography,* 635:151-154.

Karayiannis, N.I., MacGregor, J.T., Bjeldanes, L.F. (1979) Lysinoalanine formation in alkali treated proteins and model peptides. *Fd. Cosmet. Toxicol.* 17: 585-590.

Kawamura Y. and Hayashi, R. (1987) Lysinoalanine-degrading enzymes of various animal kidneys. *Agric. Biol. Chem.* 51: 2289-2290.

Lee, K. (1992) Untersuchungen zur Bilanzierung des Proteins sowie des Fructoselysins und des Lysinoalanins bei hitzebehandeltem Casein an Ratten mit der Homoarginin-Technik und an Menschen. Agrarwissenschaftliche Dissertation, Kiel, 1992, *Schriftenreihe des Instituts für Humanernährung und Lebensmittelkunde der Christian-Albrechts Universität zu Kiel, Heft 9.* ISBN: 3-926085-08-8

Reaction of Cyanide Ion with Glucose: Implications for the Maillard Reaction and its Effect on Thymidylate Synthase Enzyme Activity

L. Trézl,[1] L. Hullán,[2] V. Horváth,[1] I. Rusznák,[1] L. Töke,[1] T. Szarvas,[3] and Cs. Vida[1]

[1]TECHNICAL UNIVERSITY OF BUDAPEST, DEPARTMENT OF ORGANIC CHEMICAL TECHNOLOGY, BUDAPEST H-1521, PO BOX 91; [2]NATIONAL INSTITUTE OF ONCOLOGY, BUDAPEST H-1122; [3]INSTITUTE OF ISOTOPES, HUNGARIAN ACADEMY OF SCIENCE, BUDAPEST H-1525

Summary
Strecker reactions were extended to reducing sugars (e.g. D-glucose) because the cyanide ion could be reacted with them. As a results of such reactions, cyanohydrins were formed, and consequently the Maillard reaction was less prominent. The reaction was shown to inhibit the activity of thymidylate synthase enzyme.

Introduction

A Hungarian team from several institutes has been studying the reactions and biological effects of formaldehyde for more than two decades. The work has led to the discovery of some new reactions (Tyihak et al., 1980; Trézl et al., 1983; Trézl and Pipek, 1988; Trézl et al., 1990).

Investigations of formaldehyde reactions were extended to studies of the reactions with cyanide ion and led to the discovery of a Strecker reaction. (Houben-Weyl, 1985a,b). The Maillard reaction assumed less importance when reducing sugars (e.g. D-glucose) were applied instead of formaldehyde.

Materials and Methods

Materials:
d1-L-Tetrahydrofolic acid, 2'-deoxyuridine-5'-monophosphate, charcoal, L-lysine, L-arginine, D-glucose and D-xylose were obtained from Sigma Ltd. *Lactobacillus casei* thymidylate synthase was obtained from Biopure, Boston, MA. 5-^{3}H dUMP was purchased from the Radiochemical Centre, Amersham, UK. The active Na^{14}CN (activity: 3.159 MBq/mmol) was prepared in the Institute of Isotopes, Hungarian Academy of Science. Kieselgel 60 HF$_{254}$ was obtained from Merck. 2-mercaptoethanol, NaF, TrisHCl buffer and trichloroacetic acid were obtained from Sigma Ltd.

TLC procedure
Stationary phase: Kieselgel 60 HF$_{254}$; Mobile phase: n-propanol:water (8:2); Developing agent: Ninhydrin; Glucose spot was visualized by ninhydrin. Running time: 12 hours; Concentrations of all reagents: 0.1 mM; Details of the reactions are presented on Figure 1.

Radioactivity incorporation
Separation method: thin layer chromatography; Stationary phase: Kieselgel 60 HF$_{254}$; Mobile phase: methanol: ammonium-hydroxide (3:1); Measuring technology: LKB-Rack Beta, fluid scintillation spectrometer; Scintillator: Clinosol^{R}; Reaction temp.: 37°C; pH=7.4.

NMR measurements
^{1}NMR spectra were recorded using a Bruker AC 80 NMR spectrometer.

Quantum chemical calculations
The calculations were conducted using a Silicon Graphics Iris Indigo 2 workstation. A Gaussian 92 algorithm was used to determine the 6-31G* ESP-derived point charges. The semi empirical ESP (electrostatic potential) derived point charges were derived using the MNDO Hamiltonian. The geometries of molecules were fully optimized with the split-valence 3-21G basis set.

Enzyme assay
The commercial L. casei thymidylate synthase supplied in a frozen solution was stored in lyophilized aliquots. For each experiment, 0.1 mL lyophilized enzyme was reconstituted and diluted appropriately.

Thymidylate synthase activity was estimated as described by Roberts (1966). The reaction mixture (0.15mL) contained 13.5 nmol 5-^{3}H dUMP (22 kBq), 30 nmol tetrahydrofolate, 533 nmol formaldehyde, 1.5 μmol 2-mercaptoethanol, 7.5 μmol NaF, 7.5 μmol Tris-HCI buffer, (pH 7.5) and 0.09 mL of enzyme solution.

Reaction was started by adding the enzyme and, after incubation (0-40 min) at 37°C was terminated by addition of 0.5 mL of 3.35% trichloroacetic acid and 0.5 mL of charcoal suspension (55 mg/mL). The samples were centrifuged at 1,000 g for 15 min. Radioactivity of supernatant was counted in a liquid scintillation spectrometer. Nonadsorbable and total radioactivity were determined in appropriate controls.

Reaction velocity was linear with time under the conditions employed. The enzyme activities varied by less than 5% in the control test. To test the effect of glucose and NaCN on enzyme activity, the two compounds were added separately after the addition of all reaction participants and before the addition of enzyme. The glucose cyanohydrin reaction was started by the addition of NaCN to all samples exactly 30 seconds prior to the start of the enzyme reaction.

Results

TLC results (Figure 1) and radioactivity incorporation analysis (Figure 2) confirmed that the glucose -CN$^-$ reaction is very rapid and the glucose-cyanohydrin is a reactive molecule; it can react readily with L-lysine, forming lysine-cyanoglucose.

The glucose-cyanohydrin formation can take place at 37°C and the addition of cyanide ion to glucose is more extensive than the addition of L-lysine to glucose.

The quantum chemical calculations (Figure 3) indicated that the positive reaction centre of the glucose-cyanohydrin molecule is on the C2 atom, where the nucleophilic substitution takes place; the positive charge on the C2 atom is much higher than that of the other C atoms.

Effect of glucose and NaCN on thymidylate synthase activity.
Thymidylate synthase (EC 2.1.1.45) catalyses the reductive methylation of dUMP to dTMP. The enzyme takes part not only in *de novo* nucleotide biosynthesis producing thymidylate (a DNA precursor), but it belongs to the folate cycle as the only enzyme able to oxidize 5,10-methylene-THF.

products from the reaction of L-lysine with D-glucose cyanohydrin

L-lysine

1 2 3 4 5 6 7 8 9 10 11

Figure 1. TLC investigation of glucose-L-lysine and glucose-L-lysine-NaCN systems.
1. L-lysine test; 2. Glucose test; 3. Glucose+L-lysine 37°C 10 min.; 4. Glucose+L-lysine 80°C 10 min.; 5. Glucose+L-lysine+NaCN 37°C 10 min.; 6. Glucose+L-lysine 37°C 20 min.; 7. Glucose+L-lysine 80°C 20 min.; 8. Glucose+L-lysine+NaCN 37°C 20 min.; 9. Glucose+L-lysine 37°C 30 min.; 10. Glucose+L-lysine 80°C 30 min.; 11. Glucose+L-lysine+NaCN 37°C 30 min.;

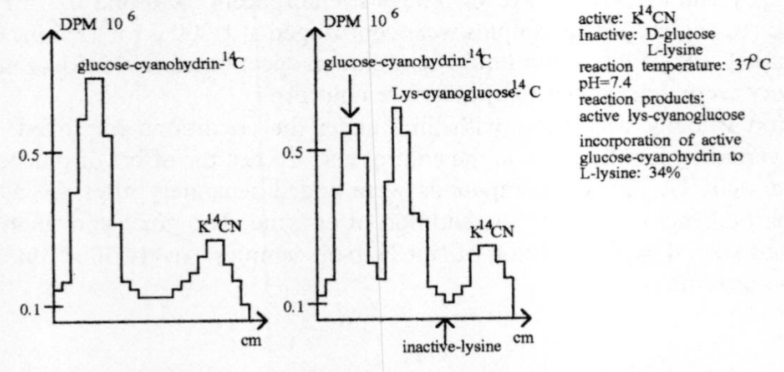

Figure 2. Radioactivity incorporation of $K^{14}CN$ into D-glucose (left) and L-lysine (right)
Solvent: Methanol: ammoniumhydroxide (3:1).

The ^{1}H NMR investigations supported the formation of both D-glucose cyanohydrin and lysine-D-cyanoglucose.

Glucose (8mM) and NaCN (0.8mM) together inhibited the activity of thymidylate synthase (see Figure 4). At the beginning of the reaction there was no effect but the degree of inhibition subsequently increased as a function of reaction time. The two compounds added separately at the above concentrations did not influence the enzyme activity (data are not given).

The results suggest that an appropriate amount of glucose-cyanohydrin for the inhibition of enzyme activity was not available until 20-30 minutes after the beginning of the reaction. In terms of mechanism, the inhibition of thymidylate synthase activity may be a consequence of an interaction between glucose cyanohydrin and the amino acid residues of the enzyme.

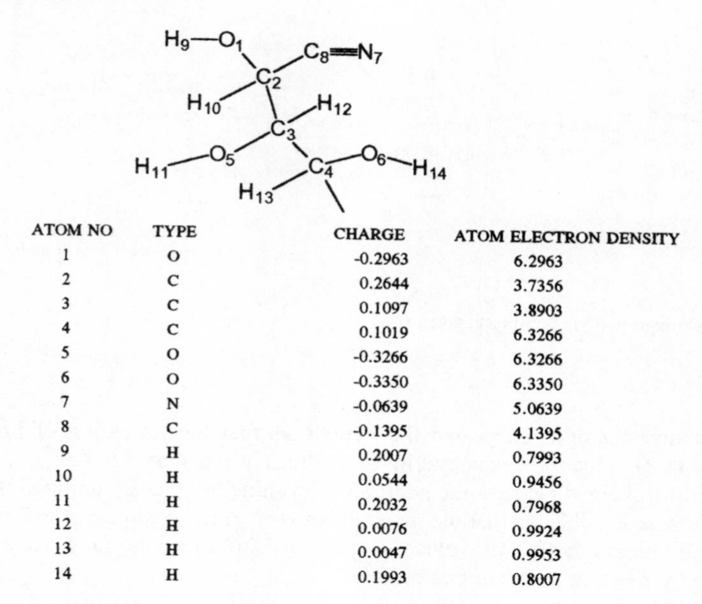

ATOM NO	TYPE	CHARGE	ATOM ELECTRON DENSITY
1	O	-0.2963	6.2963
2	C	0.2644	3.7356
3	C	0.1097	3.8903
4	C	0.1019	6.3266
5	O	-0.3266	6.3266
6	O	-0.3350	6.3350
7	N	-0.0639	5.0639
8	C	-0.1395	4.1395
9	H	0.2007	0.7993
10	H	0.0544	0.9456
11	H	0.2032	0.7968
12	H	0.0076	0.9924
13	H	0.0047	0.9953
14	H	0.1993	0.8007

Figure 3. Net atomic charges on glucose-cyanohydrin.

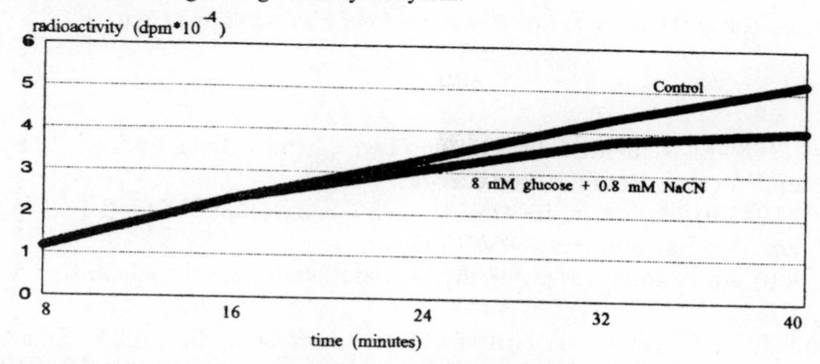

Figure 4. Effect of glucose and NaCN on thymidylate synthase activity.

Discussion

The methods used in the present study showed that the formation of D-glucose cyanohydrin from D-glucose and cyanide ion occurs rapidly. The results also verified the high reactivity of D-glucose-cyanohydrin with L-lysine. It is probable that the reactions take place according to the mechanism outlined in Figure 5.

The time required for the onset of cyanohydrin formation (see Figure 1) coincided with the time required for the onset of enzyme inhibition.

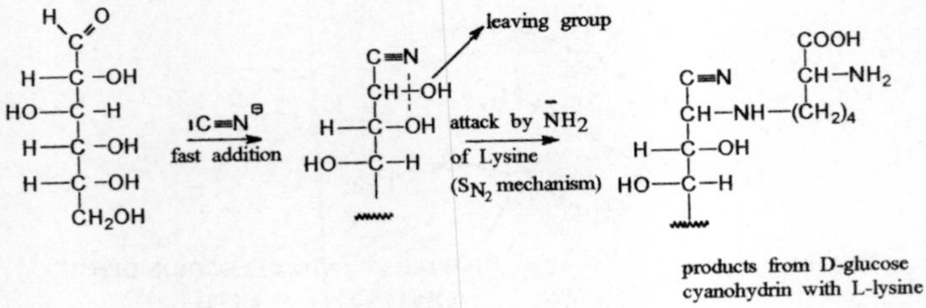

Figure 5. The suggested reaction mechanism.

Conclusions

The present investigations supported the hypothesis that the formation of product from L—lysine with D—glucose—cyanohydrin is a Strecker reaction. However, the Maillard reaction is also involved because the addition of cyanide to glucose inhibited the Maillard reaction. As there is evidence that the reaction between glucose and cyanide ion could take place in the human body, the glucose—cyanohydrin molecule may have biological significance by reacting with various proteins.

Acknowledgements
This work was supported by OTKA grant No. 7554 of the Hungarian Academy of Science.

References

Houben—Weyl (1985/a) Methoden der Organischen Chemie Band E5/Teil 2. S: 1425. George Thieme Verlag, Stuttgart, New York.

Houben—Weyl (1985/b) Methoden der Organischen Chemie Band E5/Teil 1. S: 535—37. George Thieme Verlag, Stuttgart, New York.

Roberts, D. (1966) An isotopic assay for thymidylate synthetase. Biochemistry 5: 3546-3551.

Trézl, L., Rusznák, I., Tyihak, E., Szarvas, T., and Szende, B. (1983) Spontaneous Nϵ—methylation and Nϵ—formylation reactions between L—lysine and formaldehyde, inhibited by L—ascorbic acid. Biochem. J. 214:(2)289.

Trézl, L., and Pipek, J. (1988) Formation of excited formaldehyde in model reaction simulating real biological systems. J. Molecular Structure (THEOCHEM) 170:213—223.

Trézl, L., Tyihak, E., and Lotlikar, P.D. (1990) in Protein Methylation. Eds: Paik, W.K and Kim, S. Chapter 22. CRC Press, Inc. Boca Raton, Florida.

Tyihak, E. Trézl, L., and Rusznák, I. (1980) Spontaneous Nϵ—methylation of L—lysine by formaldehyde. Die Pharmazie. 35:18.

Inhibition of Maillard Reaction by Tea Extract in Streptozotocin-treated Rats

N. Kinae,[1] S. Masumori,[1] S. Masuda,[1] M. Harusawa,[1] R. Nagai,[1] Y. Unno,[1] K. Shimoi,[1] and K. Kator[2]

[1]SCHOOL OF FOOD AND NUTRITIONAL SCIENCES, UNIVERSITY OF SHIZUOKA, 52-1 YADA, SHIZUOKA 422, JAPAN; [2]FACULTY OF SCIENCE, UNIVERSITY OF TOKYO, 7-3-1 HONGO, BUNKYO-KU, TOKYO 113, JAPAN

Summary
When male Wistar rats treated with streptozotocin (STZ) were orally administered green tea (*Camellia sinensis*) infusion, the contents of fructosylamine and advanced glycosylation end products (AGEs) in plasma significantly decreased compared with those of diabetic rats administered tap water. These data suggest that intake of tea extract may be useful for the inhibition of glycation of certain proteins during aging and diabetes of humans.

Introduction

It has been demonstrated that an amino-carbonyl reaction, the so-called Maillard reaction, occurs *in vivo* as well as *in vitro* and it is associated with the chronic complications of diabetes mellitus and aging in human beings (Monnier et al 1990). In particular, long-lived proteins such as lens crystallins (Monnier et al 1992), collagens (Monnier et al., 1992), and hemoglobin (Bunn et al., 1978) may react with reducing sugars undergoing dehydration, rearrangement, cleavage and polymerization reactions to produce Schiff's bases, fructosylamine (Amadori compound), osones and yellow fluorescent AGEs (Hayase et al., 1989). Recently, aminoguanidine has received much attention as an inhibitor of the Maillard reaction (Brownlee et al., 1986). In our previous paper, several kinds of tea extracts prepared from green tea (unfermented), oolong tea (semi-fermented), polei tea (fermented), black tea (fermented) and rooibos tea (herb) were shown to inhibit the condensation reaction between D-glucose and human or bovine serum albumin *in vitro* (Kinae and Yamashita, 1990). Upon fractionation of tea infusions, several catechins containing (-)epigallocatechin gallate were isolated as effective inhibitors of the Maillard reaction (Kinae et al., 1991). It was postulated that these catechins play a role as anti oxidants and/or as free radical scavengers (Kinae et al., in press).

In this study, we report the inhibitory effects of green tea infusion on the Maillard reaction in streptozotocin-diabetic rats; levels of fructosylamine and AGEs in plasma were examined.

Materials and Methods

Preparation of green tea extract

Green tea leaves were supplied by Shizuoka Tea Experiment Station (Shizuoka, Japan). Ten grams of tea leaves were added to 300 mL of boiling water and kept at 100°C for

3 minutes. The tea infusion was filtered through gauze and the filtrate was lyophilized. The lyophilizate was kept at 4°C before use.

Induction of diabetes by streptozotocin
Five-week-old male Wistar rats weighing 100±2g were used. Diabetes was induced by intravenous injection of 65mg/kg of streptozotocin in 0.05M sodium citrate, pH 4.5 . The diabetic rats were divided into two groups (A and B). Twenty four hours after injection, groups A (10 rats) and B(11 rats) were administered tap water and 0.05% green tea extract respectively, as drinking water. Non-diabetic rats were also divided into two groups (C and D) . Groups C (10 rats) and D (11 rats) were administered tap water and 0.05% green tea extract, respectively. All groups were fed commercial pellets CE-2 (Crea Japan, Tokyo, Japan) and drinking water *ad libitum*. Body weights and blood glucose levels were measured periodically. Glucose level of blood plasma was determined by a commercial reagent (Glucose CII-Test Wako, Wako Pure Chemical Co., Osaka, Japan). After 4 and 8 weeks rats were killed and heart tissues were excised.

Determination of lipid peroxidation level
1) Thiobarbituric acid reactive substances (TBARS)
Plasma(50 μL) was mixed with 450 μL of cold 0.9% NaCl and to which was added 1 mL of 0.7% TBA and 5 mL of n-BuOH under acidic conditions according to method of Yagi (1982). Fluorescence intensity of the n-BuOH layer was measured at an excitation wavelength of 515nm and emission wavelength of 555nm. The wavelength values selected were equivalent to those for malondialdehyde.
2) Lipofuscin-like substances in heart tissue.
Five mL of a chloroform and methanol (2:1, v/v) mixture was added to heart tissue and homogenized. Five mL of distilled water was added and the suspension was centrifuged at 2000 rpm for 5 min. The fluorescence intensity of the chloroform layer was measured at an excitation wavelength of 370nm and an emission wavelength of 440nm.

Determination of Maillard reaction products
1) Fructosylamine
Fructosylamine content in plasma was measured by using a commercial reagent (Fructosamine BMY, Boehringer Mannheim Yamanouchi Co., Tokyo, Japan) according to the method of Johnson (Johnson et al., 1982). The color intensity of formazan formed was measured at 546nm.
2) AGEs
Glycated proteins level in plasma was determined by high performance liquid affinity chromatography using a TSK gel borate column (Mallia et al., 1981). First, nonglycated proteins were removed with a solvent of 0.05M $MgCl_2$/0.02M HEPES, pH 8.5 and then glycated proteins (which were mainly AGEs) were eluted with an eluant of 0.2M sorbitol/0.02M HEPES, pH 8.5.

Results

The time course of body weight and plasma glucose levels of diabetic and non-diabetic rats with or without the green tea extract are shown in Table I and Figure 1, respectively.

The body weights of the diabetic groups (A and B) were significantly decreased compared with those of the non-diabetic groups (C and D). But there was no significant difference between groups A and B, or C and D. Similar results were observed in the glucose level in plasma. TBARS were determined as a measurement of the level of lipid per oxidation products in plasma (Figure 2).

Table I. Body weights of untreated or STZ-treated rats with or without the administration of tea extracts.

Group	Week of study -1	0	1	2	3	4	8
	Body weight in g						
A[1,3]	102.9±3.6	116.0±5.4	131.1±8.7	137.2±12	134.1±9.5	140.2±6.9	133.0±12
B[1,4]	102.7±5.4	115.2±4.2	131.9±7.1	137.6±9.8	133.4±13	137.2±11	130.9±12
C[2,3]	102.9±4.3	128.9±3.3	160.3±4.7	177.8±5.0	196.3±5.6	212.7±7.0	278.1±7.1
D[2,4]	102.8±4.5	129.3±3.7	160.7±6.2	180.3±7.5	198.7±8.3	213.7±10	289.3±12

[1] Diabetic rats; [2] non-diabetic rats; [3] rats given tap water control; [4] rats given tea extract.

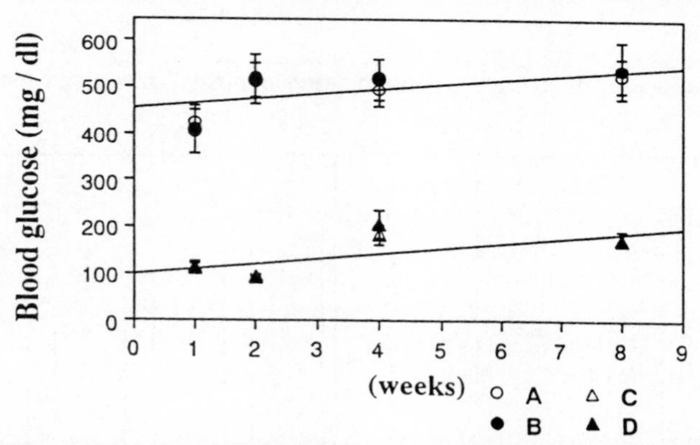

Figure 1. Time course of fasting blood glucose level of untreated or STZ-treated rats with or without tea extract (see Table I for description of groups A-D).

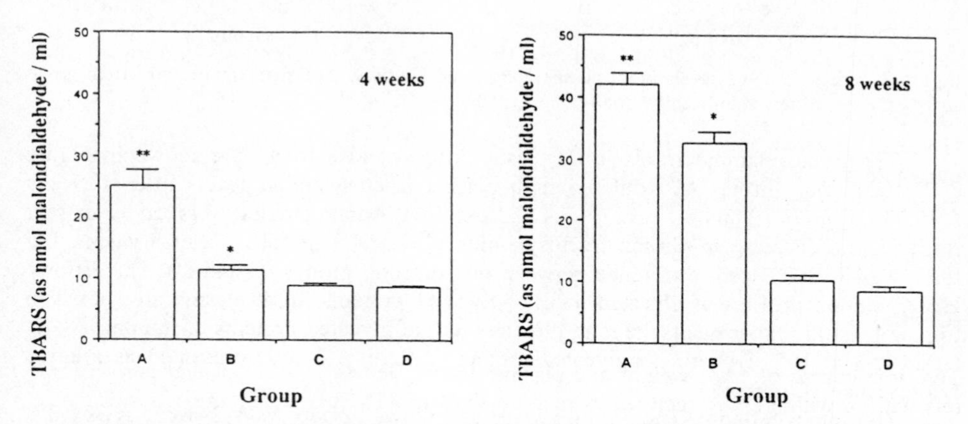

Figure 2. TBARS in rat plasma after 4 and 8 weeks; *, differs significantly from the value for A($p<0.01$); **, differs significantly from the value for C($p<0.01$) (see Table I for description of groups A-D).

Diabetic group A showed a high TBARS score (equivalent to 25.1±2.7nmol malondialdehyde/mL) after 4 weeks. But, the green tea-administered group B exhibited a low score (11.4±0.7nmol malondialdehyde/mL). After 8 weeks, values for TBARS in group A and B were 42.1±1.8nmol malondialdehyde/mL and 32.7±1.8nmol malondialdehyde/mL, respectively. The inhibitory effect of green tea extracts on lipid peroxidation was clear, especially at 4 weeks. The ability of green tea to inhibit lipofuscin formation in heart tissue was also examined (Table II). The inhibition appeared to be light under the experimental conditions used.

Table II. Content of lipofuscin-like substances in rat heart.

Group	Fluorescence int./g heart weight
A[1,3]	5.12±1.99
B[1,4]	4.90±0.68
C[2,3]	5.65±2.00
D[2,4]	4.44±0.74

[1] Diabetic rats; [2] non-diabetic rats; [3] rats given tap water control; [4] rats given tea extract.

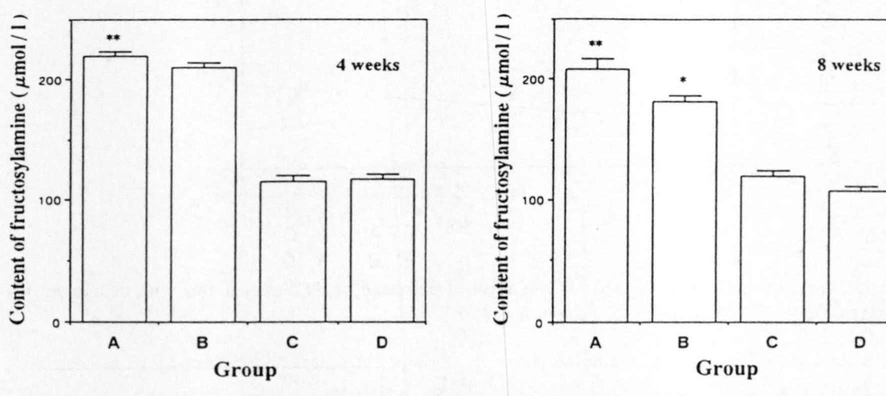

Figure 3. Fructosylamine levels in rat plasma after 4 and 8 weeks; *, differs significantly from group A($p<0.05$); **, differs significantly from group C($p<0.001$).

The fructosylamine levels in rat plasma after 4 and 8 weeks are shown in Figure 3. Diabetic rats (group A) exhibited high plasma fructosylamine levels after 4 weeks (219.7±3.7 ,μmol/L) and 8 weeks (207.4±9.3,μmol/L). Administration of green tea extract resulted in a decrease in plasma fructosylamine (181.1±4.5 ,μmol/L) after 8 weeks, but there was no significant difference between non-diabetic groups (C and D). The affinity HPLC elution profiles of glycated and nonglycated proteins in rat plasma are shown in Figure 4. In this experiment, the ratio of peak areas of glycated proteins containing AGEs (A-2, A-3 or B-2, B-3) to nonglycated protein (A-1 or B-2) was presented as relative fluorescence (Figure 5).

The relative fluorescence values were as follows: group A, A-2=1.67, A-3=0.37; group B, B-2=0.95, B-3=0.12. There results show that green tea inhibited the formation of AGEs.

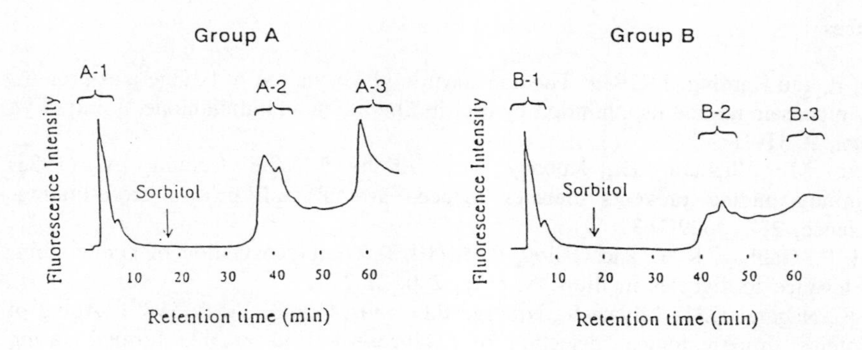

Figure 4. Elution profiles of plasma on affinity HPLC with fluorescence detection.

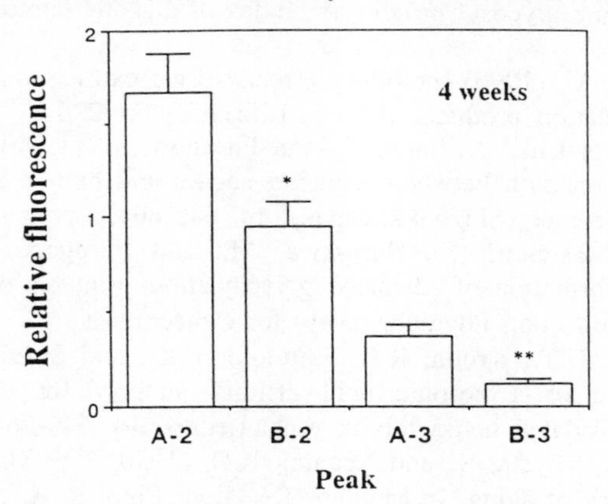

Figure 5. Fluorescent glycated protein levels in rat plasma. *, significantly different from peak A-2, $p<0.05$; **, significantly different from peak A-3, $p<0.05$.

Discussion

The levels of Maillard reaction products (fructosylamine and AGEs) in several tissues increase during aging and diabetes. Recent reports show that aminoguanidine, aspirin and rutin are candidates for the prevention of the formation of AGEs (Ajiboye and Harding, 1989; Odetti et al., 1990).

In the present study, we demonstrated a decrease in AGEs and lipid peroxidation products in streptozotocin-diabetic rats administered green tea extract. Such decreases were observed in plasma and heart tissue. As green tea extract contains polyphenolic compounds such as (-)epicatechin, (-)epicatechin gallate, (-)epigallocatechin and (-)epigallocatechin gallate, such compounds might act as antioxidants and/or radical scavenging agents in the Maillard reaction. In conclusion, green tea constituents are useful substances for prevention of the Maillard reaction *in vivo* and *in vitro*. Therefore, intake of tea extract may play an important protective role against some of the effects of aging and diabetes.

References

Ajiboye, R. and Harding, J. (1989) The nonenzymic glycosylation of bovine lens proteins by glucosamine and its inhibition by aspirin, ibuprofen and glutathione. J. Exp. Eye Res., 9, 31-41.

Brownlee, M., Vlassara, H., Kooney, A., Ulrich, P., and Cerami, A. (1986) Aminoguanidine prevents diabetes induced arterial wall protein cross-linking. Science, 232, 1629-1632.

Bunn, H. F., Gabbay, K.H., and Gallog, P.M. (1978) The glycosylation of hemoglobin: Relevance to diabetes mellitus. Science, 200, 21-27.

Hayase F., Nagaraj K.H., Miyata S., Njoroge F.G. and Monnier V.M. (1989) Aging of proteins: Immunological detection of a glucose-derived pyrrole formed during Maillard reaction *in vivo*. J. Biol. Chem., 264, 3758-3764.

Johnson, R.N., Metcalf, P.A., and Barker, J.R. (1982) Fructosamine: A new approach to the estimation of serum glycosyl protein. An index of diabetic control. Clin. Chim. Acta, 127, 87-95.

Kinae, N., and Yamashita, M. (1990) Inhibitory effects of tea extracts on the formation of advanced glycosylation products. Adv. in Life Sci., 221-226.

Kinae, N., Masumori, S., Nakada, J., Saito, T., and Furugori, M. (1990) Tea extracts inhibit the Maillard reaction between reducing sugars and human albumin. Proc. Int. Symp. on Tea Science (Shizuoka, Japan), 14, 649-661.

Kinae, N., Shimoi, K., Masumori, S., Harusawa, M., and Furugori, M. (in press) Suppression of the formation of advanced glycosylation products by tea extracts. ACS volume based on Food Phytochemicals for Cancer Prev..

Mallia, A.K., Hermanson, G.T., Krohn, R.I., Fujimoto, E.K., and Smith, P.K. (1981) Preparation and use of a boronic acid affinity support for separation and quantitation of glycosylated hemoglobins. Anal Letters, 14, 649-661.

Monnier, V.M., Sell, D.R., Miyata, S., and Nagaraj, R.H. (1990) The Maillard reaction as a basis for a theory of aging. In Maillard Reaction: Finot P. A. et al Eds. Adv. in Life Sci., 393-414.

Monnier, V.M., Sell, D.R., Nagaraj, R.H., Miyata, S., Grandee, S., Odetti, P., and Ibrahim, S. A. (1992) Maillard reaction-mediated molecular damage to extra cellular matrix and other tissue, proteins in diabetes, aging, and uremia. Diabetes, 41, 36-41.

Odetti, P.R., Brogoglio, A., Pascale, de A., and Adezati, L. (1990) Prevention of diabetes-increased aging effect on rat collagen-linked fluorescence by aminoguanidine and rutin. Diabetes, 39, 796-801.

Yagi K. (1976) A simple fluorometric assay for lipoperoxide in blood plasma. Biochem. Med. 15, 212-216.

On the Physiological Aspects of Glycated Proteins *In Vivo*

N. V. Chuyen, N. Utsunomiya,[1] and H. Kato[2]

DEPARTMENT OF FOOD AND NUTRITION, JAPAN WOMEN'S UNIVERSITY, TOKYO 112; [1]DEPARTMENT OF FOOD SCIENCE, KYORITSU WOMEN'S UNIVERSITY, TOKYO 101; [2]DEPARTMENT OF AGRICULTURAL CHEMISTRY, THE UNIVERSITY OF TOKYO, TOKYO 113

Summary

Glycated casein caused the decrease of weight gain and serum glucose of young and adult rats, while glycated soy-protein caused the decrease of weight gain and serum glucose of young rats but not that of the adult ones. Hematocrit, erythrocytes, cholesterol, triglyceride, GOT, GPT in serum of either young or adult rats were unchanged after feeding with glycated casein or soy-protein. Resemble of digestive peptide patterns in the upper part of rat small intestine was observed from glycated and native casein. The increase of peptides, free amino acids in urine were observed for all the experiments, while the decrease of urea in blood or urine of rats were observed in some cases. It is estimated that some antinutritional products were generated by arginine and lysine residues of protein in the reaction with glucose.

Introduction

Glycated proteins caused by Maillard reaction are known to be decreased in nutritive value, since amino acid residues in proteins such as lysine, arginine, tryptophan and methionine are decreased and cross-linkages are formed. Moreover, the digestibility of glycated proteins is also decreased, as well as antinutritional factors are also generated. The nutritional quality of glycated proteins was studied by various authors (Sgarbieri et al.,1973; Kimiagar et al.,1980; Lee et al., 1982), the influence of low molecular weight compounds generated by Maillard reaction was investigated by Oste et al. (1984). The metabolism of Maillard reaction products was reported by Finot et al. (1990). However, hitherto, the digestion as well as the absorption of glycated proteins were not studied in details. The physiological and nutritional effects of glycated proteins were mostly carried out on growing rats but not on adult ones.

In this study, the digestion of glycated casein in vitro, in vivo, the effects of glycated casein and soy-protein on both growing and adult rat were carried out.

Materials and Methods

Casein from milk was purified as described in a previous paper (Utsunomiya et al.,1990). Soy-protein (soy-protein isolate) was washed with diethyl ether before use. Vitamin and mineral mixture were of Harper's mixture. The other reagents were of analytical grade.

Preparation procedures of glycated proteins

Glycated casein samples were prepared under two conditions: (1) incubation of casein with glucose at 50° C and RH75%: Casein was mixed homogeneously with glucose in a ratio of 3:1 in a small quantity of water, then the whole mixture was lyophilized and ground to a powder of under 60 mesh, this powder then being incubated at 50° C and RH75% for 7 days. After the reaction, each casein reaction mixture was dialyzed against water for 3 days at 5° C to remove the reaction products of low molecular weight, and then lyophilized again to obtain the nondialyzable glycated caseins. (2) Heating of casein with glucose at 95° C in an aqueous solution. Casein and D-glucose (2:1) dissolved in distilled water were heated at 95° C with stirring for 4 hr. The reaction mixture was then dialyzed and lyophilized as just described to obtain the nondialyzable glycated caseins. Glycated soy-protein was prepared by the incubation at 50° C, RH 75% for 7 days, then dialyzed and lyophilized.

Preparation Procedures of Digestion Products

(1) in vitro, digestion products of glycated casein were prepared by the use of pepsin and pancreatin as described by Utsunomiya et al. (1990). (2) in vivo, glycated casein was administered to rat at a level of 20% (Wistar, male, about 200g of body weight) by meal feeding for 2 days. On the 3rd day, the rat was sacrificed after 30 min of feeding, then, peptides in the intestine were collected. Peptides from the digested glycated casein were separated by using an Amicon Ultrafiltration Cell. The amino acids in urine were measured by an amino acid analyzer (Hitachi 835). Peptides were analyzed by HPLC (Shimadzu, 6A) with the following conditions : Column, Asahipak ODP-50; elution solvent, 0.05%TFA in H_2O/ CH_3CN = 100/0 to 50/50 in 30 min, 0.6 ml/min); detector, UV 220nm. Analysis of the serum biochemical components was carried out as described by Utsunomiya et al., (1990).

Animal feeding and Diet.

The glycation degree of lysine residue in casein (50° C, RH 75%, 7 days or 95° C, 4 hr) or soy-protein (50° C, RH 75%, 7 days) on heating with glucose was controlledat a level of 50%. The remaining lysine was measured by amino acid analyzer and Carpenter 's method. At this glycation condition, arginine in glycated casein (CG) and glycated soy-protein (SPG) was decreased to 25% and 50% respectively, while tryptophan in CG and SPG was decreased to 20% and 10% respectively . After the lost quantity of amino acids such as lysine, tryptophan and arginine in CG and SPG was completely supplemented, these glycated proteins were fed to growing or adult rats for 1-2 months by pair-feeding.

Results and Discussion

Physiological Effects of Glycated Proteins on Rats

Effects of CG and glycated SPG on growing or adult rats are shown in Table I. CG caused the decrease of weight gain and serum glucose of young and adult rats. The administration of a mixture of CG and native casein at a ratio of 1:1 to rats also showed the same tendency. The effects of casein heated with glucose at 50° C and RH 75% for 7 days and 95° C for 4 hr in aqueous solution were also investigated on rats. No difference between the two glycated caseins was observed. On the other hand, SPG caused the decrease of weight gain and serum glucose of young rats but not that of the adult ones. Since the protein score of casein is higher than that of soy-protein, from this result, it can be estimated that by the glycation, good quality protein suffered much influence than lower quality protein. Glycated proteins also decrease the level of serum glucose, there is a possibility that some products generated after the digestion and absorption process of CG or SPG affect the activity of the enzyme system in the intestinal brush border concerning the hydrolyzation and absorption of sugar as reported by various authors.

Table I. Effect of glycated casein and glycated soy-protein on growing or adult rats fed for 1 or 2 months.

	Growing rats a)		Adult rats b)		
Glycated Casein Sample	Control (C) (native casein) (n=6)	CG-7 (n=6)	Control (C) (native casein) (n=5)	CG-7 (n=5)	CG-7:C (1:1) (n=5)
Weight gain (g)	218.6±4.6	188.3±3.5*	52.7±1.8	36.6±4.5*	44.1±4.7*
Serum glucose (mg/dl)	156±21	94±16**	121±10	85±9*	109±9.2
Glycated Soy-protein Sample	SP (n=6)	SPG-7 (n=6)	SP (n=6)	SPG-7 (n=6)	
Weight gain (g) 1 month	93.2±3.2	59.9±12.5*	24.5±31.5	11.8±10.8	
2 months	158.2±18.2	127.9±21.0**	ND	ND	
Serum glucose (mg/dl)	143±10	105±19**	158±27	170±23	
BUN (mg/dl)	16.6±2.4	8.3±4.9*	14.8±1.7	11.5±1.3**	

Figures in the Table are the average±S.D.
Wistar rats (male, body weight of 45-50g a), 292-316g b) were fed for 2 months and 1 month respestively)
* Statistically significant (p<0.01) against the control.
** Statistically significant (p<0.05) against the control.
CG-7 : Casein-glucose was heated at 50℃ and RH75% for 7 days.
SP : unheated soy-protein
SPG-7 : SP-glucose was heated at 50℃ and RH75% for 7 days.
ND : Not Done

Table II. Hydrolyzates of glycated casein hydrolyzed by pepsin-pancreatin system.

	free amino acid in urine		amino acid in urine after hydrolysis a)	
	Control	CG-7+AA b)	Control	CG-7+AA
Ala	621±70	5504±1521*	2442±251	12341±669*
Met	108±33	1101±381*	135±25	968±206*
Ile	159±17	1643±189*	697±48	1588±251*
Leu	103±14	2143±497*	1123±6	2659±12**
Phe	260±52	1115±221**	761±42	11827±827**
Lys	293±25	6515±1008*	1987±245	32195±1632**
Arg	129±12	606±97	1146±8	1929±173*

Figures in the Table are Means±S.E. (n=4)
a) Urine was hydrolyzed by 6N HCl, 110℃, 24hr.
b) CG-7+AA: Casein was modified by glucose at 50 ℃, RH75% for 7 days + supplemented Amino Acid.
*: Statistically significant (p<0.05). **: Statistically significant (p<0.01).

Results of the digestion of CG by pepsin and pancreatin are shown in Table II, native casein is much easier to be hydrolyzed by protease than CG . It is usually thought that the cleavage site of glycated protein is much different from that of the native one. Some proteases were used to determine the glycation site of protein. From the physiological point of view, when a mixture of enzyme exists, the cleavage of the glycated protein might be different. In order to clarify this problem, we analyzed the peptides collected by the digestion of CG in vitro and in vivo by HPLC. The result showed that there was difference in the quantity of each peptide, but there was mostly no difference in the pattern of these peptides (data not shown). From this result, it can be estimated that when a mixture of protease exists, the main cleavage sites of CG are mostly similar.

Table III. Amino acid composition in urine of rats fed casein modified by glucose at 50°C. RH 75% for 7 days (nmol/100gBW/day).

Molecular Weight	Nitrogen (%)					
	Native Casein			Glycated Casein		
	3[a]	8[b]	18[c]	3[a]	8[b]	18[c]
over 30,000	10.2%	5.5%	2.3%	42.2%	4.8%	4.8%
10,000-30,000	1.3	0.7	1.0	7.8	4.9	3.5
5,000-10,000	8.4	0.7	5.9	0.5	3.1	6.0
1,000- 5,000	27.5	4.0	1.4	20.1	40.1	27.5
below 1,000	52.6	89.1	89.4	29.3	46.4	58.1

a): Sample was incubated with pepsin for 3 hr.
b): Sample was incubated with pepsin for 3 hr and pancreatin for 5 hr.
c): Sample was incubated with pepsin for 3 hr and pancreatin for 15 hr.

On the Excretion of Amino Acids and Peptides in Urine

The quantity of free amino acids and peptides excreted in the urine of rats fed with CG was much higher than that fed with native casein as shown in Table III (only some representative amino acids are picked up). From this result, it can be considered that the bioavailability of CG is lower than that of the native casein. When arginine-glucose reaction mixture was administered to rats, this mixture showed similar phenomenon to CG on the urinary excretion of peptides and amino acids, therefore, it is estimated that some antinutritional products might be generated by arginine residue of casein. Besides, the excretion of urea in urine by CG fed rats was lower than that of rats fed with native casein, the BUN value also showed the same tendency. In order to explain this phenomenon, some hypothesis can be proposed : the presence of blocked peptides and amino acids in urine, imbalance due to the unavailability of some amino acids present in blocked peptides, glycated proteins products might affect the clearance of the kidney or activity of some enzymes of the urea cycle. However, further studies are necessary to clarify this problem.

References

Finot, P. A. (1990) Metabolism and physiological effects of Maillard reaction products. *The Maillard Reaction, Advances in Life Sciences*, Birkhauser Verlag Basel, pp. 259-272.

Kimiagar, M., Lee, T. C., and Chichester, C.O. (1980) Long-term feeding effects of browned egg albumin to rats. *J. Agric. Food Chem.*, 28: 150-155.

Lee, T. C., Pintaro, S. J., and Chichester, C. O. (1982) Nutritional and Toxicologic Effect of Nonenzymatic Maillard Browning . *Diabetes*, 31 (Suppl. 3) :37-46.

Oste, R.,and Sjodin, P. (1984) Effect of Maillard reaction products on protein digestion, in vivo studies on rats. *J. Nutr.*, 114: 2228-2234.

Sgarbieri, V. C., Amaya, J., Tanaka, M., and Chichester, C. O. (1973) Response of rats to amino acids supplementation of brown egg albumin . *J. Nutr.*, 103: 1731-1738.

Sherr, B., Lee, C. M., and Jelesciewics, C. (1989) Absorption and metabolism of lysine Maillard products in relation to utilization of L-lysine. *J. Agric. Food Chem.*, 37: 119-122.

Utsunomiya, N., Chuyen, N. V., and Kato, H. (1990) Nutritional and physiological effects of casein modified by glucose, diacetyl or hexanal. *J. Nutr. Sci. Vitaminol.*, 36: 387-397.

The Binding of Native and Oxidized Low Density Lipoproteins (LDL) to Collagen. The Effect of Collagen Modification by Advanced Glycosylation End Products

W. K. Lee, J. Stewart, J. Bell, and M. H. Dominiczak

LIPID/DIABETES LABORATORY, DEPARTMENT OF PATHOLOGICAL BIOCHEMISTRY, WESTERN INFIRMARY, GLASGOW C11 6NT, SCOTLAND, UK

Summary

We studied the binding of human LDL to native type I collagen and to collagen modified by advanced glycosylation end products (AGE). AGE were measured as collagen-linked fluorescence (excitation wavelength 370nm, emission wavelength 440nm). Both native and oxidized LDL bind to collagen. AGE-collagen bound less LDL than native collagen (3.1 ± 1.2 μg/mg vs. 4.3 ± 1.7 μg/mg respectively; $p=0.001$). The binding of oxidized LDL to non-modified collagen was lower than that of native LDL (2.7 ± 2.1 mg/mg vs. 4.3 ± 1.7 μg/mg respectively). The binding of oxidized LDL to AGE-modified collagen was also lower than that of native LDL (1.4 ± 0.8 μg/mg vs. 3.1 ± 1.2 μg/mg respectively). Both LDL oxidation and collagen modification by AGE affect LDL-collagen interactions.

Introduction

Low density lipoproteins (LDL) present in plasma cross the endothelial barrier and reside in the environment of the extracellular matrix. LDL interacts with matrix components. LDL was shown to bind to arterial proteoglycans (Srinivasan et al, 1982, Camejo et al., 1985) and elastin (Podet et al., 1991). Interactions of LDL with matrix components influence LDL macrophage uptake and thus may contribute to atherogenesis (Hurt et al., 1987; 1990). LDL interactions with collagen were little investigated to date.

Advanced glycosylation end products (AGE) form on arterial collagen and may also contribute to atherogenesis through their effect on macrophage and endothelial cell function (Vlassara et al., 1988; Kirstein et al., 1990; Cerami et al., 1988; Dominiczak, 1991). Our aim was to establish whether LDL oxidation and AGE formation on collagen affect LDL-collagen interactions.

Materials and Methods

Preparation of the AGE-modified collagen.

Type I calf collagen (Sigma, Poole, UK) was incubated with glucose 6-phosphate (250-1000 mmol/l) in phosphate buffered-saline (PBS) at pH 7.5 for 14 days at 37°C. After completion of incubation, AGE- collagen was washed in PBS and distilled water and freeze dried.

Preparation of LDL.

LDL was isolated from the plasma of healthy volunteers using flotation ultracentrifugation (Lipid Research Clinics Program, 1982). Plasma was centrifuged at the density of 1.019g/ml (39K at 4°C in the Beckman TI 50.4 rotor). After centrifugation the top layer was discarded. The remainder was centrifuged at 1.063g/ml (39K at 4°C for 24h). The top 2 ml constituted the LDL fraction and were used in further experiments. LDL was labeled with ^{125}I using the iodine monochloride method (McFarlane, 1958). The concentration of the iodinated LDL was estimated by measuring absorbance at 280nm and by the measurement of protein concentration (Lowry et al., 1951). After iodination, ^{125}I-LDL was purified using PD10 Sephadex columns (Pharmacia, Uppsala, Sweden).

Oxidation of LDL.
Blood samples were collected into containers with EDTA to prevent LDL oxidation. Separated LDL was dialyzed against PBS (pH 7.4) for 24h at 4°C. After dialysis $CuCl_2$ was added to a final concentration of 10μmol/L and incubated with LDL at 37°C for 18h (Esterbauer et al., 1991). To detect oxidation, the fluorescence of LDL was measured at emission wavelength 320nm and excitation wavelength 420nm. The oxidized LDL was labeled with ^{125}I as described above.

Collagen binding studies.
^{125}I-LDL was diluted to a concentration of 200μg/ml and 300μl of this was incubated with 530μg collagen for 16h at 37°C. Collagen was then washed with PBS until, in the control samples, the background radioactivity fell below 1% of the initial value. Bovine serum albumin (BSA) at a concentration of 0.5% decreased the binding of LDL to type I collagen, therefore experiments were carried out in the presence of 0.5% BSA.

The measurement of collagen-linked fluorescence (CLF).
CLF was measured as described before (Lee et al., 1993). The sample was washed with distilled filtered water. Lipids were extracted with chloroform/methanol (2:1) with gentle shaking in a cooling bath at 4°C for 24h. The pellet was centrifuged at 1300g and 4°C for 15min and washed with ice-cool methanol, distilled filtered water and 0.02M Hepes buffer pH 7.5. Collagen digestion was carried out for 24h by suspension of the pellet in 1ml of 0.02M Hepes buffer containing 250 units of bacterial type VII collagenase (EC 3.4.24.3, Sigma, Poole, UK). After overnight incubation at 37°C, the samples were centrifuged at 10000g for 5min at 4°C. Supernatant containing the digested collagen was removed for measurement of fluorescence and of hydroxyproline concentration. CLF was measured on the collagen digest at an excitation wavelength 370nm and emission wavelength 440nm using micro cells and an LS3B scanning fluorescence spectrometer (Perkin-Elmer, Beaconsfield, UK). The results were expressed in arbitrary fluorescence units per milligram of collagen (U/mg). Hydroxyproline was measured colorimetrically (Stegemann and Stalder, 1967). The amount of collagen present was calculated assuming a 14% hydroxyproline content (Hamlin and Kohn, 1971).

Results

The incubation of type I collagen with glucose-6-phosphate resulted in an increase in CLF. Figure 1 shows the increase in the fluorescence noted during collagen incubation for 14 days with variable concentrations of glucose-6-phosphate.

The binding of non-modified LDL to native type I collagen and type I collagen modified by AGE was concentration-dependent at LDL concentrations 200-1000μg/ml. In this range of LDL concentrations, the binding was non-saturable. AGE-collagen bound less LDL than native collagen (3.1±1.2 μg/mg vs. 4.3±1.7 μg/mg respectively; p=0.001). The binding of oxidized LDL to non-modified collagen was lower than that of native LDL (2.7±2.1 μg/mg vs. 4.3±1.7 μg/mg respectively). The binding of oxidized LDL to AGE-modified collagen was also lower than that of native LDL (1.4±0.8 μg/mg vs. 3.1±1.2 μg/mg respectively; Figure 2).

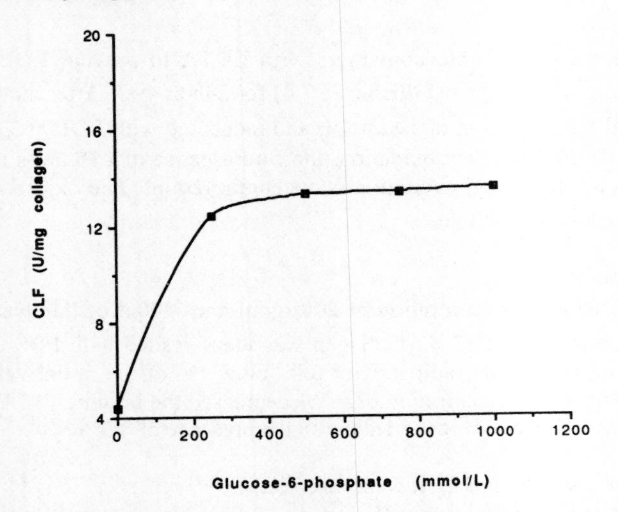

Figure 1. CLF of type I collagen incubated with indicated concentrations of glucose-6-phosphate for a period of 14 days (n=5; SD for all the points shown was < 0.79 U/mg and is omitted for clarity).

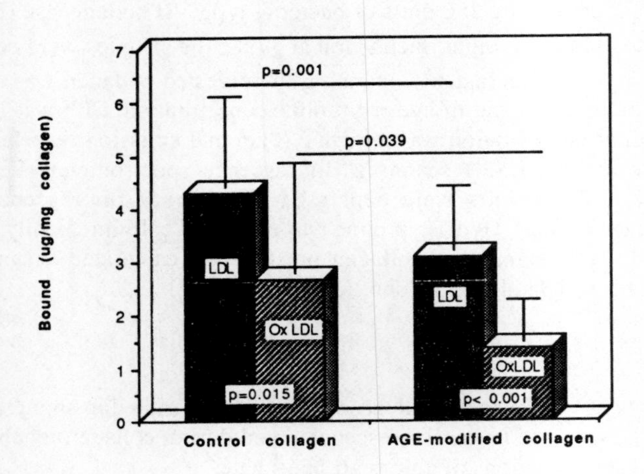

Figure 2. The binding of native and oxidized LDL to non-modified and AGE-modified collagen. For native LDL n=27 to 40; for oxidized LDL n=10 to 15.

Discussion

Structural modifications of extracellular matrix components may affect arterial deposition of lipoproteins. The collagens found in the arterial wall are predominantly of type I and type III (Trelstad, 1974). Collagen forms a significant proportion of the atherosclerotic plaque (Woolf, 1990; McCullagh et al.,1980). AGE formation on collagen might play a role in atherogenesis: it is known that AGE-collagen forms in the human arteries and that its concentration increases with patients' age (MacDonald et al, 1992). Local changes of AGE-collagen concentration occur in the atherosclerotic plaque. We have previously shown that CLF in the superficial plaques is decreased compared to that in segments of the arterial wall unaffected by atheroma (Lee et al., 1993).

This is the first study of interactions of oxidized human LDL with AGE-collagen. However our experiments were conducted using purified calf type I collagen and further studies are needed to establish whether the observed phenomena also occur *in vivo*. Our results suggest that both LDL oxidation and AGE formation on collagen affect LDL-collagen interactions with an additive effect. Our observation of a decreased native LDL binding to collagen contrasts with those of Brownlee et al. (1985) who, employing denatured, immobilized collagen, demonstrated an increased, rather than decreased, binding of LDL to AGE-collagen. Moreover, Hoover et al. (1988) observed a decreased binding of oxidized LDL to type I collagen gel. In their experiments pooled LDL isolated from plasma by precipitation was used. The above studies are not directly comparable with ours since in the present study we used intact native and AGE-modified collagens and LDL isolated by ultracentrifugation from the individual patients. Other components of the extracellular matrix also undergo qualitative and quantitative changes during plaque formation. Proteoglycan content increases (Hollander et al., 1984) and defective elastin may be synthesized (Kramsch and Hollander, 1973). LDL complexes with arterial proteoglycans (Camejo et al., 1985) and binds to elastin (Podet et al., 1991). The elastin from atherosclerotic plaques binds more LDL than normal elastin (Podet et al., 1991). Moreover, LDL pretreatment with proteoglycan leads to an increased uptake of human LDL by macrophages (Hurt et al., 1990) and to a stimulation of transformation of monocyte-derived macrophages into foam cells (Hurt and Camejo, 1987). It will now be important to investigate whether binding of LDL to native and AGE-modified collagen affects its interaction with macrophages.

Conclusions

This study suggests that modification of collagen by AGE decreases its binding of LDL. In addition, LDL oxidation decreases its binding to both native and AGE-modified collagen.

Acknowledgments
This work was supported by a grant from the Chest Heart and Stroke Association (Scotland) and from the Kidney Research Fund (UK). The authors are indebted to Miss J Gardiner and Mrs D Park for excellent secretarial assistance.

References

Brownlee, M., Vlassara, H., and Cerami, A. (1985) Nonenzymatic glycosylation products on collagen covalently trap low-density lipoprotein. *Diabetes* . 34: 938-941.

Camejo, G., Lopez, A., Lopez, F., and Quinones, J. (1985) Interaction of low density lipoproteins with arterial proteoglycans. *Atherosclerosis*. 55: 93-105.

Cerami, A., Vlassara, H., and Brownlee, M. (1988) Role of advanced glycosylation products in complications of diabetes. *Diabetes Care*. 11(Suppl. 1): 73-79.

Dominiczak, M.H. (1991) The significance of the products of the Maillard (browning) reaction in diabetes. *Diabetic Med.* 8: 505-516.

Esterbauer, H., Dieber-Rotheneder, M., Striegl, G., and Waeg, G. (1991) Role of vitamin E in preventing the oxidation of low-density lipoprotein. *Am. J. Clin. Nutr.* 53: 314S-321S.

Hamlin, C.R., and Kohn, R.R. (1971) Evidence for progressive age-related structural changes in post-mature collagen. *Biochim. Biophys. Acta.* 236: 458-467.

Hollander, W., Colombo, M., Faris, B., Franzblau, C., Schmid, K., Wernli, M., and Bernasconi, V. (1984) Changes in the connective tissue proteins, glycosaminoglycans and calcium in the arteries of the cynomolgus monkey during atherosclerotic induction and regression. *Atherosclerosis* . 51: 89-108.

Hoover, G., A., McCormick S., and Kalant, N. (1988) Interaction of native and cell-modified low density lipoprotein with collagen gel. *Arteriosclerosis.* 8:525-534.

Hurt, E., Bondjers, G., and Camejo, G. (1990) Interaction of LDL with human arterial proteoglycans stimulates its uptake by human monocyte-derived macrophages. *J. Lipid Res.* 31:443-454.

Hurt, E., and Camejo, G. (1987) Effect of arterial proteoglycans on the interactions of LDL with human-derived macrophages. *Atherosclerosis.* 67: 115-167.

Kirstein, M., Brett, J., Radoff, S., Ogawa, S., Stern, D., and Vlassara, H. (1990) Advanced protein glycosylation induces transendothelial human monocyte chemotaxis and secretion of platelet-derived growth factor: Role in vascular disease of diabetes and aging. *Proc. Natl. Acad. Sci. USA.* 87: 9010-9014.

Kramsch, D.M., and Hollander, W. (1973) The interaction of serum and arterial lipoproteins with elastin of the arterial intima and its role in the lipid accumulation in atherosclerotic plaques. *J. Clin. Invest.* 52: 236-247.

Lee, W.K., Bell, J., Kilpatrick, E., Hayes, M., Lindop, G.B.M., and Dominiczak, M.H. (1993) Collagen-linked fluorescence in human atherosclerotic plaques. *Atherosclerosis.* 98: 219-227.

Lipid Research Clinics Program (1982). *Manual of Laboratory Operations. US Dept of Health and Human Services, Bethesda : 63-70.*

Lowry, O.H., Rosebrough, N.J., Farr, A.L., and Randall, R.J. (1951) Protein measurement with the Folin phenol reagent. J. Biol. Chem. 193: 265-275.

McCullagh, K.G., Duance, V.C., and Bishop, K.A. (1980) The distribution of collagen types I, III and V (ab) in normal and atherosclerotic human aorta. *J. Pathol.* 130: 45-55.

MacDonald, E., Lee, W.K., Hepburn, S., Bell, J., Scott, P.J.W., and Dominiczak, M.H. (1992) Advanced glycosylation endproducts in the mesenteric artery. *Clin. Chem.* 38: 530-533.

McFarlane, A.S. (1958). Efficient trace-labelling of proteins with iodine. *Nature* . 182: 53-57.

Podet, E.J., Shaffer, D.R., Gianturco, S.J., *et al.* (1991) Interaction of low density lipoproteins with human aortic elastin. *Arteriosclerosis and Thrombosis* . 11: 116-122.

Srinivasan, S.R., Yost, C., Bhandaru, R.R., Radhakrishnamurthy, B., and Berenson, G.S. (1982) Lipoprotein-glycosaminoglycan interactions in aortas of rabbits fed atherogenic diets containing different fats. *Atherosclerosis* . 43: 289-301.

Stegemann, H., and Stalder, K. (1967) Determination of hydroxyproline. *Clin. Chim. Acta.* 18: 267-273.

Trelstad, R.L.(1974) Human aorta collagens: Evidence for three distinct species. *Biochem. Biophys. Res. Commun.* 57:717-725.

Vlassara, H., Brownlee, M., Manogue, K.R., Dinarello, C.A., and Pasagian, A. (1988) Cachectin/TNF and IL-I induced by glucose-modified proteins: Role in normal tissue remodelling. *Science*. 240: 1546-1548.

Woolf, N. (1990) Pathology of atherosclerosis. *Br. Med. Bull.* 46: 960-985.

Glycated Low Density Lipoproteins are Much More Susceptible to Lipid Peroxidation

T. Sakurai,[1] Y. Yamamoto,[2] and M. Nakano[3]

[1]TOKYO COLLEGE OF PHARMACY, HACHIOJI, TOKYO 192-03, JAPAN;
[2]DEPARTMENT OF REACTION CHEMISTRY, FACULTY OF ENGINEERING, UNIVERSITY OF TOKYO, BUNKYO-KU, TOKYO 113, JAPAN;
[1,3]PHOTON MEDICAL RESEARCH CENTER, HAMAMATSU UNIVERSITY SCHOOL OF MEDICINE, HAMAMATSU 431-31, JAPAN

Summary
Low density lipoprotein (LDL) from human plasma was incubated with 200 mM glucose in 75 mM phosphate buffer (pH 7.4) containing 0.5 mM EDTA at 37°C for 70hrs (200-G-LDL).Both peroxides, phosphatidylcholine hydroperoxide (PC-OOH) and cholesteryl ester hydroperoxide(CE-OOH), were estimated by HPLC with isoluminol chemiluminescence detection system. Level of CE-OOH in LDL 200-G-LDL was 4.8 times that of native LDL, while PC-OOH did not change. Both glycated and non-glycated LDL had phospholipase A_2 activity toward synthetic substrate,1-stearoyl-2-[13'-hydroperoxy-9',11'-octadecadienoyl]-*sn*-glycero-3-phosphocholine. The apparent low level of PC-OOH in 200-G-LDL may be attributed to phospholipase A_2 activity. LDL was treated with 50 mM glucose at 37°C for 70 hrs, then applied onto a phenylboronate affinity column. The relative amount of PC-OOH in the adsorbed fraction was higher than that in nonadsorbed fraction, indicating acceleration of the oxidation of glycated LDL during affinity column chromatography at pH 8.5.

Introduction

Much attention has been focused on the oxidation of low density lipoprotein (LDL), since oxidation of LDL may be involved in atherogenesis (Witztum and Steinberg,1991). However, it is not known with certainty exactly where or by what mechanisms LDL is oxidized in vivo. We have investigated glycated LDL as a possible candidate of oxidized LDL.

The reactivity of glycated protein with oxygen, iron and phospholipids has been studied by using glycated polylysine(GPL) as a model of glycated protein. The GPL can easily coordinate with ferric ion even in phosphate buffer and evoke lipid peroxidation in phospholipid liposome (Sakurai et al., 1990). LDL exposed to GPL-ferric ion is also oxidized (Sakurai et al., 1991a). Glycated LDL exposed to ferric ion, as expected, shows characteristic features of oxidized LDL (Sakurai et al., 1991b). In this paper, we report the quantity of lipid hydroperoxides, the initial products of lipid peroxidation, which are detected in glycated LDL even in the presence of EDTA.

Materials and Methods

Preparation and glycation of LDL.
Human plasma LDL was prepared by sequential ultracentrifugation. Ten ml of blood were collected in a tube containing EDTA-2Na(10mg). Plasma density was adjusted to 1.019 and centrifuged at 15°C for 18 hrs. The lower yellow part was collected and its density was adjusted to 1.061 by the addition of NaCl, followed by centrifugation at 15°C

for 18 hrs. The floating part was collected and dialyzed against 150mM phosphate buffer (pH 7.4) containing 1 mM EDTA at 4°C for 20 hrs. This was referred to as native LDL. An aliquot was kept at -40°C until assay of phosphatidylcholine hydroperoxide(PC-OOH) and cholesteryl ester hydroperoxide (CE-OOH). LDL (1mg protein/ml) was incubated with or without 200 mM glucose in 75 mM phosphate buffer (pH 7.4) containing 0.5 mM EDTA at 37°C for 70 hrs in the presence of 0.01% gentamicin. After dialysis at 4°C against 150 mM phosphate buffer to remove glucose, glycated LDL(200-G-LDL) and control LDL(C-LDL) were kept at -40° C until assay of PC-OOH and CE-OOH. After removing EDTA, aliquots of 200-G-LDL and C-LDL at the concentration of 0.5 mg protein/ml were incubated with 60 μM Fe^{3+}-ADP (Fe^{3+}: ADP=1:16.6 mol/mol) in 75 mM phosphate buffer (pH 7.4) at 37°C for 3 hrs, and PC-OOH, CE-OOH and thiobarbituric acid reactive substances(TBARS) were assayed.

Assay for PC-OOH and CE-OOH.

100 μl of LDL were mixed vigorously with 4 vol of methanol followed by centrifugation at 12000 rpm. Aliquots were analyzed for PC-OOH by HPLC with chemiluminescence detector as reported (Yamamoto et al., 1987). For CE-OOH, 300 μl of LDLs were mixed vigorously with 2 ml of methanol followed by extraction of 10 ml of n-hexane. After centrifugation, 9 ml of n-hexane extract were dried and dissolved in 270 μl of methanol/t-butanol(1:1). Aliquots were analyzed by HPLC with chemiluminescence detector (Yamamoto and Niki, 1989). Free cholesterol, cholesteryl ester and vitamin E in n-hexane extract were also assayed by HPLC with UV detector.

Assay for Phospholipase A_2 activity.

1-Stearoyl-2[13'-(s)-hydroperoxy-9',11'-octadecadienoyl]-*sn*-glycero-3-phosphocholine, which was donated by Dr. Baba (Okayama University), was used as a substrate. This was referred to as SLPC-OOH. A mixture of SLPC-OOH and sodium deoxycholate in EtOH(1:3, mol/mol) was evaporated and dissolved in 0.2 M Tris-HCl buffer (pH 7.4). Reaction mixtures of 140 μl contained 10 mM EDTA, 1 mM SLPC-OOH, 3 mM sodium deoxycholate and 100mM Tris buffer. After addition of LDL to the reaction mixture, it was incubated at 37°C. At the appropriate time, 200 μl of Dole reagent(n-heptane/2-propanol/2 N sulfuric acid: 10/40/1, V/V) and 120 μl of heptane were added, followed by centrifugation. Aliquots of heptane extracts were injected to HPLC with UV detection of 13-hydroperoxy-9,11-octadecadienic acid (18:2 OOH), the hydrolysis product of SLPC-OOH due to phospholipase A_2.

Phenylboronate affinity column.

Glycation of LDL with 25 mM glucose and 50 mM glucose was carried out under the same conditions as described for the preparation of 200-G-LDL and named as 25-G-LDL and 50-G-LDL, respectively. The 25-G-LDL and 50-G-LDL in 150 mM phosphate buffer containing 1 mM EDTA were adjusted to pH 8.5 by the addition of NaOH, and then applied on phenylboronate affinity column (Amicon Corp.,PBA-10, 1.6x10 cm) which had been equilibrated with 100 mM sodium phosphate buffer pH 8.5. The column was developed with 100 mM sodium phosphate buffer at pH 8.5 and the adsorbed fraction was eluted with 100 mM phosphate buffer containing 200 mM sorbitol at pH 8.5. Each fraction was monitored by estimation of total cholesterol content(Cholesterol Test Kit, Wako Pure Chemicals). Experiments were carried out at 4°C. Some fractions were kept at -40°C until use.

Results and Discussion

1) *PCOOH and CEOOH in glycated LDL.*
Table I shows the levels of PC-OOH and CE-OOH during glycation process in vitro. Cholesteryl ester, free cholesterol (FC) and vitamin E were also shown. At the end of 70hr-incubation in the presence of 0.5 mM EDTA(sample B & C), the level of CE-OOH/FC in 200-G-LDL(sample C) was 4.8 times that of native LDL (sample A). The increase in the level of CE-OOH/FC of C-LDL during incubation without glucose was much less than that of 200-G-LDL. The levels of PC-OOH/FC did not change during glycation process (samples A, B & C). The level of CE-OOH/FC in the 200-G-LDL was 11 times that of PC-OOH/FC. It seems that the cholesteryl esters in the core of LDL particle are more susceptible to oxidation than are phospholipids on the surface of LDL.

Table I. Formation of lipid hydroperoxides and degradation of Vitamin E.

sample	PCOOH,μM (PCOOH/FC)	CEOOH,μM (CEOOH/FC)	VE,μM (VE/FC)	FC, mM	CE, mM	TBARS,μM
(A) native LDL	0.16 (0.36x10-3)	0.38 (0.86x10-3)	6.12 (13.9x10-3)	0.44	1.14	n.d.
(B) C-LDL	0.12 (0.25x10-3)	0.59 (1.26x10-3)	5.93 (12.6x10-3)	0.47	1.40	n.d.
(C)200-G-LDL	0.16 (0.39x10-3)	1.88 (4.58x10-3)	4.93 (12.0x10-3)	0.41	1.19	n.d.
(D) Fe-treated C-LDL	0.54 (2.07x10-3)	2.96 (11.3x10-3)	3.07 (11.8x10-3)	0.26	0.77	8.75
(E) Fe-treated 200-G-LDL	0.78 (3.0x10-3)	6.69 (25.7x10-3)	2.31 (8.88x10-3)	0.26	0.79	29.0

PCOOH, CEOOH, VE, FC and CE mean phosphatidylcholine hydroperoxide, cholesterylester hydroperoxide, vitamin E, free cholesterol and cholesteryl ester, respectively.
Concentrations of protein in (A), (B), and (C) are 1 mg/ml and those of (D) and (E) are 0.5 mg/ml.
PCOOH/FC, CEOOH/FC and VE/FC are shown in the parentheses.
In (D) and (E), C-LDL and 200-G-LDL were dialyzed to remove EDTA and treated with Fe-ADP for 3 h at 37°C.
n.d. indicates not determined.

However, it has been reported that phospholipase A_2 activity is associated with apoprotein B and that lysophosphatidylcholine is detected in oxidized LDL. To confirm the phospholipase A_2 activity of LDL toward PC-OOH, 200-G-LDL and C-LDL were incubated with synthetic phosphatidylcholine hydroperoxide SLPC-OOH. 18:2 OOH is the hydrolysate of SLPC-OOH by phospholipase A_2. As shown in Figure 1(A), 18:2OOH was liberated from SLPC-OOH by treatment with C-LDL. It was increased with incubation time. 200-G-LDL showed the same enzymatic activity as C-LDL (Figure 1 (B)). The rates of hydrolysis of PC-OOH by C-LDL and 200-G-LDL, however, were very slow, compared with bovine pancreatic phospholipase A_2 (data not shown). Phospholipase A_2 associated apoprotein B may work more efficiently for PC-OOH formed in the LDL particle. The active site on the apoprotein B in the LDL could be hindered in its access to the substrate SLPC-OOH which is dispersed in the bulk solution in our assay system. Apparent low levels of PC-OOH in 200-G-LDL (Table I) may be due to its phospholipase A_2 activity. The concentration of vitamin E, a natural antioxidant, did not change in either 200-G-LDL or C-LDL during 70 hr-incubation. To investigate the effect of ferric ion on the levels of PC-OOH and CE-OOH in glycated LDL, EDTA was removed from 200-G-LDL or C-LDL sample by dialysis against phosphate buffer and dialyzed samples were used for the experiments (samples D & E). 200-G-LDL treated with Fe^{3+}-ADP showed increased levels of PC-OOH/FC and CE-OOH/FC, accompanied by a significant decrease in vitamin E content, compared with that of C-LDL treated with Fe^{3+}-ADP. Increased PC-OOH in 200-G-LDL treated with Fe^{3+}-ADP is interpreted as indication that PC-OOH production accelerated by Fe^{3+}-ADP surpassed the phospholipase A_2 activity and/or phospholipase A_2 activity was modified by degradation products of lipid hydroperoxide. Another marker of lipid peroxidation, TBARS, greatly increased after incubation of 200-G-LDL with Fe^{3+}-ADP, to levels 3 times that of C-LDL treated with Fe^{3+}- ADP. These results confirm our previous studies.

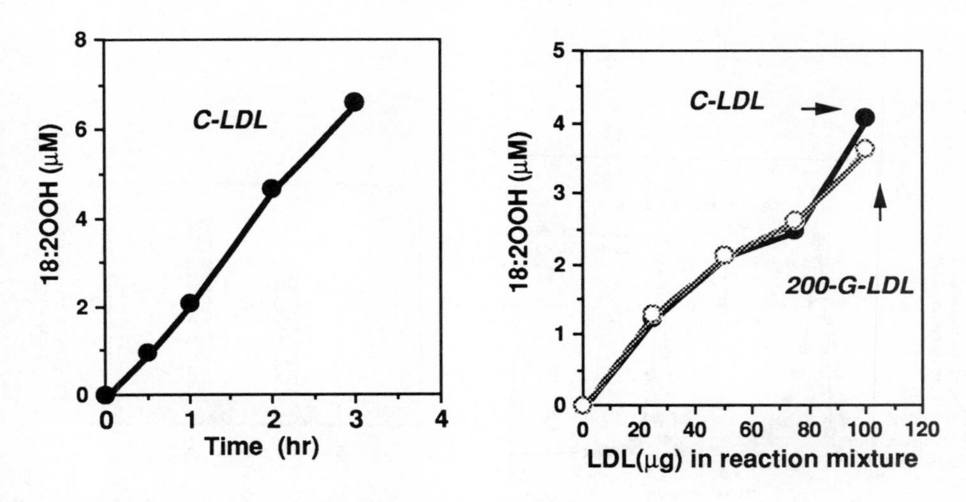

Figure 1. Time course of production of 18:2OOH(A), and effects of concentration of LDLs(B). (A) Reaction mixtures were incubated with C-LDL(50μg) up to 3hr at 37°C and 18:2OOH was analyzed by HPLC. (B) C-LDL and 200-G-LDL at a variety of the concentrations were added to the reaction mixture, followed by the incubation at 37°C for 1hr.

2) *PC-OOH in PBA-10 affinity column-adsorbed fraction.*

Figure 2 shows the elution profiles of 25-G-LDL and 50-G-LDL on PBA-10 affinity column chromatography in which glycated LDL is retained on the column. Their adsorbed fractions on PBA-10 were 18% and 38%, respectively. Figure 3 shows the relative amounts of PC-OOH, as nmol of PC-OOH per mg of total cholesterol, before and after the application of 25-G-LDL and 50-G-LDL on to PBA-10. The relative amounts of PC-OOH in the adsorbed fraction P_2 were markedly higher than those of the nonadsorbed fraction P_1 both in 25-G-LDL and in 50-G-LDL. The relative amounts of PC-OOH of P_1 fraction both in 25-G-LDL and 50-G-LDL, however, were not changed, compared with those of PC-OOH detected in LDL samples before the application on to the PBA-10. The sum of the total amounts of PC-OOH recovered in P_1 and P_2 fractions were greater than the total amounts of PC-OOH detected in the original samples, i.e.,1.7 times in 25-G-LDL and 2.3 times in 50-G-LDL. This indicates that EDTA added in the LDL samples acts as an antioxidant of LDL and that PC-OOH is produced during the column chromatography.

Such an oxidation of LDL apparently comes about in the P_2 fraction, in which EDTA had been washed out. An alkaline pH such as 8.5 of sample solutions may accelerate the formation of enediol structure of fructosamine bound to apoprotein B. Coordination of a trace amount of iron in the phosphate buffer to the enediol structure in the P_2 fraction could cause the lipid peroxidation according to our proposed mechanism (Sakurai *et al.*, 1990). Probably the level of CE-OOH in the P_2 fraction is also increased, though we did not estimate CE-OOH. Glycated LDL obtained from phenylboronate affinity column should be used carefully for experiments such as stimulation of various kinds of cells or an uptake by macrophages. It is recommended to use phenylboronate affinity column after equilibration with 1 mM EDTA-containing buffer.

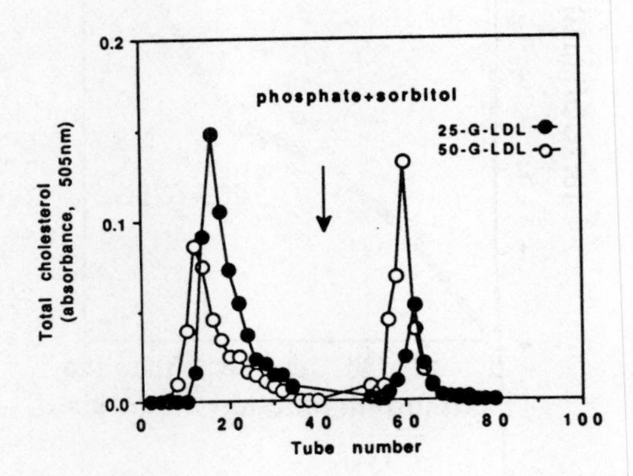

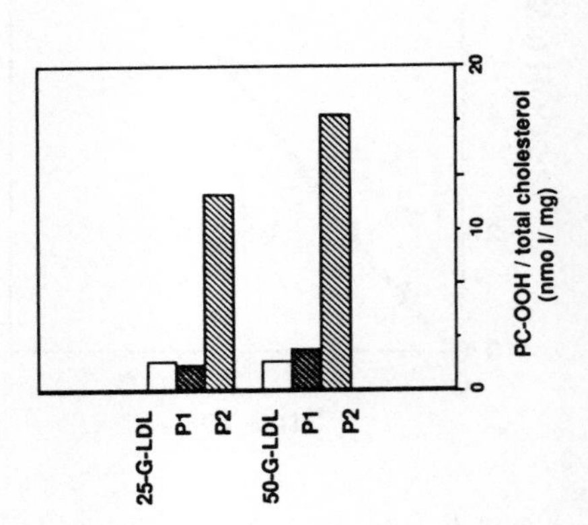

Figure 2 (left). Elution profiles of 25-G-LDL and 50-G-LDL on phenylboronate affinity column. Samples are 1mg of LDL proteins/1ml of 150mM phosphate buffer (pH 8.5) with 1mM EDTA.

Figure 3 (right). Relative amounts of PC-OOH in 25-G-LDL and 50-G-LDL before and after application on PBA-10 column. Samples shown in Figure 2 were analyzed. White bars indicate samples before application on PBA-10.

References

Sakurai, T., Sugioka, K. and Nakano, M.(1990) O_2 generation and lipid peroxidation during the oxidation of a glycated polypeptide, glycated polylysine, in the presence of iron-ADP.. *Biochim. Biophys. Acta.* 1043: 27-33.

Sakurai, T., Kimura, S., Nakano, M. and Kimura, H. (1991a) The oxidative modification of low density lipoprotein by nonenzymatically glycated peptide-Fe complex. *Biochim. Biophys. Acta* 1086: 273-278.

Sakurai, T., Kimura, S., Nakano, M. and Kimura, H. (1991b) Oxidative modification of glycated low density lipoprotein in the presence of iron. *Biochem. Biophys. Res. Commun.* 177:433-439.

Witztum, J. L. and Steinberg D.(1991) Role of oxidized low density lipoprotein. *J. Clin. Invest.* 88: 1785-1792.

Yamamoto, Y., Brodsky, M. H., Baker, J. C. and Ames, B. N. (1987) Detection and characterization of lipid hydroperoxides at picomole levels by high-performance liquid chromatography. *Anal. Biochem*. 160:7-13.

Yamamoto, Y. and Niki, E. (1989) Presence of cholesteryl ester hydroperoxide in human blood plasma. *Biochem. Biophys. Res. Commun.* 165: 988-993.

Production and Characterization of Antibodies against Carboxymethyllysine-modified Proteins

Klaus E. Gempel, Eva M. Wagner, and Erwin D. Schleicher

INSTITUTE FOR DIABETES RESEARCH, KÖLNER PLATZ 1, 80804 MUNCHEN, GERMANY

Summary
Carboxymethyllysine (CML) is the major oxidation product of fructose-lysine which results from the reaction of glucose with the ε-amino group of free or protein-bound lysine. To develop a quantitative assay for the estimation of the CML content of proteins and for possible immunohistochemical studies, CML antibodies were raised in rabbits after immunization with CML-modified proteins. The antisera obtained reacted with CML-modified proteins bound to micro titer plates but not with glycated proteins. The specificity of the antiserum was further evaluated by competition experiments using structurally related compounds. No competition occurred with concentrations of 5 mM lysine, fructose-lysine or alanine. CML competed efficiently, while sarcosine competed to the extent of 33% and 46% for CML-KHL- and CML-BSA antisera, respectively. An ELISA was developed that detected CML content below 0.1 mol/mol protein.

Introduction

Protein glycation has been shown to occur by the exposure of the protein to glucose for extended periods *in vitro* as well as *in vivo* (Ledl and Schleicher, 1990). In studies on events occurring during the later stages of the Maillard reaction, it has been demonstrated (Ahmed et al., 1986) that the carbohydrate moieties of glycated proteins are cleaved between carbons 2 and 3 in the presence of oxygen, yielding carboxymethyllysine (CML) and erythronic acid. The reaction proceeds by a metal-catalyzed, free radical mechanism (Smith and Thornalley, 1993). Subsequently, it could be shown that this reaction also occurs in. Both CML and erythronic acid have been identified in normal human urine (Ahmed et al., 1986), and urinary excretion was elevated in diabetics (Knecht et al., 1991). CML accumulates with age in the human lens (Dunn et al., 1989) and has also been detected in skin collagen (Lyons et al., 1991). The rationale for determining the metabolic consequences of carboxymethylation of low density lipoprotein (LDL) derived from the studies of Mahley et al. (1979), which documented that lysine residues of apolipoprotein B were critical for the specific recognition of LDL by its receptor. In an earlier study, we demonstrated (Gempel et al., 1993) that carboxymethylation which creates an additional negative charge on the affected lysine residue inhibited metabolism of LDL via the high-affinity receptor in fibroblasts. Since LDLs are glycated in normal subjects and glycation is elevated in diabetic patients, we reasoned that LDL carboxymethylation may also occur *in vivo*. Possible elevated
CML-modification in diabetics may contribute to the development of diabetic angiopathies. To develop a quantitative assay for estimation of the CML content of LDL or other proteins we raised antibodies to CML-modified proteins in rabbits.

Materials and Methods

Preparation of antibodies to CML-modified proteins
Bovine serum albumin (BSA) or keyhole limpet hemocyanin (KLH) were dissolved in saturated sodium tetraborate solution pH 9.1 to which 1 mg of $Na(CN)BH_3$ was added at 0°C (Saunders and Hedlund, 1984). While continuously stirring, 12 μmol glyoxylic acid was added every 10 minutes up to six times. The reaction mixture was stirred for a further 30 minutes and the proteins were extensively dialyzed before used. The molar ratio of BSA:CML was 1:10, estimated on the basis of the increase in electrophoretic mobility of the modified protein. 200 μL of immunogen was dissolved in 200 μL PBS at pH 7.4. The solution was emulsified with 200 μL of Freund's complete adjuvant and injected into rabbits subcutaneously. Rabbits received a booster injection of 100 μL of immunogen in 200 μL PBS emulsified with 200 μL of Freund's incomplete adjuvant 6 weeks later. Blood was collected and antiserum titer was determined by the ELISA as described below.

Enzyme-linked immunoassay
For preparation of radiolabelled CML-BSA for standardization of the ELISA, the reaction mixture contained [2-^{14}C]-glyoxylic acid (1.7 x 10^4 Bq/nmol). A volume of 100 μL of BSA- or KHL-conjugated with CML (1 μg/mL) was dispensed into each well of a micro titer plate as coating reagent and incubated for one hour at 37°C in a moist chamber. After washing three times with PBS, wells were blocked with crotein (27 mg/mL) (Boehringer Mannheim). Rabbit serum (50 μL) was then applied to each well and incubated for one hour at room temperature. After washing three times with buffered 0.05% Tween 20 solution containing 0.1% BSA, bound CML-antibody was determined by goat anti-rabbit IgG conjugated with alkaline phosphatase (Boehringer Mannheim) and the absorbance was determined at 450 nm with an ELISA reader (Dynatech laboratories).

Antisera and purified antibodies were checked for specific binding by incubation with increasing amounts of free hapten. Similarly, the cross-reactivity with CML analogues was tested over a range of concentrations varying from 0 to 5 mM.

Results

Production and characterization of CML-antibodies in rabbits.
The formation of CML-antibodies in rabbits that responded to CML-modified KHL and CML-modified BSA was similar with both immunogens; similar titers were obtained. Furthermore, when the antisera were tested for cross-reactivity in competition experiments using chemically related compounds, similar specificity was obtained. The specificity of CML-KHL antiserum is shown in Figure 1. Sarcosine competed weakly with both antisera while no reactivity was found with alanine. Glycine competed significantly with the binding of CML-BSA antiserum while no interference was observed when the CML-KHL antiserum was assayed. Furthermore, L-lysine, ε-fructosyl-L-lysine and alanine did not compete suggesting that in native proteins these residues are not recognized by both antisera. *In-vitro*-glycated BSA containing 13 glycated ε-lysine residues per mole was not recognized by either antiserum. Presence of 10 mM glucose in the samples did not interfere with the antigen-antibody binding.

Affinity purification of the CML antisera yielded no improvement in specificity.

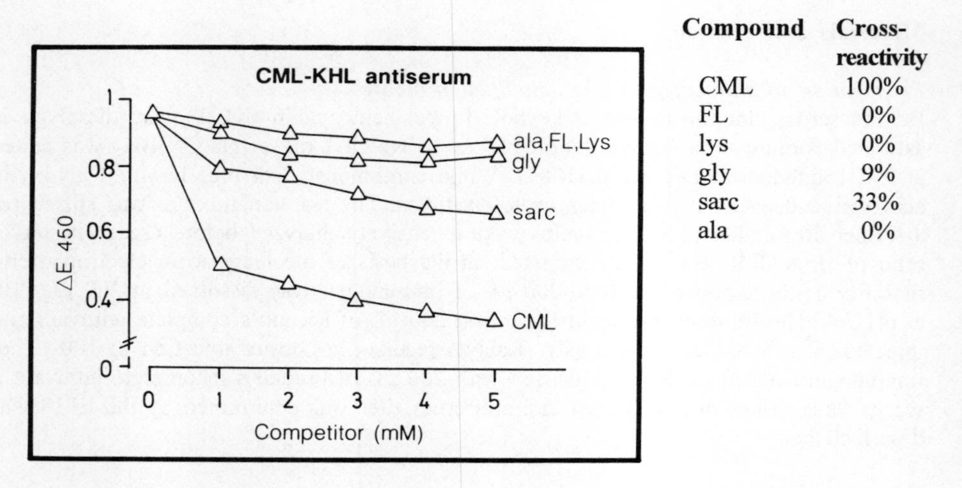

Compound	Cross-reactivity
CML	100%
FL	0%
lys	0%
gly	9%
sarc	33%
ala	0%

Figure 1. (left) Specificity of CML-KHL antiserum Microtiter plates were coated with 30 μg CML-BSA and assayed with CML-KHL antiserum obtained after immunization of rabbits with CML-KHL. Reaction mixtures contained 0-5 mM of the free hapten (CML) or other competitors. lys = lysine, FL = fructose-lysine, ala = alanine, gly = glycine, sarc = sarcosine.

Table I. (right) Cross-reactivity of CML-KHL antiserum Quantitative values were deduced from the curves of Figure 1.

Calibration curve

Figure 2 shows the calibration curve obtained with CML-modified BSA containing 0 - 2.37 CML per mole of BSA. The calibration curve indicates that the antiserum may recognize CML-modified BSA below 1 mol CML/mol BSA. The detection limit is 0.1 mol CML/mol BSA. The exact amount of CML bound to BSA was calculated from the amount of radioactivity incorporated into CML-BSA prepared with ^{14}C-labelled glyoxylic acid.

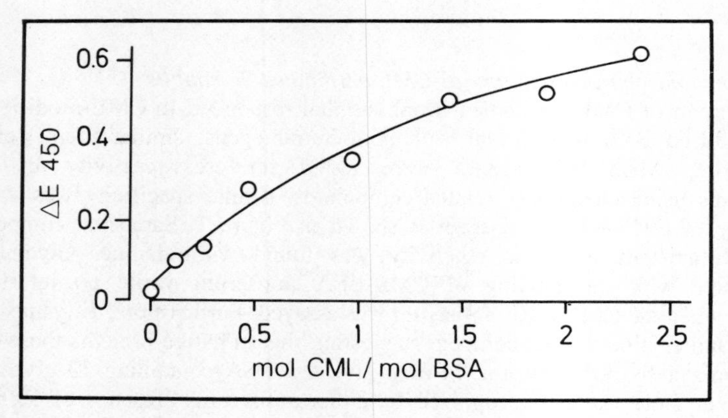

Figure 2. Calibration curve of CML-BSA with CML-KHL antiserum 1 μg CML-BSA containing increasing amounts of CML were coated on micro titer plates and assayed with CML-KHL antiserum that had been diluted 1:1000. Extent of CML modification was determined by incorporation of ^{14}C radioactivity.

Discussion

The purpose of the present study was to develop antibodies to CML-modified proteins in order to determine possible CML-modification of human plasma or tissue proteins. Two observations suggest that the epitope recognized by the antisera obtained is indeed related to CML. Firstly, the antisera could be blocked by coincubation with the free hapten carboxymethyllysine, while other chemically related haptens like sarcosine were less effective. Furthermore, the antisera did not recognize glycated lysine residues either in free or protein-bound form. Secondly, increased antibody affinity to glycated BSA was noted as a function of duration of the incubation time in the presence of oxygen. No reactivity was obtained when the incubation was performed in the absence of oxygen (data not shown). Thus, the data presented above provide strong support for the possibility to obtain CML-specific antibodies with the procedure described.

Conclusion

The results presented here indicate that we have raised specific antibodies to CML in rabbits. The antisera obtained are useful for the determination of proteins containing CML in a molar ratio of more than 0.1. Furthermore, the antisera may also be useful for the immunohistochemical detection of CML-modified proteins in tissues from normal and diabetic patients.

Acknowledgements
We thank Prof H. F. Erbersdobler, Kiel, Germany, for providing a carboxymethyllysine standard.

References

Ahmed, M. U., Thorpe, S. R. and Baynes, J. W. (1986) Identification of N_ε carboxymethyllysine as a degradation product of fructoselysine in glycated protein. J. Biol. Chem. 261: 4889-4894

Dunn, J. A., Patrick, J. S., Thorpe, S. R. and Baynes, J. W. (1989) Oxidation of glycated proteins: Age-dependent accumulation of N_ε-(carboxymethyl)lysine in lens proteins. Biochemistry 28: 9464-9468

Gempel, K. E., Gerbitz, K. D., Olgemoller, B. and Schleicher, E. (1993) *In vitro* carboxymethylation of low density lipoprotein alters its metabolism via the high affinity receptor. Horm. Metab. Res. 25: 250-252

Knecht, K. J., Dunn, J. A., McFarlane, K. F., McCance, D. R, Lyons, T. J., Thorpe, S. R and Baynes, J. W. (1991) Effect of diabetes and aging on carboxymethyllysine levels in human urine. Diabetes 40: 190-196

Ledl, F. and Schleicher, E. (1990) New aspects of the Maillard-reaction in food and in the human body. Angew. Chemie Int. 29: 565-594

Lyons T. J., Bailie, K. E., Dyer, D. G., Dunn, J. A. and Baynes, J. W. (1991) Decrease in skin collagen glycation with improved glycemia control in patients with insulin-dependent diabetes mellitus. J. Clin. Invest. 87: 1919-1925

Mahley, R. W., Innerarity, T. L., Weisgraber, K. H. and Oh, S. Y. (1979) Altered metabolism (*in vivo* and *in vitro*) of plasma lipoproteins after selective chemical modification of lysine residues of the apoproteins. J. Clin. Invest. 64: 743-750

Saunders, S. and Hedlund, B. E. (1984) Electrostatic modification of protein surfaces: effect on hemoglobin ligation and solubility. Biochemistry 23: 1457-1461

Smith, P. R and Thornalley, P. J. (1993) Mechanism of the degradation of non enzymatically glycated proteins under physiological conditions. Studies with the model fructosamine, N_{ε}-(1-deoxy-D-fructos-1-yl)-hippuryllysine. Eur. J. Biochem. 214: 729-739

Alterations of Mineral Metabolism and Secondary Pathology in Rats Fed Maillard Reaction Products

J. O'Brien,[1*] P. A. Morrissey,[2] and A. Flynn[2]

[1]DEPARTMENTS OF FOOD CHEMISTRY AND [2]NUTRITION, UNIVERSITY COLLEGE, CORK, IRELAND; [*]CURRENT ADDRESS: ROBENS INSTITUTE OF HEALTH AND SAFETY, UNIVERSITY OF SURREY, GUILDFORD GU2 5XH, UK

Summary

It is possible that some of the toxic effects ascribed to Maillard reaction products (MRPs) are secondary to gastrointestinal effects and not a consequence of direct chemical toxicity *per se*. The present studies examined the effect of feeding diets containing 5% or 10% of an MRP from a glucose-glutamate system. There was a dose-related increase in the incidence of nephrocalcinosis in animals fed the MRP, which resulted in much interstitial kidney damage in the most severely affected animals.

Mineral balance studies of rats fed 0.5% MRP suggested that the nephrocalcinosis was secondary to disturbances of calcium and magnesium metabolism and was not of toxicological significance *per se*. A more significant finding, however, was a reduction in zinc status in animals fed 0.5% MRP in a 12 d balance study as a result of increased urinary and faecal losses. Preliminary studies *in vitro* demonstrated that the glucose-glutamate MRP is capable of binding magnesium, calcium, zinc and copper.

Introduction

The literature on the Maillard reaction provides evidence for direct chemical toxicity of MRPs in a variety of organs (for review see O'Brien and Morrissey, 1989) and also describes effects, such as diarrhoea and caecal enlargement, that are similar to those produced by a host of poorly digestible carbohydrates (O'Brien and Walker, 1988). Many studies have examined the effect of MRPs on mineral metabolism but few have quantified the long-term effects on mineral status or examined the consequences of such effects for the health of animals or humans. The present studies examined the effects of glutamate-glucose MRPs because of the abundance of glutamate in the free form in food systems and because it is a major component of protein hydrolysates; it is also added to some processed foods as the flavour enhancer monosodium glutamate (MSG).

Materials and Methods

Preparation of MRP

Maillard reaction product was prepared by heating an equimolar slurry (35%, w/w, moisture) of glucose and MSG at 100°C for three hours; the reaction was stopped by cooling and the addition of water. The MRP solution was subsequently freeze dried and the dry free-flowing material produced used for diet formulation.

Protocol (90 d study)

Groups of 10 weanling male Wistar albino rats were fed semisynthetic diets containing 5% or 10% MRP for 90 days. Control groups (10 animals per group) were fed 5% unheated MSG-glucose, 10% unheated MSG-glucose or a standard semisynthetic diet. Wax-embedded formalin-fixed tissues were cut into 5μm thick sections which were stained with Ehrlich's acid haematoxylin and eosin.

Tissue morphology

Area fractions of nephrocalcinosis in the kidneys of each animal were determined using a random point counting technique on enlarged projections of 5μm thick sections. Sample sizes were determined statistically using the cumulative means test. Size distribution profiles of calcareous deposits were determined by image analysis of light photomicrographs using a MOP AMO3 image analyser (Kontron Electronics, Watford, UK).

Mineral balance study

Female Wistar albino rats (5 weeks old) were paired as littermates into two groups of eight animals and housed individually in metabolism cages for 12 days. Diets contained either 0.5% MRP or 0.5% of an unheated mixture of MSG and glucose (control). Animals were conditioned to receive diets in the form of two one-hour meals per day in conventional cages to minimize contamination of urine and faeces. Diluted samples of urine were analysed in duplicate by atomic absorption spectroscopy (Pye Unicam SP9 atomic absorption spectrophotometer) for calcium, magnesium, potassium, sodium, zinc, copper and iron. To prevent interference, samples for calcium and magnesium determination were diluted with 0.1 % lanthanum chloride solution; zinc and copper analyses were conducted using the method of standard additions. Samples of diets and faeces were analysed in a similar manner following wet ashing in 5:1 nitric-perchloric acid. Values for mineral retention were calculated by difference following the estimation of dietary intakes and urinary and faecal outputs.

Metal binding studies in vitro

The potential of the glucose-glutamate MRP to bind calcium, magnesium, zinc or copper was examined using the proton liberation method. Potentiometric titrations were conducted using a PHM-85 digital pH meter and an electrode pair consisting of a G202C glass electrode and a K401 calomel reference electrode (Radiometer Copenhagen, Denmark). Data were recorded, stored and analysed using an IBM personal computer model XT fitted with an RS-232 interface. The metal concentration used in all cases was 20mM in a 0.15mM NaCl background electrolyte. The concentrations of MRP used for the titrations were 0.8, 1.6 and 3.2 g/100mL. The values of $d[H^+]/d[A]$ at each pH were obtained by calculating the slopes of the graphs of $[H^+]$ versus the concentration of MRP ([A]).

Results and Discussion

Animals fed 5% or 10% MRP developed severe diarrhoea that persisted throughout the study. At autopsy, such animals had grossly enlarged caeca compared with control animals. All animals fed the MRP developed severe nephrocalcinosis in the region of the corticomedullary junction and, to a lesser extent, in the medulla of the kidney. Although a slight background incidence of the condition was observed in some control animals, the extent and severity of calcification was more severe in MRP-fed animals as evidenced by the area fractions (Table 1) and size distribution profiles of nephrocalcinosis. The kidneys of rats fed 10% MRP contained a greater proportion of large deposits (up to 105 μm diameter) which resulted in marked distention and damage to the surrounding tubule and necrosis of surrounding tissue.

Table I. Area fractions of nephrocalcinosis in the kidneys of rats fed MRP or control diets.

Group	Fraction of area calcified ($x10^{-3}$)
Control	3.4 ± 1.6
5% unheated MSG-glucose	1.1 ± 0.6[a,b]
10% unheated MSG-glucose	1.7 ± 1.4[c]
5% MRP	4.5 ± 1.2[b]
10% MRP	24.3 ± 11.8[a,c]

Means ± SEMs for 10 rats. Values with a letter in common differ to a statistically significant extent (Student's t-test). [a,b,c], $p<0.05$.

In the mineral balance study, there were statistically significant increases in the urinary excretion of calcium, magnesium, zinc and copper and in the faecal content of zinc in rats fed 0.5% MRP (Table II). In addition, zinc retention was significantly reduced in animals fed 0.5% MRP. Interestingly, the retention of calcium, magnesium and copper was not affected by the MRP diet and there is evidence that the renal excretion of magnesium and calcium was offset by an increased intestinal absorption.

Table II. Mineral balance data for rats fed 0.5% MRP or control diets.

Mineral	Urinary Output	Faecal output % of Intake	Mineral retention
Calcium			
Control	0.81 ± 0.21	29.1 ± 2.0	70.0 ± 2.1
0.5% MRP	1.33 ± 0.25[¶]	26.9 ± 1.1	71.8 ± 1.3
Magnesium			
Control	24.6 ± 1.8	30.5 ± 2.0	44.9 ± 1.4
0.5% MRP	40.6 ± 3.7[§]	13.0 ± 2.1[§]	46.4 ± 3.1
Zinc			
Control	0.56 ± 0.07	37.2 ± 5.0	62.2 ± 5.0
0.5% MRP	1.46 ± 0.38[*]	57.9 ± 4.5[¶]	40.6 ± 4.7[**]
Copper			
Control	1.43 ± 0.09	47.4 ± 2.2	51.2 ± 2.3
0.5% MRP	1.78 ± 0.02[¶]	51.2 ± 3.8	47.0 ± 3.9
Iron			
Control	trace	68.8 ± 2.4	31.2 ± 2.4
0.5% MRP	trace	63.8 ± 2.1	36.2 ± 2.2

Means ± SEMs for groups of 8 rats. Values with a superscript symbol differ significantly from control values. [*], $p<0.05$; [¶], $p<0.02$; [**], $p<0.01$; [§], $p<0.0001$.

These data suggest that the nephrocalcinosis observed in rats fed 5% or 10% MRP was secondary to disturbances of mineral metabolism and not a consequence of direct nephrotoxicity *per se*. An interesting feature of the mineral balance study was a significantly greater calcium x phosphorus concentration product in the urines of rats fed 0.5% MRP compared with control animals. This finding indicates that the tendency of calcium to precipitate, as calcium phosphate, in the kidneys and urinary tract was significantly increased in rats fed 0.5% MRP. However, it is unclear if a similar outcome could apply to the human ingestion of MRPs since rats are very sensitive to nephrocalcinosis and also exhibit increases in urinary calcium on ingestion of a variety of poorly digestible materials.

Increased losses via both the faecal and urinary routes contributed to the reduction in zinc retention in animals fed 0.5% MRP. In previous studies, increased urinary losses of zinc have been attributed to both intravenously and orally administered MRPs. Intravenous alimentation with solutions of amino acids or protein hydrolysates, containing MRPs as a result of autoclaving, was shown to increase the urinary excretion of zinc, copper and iron in human subjects. Similarly, there is evidence that at least some forms of diet containing MRPs may have a detrimental effect on zinc retention in humans (for review see O'Brien & Morrissey, 1989). Dietary MRPs have previously been reported to increase the urinary excretion of calcium and magnesium in rats(Adrian and Boisselot-Lefebvres, 1977; Andrieux and Sacquet, 1984). In this respect, MRPs may appear to resemble other poorly digestible materials such as xylitol or modified starches. However, in contrast to the effect of poorly digestible carbohydrates, there is evidence that MRPs may actually inhibit small intestinal calcium transport (Andrieux and Sacquet, 1984). Thus, there are clear differences between the mechanism of action of MRPs on calcium metabolism and those of poorly digestible carbohydrates.

Proton liberation as a function of pH for the glucose-glutamate MRP is described in Figure 1. Clearly, proton displacement is evident in all of the systems studied indicating that some component of the MRP was capable of binding the metal ions. The extent of proton displacement in the pH range 3.5 - 10 suggests that the strength of complexation occurred in the order: magnesium>copper=calcium>zinc. The order of binding strength differs from what would be expected from the Irving-Williams series for a mononuclear binary system (Irving and Williams, 1953): Cu>Zn>Mg>Ca. The discrepancy may be explained by the fact that some MRPs may behave as polynuclear ligands and/or that several different ligands are present in the system, each with a different complexation chemistry.

The results of the *in-vitro* mineral binding studies offer an explanation for the effect of MRPs on mineral homoeostasis *in vivo*. While it is likely that the effect of dietary MRPs on urinary zinc and copper may be due to a complexation mechanism, it is unclear to what extent complexation accounts for the increase in urinary calcium and magnesium. It is generally accepted that the action of poorly digestible carbohydrates, in facilitating intestinal calcium absorption, involves a stimulation of paracellular transport secondary to an increase in lumenal osmolality rather than a complexation mechanism. Thus, although complexation by MRPs might account for the inhibition of small intestinal transport of calcium, it is possible that the complex might be disrupted or metabolised in the lower gut facilitating a net increase in intestinal absorption as observed in the mineral balance study.

The increase in both urinary and faecal zinc in the present studies suggests either the complexation of zinc by a ligand that is partially absorbed or complexation by at least 2 distinct ligands that are selectively excreted in either the urine or the faeces. In conclusion, the present studies illustrate the importance of careful interpretation of the nutritional and toxicological effects of MRPs and suggest that further work is necessary to examine the effect of individual MRPs on mineral metabolism.

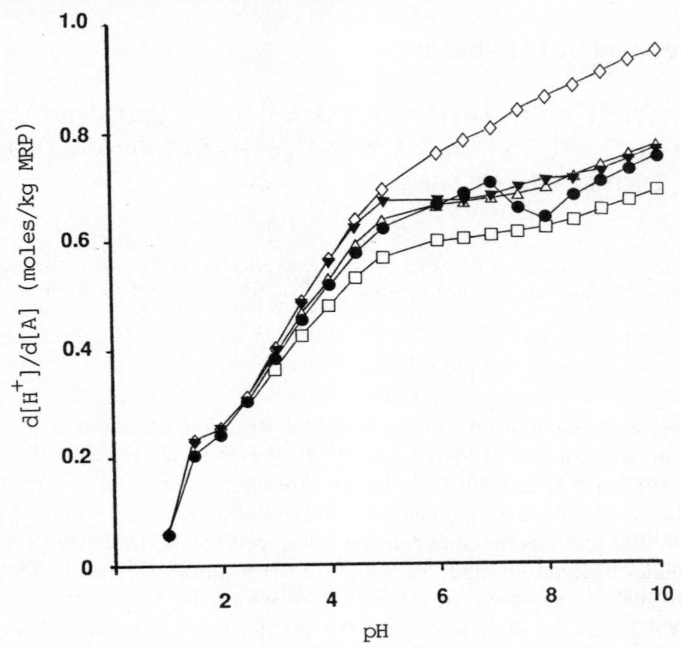

Figure 1. Proton liberation curves for glucose-glutamate MRP as a function of pH in the absence of metal ion (□), or in the presence of zinc (●), copper (▼), magnesium (◇) or calcium (Δ).

References

Adrian, J. and Boisselot-Lefebvres, J. (1977) Interactions between Maillard reaction products and calcium metabolism. *Cah. Nutr. Diet.* 12: 233-234.

Andrieux, C. and Sacquet, E. (1984) Effects of Maillard's reaction products on apparent mineral absorption in different parts of the digestive tract. The role of microflora. *Reprod. Nutr. Develop.* 23: 379-386.

Irving, H.M. and Williams, R.J. (1953) The stability of transition-metal complexes. *J. Chem. Soc.* 3192-3210.

O'Brien, J. and Morrissey, P.A. (1989) Nutritional and toxicological aspects of the Maillard browning reaction in foods. *Crit. Rev. Food Sci. Nutr.* 28: 211-248.

O'Brien, J. and Walker, R. (1988) Toxicological effects of dietary Maillard reaction products in the rat. *Food Chem. Toxicol.* 26: 775-783.

Research Provides New Insights into Non-enzymatic Browning Reactions*

John O'Brien and T. P. Labuza[1]

ROBENS INSTITUTE OF HEALTH AND SAFETY, UNIVERSITY OF SURREY, GUILDFORD GU2 5XH, UK; [1]DEPARTMENT OF FOOD SCIENCE AND NUTRITION, UNIVERSITY OF MINNESOTA, USA

Summary
The 5th International Symposium on the Maillard Reaction reviewed the new developments in nonenzymatic browning reactions and their relevance to food technology and the medical sciences.

Introduction

The 5th International Symposium on the Maillard Reaction was held at the University of Minnesota in Minneapolis in September 1993. Previous symposia in the series were held in 1979, 1982, 1985, and 1989, illustrating the accelerating pace of research in the field (Eriksson, 1981; Waller and Feather, 1983; Fujimaki *et al.,* 1986; Finot *et al.*, 1990). Almost 300 participants attended the 4-day conference to discuss aspects of the chemistry, kinetics, technology and toxicology of the reaction in foods and to review the new developments in the study of Maillard reactions *in vivo*. By an appropriate coincidence, 1993 was the 40th year since the publication of the classic review of the reaction by John E. Hodge in the first volume of the *Journal of Agricultural and Food Chemistry* (Hodge, 1953). The original review by Hodge was named a Citation Classic by the Institute of Scientific Information in 1979 and the scheme used to describe the Maillard reaction in that review is still the most satisfactory representation of the reaction (Fig. 1). The fact that the paper is the most cited ever to have been published in the *Journal of Agricultural and Food Chemistry* and one of the most cited in food science is testimony to the importance of Maillard browning reactions in food science and technology. L. De Bry (General Biscuits Belgium) emphasized during an after-dinner lecture on the anthropology of the Maillard reaction, that the potential impact of the reaction in food technology first became apparent some 700,000 years ago when Peking Man began to use fire to cook foods. De Bry suggested that, in addition to producing attractive colors and flavors, the Maillard reaction affords safe access by humans, to rich sources of kilocalories through the destruction of antinutritional factors.

Chemistry

The Maillard reaction is one of the most challenging areas of food chemistry. In the initial stages, the reaction represents a simple nucleophilic addition between the NH_2 group of amino compounds such as amino acids and proteins and the electrophilic

* *This paper was prepared for Food Technology and is included here to serve as a partial summary of the meeting.*

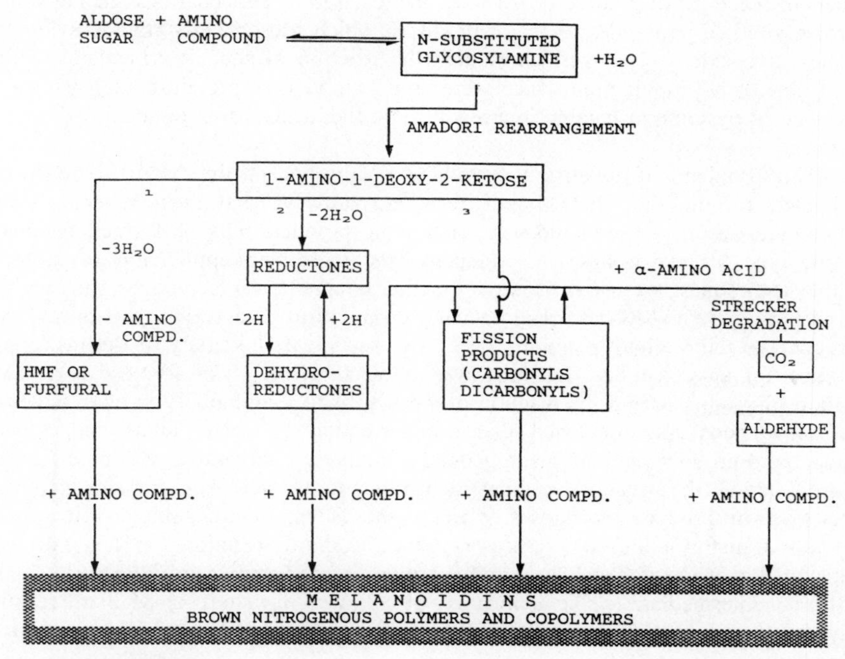

Figure 1. Scheme applied by Hodge (1953) to describe the Maillard reaction. At low pH, 1,2-enolization (1,2-E) reactions are favored whereas at high pH 2,3-enolization (2,3-E) becomes more important. Fission reactions and Strecker degradation reactions are favored by high temperatures and are important in the formation of flavor volatiles. Aldol condensations contribute to the process of chain expansion in the formation of high molecular weight melanoidin pigments.

carbonyl groups of reducing sugars.

In food technology, there is a need to better understand the reaction pathways that result in desirable attributes in finished products, especially desirable flavor profiles such as those in baked goods or in ketchup. One of the driving forces behind research in the area is the development of microwaveable and low-calorie products. Thus, knowledge of the chemistry of Maillard reactions contributes directly to new technological and biomedical opportunities arising from such reactions. The mechanisms involved in the Maillard reaction were addressed by several papers in the sessions on chemistry. M. Pischetsrieder (University of Munich) reported that while α-pyranones are typical products of reactions of glucose or fructose, the reaction of disaccharides such as maltose or lactose yields β-pyranones as major products, which subsequently transform to stable isomaltol glycosides. The workers studied the reaction of such intermediates with N-acetyl lysine or N-propylamine which were used as models of protein-bound lysine. The importance of pyridinium betaine intermediates in the mechanisms of such reactions was emphasized.

The problem of identifying synthesis pathways in the Maillard reaction is complicated by the fact that several reaction pathways and even reverse reaction pathways are operative simultaneously, and some products may be formed from more than one type of intermediate. Fortunately, the increasing application of molecular modeling techniques, the use of isotopic labeling coupled with techniques such as NMR spectroscopy, and GC-MS has greatly enhanced our ability to describe reaction pathways for products of the Maillard reaction. R. Tressl (Technical University of Berlin) reported on a series of novel studies that employed a ^{13}C label on C_1 of fructose to trace the formation of products from the reaction of fructose with γ-aminobutyric acid, isoleucine or proline. The workers showed that transformation between Heyns and Amadori rearrangement products, although energetically relatively unfavorable, was possible in the Maillard reaction; the Amadori rearrangement refers to the aldo→keto transformation step of the reaction product of aldoses whereas the Heyns rearrangement refers to the keto→aldo transformation of ketoses in the Maillard reaction. The work also demonstrated that identical labeled products may be produced from glucose or fructose via different intermediates. The application of ^{13}C labels for studying Maillard reactions clearly has much potential in the field of flavor chemistry and deserves further development.

Unfortunately, little is known of the kinetics of individual pathways of the Maillard reaction. B.L. Wedzicha (University of Leeds, UK) conducted some elegant experiments which showed that the rates of reactions of sulfite with intermediates of nonenzymatic browning can be used to characterize the kinetics of the degradation of such intermediates under various conditions.

V. Yaylayan (McGill University, Canada) reported on the determination of the open-chain forms of sugars by FTIR spectroscopic studies of the carbonyl absorption band. He suggested that the absorptions of the carbonyl groups of protein-bound Amadori products could also be monitored by FTIR which offers a new approach to the study of Maillard reactions of proteins. A relatively new technique for the study of protein glycosylation, matrix-assisted laser desorption-mass spectrometry was described by A. Lapolla (Padua University, Italy). The method was applied successfully to monitor the progressive increase in molecular weight of proteins incubated with reducing sugars.

The chemistry of Amadori rearrangement products continues to attract much attention. D.V. Zyzak (University of South Carolina) showed the phosphate-catalyzed decomposition of fructosyl glycine. It was also suggested, based on GC-MS peaks that the Amadori rearrangement might be reversible. The reversibility of the rearrangement is a most contentious issue. It is generally accepted in the literature that the reaction is irreversible. Clearly, even if the reaction was partly irreversible, it would have huge implications for evaluating the nutritional quality of foods since it is assumed that amino acids are not bioavailable from Amadori products. Conclusions regarding the reversibility of the Amadori reactions should ideally be drawn from studies that carefully characterize the composition of the system under study including the use of labels to follow individual reactions. Chemically, it is more plausible that an unknown reaction pathway rather than a reverse Amadori rearrangement could have led to the recovery of reactants from fructosyl glycine.

In line with developments in other areas of chemistry, molecular modeling techniques have found useful applications in the study of Maillard reactions. T. Kurata (Ochanomizu University, Japan) used semi-empirical molecular orbital calculations to derive information on unstable intermediates in the Maillard reaction of glucose and glycine. Information on electron densities, orbital energy levels and energy-minimized structures is critical to our understanding of structure-reactivity relationships of reactants and intermediates of Maillard reactions and will undoubtedly assist in the design of more effective synthesis strategies for the exploitation of Maillard reactions in foods. Molecular modeling was also used by D.J. Walton and coworkers (Queen's University, Canada) to study the site specificity of glycosylation of alcohol dehydrogenase; the major site of glycosylation, Lys-231, was attributed to intramolecular catalysis caused by the abstraction of the C-2 proton from the Schiffs base by the Thr-347 hydroxyl and the His-348 imidazole. Glycosylation at a secondary site, Lys-228, was catalyzed by inorganic phosphate and was inhibited by the coenzyme NAD^+.

Effect of Maillard reactions on the functional properties of proteins

Disappointingly, little work has ever been conducted to study the effect of Maillard reactions on the functional properties of proteins. It is accepted that glycosylation reactions both *in vivo* and in food systems result in changes in the functional properties of proteins although not all such reactions are detrimental. For example, S.E. Hill (University of Nottingham, UK) showed that the minimum concentration for the heat-induced gelation of bovine serum albumin was decreased by the Maillard reaction. Similarly, T. Aoki and coworkers (Kagoshima University, Japan) presented a poster that showed the improvement of several functional properties of β-lactoglobulin following Maillard reaction with glucose-6-phosphate, including heat stability and emulsifying capacity.

Effect of Physicochemical state of the food system

The effect of glass transitions on Maillard reactions has received relatively little attention to date. However, such changes in the physicochemical state of food systems can have a significant effect on Maillard reactions particularly in the region of the glass transition temperature (T_g) where the activation energy may be quite high, and even small shifts in temperature can result in large changes in reaction rate. Most of the work in the area has been conducted by M. Karel and coworkers at Rutgers University. In view of our improved understanding of the influence of glass transitions on the rate of Maillard reactions, Prof. Karel advocates the careful consideration of the physicochemical structure of food systems before applying a mathematical model to describe the kinetics of reactions. Thus, the WLF (Williams-Landel-Ferry) model will be more appropriate for some systems than the Arrhenius model and vice versa. Clearly, common sense will suggest that a curved Arrhenius plot is indicative of the lack of suitability of the model and that a model more appropriate to the physicochemical properties of the system should be applied.

Inhibition of Maillard reactions

The search for effective safe inhibitors of the Maillard reaction is stimulated both by the need to prevent the harmful *in vivo* reactions in diabetics and also to prevent undesirable Maillard reactions in foods. Additional impetus is provided by the need to find alternatives to the use of sulfites in some food products. N. Kinae (University of Shizuoka, Japan) showed that green tea extracts could inhibit Maillard reactions *in vivo* in diabetic rats which was attributed to the catechin compounds present. However, it is unclear if such an approach could also be used to inhibit Maillard reactions in foods; organoleptic considerations are a major barrier to the use of tea extracts as food ingredients. Thus, compounds such as cysteine appear to be among the best alternatives to the use of sulfite in foods at present.

The formation of free radicals in the Maillard reaction was shown to be inhibited by the naturally occurring flavones ellagic acid, gallic acid, ferulic acid and syringic acid (S. Djilas and B. Milic, University of Novi Sad, Yugoslavia). Interestingly, such flavones also showed inhibitory activity towards the formation of 2-amino-3,4,8-trimethylimidazo(4,5-*f*)quinoxaline in a model system.

Maillard reactions *in vivo*

The conference clearly established that food scientists and health scientists have much to learn from each others' work on Maillard reactions. Much time was devoted to the formation of advanced glycosylation end products (AGEs) *in vivo* and mechanisms by which the body can deal with them. Such AGEs are continuously formed in tissues both in healthy individuals and in diabetics; macrophages are responsible for recognizing and removing the glycosylated protein and for initiating the process of tissue remodeling. The drug aminoguanidine appears to have much potential in inhibiting both advanced glycosylation and crosslinking reactions (A. Cerami, The Picower Institute, New York). This differs from the activity of sulfur dioxide, a well known inhibitor of Maillard reactions, which inhibits browning but not protein crosslinking. D. Stern (Columbia University, New York) presented a particularly interesting paper on the isolation and cloning of a receptor for AGE (referred to as RAGE). It was emphasized that the RAGE is not a scavenger receptor and appears also to bind to an as yet unknown non-AGE peptide *in vivo*. The receptor has been shown histologically in samples of cerebral cortex, nerve cell bodies and axons, and in cells in culture.

Interestingly, the presence of a free reducing sugar is not necessary for advanced glycosylation reactions to proceed. H. Vlassara (The Picower Institute, New York) showed that the injection of peptide-bound AGEs into mice, rabbits or rats led to an increase in tissue AGE, tissue crosslinking and vascular permeability - classic secondary pathological sequellae of diabetes. Such findings pose a question as to the effect of dietary peptide-bound AGEs on dialysis patients.

Food Safety

The final day of the conference was devoted to food safety but several papers were presented during other sessions that provided insights into the effect of Maillard reactions on the wholesomeness of foods. The mutagenicity of cooked foods is still an area of much research activity and the number of heterocyclic mutagenic amines isolated from cooked foods and model systems is increasing steadily. K. Wakabayashi (National Cancer Research Institute, Japan) reported the isolation of three new mutagens from model systems and from cooked meat extract: an hydroxyphenyl imidazopyridine (4-OH-PhIP), a carboxymethyl derivative of dimethylimidazoquinoxaline (4-CH_2OH-8-MeIQx), and a trimethylimidazo- quinoxaline (7,9-DiMeIgQx). The paper also reported that the mutagen PhIP was the most abundant heterocyclic amine in cooked foods (0.56-69.2 ng/g) followed by MeIQx (0.64-6.44 ng/g). The human intakes of MeIQx and PhIP were estimated to be 0.2-2.6 μg/person/d and 0.1-13.8 μg/person/d, respectively, which compares with the intakes of other dietary carcinogens such as benzopyrene.

Two general approaches may be used to minimize the mutagenicity of cooked foods. One approach is to minimize the formation of mutagens by manipulation of process conditions. K. Eichner (University of Münster, Germany) reported on studies to investigate the technological conditions favoring the formation of amino imidazoazarenes during the processing of meat products. It was concluded that since the formation of MeIQx is increased in meat extract at lower water activities, the final stages of drying should be conducted at lower temperature *in vacuo* to minimize the formation of mutagens.

A second approach to minimizing the mutagenicity of cooked foods is to exploit the antimutagenic activity of some Maillard reaction products. G-C. Yen (National Chung Hsing University, Taiwan) reported that Maillard reaction products of xylose and lysine reduced the mutagenicity of IQ towards *Salmonella typhimurium* TA 100. This activity of Maillard reaction products was attributed to their interaction with proximate metabolites of the mutagen to form inactive adducts.

Future Prospects

A great advantage of the Hodge scheme (Hodge, 1953) is its simplicity. However, several papers at the recent meeting served to highlight the deficiencies in the scheme, particularly with respect to the participation of oxygen in Maillard reactions and the role of the Amadori rearrangement in the overall reaction. Perhaps the potentially most important paper of the meeting was presented by J.W. Baynes (University of South Carolina) in which he suggested, based on studies of a rat tail collagen model system, that glyoxal is the major oxidation product of glucose. It is proposed that glyoxal then reacts directly with lysine resulting in the formation of carboxymethyllysine via a Cannizaro reaction. Baynes went on to suggest that the Amadori route might contribute to less than 10% of the glycoxidation products in Maillard reaction systems. Does the chemistry of glycoxidation reactions in foods behave in a similar way? In view of such new findings, perhaps it is time to employ a paradigm shift in our understanding of Maillard reaction chemistry towards a more detailed comprehensive scheme of reactions than that which has been used successfully for the past forty years since the seminal work of Hodge.

Finally, beer drinkers may take solice from the fact that brown products of advanced Maillard reactions in dark beer are potent inhibitors of the growth of *Streptococcus mutans in vitro* and may act to prevent tooth decay *in vivo* (M. Murata and coworkers, Ochanomizu University, Japan).

References

Eriksson, C. (1981) (Ed.), *Maillard Reactions in Food, Prog. Food Nutr. Sci.* 5, 1-501.

Fujimaki, M., Namiki, M. and Kato, H. (1986) (Eds), *Amino-Carbonyl Reactions in Food and Biological Systems*, Elsevier, Amsterdam.

Finot, P.A., Aeschbacher, H.U., Hurrell, R.F. and Liardon, R. (1990) (Eds), *The Maillard Reaction in Food Processing, Human Nutrition and Physiology,* BirkhŠuser Verlag, Basel.

Hodge, J.E. (1953), Dehydrated foods: chemistry of browning reactions in model systems, *J. Agric. Food Chem.* 1, 928-943.

Waller, G.R. and Feather, M.S. (1982) (Eds), *The Maillard Reaction in Foods and Nutrition,* American Chemical Society, Washington.

Chemistry

Influence of pH on the Oxygen Consuming Properties of Heat-Treated Glucose-Glycine Systems

M. Anese*, M.C. Nicoli and C.R. Lerici; *University of Udine, Dept of Food Science, via Marangoni 97, 33100 Udine (Italy)*

The inhibiting effect of Maillard reaction products (MRP) on enzyme activity, their antibacterial properties on some food-poisoning micro-organisms and their ability to slow down lipid oxidation reaction are well known.

The antioxidant properties of MRP are generally attributed to their ability to act as chain-breakers and to chelate metals; however the effect of MRP on enzymatic activity and on microbial growth suggests other mechanisms involving oxygen consumption.

The aim of the present note was to study the oxygen consuming properties of heat-treated glucose-glycine model systems. In particular, the influence of initial pH on the oxygen uptake of glucose-glycine aqueous solutions heated at 90°C for different times was followed using a polarographic method. The oxygen uptake of MRP was related to glucose and glycine residual concentration values chosen as indicators of the extent of browning of the model system. Results confirmed the effectiveness of MRP as oxygen consumers and, in particular, a dramatic increase in oxygen uptake was observed at pH values ranging from 8.0 to 10.0.

Improvement of Functional Properties of β-Lactoglobulin by Glucose-6-Phosphate Conjugation

Takayoshi Aoki[1], Tomoko Matsumoto, Yoshitaka Kako, Yasuko Kato[2] and Tsukasa Matsuda[3]; *[1]Kagoshima University, Department of Biochemical Science and Technology, Kagoshima 890, Japan, [2]Tokaigakuen Women's College, Nagoya 468, Japan, [3]Nagoya University, Department of Food Science and Technology, Nogoya 464-01, Japan.*

Phosphate groups of casein play an important role in generating functional properties such as foaming and emulsifying properties, calcium binding and interaction with calcium phosphate. It is well known that phosphorylation of food proteins improves functional properties. However, chemically phosphorylated proteins are difficult to be accepted as food, and enzymatic phosphorylation costs a great deal since it needs ATP. In the present study, β-lactoglobulin was modified with glucose-6-phosphate (G6P) through the Maillard reaction at 50°C and 65% relative humidity for various periods (0.5-3 days). The phosphorus content of modified β-lactoglobulin was more than 1% after 1 day reaction, which is higher than that of casein. The temperature at which half of intact β-lactoglobulin precipitated at pH 7.0 was 83.5°C, while no modified β-lactoglobulin precipitated even at 100°C. Emulsifying properties of β-lactoglobulin were improved by conjugation with G6P. β-Lactoglobulin-G6P conjugates have the ability to bind calcium and phosphate, and solubilize calcium phosphate. β-Lactoglobulin-G6P conjugates also inhibited the precipitation of calcium phosphate from a supersaturated solution of dicalcium phosphate. Thus, the functional properties of β-lactoglobulin were improved by G6P conjugation.

Low Molecular Weight Coloured Compounds Formed in Xylose-Lysine Model Systems Heated with and without pH Control

Anton Apriyantono[1] and Jennifer M. Ames; *Department of Food Science and Technology, University of Reading, Whiteknights, Reading RG6 2AP, UK*

Aqueous solutions of xylose (1 m) and lysine monohydrochloride (1 m), initial pH 5.27, were refluxed for 1 h, either without pH control (final pH 2.83) or by maintaining the pH at 5.0 throughout heating by the addition of sodium hydroxide solution. The ethyl acetate extractable components were obtained from each system and analysed by TLC and HPLC with diode array detection. Five and 14 coloured compounds, respectively, were detected from the model system heated with and without pH control. Only 2 were common to both systems and one of these was identified as 2-furfurylidene-4-hydroxy-5-methyl- 3(2*H*)-furanone. Two other colored compounds were isolated from the system heated without pH control. The electronic absorption and PMR spectra indicate that these compounds possess related structures, each containing 2 terminal furan rings. The molecular formula of one of them was shown to be $C_{25}H_{19}O_7N$ by high resolution FAB-MS.

Effect of Autoxidation of Some Edible Oils in the Reaction Between Lysine and Glucose

Anna Arnoldi; *Università di Milano, DISMA, Via Celoria 2, 20133 - Milan, Italy.*

It has been suggested that fats, present in many foods, can be oxidized to aldehydes and ketones which could interact in the Maillard reaction, but only a few studies have focused on this topic. For obtaining better insight on the effect of lipid autoxidation, model systems containing glucose and lysine were heated with large amounts of edible oils in the presence of a chain radical reaction modulator: propyl gallate as an antioxidant and dibenzoyl peroxide or N,N-azobis-isobutyronitrile (AIBN) as a radical initiator. The effects were investigated through analysis of the volatiles separated by a Likens Nickerson apparatus. Models in the presence of radical chain modulators, but without any oil, indicate a direct effect especially of radical initiators on the Maillard reaction with a general increment in the formation of furans, pyrroles and pyrazines. In the presence of oils, extra virgin olive oil, commercial olive oil (i.e. a mixture of refined and virgin), and refined corn oil, the effect is less pronounced because the radical initiators are involved mainly in the autoxidation of the oil, which appears in a dramatic increase in aldehyde formation.

Diffusion and Maillard Reaction in an Agar-Microcrystalline Cellulose Matrix During Dehydration

F. Gogus, B. L. Wedzicha and J. Lamb; *Procter Department of Food Science, University of Leeds, Leeds, LS2 9JT, U.K.*

Diffusion of water and Maillard reactants (glucose and glycine), in a model food matrix consisting of 22% (w/w) microcrystalline cellulose and 1% (w/w) agar, was studied during dehydration at 80°C in a vertical flow air drier (air velocity 1.6 ms^{1}) for 30-130 min. Samples were of cylindrical geometry. Transport of water was measured gravimetrically while specific migration of the reactant was measured with the help of ^{14}C-labelled compounds. Here we report the effect of each reactant on the mobility of the other, the formation of ^{14}C-labelled reaction product retained on a Dowex 50W-X4 (H^+) cation exchange resin and the effect of sulfite on the concentrations of the latter reaction products; in all cases the data are reported as a function of radial distance within the material.

It is found that the liquid surface recedes into the solid and that solutes migrate towards the surface during dehydration indicative of a capillary-porous material. The mutual effect of the reactants on their migration rate was very small. When the reactant was ^{14}C-labelled glucose, the concentration of ^{14}C-labelled products retained on the cation exchanger increased very markedly close to the surface of the sample, particularly towards the end of the drying period. Similar behaviour was observed in the presence of sulfite, In this model system the reaction between glucose and glycine is localised within a depth of <1 mm from the surface.

Adaption of Maillard Reaction on Keratine Type Proteins with a Special Focus on the Reactions of Glucose Based Crown Ethers

L.Trezl*, P. Bakó, V. Horvath, I. Rusznak, L. Toke; *Technical University of Budapest, Department of Organic Chemical Technology, Budapest H-1521, P.O. Box 91*

During our wool-keratine treatments with crown ethers, the crown ethers penetrated into the wool-fibers by breaking up the salt linkages, by loosening and distancing from one another the peptide chains and forming complexes with certain amino acid segments (polar-NH_3 group e.g. in lysine) make wool more available for dyes.

The research team in our Department has produced unique and special crown ether; for the chemical modification of wool. A considerable new effect was noticed beside the "crown loosening" effect, forming cross linkages in the wool with certain type of glucose based crown ethers e.g. from α-D-glucopyranosido-18-crown-6.

Since the treatment with 18-crown-6 showed favourable results in wool properties (increased tensile strength-, dye uptake-, density etc.) our tests were extended to include mono and disaccharides containing glucosidic hydroxyl groups and they were used as wool modifying agents.

The investigations proved that the reducing saccharides built up into the wool-matrix (not in helical, crystal regions) and were reacted with lysine-NH_2 side groups. Below 70°C the reaction is slow but at 100°C the reaction is very fast and is accompanied by a strong browning phenomenon. The results indicate that the reaction between reducing saccharides and wool is nothing else than the reaction discovered by Maillard 80 years ago.

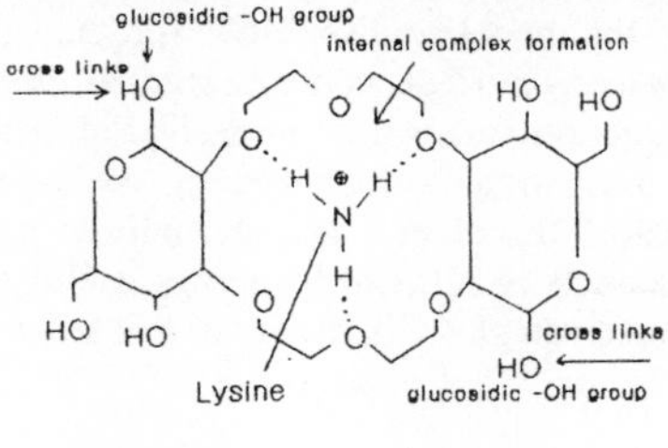

Deamination of Basic Amino acids in Protein Using Active Carbonyl Compounds Produced by the Maillard Reaction

S. Gopalan*, A.T. Gracy, A. Srinivasan; *Protein Research Unit, Loyola College, Madras, India.*

Casein, the major protein fraction of milk, interacts with oxidised ascorbic acid to yield rapid red colouration in an ethanol suspension. An apparent decrease in the free amino groups, as determined by TNBS method, reaffirms the formation of red colour. Further the electrophoretic pattern of casein changed significantly upon incubation with oxidised ascorbic acid.

The red pigment extracted from DHA-Casein system showed an absorption spectrum. TLC, Rf value and hydrolysed products were identical to those of the product produced by interaction of DHA with α-amino acid. However, the same kind of red pigmentation arises when DHA was treated with n-hexylamine.

These results indicate that the free amino group present in basic amino acids of the polypeptide chain of the protein could easily be deaminated by the amino carbonyl reaction with active carbonyl compounds (DHA).

Determination of N^{ϵ}-Carboxymethyllysine (CML) in Heated Model Mixtures and Several Foods

J. Hartkopf, Ute Albrecht, Cordula Pahlke, Inke Sievert and H.F. Erbersdobler*; *University of Kiel, Department of Human Nutrition and Food Science, Duesternbrookcer Weg 17, D-24105 Kiel, Germany.*

CML was analysed in model mixtures of lysine-monohydrochloride and glucose or ascorbic acid, that were heated at 100°C for different times. To evaluate the influence of reaction parameters on the formation of CML, additional ingredients were added to these model samples. Test mixtures were also prepared by adding glucose or sodium ascorbate and additionally sodium nitrite and sodium hydrogen phosphate to a meat homogenate normally used for sausage production. Finally, selected commercial foodstuffs were analysed for CML. In order to prevent formation of CML from fructoselysine during hydrolysis, all analyses in the model mixtures were performed without hydrolysis and in the foodstuffs after borohydride reduction. In addition to CML, furosine was also determined after hydrolysis with 7.8 M HCl using the recently available pure furosine standard. All CML values obtained by this procedure were significantly lower than CML concentrations in foodstuffs described before. In a sausage, for example, only 0.29 mg CML/l00g were found using the borohydride reduction method. Without reduction, 1.42 mg CML/100g sample were determined. The results showed an influence of ascorbic acid, phosphates and nitrites on CML formation in the model studies. However, under practical conditions, only ascorbic acid and phosphates increased the CML content. Nitrites did not increase CML in meat products when used in concentrations similar to what is applied in meat processing.

It is concluded that CML formation is enhanced by several additives like phosphates increasing in this way the expected CML content. This partially explains the relative high CML contents found in canned meat products. CML can be a very useful additional indicator of heat damage, especially in severely heated samples where furosine is known to be less reliable.

Characterization of Metal Chelating Compounds in Soluble Coffee

Seiichi Homma and Masatsune Murata; *Ochanomizu University, Department of Nutrition and Food Science, Ohtsuka 2-1-1, Bunkyo-ku, Tokyo 112, Japan.*

The inhibitory effect of coffee on non-heme iron and zinc absorption in humans has been reported. In this study metal Chelating compounds were separated from coffee and characterized.

Soluble coffee was dissolved in water, added with $ZnCl_2$ or $FeCl_2$ solution to a final concentration of 10 mM. The precipitate was centrifuged, dissolved in 1% ammonia solution, and lyophilized. The results showed that the metal content in each lyophilized material was 0.58% for Zn or 0.36% for Fe, and they were regarded as the complex of coffee with Zn(II)[Zn-Ap] and Fe(II)[Fe-Ap], respectively. The yield of the complex was 1.59% for Zn and 1.58% for Fe.

The complex was applied on Amberlite IRA-410 and then on Amberlite IR-120B columns. Metal ions of complex with low stability constant were liberated and adsorbed on the ion exchange column. Those with high stability constant were not adsorbed on the ion exchange columns, and were further purified on a cellulose column, which was eluted with n-propanol and 1% ammonia. Black-colored complex, Ap-III and Ap-IV fractions, were separated from the Zn-Ap and the Fe-Ap. The yields of Ap-III and Ap-IV fractions from 1 kg of coffee were 1.19g and 0.11g for Zn, and 0.66g and 0.03g for Fe, respectively. The molecular weight was estimated to be Ca 36KD and 50 KD by HPLC using protein standard, and the Fe content was 4.62 mg/g and 7.30 mg/g for Fe-Ap-III and Fe-Ap-IV, respectively. The addition of $ZnCl_2$ or $FeCl_2$ to their respective Ap-IIIs changed them to Ap-IVs on the cellulose column chromatography. The amount of released Fe from Fe-III was about 60% and 10% with EDTA and o-phenanthroline, respectively. Zn in the Zn-Ap-III was almost released with EDTA suggesting the stability constant was higher in the Fe-Ap-III than in Zn-Ap-III. The Ap-IIIs of both metals were chemically characterized.

Substrate Specificity Fructosyl-Amino Acid Oxidase and its Application for the Enzymic Detection of Amadori Products

Tatsuo Horiuchi, Toshiko Kurokawa and Narimasa Saito; *Noda Institute for Scientific Research, 399 Noda, Chiba 278, Japan*

We have found a new enzyme which is concerned in the decomposition of Amadori products from a microorganism. The enzyme named Fructosyl-amino acid oxidase has been isolated from the extract of *Corynebacterium* sp. 2-4-1 as rhombic crystals. Fructosyl-glycine was decomposed by the enzyme reaction to a D-glucosone and glycine concomitant with consuming an oxygen and forming a hydrogen peroxide. Optimum pH for the enzyme reaction was about 8.3 in phosphate buffer. Fructosyl-α-L-amino acid was the most susceptible substrate to the enzyme. Apparent Michaelis constant for fructosylglycine was about 0.7mM. Fructosyl-glycylglycine (peptide form Amadori product) showed fairly diminished susceptibility. Fructose in the structure of Amadori products was shown as the substrate having the most favorable configuration in comparison with that of other ketose. Fructosyl-ϵ-amino acid and fructosyl-L-amino acid were not oxidized. Enzymic detection of Amadori products was done with an oxygen electrode. In the model system consisting of D-glucose and α-L-amino acid, fructosyl-amino acid was detected easily with no interference owing to the unreacted glucose or amino acid. But with the progress of Maillard reaction, non-enzymic oxygen consumption increased its intensity level and began to disturb enzymic detection. When the analysis was applied to soy sauce, obtained value was fairly small in comparison with that of other chemical methods.

Meanwhile, an enzyme having the ability to decompose ϵ-fructosyl-L-lysine has been found from *Aspergillus*. This enzyme has not been applied for the detection of Amadori products because of the difficulty of enzyme purification. A comparison of the substrate specificity between two enzymes may be mentioned briefly.

The Influence of Free Radicals and Oxidized Fats on the Formation of Food Mutagens in a Model System

Maria Johansson* and Margaretha Jägerstad; *Department of Applied Nutrition and Food Chemistry, University of Lund, P.0. Box 124, S-221 00 LUND, Sweden*

Food Mutagens are suggested to be formed from Maillard reaction products like pyrazines, pyridines and aldehydes, which react with creatinine, during cooking of meat and fish products. The major food mutagens contain a 2-aminoimidazo-group linked to either a quinoline, a quinoxaline or a pyridine. In this study the effects of oxidized fats, iron and tocopherol content on the formation of MeIQx was investigated using a model system. The precursors; creatinine, glycine and glucose, were heated, at 180°C for 10 and 30 minutes, with the addition of five gradually oxidized deep-frying fats or different free radical donors ($FeCl_3$, $FeSO_4$ or hemoglobin ferrous). The food mutagens were extracted from the heated mixtures using the solid-phase extraction method described by Gross (1990) and were quantified using HPLC with photodiode-array detection. The mutagenic activity of HPLC fractions were determined using the Ames test and the identities were confirmed by MS. IQx, MeIQx and 4,8-DiMeIQx were formed in all heated mixtures containing any of the fats and iron or not. The amount of MeIQx formed was significantly increased when heating for 30 min at 180°C compared to 10 min. Addition of $FeSO_4$ significantly increased the formation of MeIQx by more than doubling. The oxidation degree of the fat had some effects on the formation of MeIQx. If $FeSO_4$, $FeCl_3$ or hemoglobin ferrous were added to the model system the amount of MeIQx was increased compared to the amount formed if no free radical donor was added.

Formation of Oligosaccharide Amadori Compounds and Their Heat Degradation

Lothar W. Kroh; *Technisch Universitat Berlin, Institut fur Lebensmittelchemie, Invalidenstr. 42, D-1040, Berlin.*

The formation of oligosaccharide Amadori compounds from starch and amino acids has been proved. The thermally induced decomposition (180-300°C) has been compared to those of monosaccharide Amadori compounds. The saccharide component influences the thermal process of decomposition.

Furans and furanones have been detected as predominant degradation products, the main one being 2(5H)-furanone. We suggest a reaction pathway for the formation of oligosaccharide Amadori compounds and the formation of different products via the Maillard reaction which includes 1,6-anhydroglucose.

Stereo Control of Control of Maltol Formation in Maillard Reaction

V. Yaylayan* and S. Mandeville; *McGill University, Department of Food Science and Agricultural Chemistry 21,111 Lakeshore Road, Ste. Anne de Bellevue, Quebec, Canada, H9X 3V9*

Maltol and hydroxymethylfurfural (HMF) are considered indicator compounds for the occurrence of 2,3- and 1,2-enolizations in solutions of reducing sugars. However, maltol has been mainly detected in the decomposition mixtures of reducing disaccharides and monosaccharide solutions containing amino acids heated at high temperatures but not in solutions of monosaccharides alone. On the other hand, 2,3-dihydro-3,5-dihydroxy-6-methyl-4-(H)-pyran-4-one (3), can be detected in heated solutions of reducing monosaccharides alone, with no evidence of maltol formation, compound 3 therefore cannot be considered as the main precursor of maltol. This can be due to extensive tautomerizations and unfavorable dihedral angle (67°) for E2 elimination of water. Based on molecular modeling and energy minimization studies using the MM2 force field, a stereochemical control for the formation of maltol is proposed, as shown below.

(E)-stereoisomer (2) ← R = Sugar — (1) — R = H → (R)-stereoisomer(3)

Maltol

A Study on the Structures and Reactivities of Unstable Intermediates of Amino-Carbonyl Reaction by Semi-Empirical Molecular Orbital Method

T. Kurata and Y. Otsuka[1]; *Institute of Environmental Science for Human Life, Ochanomizu University, Tokyo, Japan, and [1]Faculty of Education, Tottori University, Tottori, Japan.*

Although many studies have been carried out on the mechanisms of amino-carbonyl reaction, the detailed reaction mechanisms are still not clear owing to the difficulties in the isolation and chemical characterization of the important unstable reaction intermediates. For instance, the enol form of Amadori rearrangement product, one of the most important intermediates of amine-sugar reaction, have never been isolated and, therefore, we have not yet been able to get enough and reliable information about its structure and reactivities from experimental data. In this study, we used a molecular orbital (MO) method to get an insight into the molecular structure and orbitals of unstable intermediate compounds of amino-carbonyl reactions, including the enol form of Amadori rearrangement products and others. Calculations were made by using a semi-empirical MO program, MOPAC ver.5.0 (or ver.6.0), and PM3 hamiltonian was mostly used. From the results of MO calculation made on the enol form of fructoseglycine, Amadori rearrangement compound produced by the reaction of glucose with glycine, its heat of formation was estimated to be about -344 Kcal/mol. The optimized structure, energy levels of orbitals, electron densities of atoms of this molecule were also obtained. In the case of the enol form of Amadori-compound produced by the reaction of glyceraldehyde with glycine, the heat of formation and energy level of Highest Occupied MO were estimated to be ca. -165 Kcal/mol and -9.92 eV, respectively. Its optimized structure showed that both oxygen and nitrogen atoms of the enaminol group took a coplanar orientation.

Structures and Reactivities of Reaction Products Formed by the Reaction of L-Ascorbic Acid with an Amino Acid in the Presence of Oxygen

T. Kurata, N. Miyake and Y. Otsuka[1]; *Institute of Environmental Science for Human Life, Ochanomizu University, Tokyo, Japan, and [1]Faculty of Education, Tottori University, Tottori, Japan.*

The oxidation process of an ene-diol to its corresponding di-oxo compound, such as the oxidation of L-ascorbic acid (ASA) to dehydro-L-ascorbic acid (DHA), is a very popular reaction process observed in various biological systems and also in foods. The resulted di-oxo compounds are usually typical electrophiles and react with an amino acid very easily. Thus, in the presence of oxygen, ASA and its degradation products, such as DHA and 2,3-diketo-L-gulonic acid (DKG), the hydrolysis product of DHA are known to be involved in amino-carbonyl reactions observed in both food and biological systems, however, the details in their reaction mechanisms are still not well understood. In this study, we used a semi-empirical molecular orbital (MO) method, to get more information about the structures and reactivities of various reaction products, especially those of the rather unstable reaction intermediates, such as, monodehydro-AsA, oxidized form of L-scorbamic acid (SCA), and others. Most calculations were made by using a semi-empirical MO program, MOPAC ver. 5.0 (or ver. 6.0), and PM3 hamiltonian was mostly used. Some of the reaction products, such as K-salt of DKG, SCA and others were synthesized by the usual methods reported in our earlier papers. The optimized structure of AsA obtained by MOPAC calculation was very similar to the one calculated by an initial MO method, and the heat of formation of AsA was estimated to be ca. -234 Kcal/mol. The optimized structure of ASA monoanion, and monodehydro-ASA showed some similarities to that of DHA. The energy levels of orbitals, ionization potentials and optimized structures of other unstable reaction intermediates, such as monodehydro-AsA, dehydro-SCA were also obtained.

A Preliminary Study on the Products Arising from the Reaction of Protected Lysine and Glucose

A. Lapolla*, D. Fedele, C. Gerhardinger, L. Baldo, G. Crepaldi, [1]R. Seraglia, [1]S. Catinella, [1]P. Traldi, [1]R. Bertani, B.E. Rizzi; Inst. of Int. Medicine, [1]C.N.R., B Italfarmaco, Padova, Italy.

The interaction between glucose and ϵ-amino groups of lysine present in the protein chains leads to the formation of Advanced Glycation Endproducts responsible for the tissutal modifications characteristic of diabetic complications. The structural identification of such products is a topic of wide interest either from the biochemical or from the medical point of views, so we undertook a systematic study on the products arising from the reaction of protected lysine with glucose by means of different analytical techniques (gel permeation chromatography, silica gel column chromatography, FAB mass spectrometry, HPLC/MS,^{1}H and ^{13}C-NMR). By gel permeation chromatography, a molecular weight distribution up to 200,000 Da was found: a major peak with components corresponding to mean molecular weights of 870, 1228 and 1744 Da is evidenced. Other quite abundant components of the reaction mixture have mean molecular weights of 172 and 406. The higher molecular weight components are significant, proving the occurrence of extensive polymerization. By further silica gel column chromatography, six different fractions were obtained and we focused our attention on four of them, exhibiting spectroscopical properties typical of the advanced glycated proteins. FAB mass spectrometry, HPLC/MS and ^{1}H and ^{13}C-NMR allowed identification of a number of possible structures, all containing furane and tetrahydrofurane moieties, which are the object of the present communication.

Cu(II) Chelating Activity of Glucose-Lysine Model Melanoidin

Gapsoon Moon*, [1]Masatsune Murata and [1]Seiichi Homma; *Kunie University, Department of Nutrition and Food Science, 18-3 Ubang Dong, Kimhae, Kyong Sanganam-Do, Korea.* [1]*Ochanomizu University, Department of Nutrition and Food Science, Ohtsuka 2-l-1, Bunkyo-ku, Tokyo 112, Japan.*

The effect of chelating activity of melanoidin on antioxidative activity was investigated. Model melanoidin was prepared by reflux on heating of 2M glucose and 2M lysine for varied times and separated into dialyzable and nondialyzable fractions. The melanoidins were chromatographed on a Cu(II)-chelating Sepharose 6B column, and underwent antioxidative assay against linoleic acid using absorbance at 234nm as an oxidative index of the fatty acid.

The model melanoidin was separated into three fractions by Cu(II)-chelating Sepharose 6 column chromatography by falling pH gradient. The first and second melanoidin components appeared in pH 7.9 and 7.9-3.4 eluates, respectively and the third was in the eluate with EDTA. The non-dialyzable melanoidin was found to be eluted only with EDTA. Each melanoidin component in the eluates was found to be combined with Cu. Since the first melanoidin component was not retained on the Cu-column, it was indicative of the presence of melanoidin with stronger chelating activity than imino acetic group on the chelating Sepharose. The longer the heating time of melanoidin, the larger the amount of combined Cu was. Gel filtration of each melanoidin component showed that molecular weight of the third melanoidin component was larger than the first and second melanoidin components. Antioxidative activity of the model melanoidin was stronger in the low molecular (m.w.4000-7000) fraction than in the high.

The Relationship Between the Coloured Compounds Present in the Pressed Liquor of Cane Sugar and Those Formed in Maillard Reactions, in Alkaline Degradation of Sugars, and in Caramelisation

Javad Reramat and Harry E. Nursten; *University of Reading, Department of Food Science and Technology, Whiteknights, PO Box 226, Reading RG6 2AP, UK.*

The coloured compounds of pressed liquor (PL) have been compared with those formed in Maillard reactions from alanine and glucose (MAG) and alanine and fructose (MAF), in the alkaline degradation of glucose (ADG) and of fructose (ADF), and in caramelisation of sucrose (CAR) under conditions simulating those obtaining in carbonation. The coloured compounds were isolated by adsorption on non-ionic resins (Amberlite XAD-7 and XAD-16, in series) at pH 2.5, the columns being washed with 0.01 N HCl, and eluted with 0.01 N HCl in methanol. The coloured extracts were evaporated to dryness and analysed by spectrophotometry and TLC. Distribution of absorbance units measured at 420 nm between unabsorbed material and XAD-7 and XAD-16 extracts showed CAR to be closest to PL, but the recovery of absorbance units measured at the maximum in the range 300-320 nm for the XAD-7 extracts were PL, 54%; MAG, 62%; and MAF, 20%; ADG, ADF, and CAR have no maximum in this range. The XAD-7 extracts of ADG and ADF were exceptional in exhibiting maxima at 440 nm. On TLC of the XAD-7 extracts in BAW, PL gave 10 bands, of which 7 appeared also in MAF and in MAG and 4 in ADF,in ADG and in CAR. Overall it seems that the coloured compounds in PL bear most resemblance to those formed in Maillard reactions.

Evaluation of the Advanced Maillard Reaction in Dried Pasta

Pierpaolo Resmini and Luisa Pellegrino; *University of Milan, Department of Food Science and Technology, Via Celoria, 2, Milan 20133, Italy*

Among water soluble Maillard compounds 2-acetyl-3-D-glucopyranosyl-furan (AGPF) was quantified in dried pasta by IP-RP HPLC with 280 nm detection. Peak identification was achieved both by using the [^{14}C] labelled molecule obtained in model solutions of maltose with free amino acids and by FAB-MS. Being negligible (<1 ppm) in spaghetti dried under low or medium temperature conditions but significant (up to 20 ppm) in some commercial samples, AGPF formation is related to severity of heating and availability of free amino acids, mainly glutamine. Relationships were found between amount of the colourless AGPF and redness (a*red index) developed in pasta dried under either experimental ($r^2=0.99$) or industrial ($r^2=0.81$) conditions. Taking into account the properties of AGPF precursor 1-desoxyketulose, the extent of this type of advanced MR could explain undesirable sensorial changes appearing in HT dried pasta.

The advanced MR involving protein-bound lysilketoses degradation was studied on model doughs of purified gluten with [^{14}C] labelled sugars added and on industrial pasta. Despite the lower rate of furosine accumulation in the last steps of HT drying, only small amounts of free radioactive compounds were split from labelled maltulosyllysil- or fructosyllysil-residues of gluten. Lysilpyrrolealdehyde (ϵPL) was recognized as a main derivative of lysilketoses residues. Under the adopted chromatographic conditions ϵPL is eluted at 15-16 min, 9-10 min being the retention time of tryptophan. This interference-free separation allows routine quantification of ϵPL in several cereal products. Due to the close response factor, the commercially available 2-acetylpyrrole can be used as an external standard. ϵPL formation is highly enhanced if pasta drying includes temperatures close to 80°C or higher, when pasta moisture is close to 15% or lower (0.8-0.7 aw). Concentration of reducing sugars and glucose/maltose ratio in semolina and storage conditions of pasta are additional factors involved. Considering the anti-nutritional properties reported for ϵPL and the wide range of values (0-40 ppm) found in commercial spaghetti, a better control of the advanced MR occurring in this basic food is advisable.

The Maillard Reaction in Pasta Drying: Study in Model Systems

A. SenSidoni[1], C.M. Pollini[2], D. Peressini[1], P. Sari[2]; [1]*Department of Food Science, University of Udine, via Marangoni, 97-33100 Udine Italy;* [2]*Pavan-Mapimpianti S.p.a., via Monte Grappa, 8-35015 Galliera Veneta (PD) Italy*

Colour and nutritional pasta quality can be reduced by nonenzymatic browning during drying. The Maillard reaction was studied in model systems starch/glycine (33.3% glycine), starch/glucose /lysine (2% glucose; 1% lysine) and flour/water in order to find out which factors may be important during pasta processing. The main variables of drying temperature, time and water activity were considered.

The rate of Maillard reaction was influenced by initial water activity of model system and by water activity changes during heat treatment. Different indicators such as headspace CO_2 and O_2, volatiles concentration, colour, optical density and pH showed a rise of browning as initial water activity increased (from 0.30 to 0.95). A control of reducing sugars formation during kneading and extrusion could improve pasta quality.

Influence and Impact of Nonenzymic Browning Reaction on Proteins in Milk and in Indigenous Dairy Products of India

Akur Srinivasan* and S. Gopalan; *Ramabadran-Protein Research Unit Loyola College, University of Madrea, Madras-6000 034, India*

Milk and dairy products serve as best substrates and good biochemical media for undertaking studies on nonenzymic browning reaction. Maillard browning due to the interaction of whey proteins and lactose alters physicochemical properties of processed milk, nutritional status of milk proteins and affect shelf life of dairy products. In India, where cow and buffalo milk is consumed after boiling and dairy products like Koah and Koah based sweets are prepared from continuously heated and evaporated milk in presence of sweetening agents, the problem of Maillard Browning assumes special significance and importance. Attempts have been made to evaluate the intensity of browning in heated cow and buffalo milk samples in presence of sweetening agents like glucose and sucrose of different concentrations to study the influence on physicochemical properties of milk and on the isolated milk proteins and their roasted samples at 393K at different intervals of time. Experimental results indicate that the intensity of browning increases with increase in temperature, time of heating and concentration of sweetening agents. Buffalo milk showed greater intensity of browning and the presence of glucose and milk lipids enhanced the intensity of browning in heated milk samples. Browning was found to reduce protein content and its nutritional status. Reddish brown Melanoidin pigments from milk proteins containing sweetening agents were found to exhibit metal ion absorbing properties and antioxidant effects. Roasted samples of milk proteins were analysed for browning and composition.

Foodstuffs Melanoidines Quantitative Isolation Methods

Margarita J. Kintcheve, Tzvetan D. Obretenov; *Higher Institute of Food and Flavour Industries, Department of Organic Chemistry, 26 Maritza Blvd., 4000 Plovdiv, Bulgaria*

The isolation of melanoidines from foodstuffs is a current problem in food chemistry. This is required to investigate their formation and study the influence of processes on melanoidine structure and characteristics. The essence of the methods worked out by us are to combine different analytical approaches for extraction and for purification of brown substances. Melanoidines have been isolated from heat processed meat, cooked products, bread, malt, beer, roasted and instant coffee. They are characterized by an approximately high results reiteration degree and approximately low relatively standard deviation (from 1,2% to 5,3%).

Pharmacology/Biochemistry

Prevention of DM-Complications by Organic Germanium Compounds and Their Mode of Action in Reversible Solubilization of Maillard Products

Kunie Nakamura*, Keiko Nomoto, Kazutoshi Kariya, and Norihiro Kakimoto; **Molecular Biol. Lab., and Dept. Biochem, Kitasato Univ. School of Med., 1-15-1 Kitasato, Sagamihara, Kanagawa, 228 Japan, and Asai Germanium Institute, 1-6-4, Izumi-Honmachi, Komae-shi, 201 Japan.*

Organic germanium compounds (2-Carboxyethylgermanium sesquioxide: Ge-132, and 2-Carboxy-2-aminoethylgermanium sesquioxide: Ge-385) were found effective in reducing turbidity of cataract lens in SAM mice and DM-rats induced by streptozotocin (STZ). The main mechanism of the action of the drugs was elucidated to disconnect amino-carbonyl conjugate of Advanced Glycation Endproducts (AGE) to glucosone and amino-residues. These observations led to experiments to examine preventive effects of the drugs in the progress of clinical complications in DM-rats. Both drugs were perorally administered to STZ-DM-rats, and clinico-pathological examinations revealed that glycated serum-albumin, cataract (turbidity of lens), retinopathy, neuronal conductivity, peripheral circulation were remarkably improved.

Ge-compounds could not prevent the formation of Amadori-rearranged substances, but inhibited further progress to Advanced Glycation Endproducts (AGE) in a model reaction between arginine and ribose. The mode of action of Ge-compounds in the interaction with glucose was analysed by NMR, elucidating the formation of Ge-glucose complexes through oxygens of glucose at 1 and 2 carbon sites. This conjugate results in reversible disconnection of amino-carbonyl products to glucosone and amino residues. Ge-compounds, specifically Ge-385 which has two amino residues in its molecule, appear to be effective in the prevention of AGE formation by making carbonyl-(NH_2-Ge-385) complex as the substitute of reactive amino residues on functional polypeptides. Ge-compounds are the first candidates that prevent and reversibly solubilize the Maillard products which induce severe clinical complications in the patients with diabetes mellitus.

Maillard Polymers and Regulation of Activity of Adenylate Cyclase

Ru Duo Huang* and Milton S. Feather; *Anhui University, Department of Biology, Hefei, Anhui 230039, China; and Univesity of Missouri, Department of Biochemistry, Columbia, MO 65211, USA*

The Maillard reaction represents a reaction common to many types of food processing, including the formulation of traditional Chinese medicines. During the formulation of a Chinese herbal product derived from Wild Bluish Doghane, a water soluble Maillard polymer is produced along with the desired product(s). This particular herbal medicine is used for reducing blood pressure and glycerides in blood. We have found that this polymer influences the activity of forskolin-activated adenylate cyclase. The presence of the polymer has an inhibitory effect on the enzyme system, which was isolated from either corpus luteum or from fat cell membranes. Data collected to date (based on double reciprocal plots) suggest that the inhibition is noncompetitive in nature.

Immunochemical Characterization of the Major Fluorescent Compound Isolated from Advanced Maillard Reaction Products

Norie Araki* and Seikoh Horiuchi; *Deparunent of Biochemistry. Kumamoto University School of Medicine, Honjo, 2-2-1, Kumamoto 860. Japan*

The Maillard reaction proceeds through a series of reactions from early stage products to the advanced glycation end products (AGE) exhibiting fluorescence, brown color and cross-linking. To elucidate the mechanism of the advanced stage of the Maillard reaction, the chemical structure of the AGE-products must be determined. AGE-structures postulated so far include FFI, pyrrole aldehyde, 3-deoxyglucosone, pentosidine and crosslines. Although the *in vivo* presence of these compounds is also claimed, the chemical structure of the main AGE-product remains unidentified. Our previous immunological studies using anti-AGE antibodies suggested the presence of a common structure in AGE-products. In the present study, we isolated the main fluorescent compound (named as X1) from AGE-modified α-tosyl-L-lysine methyl ester. Although its chemical structure is being determined, we made an immunological approach to characterize X1 structure by its antibody raised in rabbits. [Experimental Procedures] The main fluorescent product (X1) was isolated to a homogeneity from AGE-modified α-tosyl-L-lysine methyl ester by reverse-phase HPLC. The anti-X1 antibody was raised in rabbits by immunizing X1-coupled with keyhole limpet hemocyanin. The titers were determined by ELISA [Results] (1) X1 showed the positive reaction toward the anti-AGE antibodies which were characterized previously (J. Biol. Chem. 266, 7329-7332, 1991). (2)The anti-X1 antiserum reacted with AGE-proteins obtained from bovine serum albumin, RNase and hemoglobin, but not with unmodified proteins. (3) The anti-X1 antiserum reacted with the extracts from human lens (80 years of age) but no reaction was observed with human lens proteins obtained from 4 year of age. (4) FFI, pyrrole aldehyde and pentosidine, AGE structures proposed so far, did not react with the antibody. The cross-reactivity of the antibody to crosslines is not yet known. [Conclusions] These data suggest that the fluorescent compound isolated from AGE-modified α-tosyl-L-lysine methyl ester is a new AGE-product and constitutes a major fluorescent compound among the fluorescent AGE-products.

Antioxidative Mechanism of Maillard Reaction Products in vivo

Nguyen van Chuyen*, Nobuko Utsunomiya[1] and Hiromichi Kato[2]; **Dept of Food & Nutrition, Japan Women's University, Tokyo 112, [1]Dept of Food Science, Kyoritsu Women's University, Tokyo 101, [2]Dept of Agricultural Chemistry, The University of Tokyo, Tokyo 113, Japan.*

In this study, amino acid-glucose reaction mixture (amino acid MRP), peptide-glucose reaction mixture (peptide MRP), and miso (a Japanese soy-paste containing natural MRP) were used to investigate the antioxidative mechanism of MRP *in vivo*. A mixture of arginine:histidine:glycine MRP (3.5:3.5:3), or milk casein hydrolysate (peptide MRP) or miso were fed to rats respectively for 6-8 weeks. The results showed that rats fed with these MRP showed lower liver TBA values and chemiluminescence values than that of the control i.e., amino acid MRP, peptide MRP and miso MRP exhibited antioxidative activity *in vivo*. Absorbed MRP of amino acid, peptide or miso were prepared by the using of rat small intestine. When the absorbed MRP was incubated with vitamin E deficient rat liver homogenate and H_2O_2, it was observed that absorbed MRP inhibited the lipid peroxidation caused by H_2O_2. When 3,7-bis-dimethylamino-10-N-methylcarbamoyl phenothiazin, a precursor of methylene blue, was mixed with absorbed MRP in the presence of alkoxy, hydroxy radical or peroxy radical, absorbed MRP inhibited the generation of methylene blue i.e., the scavenging of these radicals by absorbed MRP was confirmed. From these above results, it was estimated that MRP from amino acid, peptide and miso exhibited antioxidative effect *in vivo* by scavenging hydroxy, alkoxy, peroxy radicals and blocking the reactive oxygen species.

Enzymatic Degradation of Glycated-ϵ-Aminocaproic Acid and Glycated-Lysine by a Mucoid Soil Strain of ps. Aeruginosa

Chiara Gerhardinger*, Susan M. Marion, Vincent M. Monnier; *Case Western Reserve University, Institute of Pathology, 2085 Aderbert Road, Cleveland, Ohio 44106*

Searching for new approaches to reverse the glycation reaction, we have studied the ability of soil organisms to utilize a synthetic Amadori compound as carbon source. A microorganism identified as a mucoid strain of *Ps. Aeruginosa* was found to grow on minimal medium supplemented with glycated-ϵ-aminocaproic acid as only carbon and nitrogen sources. Oxygen consumption was observed when the cell-free bacterial extract was incubated both with glycated-ϵ-aminocaproate and glycated-α-t-boc-lysine but not with the unglycated amines or glucose. Borohydride reduced glycated substrate had 20% activity compared to unreduced control. The reaction was inhibited by heat treatment (100°C, 10 min) but not by dialysis supporting it enzymatic nature. An apparent Km of 0.44 mM for Glycated-ϵ-aminocaproate and 0.105 mM for glycated-lysine were calculated. A preliminary purification step showed that the ability to oxidize the Amadori compound is retained in the 50% ammonium sulfate precipitable fraction. This enzyme may be similar to the one described by Horiuchi et al. (Agric. Biol. Chem. 53, 103-ll0 1989) i.e. Fructosyl-amino Acid Oxidase (E.C. 1.5.3.).

Triazine Derivatives Formed with Aminoguanidine During Degradation of Glycated Proteins

Marcus A. Glomb, Martin Grissom, Vincent M. Monnier; *Case Western Reserve University, Institute of Pathology, 2085 Adelbert Road, Cleveland, Ohio 44106*

α-Dicarbonyl compounds are of major importance in the chemical pathways of the Maillard reaction. Due to high reactivity of these structures, it is almost impossible to isolate them from reaction mixtures. Aminoguanidine reacts with α-dicarbonyl compounds and α-ketocarboxylic acids to form relative stable 3-amino-1,2,4-triazine derivatives, which can be separated from reaction mixtures and detected after derivatization by a GC/MS system. All triazines monitored in this study were either synthesized independently or isolated from the reaction mixture and the structure established by their spectroscopical data. First the degradation of the Amadori product of propylamine and glucose in the presence of aminoguanidine (80°C, phosphate buffer pH 7) has been investigated. After 12h the triazines formed by glyoxal, methylglyoxal, diacetyl, pyruvic acid, 2,3-dioxobutanol and 3-deoxyglucosone could be identified. During degradation of glycated bovine serum albumin (BSA) under physiological conditions for 1.5 months the triazines derived from glyoxal, methylglyoxal, and 1-deoxyglucosone were present but the one from 3-deoxyglucosone was present only in trace amounts. Incubating the glycated BSA with aminoguanidine under anaerobic conditions and in the presence of EDTA showed almost no degradation. The detection of triazine derivatives in these model systems may help to explain the reported positive effects of aminoguanidine treatment on the long-term complications of diabetic subjects.

Evaluation of Urinary Excretion of Fructoselysine in Different Groups of Diabetic Patients

Anne Gross and Helmut F. Erbersdobler*; *Institut für Humanernährung und Lebensmittelkunde, Kiel University, Duesternbrooker Weg 17, D-24105 Kiel, Germany*

Urine samples which were collected for 24 h from a total of 56 diabetic patients (24 type I and 32 type II; 34 female and 22 male) and total blood samples from 18 patients were analysed for fructoselysine (FL) by the furosine method. The urinal FL originating from heat damaged foods (exogenous FL) was calculated from balance data determined in healthy volunteers and by using food questionnaires of the patients. The results were compared and correlated with the blood glucose or HbA_1 -values available from the hospital.*

The average of the urinary fructoselysine excretion was 1.3 times higher in diabetics compared to non diabetics However, among both diabetic types no correlation between fructoselysine in the urine and blood glucose values was found. On the other hand for the type I patients a correlation was found between FL in the urine and blood glucose, being highest ($r=0.69$) for the blood glucose levels from the 6th week prior to the urine collection. For the type II diabetics no glucose levels were available from such a long period of time but the correlations were generally lower ($r=0.34$ for the glucose in the 3rd week before the test) The correlations between the FL excretion and the HbA_1 values, however, were highest in the type II patients ($r=0.82$), whereas there was no correlation for the type I diabetic patients. Consequently no correlation was found between blood glucose and HbA_1 for this group. This finding suggests that the HbA_1 determination was somewhat questionable in the type I patients, which were mostly outpatients. The FL in the blood was 3 times ($p>0.001$) higher in the diabetics than in healthy persons.

Thus, it is concluded that under practical conditions the prediction of the metabolic state in diabetics is more difficult than reported from studies with selected groups of patients.

**We thank the City Hospital (Second Department of Internal Medicine of Kiel University) for providing us with urine and blood samples and the blood glucose and HbA_1 data.*

Phosphorylated Amadori Products in Mammalian Kidneys

F. Kappler*, B. Su, B.S. Szwergold, and T.R. Brown; *Fox Chase Cancer Center, Dept. of NMR & Medical Spectroscopy, 7701 Burholme Avenue, Philadelphia, PA 19111.*

We have identified phosphorylated Amadori products in mammalian kidneys. These are a new class of potentially reactive compounds which are present in normal kidney in small amounts and elevated by hyperglycemia.

We have examined two rat models for hyperglycemia, galactose fed and STZ diabetic, and have isolated and identified tagatose lysine 3-phosphate as the major new metabolite in the galactosemic kidney and have tentatively identified fructoselysine 3-phosphate in the diabetic kidney. These compounds are structurally similar to fructose 3-phosphate, a potent glycating agent, suggesting they may also possess a high glycation potential.

The structure of these compounds is a surprise, since current thinking is that Amadori products are not utilized by mammalian organisms. In addition to the kidney, we have found that human RBCs also phosphorylate Amadori products.

In conclusion, Amadori compounds are not simply deadend products, but are metabolized by mammalian tissues to potentially reactive compounds which may be precursors to advanced glycosylation endproducts.

Characterization of Model Melanoidin by Lectin Affinity and Immunochemistry

Seiichi Homma*, Masatsune Murata, Masako Fujii and Young Soon Lee; *Ochanomizu University, Department of Nutrition and Food Science Ohtsuka 2-1-1, Bunkyo-ku, Tokyo 112, Japan*

Model melanoidin was characterized by lectin affinity and immunochemistry. Non-dialyzable model melanoidin prepared from glucose and glycine was fractionated into test samples [molecular weight (m.w.) 17,000 to 156,000] by GPC-HPLC. The melanoidins were applied on a Con A-Sepharose 4B column, and the results showed that the amount of affinity melanoidin increased proportionally with the m.w. of melanoidin. The ratio of affinity fraction was 2.0% for m.w. 17000 melanoidin and 10.5% for m.w. 156,000. The affinity of the melanoidins to Con A was determined by dot blotting. The larger the m.w. of melanoidin, the more intensified the Con A affinity per mole of the melanoidin was. The minimum amount for detection was 2.8 pmole for m.w. 17,000 melanoidin and 0.6 pmole for m.w. 156,000. The non-dialyzable melanoidin also showed affinity to other lectins such as WGA, RCA-120 and PHA-E_4. Sugar in the melanoidins was determined with HPLC after hydrolysis with 0.5 N trifluoroacetic acid. The melanoidins were found to have glucose, mannose, xylose, ribose and others. The total amount of sugar determined in m.w. 156,000 melanoidin was 48 mg/g for the non-affinity fraction and 110 mg/g for the affinity.

Furthermore, polyclonal antibody was formed by injecting m.w. 156,000 melanoidin into mice with Freund's complete adjuvant. Although the antibody value was low, the antiserum responded with the melanoidins and polysaccharides such as glycogen on ELISA. The response of the antiserum to the melanoidins was higher in high m.w. than in low m.w. melanoidin.

Comparison of melanoidin in soy sauce and fish sauce by the Con A affinity chromatography showed that the amount of the affinity melanoidin was found to be larger in soy sauce than in fish sauce, and soy sauce melanoidin showed similar chromatographic pattern to the model one.

Aldosine, Difunctional Amino Acid Derived from Aldol Crosslink of Elatin and Collagen; Effect of Aging and Two Models of Hyperglycemia

Fumihiko Nakamura* and Kyozo Suyama; *Molecular Tech. Animal Prod., Fac. of Agric., Tohoku Univ., Sendai 981, Japan*

One of the most predominant crosslinking amino acids, named aldosine, was isolated from acid hydrolysates of bovine elastin and collagen, and identified by ^{1}H-and ^{13}C-NMR spectroscopy. The mass spectral analysis indicated a parent compound with a mass of 256 ($C_{12}H_{20}N_2O_4$). It was deduced from the structure that aldosine was derived from aldol crosslink which was a difunctional crosslinking amino acid formed by aldol condensation of allysine. Determination of aldosine was carried out by an ion-pair reversed-phase HPLC as pyridine derivative, 6-(3-pyridyl) piperidine-2-carboxylic acid, which was synthesized theoretically by oxidative decarboxylation of aldosine using Fe^{3+}. The age-related changes in the concentration of aldosine was followed for aorta from both a short-lived species (rat) and a long-lived species- (bovine). The content of aldosine in new born rat was very low but increased markedly with growth. After reaching maturity, the aldosine content decreased. On the other hand, in bovine aorta, the content of aldosine decreased gradually from 7 months to 16 years. Aldosine was also quantified in aorta and tail tendon of rat in two models of hyperglycemia, diabetes and galactosemia. Hyperglycemia had a significant effect on aldosine content; aldosine remarkably decreased in both experimental diabetic (one-half) and galactosemic (one-sixth) animals relative to controls.

Ocular Albumin Permeability in Experimental Diabetes: Effects of Aminoguanidine and Inhibitors of Oxidation and Polyol Pathway

G. Jerums*, T. Lim Joon, T. Gin, V. Lee, N. Carroll, S. Panagiotopoulos and H. Taylor; *Endocrine Unit, Austin Hospital and Melbourne University, Melbourne, Australia*

Increased vascular permeability to albumin is observed in both experimental and human diabetes. Candidate mechanisms for this process include advanced glycation, oxidative stress and the polyol pathway. The present study has assessed the effects of inhibition of advanced glycation (aminoguanidine), oxidative stress (butylated hydroxy toluene, [BHT]), and the polyol pathway (Statil) on ocular albumin permeability in the diabetic rat.

Diabetes was induced in male Sprague-Dawley rats aged 8-10 weeks with streptozotocin (50mg/kg iv). Animals were randomly assigned to the following groups: control, diabetic untreated, diabetic + aminoguanidine 500 mg/L in drinking water, diabetic + BHT 0.4% w/w, and diabetic + Statil 0.03% w/w. There were no significant differences in body weight or blood glucose levels (25-35mM) among the four diabetic groups. Ocular vascular permeability to albumin was assessed at 16 weeks, using a double isotope technique. Retinal albumin vascular clearance was increased significantly in diabetic rats (control 137 ± 42, diabetic 573 ± 205 nl/min, mean $\pm$SEM, $p<0.05$). Diabetes-related increases in retinal vascular clearance were prevented by aminoguanidine (55 ± 7nl/min, $p<0.05$) and by Statil (185 ± 5nl/min, $p<0.05$) but not by BHT (316 ± 92 nl/min). Similar results were observed in the anterior and posterior uvea.

These data support the concept that advanced glycation and the polyol pathway participate in the pathogenesis of diabetes-related increases in ocular vascular permeability. Although inhibition of oxidation showed a similar trend, this did not reach significance, implying that oxidative stress does not play a major role in the above process.

Cyclopentenosine, Trifunctional Crosslinking Amino Acid of Elastin and Collagen; Structure, Characterization and Distribution

Kyozo Suyama*, Kazuyuki Yamazaki and Fumihiko Nakamura; *Molecular Tech. Animal Prod., Fac. of Agric., Tohoku Vniv., Sendai 981, Japan*

A trifunctional crosslinking amino acid having a cyclopentenone skeleton, named cyclopentenosine, was isolated from the hydrolysates of bovine aorta elastin and the bone collagen, and the spectroscopic properties were characterized. The mass spectral analysis indicated a parent compound with a mass of 381 ($C_{18}H_{27}N_3O_6$). From the structure, cyclopentenosine was formed from three allysine residues by amino-carbonyl reaction accompanied with condensation. The UV maximum was 265 nm (O.1N HCl), and the strong fluorescent exhibited (excitation/emission maxima at 325/395 nm in O.1N HCl). Cyclopentenosine was detected and determined in elastin-rich tissues such as ligamentum nuchae, aorta, lung and spleen, and also in collagen-rich tissues such as bone and Achilles tendon by an ion-pair reversed-phase HPLC. The content of cyclopentenosine of such protein was more than or equal to those of such pyridinium crosslinks as desmosine and pyridinoline. Effect of aging and diabetes on cyclopentenosine of such proteins were also investigated. The cyclopentenosine level of aorta in newborn rats was very low. However, it increased markedly with the growth of rats. This marked increase was completed by the time rats reached physiological maturity, i.e., about 15 weeks of age. No change on the content of cyclopentenosine was found after the maturity from 15 to 27 weeks of age, in the case of the bovine aorta as well (from 7 months to 16 years of age). The cyclopentenosine level in such protein decreased in streptozotocin-induced diabetic rats relative to controls.

Chemistry of Crosslines

K. Nakamura, T. Hochi, Y. Nakazawa, Y. Fukunaga and K. Ienaga*; *IBAS, Nippon Zoki Pharm. Co. Ltd., Yashiro-cho, Kato-gun, Hyogo 673-14, Japan*

A pair of fluorophores, N-diacetates of crosslines (XLs) A and B (Fig.), were isolated as major products in the Maillard reaction of α-N-acetyl-L-lysine and D-glucose. Since 2 molecules of amines and 2 molecules of glucose form XL derivatives with cyclization and 6 molecules of water dehydration, contribution of oxidation and/or reduction in the formation of XL is doubtful. Fluorogenesities of XLs are similar to those of fluorophores, advanced glycation end products (AGEs), in age- and diabetes-related cross-linked proteins. Furthermore, since specific antisera against the pair of XLs, using KLH as a carrier protein, never recognized serum albumins (SAs) but AGE-SAs, we concluded that AGEs should include XL-like structures. Neither pyrraline nor pentosidine showed immunocrossreactivities. Not only immunochemical evidence but also immunohistochemical evidence using animal models indicated the diagnostic importance of XLs as markers for diabetic complications.

Figure. XLs A and B are epimers at *C9. R = $-(CH_2)_4CH(NHAc)CO_2H$

Aminoguanidine Decreases Atherosclerotic Plaque Formation in the Cholesterol-Fed Rabbit

S. Panagiotopoulos, R.C. O'Brien, M.E. Cooper, G. Jerums and R. Bucala*; *Endocrine Unit, Austin Hospital, Australia and Picower Institute for Medical Research, NY, USA**.

Advanced glycation endproducts (AGEs) may be implicated in the vasculopathy of diabetes and aging. Monocytes which have AGE specific receptors, migrate and release growth factors in response to AGEs. Since aminoguanidine inhibits the accumulation of AGEs, we explored its effects on the development of atherosclerosis in the cholesterol-fed rabbit.

Male New Zealand white cross rabbits fed a high cholesterol (1% w/w) diet were randomized to control or aminoguanidine treatment. Three dose levels of aminoguanidine were mixed into the diet: 26, 50 or 100 mg/kg body weight/day. The animals were sacrificed at week 12. Sudan IV was used to stain the lipid containing plaques of the aortic arch, thoracic and abdominal aorta. The surface area (%) occupied by atheromatous plaques is shown in the table (mean ± SEM, $*p<0.05$, $**p<0.01$):

Dose of	Control	26mg,n=3	50mg,n=3	100mg,n=5	p value
aminoguanidine					vs.
(per kg body wt/day)					cont.
Aortic arch	92.1±2.4	83.3±6.5	82.2±3.2	65.9±8.6	*
thoracic aorta	48.4±3.1	49.2±12.6	34.0±16.5	17.7±3.9	**
Abdominal aorta	39.3±7.0	33.5±11.1	20.6±7.5	15.3±6.6	*

The reduction of plaque formation by high dose aminoguanidine treatment was independent of significant cholesterol-lowering effects. These data suggest that glycation may participate in atherogenesis and raise the possibility that inhibitors of glycation may retard this process

These studies were funded by Alteon Inc.

Glycation and Inactivation of Rat Aldehyde Reductase, a Major 3-Deoxyglucosone Reducing Enzyme

Motoko Takahashi, Junichi Fujii, Tadashi Teshima[1], Keiichiro Suzuki, Tetsuo Shiba[1] and Naoyuki Taniguchi; *Department of Biochemistry, Osaka University Medical School, 2-2 Yamadaoka, Suita, Osaka 565, Japan, [1]Peptide Institute, Protein Research Foundation, 4-1-2 Ina, Minoh, Osaka 562, Japan*

We have purified a rat liver enzyme that catalyzes the NADPH-dependent reduction of 3-deoxyglucosone (3-DG), a major intermediate in the Maillard reaction and a potent cross-linker responsible for polymerization of proteins. Comparison of the amino acid sequences of peptides obtained from the rat 3-DG reducing enzyme by lysyl-endopeptidase digestion with the sequence of human aldehyde reductase strongly suggested that the purified enzyme was rat aldehyde reductase. We cloned the cDNA encoding aldehyde reductase from a rat kidney cDNA library. All the peptide sequences obtained from the rat 3-DG reducing enzyme were included in the deduced amino acid sequence of the rat aldehyde reductase cDNA. Moreover, the protein expressed in COS-1 cells transfected with the rat aldehyde reductase cDNA exhibited NADPH-dependent 3-DG reducing activity and cross-reacted with antiserum raised against the purified rat 3-DG reducing enzyme. All the above data clearly indicated that the 3-DG reducing enzyme was identical with aldehyde reductase. Northern blot analysis of total mRNA from a variety of rat tissues showed fairly high levels of expression of aldehyde reductase mRNA. This suggests that sufficient aldehyde reductase is present to detoxicate 3-DG when it is formed through the Maillard reaction *in vivo*. The presence of glycated ALR *in vivo* has been confirmed by boronate column chromatography and anti-hexitollysine antibody. The specific activity of the glycated form was lower than the nonglycated form, indicating that the glycation of the enzyme led to a low active form. This suggests that aldehyde reductase undergoes glycation and lose its activity under hyperglycemic conditions.

Effect of Glucose and Glycation on Protein Oxidation

Mary C. Wells-Knecht*, Min-Xin Fu, Suzanne R. Thorpe, John W. Baynes; *University of South Carolina, Dept of Chemistry and Biochemistry, Columbia, S.C. 29208*

Oxidative stress and resultant oxidative damage to biomolecules may play a role in the aging of tissues and the development of complications associated with diabetes. The possibility that glucose and glycation could cause oxidative damage to protein was studied using collagen incubated with varying glucose concentrations or for varying time under oxidative and antioxidative conditions. Additionally, to determine how the Maillard reaction influences oxidation of protein, collagen which was preglycated under antioxidative conditions was incubated under oxidative conditions for varying periods of time in the absence of glucose. Pentosidine and o-tyrosine were measured as indicators of oxidative damage to protein. Pentosidine is a glycoxidation product formed by reaction of carbohydrate with protein during the Maillard reaction, while o-tyrosine is formed by reaction of hydroxyl radical with phenylalanine. Neither pentosidine nor o-tyrosine were formed under antioxidative conditions (N_2 atmosphere and metal chelators). In collagen incubated under oxidative conditions, both pentosidine and o-tyrosine increased as a function of glucose concentration and as a function of time at constant glucose concentration. Levels of pentosidine and o-tyrosine also increased with time in preglycated collagen incubated under oxidative conditions in the absence of glucose. Our results show that glucose and/or the glycation reaction as well as Amadori adducts on glycated proteins can cause oxidative modification of amino acids in protein. We hypothesize that other amino acid oxidation products in addition to o-tyrosine are formed in proteins during glycation reactions; and that, like pentosidine, they may be useful biomarkers of oxidative damage to protein in aging and disease.

Toxicology/Aging

Reactivity of Age Collagen in the Radioreceptor Assay for Age Modified Proteins

S. Hepburn*, W.K. Lee and M.H. Dominiczak; *Department of Pathological Biochemistry, Western Infirmary, Glasgow G11 6NT, Scotland, U.K.*

A specific high affinity macrophage surface receptor for advanced glycation endproduct (AGE) modified proteins has previously been used to develop a whole cell radioreceptor assay for AGE using RAW 264.7 cells. Using the radioreceptor assay we have evaluated the specificity and affinity of the receptors in question. Our data support the presence of a saturable high affinity (3.6 $e10^7$ M^{-1}) receptor for AGE BSA. The receptors become saturated at around 10ug/mL AGE BSA standard and 50% competition is achieved using 33ug/mL AGE BSA competitor. However, *in vitro* glycated type 1 calf skin collagen (14 days at 37°C in 250mM glucose-6-phosphate, 18.3 U/mg at 370nm excitation, 440nm emission) digested for 24 hours with bacterial collagenase does not compete for binding sites with the AGE BSA standard (44 days at 37°C in 500mM glucose-6-phosphate, 115 U/mg at 370nm excitation, 440nm emission) at the range cf concentrations tested (60-500 ug/mL collagen). It has previously been reported that AGE collagen is reactive in this assay system. Our results show that the AGE formed by *in vitro* glycation of collagen forms AGE which are fluorescent but are not reactive with the macrophage receptor. The epitope which the macrophage receptor recognizes may not be present in high concentrations in this AGE collagen.

Inhibition of Cariogen Glucan Synthesis by Maillard Reaction Product of Beer

Masatsune Murata*, Yukako Nakajima, and Seiich Homma; *Ochanomizu University, Department of Food Science and Nutrition, 2-1-1 Otsuka, Bunkyo-ku, Tokyo 112 Japan*

The extracellular insoluble glucan (IG) synthesized by *Streptococcus mutans* and *S. sobrinus* from sucrose plays an important role in the induction of dental caries of teeth. We investigated inhibitory activities of several brownish foods and beverages against IG synthesis by glucosyltransferase of cariogenic bacteria, *S. mutans* MT8148 and *S. sobrinus* 6715, were grown in BHI broth at 37°C. The supernatant of the cultured broth was used as a crude enzyme. The reaction mixture for standard assay contained crude enzyme, food extract or beverage, 0.01% sucrose, 0.07mg dextran, 0.02% NaN_3, and O.1M phosphate buffer (pH6.5). The final volume was adjusted to 1.75mL. Incubation was carried out for 15hr at 37°C. The enzyme activity was determined by measuring the amount of IG by turbidimetry.

Among 15 foods and beverages, dark beer, green tea, and oolong tea definitely inhibited IG synthesis. Dark beer showed the strongest inhibitory activity, compared by volume as beverages. Black tea only inhibited IG synthesis of *S. sobrinus*. Soy sauce, caramel, and cocoa did not show the inhibition. Pale beer showed weak activity . The active material in dark beer was purified by ethanol precipitation and the chromatography of Sephadex G-200 and Sepharose 4B. The active fraction was obtained as a brownish amorphous powder. It was mainly composed of sugars and contained nitrogen. The molecular weight was estimated to be ca 200kD as dextran. I_{50} was 30 and 40ug/mL against the enzymes of *S. mutans* and *S. sobrinus*, respectively. Dark beer also inhibited the adherence of the bacteria to surface of glass in the presence of sucrose. This material was considered to be produced by the Maillard reaction or caramel reaction during kilning of malts.

Human β_2-Microglobulin Modified with Advanced Glycation End Products in Hemodialysis-Associated Amyloidosis

Toshio Miyata*, Reiko Inagi, Yoshiyasu Iida, Norie Araki, Norio Yamada, Seikoh Horiuchi, Naoyuki Taniguchi, Taroh Kinoshita, and Kenji Maeda; *Dept. Immunoreg. Res. Inst. Micro. Dis., Osaka Univ., Dept Bacteriol. & Biochem. Osaka Univ. Med. Sch., Dept. Int Med., Nagoya Univ. Sch. Med., Dept. Biochem., Kumamoto Univ. Sch. Med. Japan*

Hemodialysis-associated amyloidosis (HAA) is a long-term complication among chronic dialysis patients. Its clinical features include carpal tunnel syndrome (CTS), erosive arthropathy and lytic bone lesions. Recent studies demonstrated that β2-micro- globulin (β-MG) is a major protein constituent of the amyloid fibrils of dialysis patients. Serum β-MG levels are known to increase markedly (usually 30 > fold) in these patients than in normal subjects, but statistical correlation between serum concentrations and the occurrence of HAA is unclear. To elucidate the pathological role of β-MG in HAA, the present study was undertaken to compare the acidic form of β-MG with normal β-MG in their physicochemical and immunochemical properties. [Results] Amyloid fibril proteins were isolated from connective tissues forming carpal tunnels in hemodialysis patients with CTS. Two-dimensional polyacrylamide gel electrophoresis and Western blotting demonstrated that most of β-MG forming amyloid fibrils exhibited a more acidic pI value than normal β-MG. This acidic β-MG was also found in a small fraction of β-MG in serum and urine from these patients, whereas heterogeneity was not observed in healthy individuals. We purified acidic and normal β-MG from the urine of long-term hemodialysis patients, and compared their physicochemical and immunochemical properties. Acidic β-MG, but not normal β-MG, was brown in color and fluoresced, both of which are characteristics of advanced glycation end products (AGEs) of the Maillard reaction. Immunochemical studies showed that acidic β-MG reacted with anti-AGE antibody and also with an antibody against an Amadori product, but normal β-MG did not react with either antibodies. Incubating normal β-MG with glucose in vitro resulted in a shift to a more acidic pI, generation of fluorescence and immunoreactivity to the anti-AGE antibody. The β-MG forming amyloid fibrils also reacted with anti-AGE antibody. [Conclusions] The present data provided evidence that AGE-modified β-MG is a dominant component of the amyloid deposits in HAA, implicating the potential link of AGE-modification of β-MG to the pathogenesis of HAA.

Cyto- and Genotoxic Activity of Heat Processed Cooked Products on in-vitro Test System. 5178Y Mouse Lymphoma

M. Draganov, M. Kuntcheva, S. Ivanova, N. Popov, Tzv. Obretenov; *Higher Institute of Food and Flavour Industries, Department of Organic Chemistry, 26 Mariza Blvd., 4000, Plovdiv, Bulgaria*

Cyto- and genotoxic activity of heat processed cooked products have been investigated by a cell line mouse lymphoma cells L 5178Y (wild type) test.

The melanoidines have been applied without exogenous activation with an increasing concentration. IC_{50} level has been time-tested by reading the cell proliferation and cloning efficiency for 4, 24, 48 hours. A test for mutagenicity of melanoidin has been done without an advance metabolic activation. The genes for thymidinekinase (TK) hypoxanthine-guanine phosphoribosyltransferase (HGPRT) $N_a+/K+$ dependent ATPase have been analysed by $5BrdUrd^{res}$ $6TG^{res}$ mutants selection.

Physiological Significance of Glycation of Myofibrullar Proteins in Mice and Rats, and in Cultured Muscle Cells (16)

N. Nishizawa*, H. Watanabe and W. Sato; *Iwate University, Department of Bioscience and Technology, Faculty of Agriculture,*
Morioka, Iwate 020, Japan

Skeletal muscle is major tissue in human and animals, and has the largest intracellular protein, myofibrillar protein. It plays an important role in the overall regulation of whole-body protein and carbohydrate metabolism. However, in spite of this importance of muscle, studies on glycation of muscle proteins have been hardly done. We reported identification of Nε-fructoselysine in myofibrillar protein, and that the amount of Nε-fructoselysine increases in aged mice and rats.[1] We confirmed the presence of Nε-fructoselysine in myofibrillar proteins with immunoblotting procedure using monoclonal antibody. The content of Nε fructoselysine in actomyosin increased in streptozotocin-induced diabetic rats and with administration of insulin to rats, respectively. Studies with cultured muscle cells (L6) showed that the glycation of the myofibrillar protein is enhanced by addition of glucose and insulin or an inhibitor of glucosephosphate isomerase to the medium. These results showed that glycation product of large amount is released from muscle tissue after protein degradation and would appreciably affect diseases such as diabetes or aging. They also suggested a possibility that glucose-6-phosphate would effectively act as a carbohydrate source for glycation of muscle protein rather than glucose within cells.

[1]H. Watanabe *et al., Biosci. Biotechnol. Biochem.*, 56:1109 (1992).

Plasma Advanced Glycoxidation Endproduct and Malondialdehyde in Maintenance Hemodialysis Patients

P. Odetti*, L. Cosso, D. Dapino, M.A Pronzato, G. Noberasco, and G. Gurreri; *University of Genova, DI.M.I. and Inst. of Gen. Pathology, Italy.*

An increase of pentosidine (P), a well known advanced glycation endproduct, has been described in patients with end stage renal failure. Nevertheless the cause of this increase is still poorly understood. The recent hypothesis that P is a product of the "glycoxidation", because it is formed with sequential steps of oxidation and glycation, opened new possibility of interpretation.

Fifteen patients (age 24-78 yrs) undergoing periodic hemodialysis (HD) were studied; the patients were on maintenance HD for an average of 32 months (range 12-71 months), serum creatinine ranged between 415.5 and 1591.2 μmol/L. Plasma levels of P and malondialdehyde (MDA), a side product of lipoperoxidation damage, were evaluated before and after a HD session. It has been used for both parameters a HPLC method. The results of this study showed that in plasma levels of P were, as expected, much higher in uremic patients than in age-matched healthy controls (23.7 ± 3.6 vs 1.2 ± 0.14 pmol/mg protein; $p < 0.001$). The level of P was unaffected by the acute HD treatment or by the type of HD treatment (hemofiltration, acetate-free biofiltration and bicarbonate HD). The time on HD and serum creatinine did not correlate with the level of plasma P. Plasma MDA concentration was higher in uremic patients than in healthy controls (1.9 ± 0.12 nmol/mL). The concentration of MDA was slightly increased by the HD session (2.78 ± 0.21 vs 3.03 ± 0.23 nmol/mL). A significant correlation between MDA and plasma P was observed and, notwithstanding the small number of subjects ($n=15$), was significant ($r=0, 53$; $p < 0.05$). These results suggest that HD increases the oxidative stress, indirectly evaluated by MDA. The relationship between MDA and P support the hypothesis that oxidative stress may contribute, besides the main role of reduced renal clearance, to the formation of plasma P during end stage renal failure.

The Changes of Collagen Cross-Links in the Connective Tissues of Guinea Pigs Supplemented with L-Ascorbic Acid

M. Otsuka*, M. Kim, E. Shimamura, T. Kurata & N. Arakawa; *Ochanomizu University, 2-1-1, Otsuka, Bunkyo-ku, Tokyo Japan 112*

The oxidation of L-ascorbic acid (AsA) in human lens protein has become of interest in recent years. It is assumed that AsA participates in the formation of some cross-links *in vivo*. AsA is well known to be required for the enzymatic hydroxylation of proline and lysine in collagen fibrils. Therefore, the formation of cross-links in the connective tissues may be influenced by the supplementation of AsA that enhances collagen synthesis. We investigated changes of collagen cross-links in the connective tissues of guinea pigs supplemented witb AsA. Pyridinoline, a mature cross-link of collagen which is presumably formed through dehydration and condensation of Schiff bases derived from hydroxylysine residues, remarkably increased with growth in cartilage of guinea pigs in a similar manner as humans. Depletion of AsA for 4 weeks in the growing process of the animals resulted in the increase of pyridinoline content in cartilage, indicating that AsA may indirectly regulate the excess formation of pyridinoline. It was also supported by the results of *in vitro* aging in bone and cartilage. They revealed that AsA in oxidized form suppressed the formation of pyridinoline in its early stage. The effect of AsA on the formation of other cross-links related to pyridinoline in the connective tissues of guinea pigs will be discussed.

Identification of 2-Lactylamido-Guanine (LAG) as a DNA-Advanced Glycosylation Endproduct

Andrew Papoulis, Anthony Cerami, and Richard Bucala*; *The Picower Institute for Medical Research, 350 Community Drive, Manhasset, New York 11030*

Physiological aging is accompanied by a decline in the functional integrity of the genetic material and is evidenced by a decrease in cellular viability and by an increase in the frequency of oncogenic transformation. The presence of primary amino groups on nucleotides led us to postulate that long-lived DNA molecules might serve as pathophysiologically important targets for advanced glycosylation (Maillard) reactions. In model studies, plasmid DNA that was incubated with glucose or glucose-derived reactive intermediates was found to undergo mutation and DNA transposition when re-introduced into prokaryotic or eukaryotic cells. To begin to address the chemical basis for DNA modification by advanced glycosylation endproducts, the model nucleotide 9-methylguanine (mG) was reacted with glucose, 1-n-propylamino-N-D-glucoside (a Schiff base), or 1-n-propylamino-N-D-fructose (an Amadori product) at physiological pH. In each experiment, a similar guanine-derived reaction product was identified. After purification by flash chromatography and preparative TLC, structural analysis by ^{1}H-NMR, FT-IR, and mass spectroscopy revealed the modified guanine residue to be 2-lactylamido-9-methylguanine (LAmG). Independent synthesis of LAmG from methylglyoxal and mG confirmed this structural assignment. We propose that this guanine modification results from the addition of methylglyoxal that forms by fission of an Amadori product to 1-deoxyglucosone. 1-Deoxyglucosone then undergoes retroaldol fragmentation to generate methylglyoxal. Nucleophilic addition of the N-2 of guanine to the aldehyde moiety of methylglyoxal is followed by a hydride shift to generate stable LAmG. Efforts are underway to characterize the mutagenic activity of LAG *in vivo* and to quantitate the frequency of LAG nucleotides in normal and aged human tissues.

3-Deoxyglucosone is a Decomposition Product of Fructose-3-Phosphate

S. Lal, B. Szwergold, F. Kappler, T.R. Brown; *Fox Chase Cancer Center, Dept. of NMR, Philadelphia, PA 19111.*

Fructose-3-phosphate (F3P), a potent crosslinking agent is produced in significant amounts in diabetic rat lenses. Data from these lenses indicate that F3P achieves steady state levels in a matter of days after induction of diabetes suggesting that it may be decomposing or is further metabolized in the lens. One possibility for the decomposition of F3P, is a β-elimination of its phosphate group to form 3-deoxyglucosone (3dG) which is another powerful crosslinking agent. To test this hypothesis F3P was incubated at 37°C in a buffer, with lens crystallins (10mg/mL) or other model amino compounds and the incubations were examined by ^{31}P NMR. Data from these experiments suggest that at a neutral pH F3P by itself in solution is stable but in the presence of crystallins and amino compounds it decomposes relatively rapidly (ca. 0.2%/hr @ 10 mg/mL crystallin). This rate is sufficient to account for the steady state levels of F3P observed in the diabetic rat lens.

The decomposition products of F3P were characterized by ^{13}C NMR and further identified by TLC and Electrospray Mass Spectroscopy following derivatization with 2,4 dinitrophenyl hydrazine. Results of these analyses demonstrate conclusively that the major decomposition product of F3P in the presence of amines is 3dG.

In conclusion, our data suggests that F3P and 3dG may be important contributing factors in cataractogenesis in the diabetic rat.

Antibodies to Pentosidine and Pentose-Derived Advanced Maillard Products

Shinji Taneda and Vincent Monnier; *Case Western Reserve University, Institute of Pathology, 2085 Adelbert Road, Cleveland, OH 44106*

Increasing evidence points toward the involvement of pentose in post-synthetic modification and crosslinking of tissue proteins in diabetes. For quantitation and immunolocalisation of pentosidine, monoclonal antibodies to pentosidine and pentose derived advanced Maillard products are being generated. Acetylated pentosidine (AP) coupled with keyhole limpet hemocyanin (KLH) prepared using the carbodiimide method was injected to 6 Balb/c mice. The molar ratio of hapten: lysine in KLH was about 9:1 prior to fusion immune serums recognized pentosidine and AP. Spleen cells from the mouse whose antiserum showed highest titer and highest sensitivity against pentosidine were fused with mouse myeloma cells of the SP 2/0 cell lines at the cell ratio of 5:1. The cells were cultured with HAT-supplemented medium. Hybridoma clones prepared by limiting dilution were screened by 2 steps in ELISA system as follows; 1) the antibody positively reacted with bovine serum albumin (BSA)-AP conjugate as a coating agent, 2) the reaction to the BSA-AP congugate a coating agent was well inhibited by free pentosidine. Two different cell lines were obtained, one secreting IgM which recognized mainly pentosidine and AP, the other secreted IgM which recognized protein bound pentose prepared by incubating BSA with ribose, as well as AP. These hybridomas are now being expanded and ascites is produced by injecting these hybridomas to Balb/c mice. These antibodies are expected to be a useful tool for elucidating the role of pentose in post-synthetic modification of key protein in tissues in diabetes.

Existence of Maillard Reaction Products in Tumor Bearing Mice

Kazuhiko Takamiya*, Emi Tsukamoto, Mikako Muraga; *Kyoritsu Women's University, Dept. of Nutrition and Food Sciences, Hitotsubasi, Chiyoda-ku, Tokyo, Japan.*

Four days after the inoculation of Ehrlich tumor cells into mouse intraperitoneal cavity, ascites fluids accumulate and the number of tumor cells increase remarkably. Decreases of liver catalase activity soon after the tumor inoculation helps the formation of free radicals, and leads to cachexia and promotes the tumor proliferation.

A compound having 410 nm absorbance was found in the bleeded ascites fluids, which suggested the existence of some Maillard reaction products *in vivo*. By the successive administration of aminoguanidine (AG), the absorption of 410 nm of the ascites fluid disappeared. Administration of AG shortened the survival days of the tumor bearing mice, and promoted solid tumors inoculated subcutaneously. Administration of AG also decreased the hepatic and the tumor cell cholesterol contents. The component which was formed in the ascites fluid and showed absorption of 410 nm was suggested to inhibit the cachexia and to show some antitumor activities.

Mesangial and Retinal Microvascular Phenotypic Cell Changes in Response to Glycosylated Matrix Components

E.C. Tsilibary*, S.S. Anderson, S. Setty, T. Kalfa, M. Gerritsen, Y. Kim and E. Wayner; *Univ. of Minnesota, Depts. of Lab. Med. Pathology and Pediatrics, 420 Delaware Str. SE., Minneapolis, and Miles Labs., New Hawn.*

Type IV collagen (tIV) promotes the adhesion and spreading of various cell types including human mesangial cells (HMC) and retinal microvascular cells (retinal endothelial cells:RER and retinal pericytes:RP). In long-standing diabetes mellitus, non-enzymatic glycosylation of tIV occurs and there are distinctive changes in the mesangium and the retinal microvasculature as well. We examined the phenotypic cell response of each of these cell types, upon interaction with control and glucose incubated tIV, which were incubated with either 50-or 500-mM glucose and had incorporated 1 or l0 moles of glucose per mole of tIV, respectively. HMC incubated on a substrate of glucosylated tIV (tIV 50, tIV 500) were 19% and 39% less adherent, respectively, when compared to control tIV (tIV-0). HMC also spread less on tIV-500 with 44% smaller average surface area and 24% less average perimeter area compared to HMC on tIV-0 after 30 min of incubation. Because integrins are known to mediate adhesion to tIV, we tested the presence of a1-, a2- and b1- subunits of integrins on HMC grown for 8-15 days in 5-or 25-mM DMEM. By flow cytometry and immunoprecipitation (which made use of specific monoclonal anti-integrin antibodies) there was an apparent increase in the a2 and b1 subunits, when HMC were grown in 25 mM DMEM. The proliferation rate of HMC grown on tIV-50 tIV-500 and even laminin-50 or-500 did not vary singificantly. However, the proliferation rate of REC in response to tIV-50 was increased by $\approx$13%. In contrast, the growth rate of RP was substantially decreased (by $\approx$ 38% on tIV-500 and $\approx$ 20% on laminin-500). The observed growth rates of each of the cell types in response to glycosylated BM components resembles the *in situ* findings in diabetic nephropathy and angiopathy. We conclude that the phenotypic cell response to glycosybted BM varies depending upon the cell type, and that glycosylation of tIV and laminin could result at least in part to the altered cell phenotype.

Aldehyde Reductase and Aldose Reductase May Act as Scavengers of Aldosuloses Formed from Glycated Proteins

Donald J. Walton*, Terrance J. Kubiseski, Kim A. Munro, T. Geoffrey Flynn, and Milton S. Feather[1]; *Departments of Biochemistry, Queen's University, Kingston, Ontario, Canada K7L 3N6, and [1]University of Missouri, Columbia, MO 65211, U.S.A.*

The monomeric, NADPH dependent enzymes, aldehyde reductase (ALR1) and aldose reductase (ALR2), catalyze the reduction of a large variety of aldehydes. Because of their broad substrate specificity and widespread tissue distribution it has been difficult to assign a special physiological role for these enzymes. In many mammalian tissues significant quantities of aldosuloses (osones) are formed by elimination of carbohydrate residues from glycated proteins [Knecht, Feather & Baynes (1992) *Arch. Biochem. Biophys.* 294, 130-137]. The aldosuloses are very reactive, and are capable of damaging tissues by crosslinking of proteins. One of them, 3-deoxy-D-*erythro*-hexosulose (3-deoxy-D-glucosone), is reduced, *in vivo*, to the less reactive 3-deoxy-D-fructose. Our hypothesis is that a major role for the reductases is catalysis of reduction of aldosuloses by NADPH, and that this accounts for their ubiquity. This is supported by the results of a substrate specificity study, in which we have established that D-*arabino*-hexosulose (D-glucosone), D-*threo*-pentosulose (D-xylosone), and their 3-deoxy- analogues, are good substrates for ALR1 (porcine kidney), ALR2 (porcine striated muscle) and ALR2 (human striated muscle). These enzymes may have, therefore, have an important role as scavengers of aldosuloses that are leached from glycated proteins in many tissues of the body.

(Supported by Canadian Diabetes Association, Medical Research Council of Canada and Juvenile Diabetes Foundation International).

Subject Index